Change, Choice and Inference:

A Study of Belief Revision and Nonmonotonic Reasoning

HANS ROTT

CLARENDON PRESS • OXFORD

2001

OXFORD

UNIVERSITY PRESS

Great Clarendon Street, Oxford OX2 6DP

Oxford University Press is a department of the University of Oxford.
It furthers the University's objective of excellence in research, scholarship,
and education by publishing worldwide in

Oxford New York

Athens Auckland Bangkok Bogotá Buenos Aires Cape Town
Chennai Dar es Salaam Delhi Florence Hong Kong Istanbul Karachi
Kolkata Kuala Lumpur Madrid Melbourne Mexico City Mumbai Nairobi
Paris São Paulo Shanghai Singapore Taipei Tokyo Toronto Warsaw

with associated companies in Berlin Ibadan

Oxford is a registered trade mark of Oxford University Press
in the UK and in certain other countries

Published in the United States
by Oxford University Press Inc., New York

British Library Cataloguing in Publication Data

Data available

Library of Congress Cataloging in Publication Data

Data available

ISBN 0 19 850306 7

1 3 5 7 9 10 8 6 4 2

Typeset by the author in LaTeX
Printed in Great Britain
on acid-free paper by
Biddles Ltd, Guildford and King's Lynn

Let me not to the marriage of true minds
Admit impediments. Love is not love
Which alters when it alteration finds,
Or bends with the remover to remove:
O, no! it is an ever-fixed mark,
That looks on tempests and is never shaken;
It is the star to every wandering bark,
Whose worth's unknown, although his height be taken.
Love's not Time's fool, though rosy lips and cheeks
Within his bending sickle's compass come;
Love alters not with his brief hours and weeks,
But bears it out even to the edge of doom.
    If this be error, and upon me prov'd,
    I never writ, nor no man ever lov'd.

Shakespeare, Sonnet 116

For my parents

Jutta Karin Rott and Johann Ludwig Rott

with love and gratitude

Thanks

Early versions of parts of this book were written while I was visiting Australia and Sweden. I thank the DAAD (German Academic Exchange Service) for supporting a number of Swedish-German exchanges in belief revision, and in particular a research visit of mine to the Department of Philosophy at the University of Uppsala in October 1994. I want to thank Mary-Anne Williams, M.-A., Norman Foo and the DFG (German Research Council) for supporting a research visit to the Basser Department of Computer Science at the University of Sydney and the Department of Management at the University of Newcastle from February to April 1996. It is a great pleasure to express my gratitude to Grigoris Antoniou, John Cantwell, Norman Foo, Peter Gärdenfors, Aditya Ghose, Sven Ove Hansson, Sten Lindström, Sebastian Melmoth, Abhaya Nayak, Erik Olsson, Maurice Pagnucco, Pavlos Peppas, Wlodek Rabinowicz, Abdul Sattar, Krister Segerberg, Mary-Anne Williams and all the other friends and colleagues in Australia and Sweden for many valuable discussions and for their warmth and hospitality. They all made me feel that the belief revisionists form a very nice family indeed.

I owe very special debts to Isaac Levi and David Makinson for their infallible, unfailing help and advice on belief change and related matters over all the years, and to Jürgen Mittelstraß and Johan van Benthem for their sustained support of this project during my times at the Universities of Konstanz and Amsterdam.

It is a pleasure to acknowledge numerous instructive exchanges of ideas with Carlos Areces, Johan van Benthem, Eduardo Fermé, Ulf Friedrichsdorf, André Fuhrmann, Andreas Herzig, Wolfram Hinzen, Hans Kamp, Bernhard Nebel, David Pearce, Peter Schroeder-Heister, Frank Veltman, Heinrich Wansing, Renata Wassermann and Emil Weydert.

An early version of this book was submitted to the Philosophical Faculty of the University of Konstanz in October 1996 as a *Habilitationsschrift*. Sven Ove Hansson, Erik Olsson and Wolfgang Spohn read that version very carefully and have given me valuable comments. I thank them very much for that, as well as for continuing the discussion and doing excellent research in their DFG research group at Konstanz.

I thank Brigitte Weininger for her efficient help in manufacturing the final version of the manuscript.

Finally and most of all, I want thank my family – my parents, my wife and my children – for all their support and understanding.

Regensburg, 8 August 2001
H.R.

CONTENTS

FIGURES

TABLES

0

OVERVIEW

... die Wahrheit liegt bei einem solchen Gegenstand nicht in der Mitte, sondern rundherum wie ein Sack, der mit jeder neuen Meinung, die man hineinstopft, seine Form ändert, aber immer fester wird.

Robert Musil, 'Das hilflose Europa oder Reise vom Hundertsten ins Tausendste' (1922)

General theories of *belief change*—or, more aptly, of *opinion change*—and of *non-monotonic reasoning*—or, more positively, of *defeasible reasoning*—have been flourishing for 20 years now.[1] Today they are actively studied by researchers in philosophy, logic and artificial intelligence. Though not devoid of toilsome logical details, the present book is primarily meant to be a philosophical contribution to the field. Being a philosopher means offering new perspectives.[2] I accordingly attempt to open up two new perspectives on the formation and transformation of belief. The first view focuses on the problem of how to generate, with the help of sophisticated inference operations, coherent belief systems from information bases which may be both incomplete and inconsistent. This aspect is static in so far as no pieces of information are added to, or retracted from, the current stock of beliefs in response to some 'perturbation' from an external source. The sophisticated statics makes it possible to capture dynamic processes by very simple means. The alternative, second view is based on simple, essentially classical inference operations and lays the emphasis on the dynamic side, on non-trivial coherence requirements for the choices which are necessary to resolve incompatibilities between old and new information. From both perspectives close connections may be exhibited with the area of nonmonotonic reasoning in the abstract style of D. Gabbay (1985), D. Makinson (1989) and S. Kraus/ D. Lehmann/M. Magidor (1990). The drawing of inferences represents the *static aspect*; learning something or calling something into question (accepting or rejecting something) represents the *dynamic aspect* of the topics covered in this book. Due to the choice-theoretic perspective taken in major parts the book, one might say that it is about the *making up* and the *changing of one's mind*, where these idiomatic phrases are being (re-)interpreted here as respectively referring to the static and the dynamic aspects of the change of doxastic states.

---

[1] Detailed surveys of the research conducted in about the first dozen of years are offered by Gärdenfors and Rott (1995), Hansson (1999a) and Makinson (1994).

[2] This is not beyond controversy, however.

As a preparation for further analyses, I begin by contrasting doxastic attitudes like believing, being convinced, accepting, having an opinion, assuming,
expecting etc. in Chapter 1. I argue that 'acceptance' is the most appropriate
term for the purposes of the present investigations, since it best expresses the
idea of a 0-1 decision vis-à-vis a certain proposition. For terminological continuity
with existing work and for ease of formulation, however, I keep on using the word
'belief'. No confusion is likely to arise from this slight inaccuracy. We just keep in
mind that we don't require a high degree of certainty to consider a belief, or the
acceptance of a proposition, as legitimate. Belief and acceptance may be—and
in general will be—provisional. Since choice functions will play a central role in
later parts of the book, I ask whether it is in fact possible to *choose* or to *decide*
whether to believe or not to believe something. Such a doxastic voluntarism is
found to be more plausible than its bad reception by philosophers would suggest,
especially when applied to the retracting rather than to the acquiring of a belief.
But we do not have to commit ourselves to an act of free will with respect to
opinions or beliefs, if the only thing we do is to employ abstract choice functions
for processes of belief change.

Having addressed these questions of interpretation, I discuss the format in
which systems of beliefs should be represented in the second part of Chapter 1. It
is argued that it is important to distinguish belief bases, which are typically incoherent, from coherent systems of belief which are called 'belief sets' or 'theories'.
Belief sets are derived from bases by a process of forming the inferential closure.
In general, inference relations suitable for this task must be more sophisticated
than standard logics in that they are required to support both nonmonotonic
and paraconsistent reasoning. Such reasoning is performed on the background of
implicit expectations of the reasoner, and it makes use of operations for changing
expectations and beliefs. The picture is rounded off by orderings of priority or
entrenchment that may be associated with belief bases or belief sets and help
governing inference and revision.

Chapter 2 advocates the thesis that problems of epistemology and problems
of belief change are tightly interwoven. Our aim is to show that a successful
analysis of knowledge depends on a proper solution of the problems of belief
change. What an agent knows at a certain time instant is dependent on how
he or she is inclined to change his or her beliefs in time. This is so at least if
one follows the conception of knowledge set forth in the theory of the influential
contemporary epistemologist K. Lehrer (1990). I briefly recall the architecture of
his system and criticize two of his fundamental concepts, *personal justification*
and *undefeated justification*, as being insufficiently clarified. Both concepts leave
open important problems, and systematic solutions to them must have recourse
to a well-developed theory of belief change.

This argument tries to establish that the theory of (the static concept of)
knowledge needs help from the theory of the dynamics of belief. In the reverse
direction, I argue that the methodological approach to belief revisions one adopts
strongly depends on the fundamental epistemological position that one takes. For

the rest of the book I take advantage of the distinction between foundationalist and coherentist theories that has become central in contemporary epistemology, and turn it into a framework for the analysis of belief representation and revision.

Elaborating on this topic, Chapter 3 opens with the identification of the two dimensions along which processes of belief change are characterized. Five general regulative maxims for belief change are then listed. I distinguish two logical maxims for the statics of belief (consistency and closure, which taken together make up the concept of inferential coherence), two economic maxims for the dynamics of belief (principles of minimal change or conservatism, making up a concept of diachronic coherence), and finally a fifth maxim concerning the coherence of choices involved in non-trivial dynamic processes. Then I introduce two fundamentally different, and as it were orthogonal, perspectives on belief change. From the *vertical perspective*, the static inference operations are interesting and sophisticated, while the dynamic revision operations can in compensation be conceived of as very simple. From the *horizontal perspective*, which underlies the prevailing paradigm of C. Alchourrón, P. Gärdenfors and D. Makinson (1985), the static inference operations are comparatively trivial, but the dynamic revisions operations are interesting and sophisticated. These perspectives determine the further organization of this book. They correspond, in the prototypical cases, to the epistemological distinction between foundationalism and coherentism. What we will call the logic-constrained mode of belief base revision, however, shows that it is possible to combine the horizontal perspective with a foundationalist philosophy.

The third chapter further includes a discussion of the connections between belief revision and nonmonotonic reasoning—i.e., ampliative reasoning that takes into proper account the notorious incompleteness and uncertainty of the information explicitly available to human reasoners. The 'direction' of the connection is dependent on the perspective taken. From the vertical perspective, belief revision results as a by-product of trivial base changes and sophisticated inferences. From the horizontal perspective, one can interpret nonmonotonic reasoning as a process of revising one's implicit background theory of the normal state and development of the world by coherently assimilating pieces of explicit information (D. Poole 1988; Makinson and Gärdenfors 1991), or as a process of withdrawing 'the inconsistent belief', $\bot$, from a suitable combination of expectations and information. Roughly speaking, the vertical perspective reduces the dynamics of belief to problems of the statics of expectation and information, while the horizontal perspective explains the statics of belief as a dynamic process of changing one's expectations.

Chapter 4 presents and comments on three long lists of rationality postulates that have either played a major role in the literature or will play such a role in later parts of the present book. I use two identities (due to I. Levi 1977 and to Makinson and Gärdenfors 1991) as conceptual bridges that help in exhibiting a one-to-one correspondence between postulates for belief contraction, postulates for belief revision and postulates for non-monotonic inference operations. Existing work is further developed by extending the list of postulates considered and by pinning down case-by-case the exact parallels between the lists of postulates. All these postulates may be called postulates of theoretical reason that act as regulative principles for the drawing of inferences (static dimension) and for the processing of changes of information (dynamic dimension).

The central foundationalist part of the book is expounded in Chapter 5. Here I distinguish an information Base with a (possibly empty) priority ranking of its elements from the full set of accepted sentences (opinions) that can be derived from the base. For the derivation to make sense, we cannot apply a standard logic, but have to use an inference operation that is capable of modelling everyday reasoning. Such reasoning is in general nonmonotonic (that is, bold, credulous, ampliative, information-tropic) and paraconsistent (that is, cautious enough not to let classically inconsistent bases burst into a doxastic explosion[3]). The inference relations considered are thus stronger in some respects and weaker in other respects than classical logic. Belief revision in this mode is performed by direct, more or less trivial changes on the base level, with a precise understanding of 'trivial': Such changes do not require the agent to solve any problems of choice. Through the sophisticated inference relation, the trivial base change operations generate, as a by-product as it were, non-trivial change operations on the level of coherent belief sets. Two immediate ways of making this idea precise using the revision-of-expectations idea of Gärdenfors and Makinson fail due to fundamental problems. A third alternative based on the notion of consolidation is proposed, and its properties are investigated. Its most striking, unusual, but only seemingly counterintuitive, feature is that it violates a very basic postulate of belief preservation.

Chapter 6 leaves foundational belief change and returns to the fifth coherence maxim of the third chapter. I rehearse and discuss the *general theory of rational choice* as initiated in the middle of our century by leading economists like P. Samuelson, K. Arrow and, somewhat later, by A. Sen. In its classic form, it puts up criteria of coherence stating how decisions in different choice situations must hang together in order to be 'rationalizable' by an underlying, context-independent pattern of (hopefully transitive or modular) preferences. I slightly extend the crystalline picture of the classic theory by paying special attention to agents that do not really have to choose (because there are absolutely satisfactory options), to agents that refuse to choose (because all options are taboo) and agents whose choices are not rationalizable in the above sense, but still show a form of rationality which is captured by a condition of 'path independence'. No mention of inferences or revisions is made in Chapter 6; it deals exclusively with an abstract and highly idealized, but empirically contentful theory of practical reason.

Applying the theory of rational choice to the problems treated before, Chapter 7 presents the central coherentist part of the book. It confronts the (idealized) account of practical reason given in Chapter 6 with the characterization of belief

---

[3]For definitions of paraconsistency and explosiveness, see Section 1.2.5 below.

change and the (slightly de-idealized) logic of everyday, nonmonotonic reasoning as laid down in Chapter 4. The theoretical problem of belief revision is thus understood as a problem of rational choice. More exactly, the problem consists in the general specification *which* parts of the background theory must be given up in response to *which* informational input. The coherentist perspective is taken on both a semantic and a syntactic level: Which possible worlds are chosen as the most plausible ones, relative to the context of a required belief change? Which beliefs are chosen as the ones that are most readily given up, again relative to the context of a required belief change? I define suitable rules of constructing belief contractions and revisions based on the relevant choices, and study how the logical rationality postulates for belief change and nonmonotonic reasoning relate to the coherence criteria for rational choice.

The central and most surprising result of this chapter is that all postulates of theoretical reason are derivable from more general, practical principles of rational choice. This can even be done in a such way that the former postulates correspond to the latter principles—which are motivated by entirely independent considerations—in a one-to-one fashion. And for all this, it is irrelevant, with the exception of a single case, whether we work on the semantic or the syntactic level! Technically, I substantiate this derivation by a number of *representation theorems*: All operations of belief change or nonmonotonic logic that are generated by rational choice functions, with the choices satisfying certain coherence constraints, satisfy corresponding rationality postulates ('Soundness'). And conversely, all operations of belief change or nonmonotonic logic that satisfy certain rationality postulates can be represented as operations that are generated by rational choice functions, with the choices satisfying corresponding coherence constraints ('Completeness'). Philosophically, I take this to be a strong indication of the *unity of practical and theoretical reason*. It remains to be clarified in future work, however, how the parallelism of postulates can be fleshed out to actually support such a conclusion.

There is, however, a slight shift in the relative plausibilities of the various postulates. The choice-theoretic perspective favours relationalizability as the main criterion of rationality, but in the general case (in which the preference relation is not necessarily transitive) this results in a quite unusual logical system. From the logical point of view, the principle of path independence for choices turns out to characterize a more agreeable and uncontroversial concept of rationality. As a limiting case, ordinary monotonic logics turn out to be logics based on 'satisficing' choices in the sense of H. Simon.

In the final part of this chapter on coherentist belief change, I give an explanation for the surprising observation that the same correspondences between choice criteria and logical postulates hold on the semantic level as we find them on the syntactic level. This is due to the fact that a choice function on the syntactic level may be seen as generating an equivalent choice function on the semantic level *under preservation of all choice-theoretic features.* The same is true for the reverse direction, except for one notable case.

In Chapter 8 I continue the theme of the previous chapter and study the associated theory of *revealed preferences* that may be considered as an integral part of theory of rational choice. The semantic level would not offer any new insights. But the syntactic level is interesting, since there not only choice-theoretic constraints, but also two logical constraints are to be taken into account. The theory of *epistemic entrenchment* as a basis for belief change has been developed by Gärdenfors and Makinson (1988) and generalized by Rott (1992b). The axioms of entrenchment relations were rather selected for their elegance and simplicity, independently of any considerations of rational choice. In this final chapter I show that if we take an entrenchment relation to represent the preferences revealed or manifested by a syntactic choice function as introduced in Chapter 7, then the theory of entrenchment can be fully reconstructed and decomposed into modules of varying degrees of plausibility. A carefully motivated translation procedure demonstrates which conditions for epistemic entrenchment may be derived from conditions for rational choice. I separate structural, purely choice-theoretic and purely logical components in the theory of entrenchment. This procedure results, I think, in the most clear-cut and natural interpretation of the concept of epistemic entrenchment available. I close this chapter with sections on the semantics of entrenchment relations, on two quite different methods of applying them in the construction of belief changes, and on their use in the analysis of a sceptical approach to the so-called relational belief change.[4]

This book is intended to satisfy the needs of the philosopher interested in general conceptual analyses as well as the more specialized logician or AI researcher. It is perhaps best to read it as an essay on the notion of *rationality*, which in turn is explicated mainly by means of various concepts of *coherence*. I shall try to elucidate carefully the underlying motivation of my formal framework and locate this work in the broader context of epistemological theories. At the same time I set a high value upon demonstrating rigorously that my more sweeping philosophical claims—like 'Principles of theoretical reason are derivable from principles of practical reason'—are capable of a precise formulation and verification. Thus I have attempted to be more careful in justifying and motivating what I do than logic texts usually are, and to be more precise than philosophy texts usually are. I hope that the reader, whether philosopher, logician or AI researcher, will find some new contribution to the subject. It may also be interesting for economists to see how their ideas about rational choice

---

[4] The enterprise of Chapter 8 is reconstructive in nature. But of course one may seek alternative *constructive interpretations* of the relation of epistemic entrenchment. In the special context of (prioritized) belief bases, such interpretations are offered and analysed in Rott (2000a) which may serve as a good supplement to our final chapter. In that paper a positive idea based on the quality of 'proof sets' for the sentences to be compared is distinguished from a negative idea based on the quality of 'remainder sets', and it is shown that the negative relation is in fact a refinement of the positive one. It turns out that only the negative idea of epistemic entrenchment is suitable for a reproduction of a foundationalist approach to belief change on the level of coherent theories, and even this reproduction can serve only as an approximation.

can be put to use in the field of epistemology and logic. Papers by theoretically oriented economists like Arrow, Chernoff, Samuelson, Sen or Simon contain sophisticated and far-reaching reflections on the concept of rationality that should not be neglected by any philosopher working on that concept.

I have decided to separate the more philosophical aspects of this book from the more technical ones. For this reason the reader will find quite a number of announcements and promises in the first three chapters which may leave him in temporary frustration. I hope, however, that the concrete constructions and results in later chapters show that the promises are actually kept. This may involve some inconvenience in moving back and forth in the text for the reader. But I found it desirable to have some informal reflection my approach in the first part of the book that is not pervaded by formal details. I have relegated almost all proofs (for Chapters 4–8) to appendices. This is done in order to make the line of argument in the main text more transparent. Needless to say, the reader is invited to study the proofs in detail; only then will he or she understand what is really involved in the arguments.

This book also contains a number of tables and figures. The tables in particular are inserted as a means of helping the reader to keep track of the parallels that hold between contractions, revisions, nonmonotonic inference operations, choice functions and revealed preference relations (*alias* entrenchment relations). At the end of the book the reader may find an index of symbols with brief explanations of their meanings.

In various parts of Chapters 6–8 the reader will find traces of Rott (1992b, 1993, 1994). However, I have extended and revised not only the main line of argument but also almost every detail of the material to be found there. So although there is a certain amount of overlap with previously published papers, I think that it is fair to say that the book is an entirely new piece of work. A short report on the central results of Chapter 7 appeared as Rott (1998).

Recent years have seen much good work in the fields covered in this book. Of the many things I could mention, I first want to refer to Sven Ove Hansson's authoritative *Textbook on Belief Dynamics* (1999a). Then there are three recent products particularly relevant to the topics treated here that were written largely in parallel with the present work. This is John Cantwell's Uppsala dissertation *Non-Linear Belief Revision* (2000), Alexander Bochman's forthcoming book *A Logical Theory of Nonmonotonic Inference and Belief Change* (2000c) and the collection *Frontiers in Belief Revision* edited by Mary-Anne Williams and myself (2001). I would have liked to think about what is contained in these works and relate it to my own line of thought. As time was too short to do that, I can only recommend warmly to anyone interested in belief dynamics and nonmonotonic reasoning to have a look at these books.

In most of this book, I shall refer to agents, reasoners etc. with the male anaphora 'he', 'him' and 'his' only. I have found that this is the easiest and most natural way of expressing myself, but it is of course understood that the capacities to act, to choose, to reason etc. are as well-developed in women as they are in men.

1

DOXASTIC STATES AND THEIR REPRESENTATION

1.1 The menagerie of doxastic states

How do we make up our minds? How should we make up our minds? How do
we change our minds? How should we change our minds? This book attempts
to give answers to these questions, descriptive and normative questions about
the *formation* and the *transformation* of mental states. Only particular types
of mental states, however, are dealt with in this book, states which are most
commonly referred to as *beliefs*. I shall not be concerned with other mental
states, in particular not with desires or intentions. So the questions put above
may seem somewhat misleading, since they suggest that the book deals with
the formation of decisions, intentions or plans for action, whereas in fact it only
deals with beliefs, or 'belief-like' states of mind. But the allusion to decisions and
choices is deliberate. It is motivated by the fact that a large part of the following
is based on the formal concept of choice or selection functions—and the talk
about making up one's mind and changing one's mind above just indicates that
the mental states we are concerned with, beliefs, are something which can be
chosen or decided about.

The idea of explicating belief and belief change in terms of choices is certainly
in need of justification. In the present chapter I shall explain my use of the term
'belief'. Several dimensions are listed which allow us to distinguish a wide variety
of doxastic concepts. Our treatment of these topics will have to be superficial—
an adequate discussion of the many subtleties involved would require a book of
its own. But having identified and demarcated various doxastic concepts, I shall
feel free in being more sloppy later on and talk of beliefs in contexts where it is
doubtful whether that term is justly applied. When I do so, I do it only for the
sake of easy and comprehensible expression. Whenever I speak of beliefs I really
mean mental states, doxastic states, which resemble beliefs in some important
respects but are not necessarily most adequately referred to as beliefs. It is the
purpose of the present chapter to help clarifying this issue.

Belief as I will understand the concept is a propositional attitude. When I
talk about beliefs, I have in mind *believing that* something is the case. I shall
exclude from my discussion *beliefs in* somebody or something, as well as *beliefs
of* somebody or something (*de re* beliefs and *de se* beliefs). If Pierre believes of
London that it is beautiful, there is a relation between the person Pierre, the
city of London and the property of being beautiful. Such relations are beyond
the scope of the present book. I restrict myself to cases of belief where the object
of belief is a proposition. A *proposition* is something that can be expressed by

a declarative sentence (a statement) in a certain language. It is the meaning or
the content of such a sentence, and I identify it with the content of the beliefs
themselves. I take it that a pre-theoretical characterization of propositions is
sufficient for my purposes. I assume that the object language is given, and I
shall not analyse the internal structure of the sentences of this language at a
level that is finer than the level of analysis met within propositional logic. These
assumptions are made not because I think that a non-trivial analysis of the
subsentential level is uninteresting or even unnecessary, but rather because I
have found the problems involved at the propositional level substantial enough
to justify a treatment of their own.

How do we know what a person believes? When somebody believes that ϕ is
the case, it is certainly neither necessary nor sufficient that she openly states or
asserts that ϕ is the case. She might be mute, or she might be lying. But even
if we suppose that she makes an honest statement as to what her beliefs are,
there is no guarantee that it is in fact her beliefs that she is describing. First,
a person may be the victim of self-deception or unable to recognize explicitly
and in public something that she has long been recognizing for herself. Second,
a person may be unable to find the right words for what she believes. Even the
most honest report of one's beliefs may be grossly inaccurate if the person does
not fully master the language in which she formulates her beliefs.

From a third-person perspective, there is no other way than to frame hypotheses about another person's beliefs on the basis of a great variety of clues
which need not all point in the same direction. We surely pay much attention to
what a person utters, to her plain statements as well as to her statements about
her own beliefs. But not in every case shall we take these statements at face value.
We compare them and try to bring them in harmony with each other and with
further information we have about the person, with the statements she made
earlier and in particular with what we observe about her behaviour. Ascribing
beliefs is part of a larger project: that of making sense of the agent, on in other
word, of understanding the agent as a 'rational' being. If Harry declares that he
does not at all believe that there is a spider in the kitchen or that his former girl
friend is going to ring him up, but he continues to look concerned at the walls
and ceilings of the kitchen and cast expectant looks at the telephone, then we
at least doubt whether Harry has given adequate expression to his beliefs. The
same is true if he complains about his arthritis and tries to get relief from his
pains by massaging his calf muscles. There is no immediate and foolproof way
of 'finding out' what a person 'really' believes. It is only as a result of a complex
and varied process that we *ascribe* beliefs to a person. Even the most careful and
successful examination of a case of killing in alleged self-defence cannot claim to
establish definitely whether the sued person had in fact believed that the burglar
carried a gun when he shot the burglar. The testimony of the defendant is not
a very reliable indicator, even if it is made without any conscious distortion of
the truth. It is only after all things have been considered that a judgement can
be passed about the defendant's beliefs.

The ascription of beliefs then is a holistic process, that has to take into account linguistic as well as extra-linguistic evidence and cannot be separated from the ascription of other propositional attitudes of an agent. This process is an integral part of what has been discussed under the heading 'radical interpretation'.[5] In everyday life we do not easily recognize that we face the task of radical interpretation, since we usually know quite a lot about the people we meet and have conversations with. But this fact does not change the holistic character of the task, it only limits down the number of alternative interpretations. The talk about mental states must be chosen carefully. It does not seem appropriate to say that a person at a certain time is in a multitude of mental states simultaneously. It makes more sense to say that a single mind at a certain time is in a unique mental state that supervenes on its unique physical state. Aspects of this state can be captured by ascribing propositional attitudes to the person, and it is only through the ascription that the multiplicity which is typical of complex minds like ours enters the stage.

That the ascription of beliefs is always dependent on a large range of factors does not entail that nothing can be said about beliefs taken in isolation. In fact the validity of much of what I say in the present book depends on the thesis that if all these factors remain constant—*ceteris paribus*—then one can say something about the workings of belief in processes of inference and change. Although a very wide context is necessary for an external, 'objective' observer to find out about (or rather: to form sensible hypotheses concerning) the beliefs of somebody, we may be confident that we are able to discern a logic of belief and changes of belief—as long as that context of interpretation does not change. This discerning is not a discovering, but again a matter of ascription. What we may be confident of is that we find a holistic interpretation of the agent that satisfies the constraints imposed by a good logic of belief and belief change.[6]

Belief, then, is part of a larger pattern of attitudes ascribed to an agent by external observers that try to rationalize the agent's behaviour. It follows that the belief component itself comes out as *rational* and *coherent* and *well-balanced* to a degree that would never be obtained if belief were only what the agent explicitly declares to believe, or would explicitly declare to believe if persistently interviewed. It would be closer to the truth (though not altogether true) to say that the beliefs we take the agent to hold are those he is committed to, given his declarations of belief. In a different (and again not fully accurate) terminology, we might say that we are concerned with the competence, not with the performance, of believers. For the most part of this book I base my arguments on the understanding that what the agent 'really' believes (if only implicitly) is *logically consistent* and *logically closed*. I shall say more about what I mean by 'logic' here in Sections 1.2.4 and 1.2.5. Notice that in this understanding

---

[5]Davidson (1973) and Lewis (1974) list factors that must be taken into account when interpreting the utterances and actions of a person.

[6]Thanks to Wolfgang Spohn for helping me to express this point more clearly.

consistency and closure are neither descriptive idealizations of, nor normative standards for, the more irregular belief states of real-life believers; it is rather the notion of belief I use that makes these properties natural properties of belief states—provided that the agent's behaviour can be rationalized at all. Logical consistency and closure of the set of ascribed beliefs are thus taken to be partial explicates of the explicanda of the rationality, coherence or well-balancedness of the agent's mental state.

Why should we want to ascribe beliefs? Because we want to use beliefs for the explanation and prediction of human behaviour.[7] Together with desires (and possibly other factors), beliefs form the basis of many successful accounts of why humans act in the way they do act. Taking 'the intentional stance' is very often the only way of making sense of our behaviour. Ascribing a coherent (well-balanced) and stable (situation-independent) pattern of beliefs and desires to a person that can account for the observable motions of her body turns mere *behaviour* into purposive *action*. From a scientific point of view, it would be equally satisfactory, and perhaps preferable, to understand why we do what we do in terms of purely causal mechanisms, to give a purely physical explanation of our behaviour. However, only a tiny fraction can be captured by such explanations. The vast majority of our deeds can be understood only in intentional, psychological terms—if it can be understood at all. It may well be that a Laplacian demon could predict the course of the world in all accuracy, using only descriptive means from the repertory of the natural sciences. But there are principal bounds for humans in this respect, bounds that will never be transcended (or so it seems). Due to the chaotic constitution of the world, the tiniest variations in initial conditions often lead very soon to the most striking divergencies in future developments of a certain situation. This seems all the more true for such an incredibly complex system as the human brain. So it may be advisable to refrain from attempting a causal-physical explanation right from the start, and look at fundamentally different similarity classes that might be used in the explanation and prediction of human activity: those offered by the intentional concepts.

In such a picture, believing that ϕ is true is very much like acting as if ϕ were true or, more accurately, showing a disposition to act in a way that could be construed as rational if it were taken to be based on the assumption that ϕ is

---

[7] I do not see any reason why higher animal behaviour should not be successfully explainable and predictable by the same kind of psychological ascriptions. For lower forms of life, scientists may well be successful with purely causal, physical explanations and predictions, which would then be preferable to less clear-cut explanations and predictions in terms of beliefs and desires. But it may even make sense to describe simple inanimate devices in terms of psychological states. See, for a provocative example, McCarthy (1979, Section 4.1) about thermostats: 'The simplest belief predicate ... ascribes belief to only three sentences: "The room is too cold", "The room is too hot", and "The room is OK"—the beliefs being assigned to states in the obvious way. We ascribe to it the goal, "The room should be OK". When the thermostat believes the room is too cold or too hot, it sends a message saying so to the furnace.' This is not supposed to be metaphorical talk.

true. In this way belief gets intimately tied up with action. Believing something, however, is not itself performing a certain kind of action. And there need be no occurrence of an event (in the mind or in the brain) that may be referred to as the belief-that-ϕ.

There are a number of doxastic concepts that are much more like actions than belief itself. All of these concepts happen to begin with the letter 'a'. An agent can assert a proposition, he can affirm it, or assent to it, or accept it. Like a statement, an *assertion* is usually understood to be an explicit speech-act. It is certainly to be classified as an action and can thus be analysed by action-theoretic means. In the standard cases statements and assertions give expression to a belief of the speaker, and it is not the particular choice of the words or the syntactic structuring of the sentence uttered that is important, but rather its content. *Affirming* a proposition or *assenting* to it usually does not involve uttering a sentence expressing that proposition. The typical situation would rather be that the agent is confronted with an occurrence of the sentence and gives some signal of approval. This may, but need not, consist in the uttering of words or sounds. A nod, a wink or thumbs up would do just as well. In any case, however, affirmation and assent involve an action, a manifest movement that would most naturally be accounted for in non-physical terms.

Acceptance is perhaps the most iridescent and versatile term in the family of doxastic concepts. Unlike assertion, affirmation and assent it does not necessarily involve any observable behaviour. But it does involve some kind of 'inner action', some jerk that makes the agent pass from an earlier, somehow unresolved state of mind to another state in which something has been settled. In many cultural and institutional contexts acts of acceptance are performed in accordance with a formalized procedure. In the court-room the judge decides which pieces of evidence to admit. For instance, it may be formally accepted that the fingerprint, say, shows that the defendant was at the scene of the crime on the day when it was committed. Statisticians know many very elaborate rules for accepting or rejecting statistical hypotheses.[8] In scientific practice rules for accepting or rejecting theories are not expressly stated, and some philosophers of science have held that scientists should never commit themselves to either the truth or the falsity of a scientific hypothesis or theory. That does not mean, however, that theories are not accepted as tools or auxiliary devices in practical contexts or when other threads of the web of scientific theories are being tested. Scientists use theories, as if they were true, as long as there are no better theories around.[9]

---

[8] For rules of acceptance and rejection in statistics, see Berger (1985) and Lehmann (1986). A formal unification of the theories of estimation and hypothesis testing was achieved in Abraham Wald's 1950 general theory of decision procedures. See also Hilpinen (1968).

[9] Clearly, the term 'acceptance' is multi-faceted even within the context of philosophy of science. In the year 1968, for instance, each of the following philosophers of science has distinguished three different ways of interpreting the 'acceptance' of a scientific theory: Bar-Hillel (1968, p. 124), Giedymin (1968), and Lakatos (1968, Section 6). It is striking that in scientific methodology as well as in inductive logic in the 1960s there has been an intensive discussion about 'rules of acceptance' but very little was said in this field about 'rules of rejection' of

In all these cases there has to be some *decision* that a certain piece of evidence, a certain hypothesis or a certain theory will not be questioned for the time being, and a quite clearly identifiable point in time which marks the acceptance of it. Acceptance is thus much more like an action than just a disposition to exhibit some special behaviour in some special contexts. Acceptance manifests itself in some kind of occurrent event, in some observable or inner act; belief does not necessarily do so. I have been believing, if only implicitly, that 365 plus 366 equals 731 for many years, but I have never based any further thoughts or acts on my belief that this particular proposition is true. This is why it would seem strange to say that I have accepted that 365 plus 366 equals 731. There is no such thing as a merely implicit, dispositional acceptance of a proposition.

Insofar as a doxastic concept involves some action, it involves voluntary components. Actions are different from behaviour: the agent has control over the action, he may perform it or refrain from performing it, and he is held responsible for the action and its (immediate) consequences. Belief itself, in its most typical instances, is not voluntary. I cannot believe that a pink elephant is sitting on my desk or that Harry is my brother, even if I wished to believe it.[10]

Quite a number of philosophers have disagreed. If belief were involuntary, then nobody could be reproached for believing or not believing something. Still we sometimes think that somebody should have known (and thus, believed) ϕ or that he should not have believed ψ. In fact, philosophers have suggested that there is an *ethics of belief*. Most well-known for his doctrine that belief can be controlled by the will is Descartes, but also Locke and Hume were involved in the discussion whether there can possibly be an ethics of belief.[11] In the

previously accepted hypotheses.

[10]The most eminent philosopher denying that belief can be voluntarily adopted is of course David Hume: 'belief consists ... in something, that depends not on the will, but must arise from certain determinate causes and principles, of which we are not masters.' (Hume 1978, p. 624). More recently, Bernard Williams (1970) has claimed that although one can of course decide to assert something, one cannot decide to believe something. Still Williams holds that the concepts of assertion and belief are intimately related: 'our very concept of assertion is tied to the notion of deciding to say something which does or does not mirror what you believe. ... for belief, full-blown belief, we need the possibility of deliberate reticence, not saying what I believe, and of insincerity, saying something other than what I believe.' (Williams 1970, pp. 105–106) Incidentally, this is why he thinks that computers may be *knowers*, but can never be *believers*. His argument for this, however, is begging the question, since it is based on the premise (p. 104) that computers cannot lie. The advance of slightly more 'intelligent' systems, I take it, has invalidated that premise. One can easily program computers to tell 'conscious' lies, if they judge it expedient to do so.

[11]See Descartes in the *Principles* (1985, Part 1, Principle 34, p. 204): 'In order to make a judgement, ... the will is also required so that, once something is perceived in some manner, our assent may then be given.' and in the fourth *Meditation* (1984, p. 60): 'If ... I simply refrain from making a judgement in cases where I do not perceive the truth with sufficient clarity and distinctness, then it is clear that I am behaving correctly and avoiding error. But if in such cases I either affirm or deny, then I am not using my free will correctly. ... In this incorrect use of free will may be found the privation which constitutes the essence of error.'

Descartes was immediately criticized for this doctrine by Hobbes (see Descartes 1984, Third Set of Objections, 13th Objection, p. 134): 'it is not only knowing something to be true that is

nineteenth century, there was a heated discussion whether religious faith can be reconciled with the critical method of the advancing sciences, culminating in William James's notorious 1897 paper on 'The Will to Believe'.[12] In our century, C.I. Lewis, R. Firth, R. Chisholm, and B. van Fraassen are among those that have endorsed, in varying forms and degrees, doxastic voluntarism. K. Lehrer draws a distinction between belief and acceptance: 'Acceptance, unlike belief, is a matter of choice.'[13]

The common sense view seems to be somewhat like this. In a particular situation it is not—or, at best, hardly—possible to decide whether to adopt or not to adopt a belief.[14] However, beliefs or non-beliefs may be cultivated by systematically seeking favourable and avoiding unfavourable evidence. Believers have a duty to make sure that they believe the truth, the whole truth and nothing but the truth. The efforts they put into their investigations into the truth should be well-proportioned in relation to the importance of the issue. If we risk only a drink in the pub, we need not be too fussy about the credentials of a belief. But if somebody's life is at stake, we must be extremely careful in accepting uncertain information. In any case, if somebody neglects his duty to take account of the available evidence or seek further evidence, but cultivates his belief (or non-belief) in an irresponsible manner, then he may be blamed for believing (respectively, not believing) what he ought not to believe (respectively, what he ought to believe).

I have no qualms about this view, but I want to supplement it. The point I want to make is that the acquisition of our first beliefs is more or less automatic. For one thing, children have some natural credulity. For another, even if they were more sceptical, children would have no alternative to ascribing true beliefs to the adults they talk to during the long process of acquiring their mother tongue. By the time they are teens, they have simultaneously acquired a language and a huge corpus of beliefs (most, but of course not all, of which are true). So it appears that during this initial phase of the cognitive development there is very little to choose or decide. A more appropriate model seems to be provided by the behaviourist conditioned-response model described in Quine's writings on the philosophy of language. Only after having mastered the language and—by the same token—

---

independent of the will, but also believing it or giving assent to it. If something is proved by valid arguments, or is reported as credible, we believe it whether we want to or not.' (Descartes's reply is curt and obscure.) According to Locke, there is an ethics of belief, according to Hume there is none. Illuminating analyses on the ethics of belief in Descartes, Locke and Hume may be found in Curley (1975), Passmore (1986) and Passmore (1980) respectively. See also Price (1969, Series I, Lecture 10) and Wolterstorff (1996).

[12] James (1896/1979). James later thought that 'The Right to Believe' would have been a much better title for his book. A good account of James's doctrine is given by Wernham (1987). A very useful reader of the ethics-of-belief debate in the nineteenth century, including James and his context of discussion, is available in McCarthy (1986).

[13] Lehrer (1975, p. 149). Lehrer's concept of acceptance is put in perspective and subjected to a close investigation by Piller (1991). A view similar to Lehrer's has recently been taken and traced out in considerable detail by Cohen (1992).

[14] It is not the issue here to say when such a decision may be called rational or irrational.

having acquired knowledge about the basic facts of life, persons start reflecting about their own beliefs and about the information they receive from all kinds of sources. To some extent, they are then capable of deciding which new beliefs to *acquire*. But here the degrees of freedom seem to be rather limited.[15] It is much more plausible to locate crucial decisions in the *elimination* of beliefs. In the first, credulous phase just described, as well as on many occasions later on, one acquires, as a rule, too many beliefs and will later find quite a number of them incompatible with one's experiences.[16] In the face of recalcitrant experiences made later, it will usually not do just to give up a single item of belief, since the individual beliefs of a believer in a well-balanced state of mind are usually backed up by quite a number of 'surrounding' beliefs. Some of these have to be given up along with the falsified item, some of them may stay, and the believer has to make a choice as to *which of them* should be given up and which can stay. The resolution of logical conflicts is *not* a purely logical task. If I find that the radio-receiver does not work tonight, I not only withdraw my original belief that the radio works well. I also have to withdraw some of the (truly naive) beliefs which taken together entail that the radio is not damaged. The thesis I want to advance here is that doxastic voluntarism applies to the elimination rather than to the acquisition of beliefs. This view is of course not new. In fact it may be traced back to Descartes. Let us quote from Price:

... when Descartes maintains that assent is an act of will, what chiefly interests him, of course, is the negative side of this doctrine: not so much that it is in our power to assent as we choose, but that it is in our power to withhold assent as we choose. The important point about his 'freedom of judgement', in his eyes at least, is that it is a freedom to suspend judgement.[17]

Roughly, we may say that although believing something is not itself performing an action, deliberately changing one's beliefs is.[18]

---

[15] William James's (1896/1979, p. 20) classical statement of doxastic voluntarism applies to belief acquisition: "Our passional nature not only lawfully may, but must, decide an option between propositions, whenever it is a genuine option that cannot by its nature be decided on intellectual grounds; for to say, under such circumstances, 'Do not decide, but leave the question open,' is itself a passional decision—just like deciding yes or no—and is attended with the same risk of losing the truth." (The whole passage is italicized by James.)—Acquiring new beliefs is not as simple as one might think. One has to harmonize new beliefs not only with old beliefs, but also with background presumptions, presuppositions or prejudices which may be partly subconscious. New information is very likely to generate conflicts with old background assumptions, and this is precisely why we need a dynamic model for belief formation. Compare Chapter 5.

[16] Perhaps credulity is not limited to youthful naïveté. Peirce's (1982, p. 66) famous dictum seems to come to that effect: 'Doubt is an uneasy and dissatisfied state from which we struggle to free ourselves and pass into the state of belief; while the latter is a calm and satisfactory state which we do not wish to avoid, or to change to a belief in anything else.' Peirce, however, is advocating inquiry, not credulity as the road to belief.

[17] Price (1969, p. 224). Popper's (1976, p. 87) dictum 'Belief, of course, is never rational: it is rational to *suspend* belief.' (Popper's emphasis) seems to be relevant, but in fact he just rephrases his falsificationist doctrine that empirical findings only have a negative force vis-à-vis scientific hypotheses or laws.

[18] Logicians have discovered actions as a field of research. Presently, 'dynamic logic' (Harel 1984) is the most prominent framework for a logical modelling of actions. Building on work in

I have said that I believe that 365 plus 366 equals 731 but I do not accept that proposition, because usually there is no manifestation of this belief. (Let us disregard the fact that *now*, in writing these lines, I make that belief manifest and thus perhaps do accept the proposition.) There are more reasons why it would seem strange to say that I have accepted that 365 plus 366 equals 731. When we speak of acceptance we usually imply that there was a possibility of rejection, however remote; that there are some reasons which could at least possibly make one decide not to accept the hypothesis in question. If there is nothing to choose between, there is hardly room for acceptance. Nobody has ever had serious doubts that 365 plus 366 really equals 731, so nobody has felt the need to accept that equation.

This leads us to another dimension along which to distinguish doxastic states. The concept of acceptance encompasses propositional attitudes weaker than beliefs. I may accept the predictions given by the weather forecast, although I would be hard pressed if I had to say that I really believe that what it says is true. If I say 'I think that ϕ', this leaves room for various strengths of endorsement of the proposition ϕ. This can range from vague expectations and uninformed guesses, via various sorts of opinions, to fully-fledged beliefs and firm convictions. There is no clear line to be drawn between these kinds of doxastic concepts, but they clearly differ in the degree to which an agent is willing to commit himself to the truth of the things accepted. It will be a major point of the present book that it focuses not only on what can count as a fully endorsed belief but also on weaker, less firm notions of acceptance.

Robert Stalnaker (1984, p. 79) uses 'acceptance' as a 'generic propositional attitude concept with such notions as presupposing, presuming, postulating, positing, assuming and supposing as well as believing falling under it'. To accept a proposition, he says, is 'to treat it as a true proposition in one way or another— to ignore, for the moment, the possibility that it is false (p. 79), to 'act ... as if one believed it' (p. 80). In his view, acceptance may be passive, provisional, compartmentalized and (self-)imposed (pp. 80–81). Stalnaker's principal insights deserve to be taken seriously. Traditionally, philosophy and philosophical logic have tended to focus on particularly *strong* forms of beliefs, namely, the kind of true justified belief we call knowledge.[19] Stalnaker claims that it is of equal importance for the understanding of human behaviour to analyse *weak* forms of belief.[20]

Belief is not the only attitude that is relevant to the cognitive situation of inquirers. Inquirers make posits, presumptions, assumptions and presuppositions as well. These

---

this field, Segerberg (1995, p. 536) and van Linder, van der Hoek and Meyer (1995) speak of '(doxastic or epistemic) actions' and 'actions that make you change your mind'.

[19] We are going to treat knowledge and the very special way in which knowledge involves strong forms of belief in Chapter 2.

[20] Weak forms of beliefs are of course dealt with extensively in Bayesian theories. The charge of negligence in this respect can only be directed against non-probabilistic approaches. Recently, the discovery of how important nonmonotonic reasoning is for AI systems has confirmed Stalnaker's claim. See Sections 1.2.3 and 1.2.5 below.

methodological attitudes may diverge from belief in various ways, giving rise to additional complexity in a representation of a doxastic situation. But, I suggested, the cluster of propositional attitudes which were grouped together under the label *acceptance* share a common structure with belief. (Stalnaker 1984, p. 99)

I agree with Stalnaker and shall offer a precise explication of the common structure alluded to by Stalnaker.

Stalnaker imposes logical constraints on acceptance states, but he removes much of the effects of his idealizing assumption by holding that an agent usually is in several acceptance states at the same time—states that may well be logically incompatible.[21] These acceptance states represent different compartments of the agent's mind. But there are at least two problems. In practical reasoning, Stalnaker's inquirers must somehow strive to integrate conflicting compartments of mind, and it is not clear how this should be done. And second, it is difficult to see how the compartments-of-mind account fits together with the idea that we want to explain and understand human behaviour through the ascription of a coherent pattern of acceptance and belief. In our modelling, agents will always be understood to be in a single doxastic state.

Quite a lot of work has been done in the past two decades on a particularly weak form of propositional attitude. It has been recognized by researchers working on practical applications in Artificial Intelligence (notably in the construction of so-called expert systems) that an extremely important role in practical reasoning tasks is played by implicit *default assumptions*. These are assumptions about the *normal* or *typical* state and development of the world in general, or about the *normal* or *typical* features of relevant situations or individuals in particular. Normally or typically, the persons we meet in the corridor are not crazy, cars that have just been parked around the corner are still there, birds can fly, students do not have a high salary, and so on and so on.

One would usually not become conscious of such pieces of information, nor would one think of explicitly entering them into some database—this kind of information seems just too trivial for us humans to ever mention them (except perhaps in some conversations with little children). Still it turns out that in order to deduce formally which action should be taken in a situation one has to represent and process the default assumptions. Such assumptions are very weak in the sense that they are easily defeated. Should anybody provide us with the information that there are a number or crazy colleagues in the department, that the police is removing cars from improper places, that the bird is a penguin, that the student is a business consultant, then we are ready to cancel our initial assumptions. Before the advent of Artificial Intelligence, default assumptions have been called presumptions, expectations, presuppositions, or prejudices. In all normal or typical situations, these are just the propositions that are taken for granted and acted upon without reflection. In exceptional cases, however, they are—and should be—given up as easily as they are relied upon in the non-

---

[21] Stalnaker hereby aims at circumventing the 'problem of logical omniscience'.

exceptional ones. For a more extensive discussion of default assumptions I refer the reader to Section 1.2.3.

There is yet another aspect which casts an illuminating light on different sorts of doxastic attitudes. It concerns the question as to whether beliefs come in various degrees, or whether there are only belief, disbelief and a neutral abstention from taking a doxastic stance. Is belief (or are its siblings) a continuous and analogical, or is it rather a discrete and digital matter? This question is closely connected with the question of whether a doxastic concept is more like a *state*, or more like an *action*, *decision* or *intention to act*. There is no such thing as a partial action or decision. Either one performs an action or one does not, one either makes a decision, forms an intention or one does not. It is not possible to act or decide or intend partially. In contrast, it makes perfectly good sense to speak of partial belief. A believer can believe something in various degrees, just as he can be tired, nervous and intelligent in various degrees. Many authors have argued that the degree of belief that a proposition is true can be, or should be, proportional to the evidence that speaks in its favour. There is a huge literature on subjective probability theory which interprets subjective probability precisely as partial belief ('believing ϕ to degree r').[22] Interestingly enough, partial belief need not necessarily be treated in numerical terms. One can develop theories about comparative probability ('believing ϕ to a higher degree than ψ'),[23] or one can even talk about partial belief in purely categorical terms ('partially believing ϕ').[24]

I said that doxastic concepts which are more like actions than like states, such as assertion, assent and acceptance, do not admit distinctions by degrees. However, there is an aspect to them that tends to blur these apparently clear-cut distinction. Intuitively, an agent may assent to, assert or accept a proposition ϕ may be more or less whole-heartedly—even though the lack of vivacity or emphasis of an act of assent/assertion/acceptance does not make it less of an assent/assertion/acceptance. We shall see that it is possible to capture some aspects of the degree of vivacity and emphasis in precise terms. Even though assent/assertion/acceptance are all-or-nothing terms, there may be a lesser or greater readiness to turn acceptance into rejection in the face of evidence to the contrary. So even though the set of accepted propositions is perfectly homogeneous in so far as all elements are equally accepted, the set forms a structured system in so far as the elements are likely to be discarded to varying degrees. This 'degree of entrenchment', as it is called, should, however, be carefully dis-

---

[22] Probabilistic modellings of belief have problems in giving an explication of *plain* belief and *plain* ignorance. Advocates of subjective probability like Carnap and Jeffrey argue, however, that there is no need for the concept of plain belief or acceptance, since all that matters are probability assignments. For a recent collection covering numerous aspects of subjective probability theory, see Wright and Ayton (1994).

[23] See for instance, Suppes (1994).

[24] See Price (1969, Series II, Lecture 4) on 'half-belief'.

tinguished from the 'degree of belief' as it is modelled in probability theory.[25]

There is another sense that has been given to the term 'acceptance' which should be mentioned. Lehrer (1990, pp. 10–11) separates acceptance from belief by reserving acceptance a special 'epistemic purpose' which is the one relevant for knowledge: 'A person need not have a strong feeling that something is true in order to know that it is. What is required is acceptance of the appropriate kind, acceptance in the interest of obtaining a truth and avoiding an error in what one accepts.' Lehrer contends that we sometimes *believe* that ϕ for the sake of felicity or the pleasure of believing so, but we would not *accept* that ϕ for these reasons.[26] Although they often go hand in hand, he thinks that there is always a potential conflict between the 'ancient system of perceptual belief' which is the 'yield of habit, instinct, and need' and the 'truth-seeking ... scientific system of acceptance'. The former is an 'automatic input system', the latter is 'capable of ratiocination'. (Lehrer 1990, pp. 113–114) As indicated above, it seems to me that there is a difference in the concepts of belief and acceptance in that the former is more amenable to explanation in terms of causes and the latter in terms of reasons. I do not see, however, that this difference is necessarily linked up with the issue of truth. Throughout this book, I work on the simplifying assumption that all doxastic attitudes are 'aiming at' truth. In the case of acceptance, this may be a deliberate orientation, while in the case of belief it may happen to be so because truth is of a great survival value. Nevertheless, if an agent either believes or accepts that ϕ, this means that he *holds ϕ true*. This does not mean, of course, that there is an easy way from belief or acceptance to knowledge, even if the proposition held true is in fact true. I shall examine one account of how belief and acceptance can attain the status of knowledge in the next chapter. There I analyse Lehrer's *Theory of Knowledge*, but I shall not follow him in his use of the term 'acceptance'.

Let us summarize the above and relate it to the analyses that are going to be presented in the course of this book. We have seen that there is a menagerie of doxastic concepts that differ in various ways and aspects. It is beyond the scope of the present book to work out all the subtle differentiations that may be discovered here in ordinary language and philosophical writings. Moreover, it does not seem necessary to do so.

When I am speaking of beliefs in later parts of this book, this is always to be taken with a grain of salt. In terms of our classifying characteristics of doxastic concepts, the mental states I am going to discuss are like beliefs in that they are

---

[25] We shall see in Chapter 8 that the 'laws of entrenchment' are quite different from the laws of (quantitative or qualitative) probability theory. This point has long been made by Isaac Levi, for instance in Levi (1983, p. 165): 'we can talk of differences within the corpus with respect to grades of corrigibility. It is tempting to correlate these grades of corrigibility with grades of certainty or probability. According to the view I advocate, that would be a mistake. All items in the initial corpus K_1 which is to be contracted are, from Xs [the person's or community's, H.R.] point of view, certainly and infallibly true. They all bear probability 1.'

[26] Contrast this with Bernard Williams's (1970, pp. 109–111) discussion of the notion of 'wanting to believe' for 'truth-centred' and 'non-truth-centred motives'.

states, *attitudes*, or *dispositions* to utter certain linguistic expressions or to act in certain ways, rather than instantaneous acts or occurrent events.

But our mental states are like acceptances in that they do not come in degrees, and they are explained with reference to conceptions of rationality rather than of causality. Though there is no partial accepting of a proposition, there is a sense of 'degrees of acceptance' that will prove to be fully appropriate for our mental states. I shall make use of relative degrees of retractability (*alias* degrees of firmness, certainty, corrigibility or entrenchment) which express how severe an incision it would mean to the agent's doxastic state to give up a certain proposition, and I include very weak forms of acceptance or belief, in a sense which includes propositions that are very easily retracted.

Non-trivial changes of our mental states are analysable in terms of rational choice, but the mere possibility of reconstructing some processes in terms of choice does not mean that there is really free choice or voluntary action involved. A chess computer always 'chooses' to make the best move it has found within a (more or less) prefixed time interval, but it would seem unnatural to say that the computer executes acts of free choice.[27] If we avail ourselves of the device of choice or selection function in the analysis of belief change, then, this does not entail that we actually commit ourselves to doxastic voluntarism.[28]

By and large, I shall most freely and frequently use the talk about *belief*, for the simple reason that it is this most convenient word to use. When I wish to emphasize that I want belief to be seen as defeasible and open to revision, I shall talk of opinions or of propositions which are accepted. When I refer to propositions that are explicitly represented and comparatively well entrenched, I shall talk of beliefs or pieces of information. When I refer to propositions that are likely to remain implicit and are easily overturned in the face of conflicting explicit information, I shall talk of expectations or default assumptions.

In his penetrating study of the differences between belief and acceptance (and numerous related concepts), L. Jonathan Cohen (1992, p. 28, note 28) expresses

---

[27] The computer executes moves that are 'best' in terms of an in-built evaluation of positions, which in some way expresses the prospects of reaching the ultimate aim, i.e., of winning the game. The boundary between 'stupid' utility maximizers and 'intelligent' systems showing genuine choice behaviour, however, is vague. For instance, if we discover that the computer has a very sophisticated program for deciding how much time to spend on the search for, and evaluation of, further ramifications of the game tree, and thus for budgeting its time resources, we become quite inclined to ascribe conscious action to it.

[28] Compare Price (1954, pp. 15–16): ' ... something rather like preferring or choosing does quite often occur as a stage in the process by which a belief is formed, especially when we acquire our belief in a reasonable manner, after careful consideration of the evidence *pro* and *con*. ... this assenting, and especially the initial assent, has a *preferential* character. ... Now because of this preferential element in it, assent may look rather like voluntary choice. But the appearance is deceptive. It is not a free choice at all, but a forced one. If you are in a reasonable frame of mind (as we are assuming that you are in this case) you cannot help preferring the proposition which the evidence favours, much as you may wish you could. ... It just is not in your power to avoid assenting to the proposition which the evidence (your evidence) favours, or to assent instead to some other proposition when the evidence (your evidence) is manifestly unfavourable to it.'

the suspicion that

discussions [of the formal structure of reasoning about revisions in what we believe or accept] ... are arguably premature, and risk confusion, in a context in which the fundamental distinction between belief and acceptance has yet to be clarified.

The present book is concerned precisely with discussions of the kind referred to by Cohen. However, Cohen does not substantiate his suspicion, and I have not become aware of any confusion or mistake that arises from a somewhat sloppy way of using doxastic terms. For this reason, I will not attempt to be overly pedantic about this.

1.2   Representing doxastic states

The most common representations of belief states in computational contexts are *sentential* or *propositional* in the sense that beliefs are coded as formulas representing propositions. Even if we stick to propositional representations of belief systems, there are many options open. First of all, we must choose an appropriate *language* to formulate the sentences expressing beliefs.

We will be working with a language L which is based on propositional logic. It is assumed that L is closed under applications of the Boolean operators $\neg$ (negation), $\wedge$ (conjunction), $\vee$ (disjunction), and $\supset$ (material conditional). p, q and r are the atoms (propositional variables) of the language. We will use ϕ, ψ, χ as metavariables for sentences in L, and H (and sometimes F and G) as metavariables for sets of sentences in L. Sets of sentences meeting special constraints (see Section 1.2.2) will be denoted by K. All these symbols may occur with primes or indices. It is convenient to introduce the symbols $\top$ and $\bot$ for the two propositional constants 'truth' and 'falsity' into our language. When there is no danger of confusion, 'L' will also be used to denote the set of all sentences of the language.

1.2.1   *Belief bases: Independent items of information*

The simplest way of representing a belief state is to represent it by a *set* of sentences from L. The interpretation of such a set is that it contains all the sentences ('beliefs', 'opinions') that are *accepted* in the belief state so represented. Consequently, when ϕ is in H we say that ϕ is *accepted in H*; when $\neg\phi$ is in H we say that ϕ is *rejected in H*; otherwise the agent *suspends judgement* with respect to ϕ (ϕ is *undecided* or *left open*). A sentence's being accepted does not imply that it has any form of justification or support.

A belief base is a partial description of the world. Formally, this idea is modelled by calling any arbitrary set H of sentences in L a *belief base*. Given some inference operation *Inf*, a base H *generates a belief set K*, or is a base *for K*, if and only if $Inf(H) = K$. The idea is that the elements of H are *basic beliefs* with an independent warrant, whereas the elements of $K - H$ are called *derived* *beliefs* which are there only because they can be derived from the basic beliefs.[29] A belief base in this sense is not just a convenient finite, or at least decidable, representation ('axiomatization') of a theory. The fact that some element in K is included in the base H for K is supposed to have important epistemological implications. Such an element has a preferred status:

(**Maxim B**) The elements of a base are explicitly given beliefs. They comprise beliefs, and only beliefs, which have some kind of independent standing, i.e., which are not derived by processes of inference from other beliefs. They serve as a basis or foundation of the belief set that can be 'built on' the base.

No constraints whatsoever are put on the structure of belief bases or on the truth of its elements. It is not my concern in this book to discuss the origins or contents of belief bases. I shall take bases as *somehow* given, no matter how. What is of interest to us is the reflection that an agent may or should engage in, when he or she is given some such base as the foundation of his doxastic state. He should start to draw inferences from what is given in order to reach a cognitive equilibrium. Only well-balanced states, states in cognitive equilibrium may be expected to satisfy substantial constraints of coherence and rationality.

If we want to emphasize that belief bases need to be processed before their elements are finally accepted as genuine beliefs or opinions, we call them *information bases*. In computer science they are called *databases*, in the theory of knowledge advocated by Keith Lehrer (1990) *acceptance systems.* In real-life cases of applications, belief bases may safely be supposed to be finite.

1.2.2 *Belief sets: Coherent theories*

Being directly and explicitly given, the items in a belief base are consciously accessible to the agent. An ideal reasoner, however, should not rest content with what is immediately before his mind. He should try to work out, by processes of inference, what is implied by the base, in other words, what acceptance of the base commits him to. Only if all logical consequences are traced out and called to consciousness, the ideal reasoner will have reached a cognitive or reflective equilibrium.[30]

We call sets of beliefs which are coherent in this sense *belief sets*. Formally, belief sets are *theories* in the logician's sense, sets of sentences in L that are closed under an appropriate inference operation. Intuitively, a belief set or theory contains what the agent 'really believes', what he or she believes 'implicitly' or

---

[29] In standard accounts, H is a subset of K. In some of the models considered in Chapter 5, however, basic belief will be treated as fallible or defeasible.

[30] Evidently, this is not the sense of 'reflective equilibrium' made popular in recent metaphilosophy by Goodman and Rawls. What we have in mind is more general than the mutual adaptation of abstract, general principles and well-considered judgements concerning more concrete questions. But compare Ellis's laws of rationality which are '*equilibrium laws*, because a rational belief system is conceived to be one which is in equilibrium under all pressures of internal criticism and discussion'. (Ellis 1979, p. 84)Ellis

is 'committed to believe' (the latter term is Isaac Levi's favourite). What is—or at any rate, should be—accepted in a belief state is not only the sentences that are explicitly laid down in the database, but also what may legitimately be *inferred* from these basic beliefs. Hence, an important factor which has to be decided upon when representing a belief state is what inference relation governs the beliefs.

I shall follow two ways of explicating the inferential moves an agent is allowed to take, either by means of applying a traditional Tarskian consequence operator or by means of applying a more sophisticated, typically nonmonotonic and paraconsistent inference operator. This distinction will be elaborated in Sections 1.2.4 and 1.2.5.

Logical closure may be seen as a coherence property of sets of beliefs. If you believe that ϕ and that $\phi \supset \psi$ then you should also believe that ψ; otherwise your stock of beliefs may justly be called incoherent. The former beliefs commit you to the latter one, you implicitly, or 'really', do believe that ψ, even if you are not aware of it. Taken together with consistency, which is more commonly interpreted as coherence, closure constitutes a notion of *logical* or *inferential coherence*.[31]

Logic furnishes the concepts of closure and consistency, and thus helps to define a concept of inferential coherence. If your body of beliefs is not closed or not consistent, then there is an *internal problem* with your beliefs. I shall refer to this as *the static problem* of making up one's mind. Logical reasoning is supposed to solve the problem of establishing a reflective equilibrium. *External* or *dynamic problems* are created when the need to add or subtract beliefs arises, in other words, when you have to change your mind. The insertion of the new belief or the cancellation of the old belief leads to a perturbation of the initial equilibrium. If you add a single belief to a coherent belief set, you violate the logical closure condition; but if you close after the addition, you may well violate the consistency condition (so you do not reach your aim). If you retract a single belief, you similarly violate the closure condition; but if you close after the retraction, you may well rederive the very belief you wanted to retracted (so you did not achieve anything). This task of maintaining the coherence of belief sets while effectively adding new or retracting old beliefs is precisely what makes *the problem of belief change* interesting and non-trivial.

If a collection of beliefs is understood to be logically closed, that is, if it is a *belief set* or a *theory* in the technical sense of these terms, we denote it by the metavariable K rather than by H which we reserve for sets (belief or information bases) which are not supposed to be closed.

---

[31] Similarly, an *acceptance state* in Stalnaker's (1984, p. 82) sense has to satisfy three 'deductive conditions': It must be closed under consequences of singletons, closed under conjunctions, and it must be consistent. For Dennett (1987, p. 21), consistency and closure make up the 'ideal of perfect rationality'. A concurring view is expressed by Levi (1995). Forceful counterarguments are given by Kyburg (1970). See also Section 3.2.1 below.

Against representing belief states as belief sets it has been argued that some of our beliefs have no independent standing but arise only as inferences from our more basic beliefs.[32] This distinction is not possible to express in a belief set since there are no markers for which beliefs are basic and which are derived.

Even if we agree that doxastic states should be understood as comprising what the agent 'really' believes, we have to bear in mind the impact of Maxim B of the last section. Even if we aim at turning a belief base into a coherent theory when trying to find out what we currently believe, the belief base may become important when our beliefs have to be changed. When we perform revisions or contractions of belief, very often we do not operate on the belief set itself (which contains an infinite number of derived elements), but rather directly to the set of basic beliefs. Instead of introducing revision and contraction operations that are defined on belief sets, one can insist that belief sets are inferentially generated from belief bases and that therefore change operations should be defined *on the bases*—or at least *with essential reference to the bases*. Operations of that kind will be called *base revisions* and *base contractions*. This approach results in an assignment of varying belief revision strategies to a single belief set K, since we may have two distinct bases H and H' for K (i.e., $K = Inf(H) = Inf(H')$) which will in general lead to distinct revision behaviour. This can happen either for the reason that the revisions of the bases H and H' lead to new bases that expand to different theories, or for the reason that the theory K itself is revised in different ways that crucially depend on the structure of the base from which the theory has been generated. Explanations and examples for these variants of 'foundationalist' belief change will be given in the course of the discussion in Chapter 3 (see particular Section 3.4.1) and Chapter 5.

There is no general answer to the question which representation is the better, full belief sets or bases. This depends on the particular level of analysis or purpose of application. Bases, whether prioritized or not, are easier to handle since they are finite structures; theories lend themselves more naturally to a semantic investigation.

1.2.3 *Acceptance: Beliefs and expectations*

Earlier I have mentioned default assumptions as an interesting species of the menagerie of doxastic states, and I now want to discuss how they function in the process of belief formation. The idea of default assumptions has come into the focus of research only fairly recently, but it does of course have its precursors. Since this is not widely known, let us pause a while with the systematic presentation and have a glance at the relevant teachings of some major figures in the history of philosophy.

Gottfried Wilhelm Leibniz (1765/1981, pp. 465–466) spoke with approval of the practices among jurists which are aware of 'many degrees of conjecture and of

---

[32] This has been put forward, amongst others, by Alchourrón and Makinson (1982), Makinson (1985), Hansson (1989), Fuhrmann (1991) and Hansson (1992), all of them referring to the case of a standard Tarskian consequence operation.

evidence'. Among them 'there are *presumptions*, which are accepted provisionally as complete proof—that is, for as long as the contrary is not proved'.[33] When Leibniz went on to say that the 'entire form of judicial procedures is, in fact, nothing but a kind of logic, applied to legal questions', he referred to the then beginning theory of probability, not to a logic or inference relation of the kind we shall be dealing with in later chapters.

Thomas Reid (1788/1985b, p. 620) transferred this observation about the practices of jurists into the realm of genuine philosophy. He held that we may trust the common sense. For instance, he argued that our beliefs that we are acting freely and that the persons with whom we converse are thinking, intelligent beings are justified. They have, he said, 'what lawyers call a *jus quæsitum*, or a right of ancient possession, which ought to stand good till it be overturned'. Being 'natural convictions', they should not be thrown away gratuitously: 'I should think it reasonable to hold the belief which nature gave me before I was capable of weighing evidence, until convincing proof is brought against it.'[34] In Lehrer's (1990, pp. 65 and 176–177) felicitous phrase one can say Reid held that many beliefs, including perceptual beliefs and beliefs of common sense, are 'innocent until proven guilty'.

But as Reid was ready to admit, even our most cherished beliefs *can* be proven guilty. Quite frequently we have to realize that we are not infallible. All too often it turns out that what we have taken for granted *is* wrong, or at least that it *may be* wrong. Then we have to 'change our mind' by retracting some presumptions, and possibly replacing them by their negation.

Throughout the 40 years of his teaching career, Immanuel Kant lectured on logic, and it is striking to see that this master of rigid reasoning was very well aware of the valuable role of judgments that might well be termed 'presumptions' or 'defaults':

A provisional judgment is one in which I represent that while there are more grounds for the truth of a thing than against it, these grounds still do not suffice for a *determining* or *definitive* judgment, through which I simply decide for the truth. Provisional

[33] Also compare p. 457: 'As for "presumption", which is a jurists' term, good usage in legal circles distinguishes it from "conjecture". It is something more than that, and should be accepted provisionally as true until there is proof to the contrary; whereas an indication, a conjecture, often has to be weighed against another conjecture. ... In this sense, therefore, to *presume* something is not to accept it *before* it has been proved [Locke's formulation, H.R.], which is never permissible, but to accept it *provisionally* but not groundlessly, while waiting for a proof to the contrary.' The most prominent presumption which is still an important principle of many legal codes is the 'presumption of innocence'.

[34] Reid's philosophy relies very heavily on the first principles of common sense given to us 'by Nature'. However, he is ready to extend his fallibilism even to these very First Principles: 'We do not pretend that those things that are laid down as first principles may not be examined, and that we ought not to have our ears open to what may be pleaded against their being admitted as such. Let us deal with them as an upright judge does with a witness who has a fair character. He pays a regard to the testimony of such a witness while his character is unimpeached; but, if it can be shown that he is suborned, or that he is influenced by malice or partial favour, his testimony loses all its credit, and is justly rejected.' (Reid 1785/1985a, p. 234) At many places, though, Reid says that denying first principles is absurd.

judging is thus merely problematic judging with consciousness. ... Provisional judgments are quite necessary, indeed, indispensable, for the use of the understanding in all meditation and investigation. For they serve to guide the understanding in its inquiries and to provide it with various means thereto. ... We can think of provisional judgments, therefore, as *maxims* for the investigation of a thing. We could also call them *anticipations*, because we anticipate our judgment of a thing even before we have the determining judgment. Judgments of this sort have their good use, then, and rules can even be given for how we ought to judge provisionally concerning an object. (Kant 1800/1988, pp. 577–578)

Kant distinguishes between preliminary judgements ('vorläufige Urteile') and prejudices ('Vorurteile') by saying that the latter are taken as principles ('Grundsätze') which makes them the source of erroneous judgements.

In a more emphatic vein, but perhaps gluing together truth and acceptability too firmly, William James (1907/1975, p. 100) holds that 'Truth lives, in fact, for the most part on a credit system. Our thoughts and beliefs "pass", so long as nothing challenges them, just as bank notes pass so long as nobody refuses them.'

Default assumptions or expectations have two important characteristics. First, although they are almost universally at work in everyday reasoning processes, they do their work in secret as it were. It is quite hard, if not impossible, to expose (even by way of ascription) all and only the hidden premises of our ordinary inferences. The second feature of default assumptions is that they are, or should be, easily overturned. In case of conflict, they should give way to explicitly documented items of information. It is not obvious that these two traits always go hand in hand. And it is evident that some prejudices are extremely hard to exterminate. Prejudices in the broad used here are neither harmful nor unjust; they are ordinary, value-neutral assumptions about just anyone and anything. They start getting disagreeable or dangerous only if people cling to them too stubbornly. So the fact that prejudices are hardly made explicit does not detract from their tenacity. Another case making abundantly clear that implicitness does not equal defeasibility is the case of explicitly documented weak beliefs. It may well happen that somebody tells us expressly that it is dangerous to go to Redfern Station, but we may be very ready to give up such a belief upon learning that there are always many security people around. In this book the main trait of defaults is not seen in their being liable to be given up easily—this is also true of full beliefs—but rather in their being *only implicitly assumed or taken for granted*. As a simplification, however, I shall sometimes[35] proceed on the premise that implicit assumptions are more likely to be discarded than explicitly given data.

Default assumptions are assumptions we are used to take for granted. This feature about default assumptions has also been made a topic in philosophy for quite some time. Walter Bagehot, in his paper 'The Emotion of Conviction' (1871/1898), considered unquestioning acceptance to be a form of prim-

[35]In Chapter 5.

itive credulity; John Cook Wilson stressed the importance of things which we *take for granted* or *under the impression* of which we are; Harold A. Prichard contemplated the state of *thinking without question* that ϕ or of *accepting ϕ as a working hypothesis*. When we take for granted that ϕ or when we think without question that ϕ, this does not mean that we would never give up our presumption the ϕ is true. It just means that ϕ is not going to be put on the agenda for the time being, it is *used*, but not explicitly *discussed*. We need not even be conscious that we actually use ϕ in our deliberations. Should we find occasion to focus our attention on ϕ, then ϕ is not any safer than any other belief we happen to have.[36]

Like beliefs, default assumptions may be *ascribed* to humans in the process of explaining their behaviour. Poole (1988, p. 28) refers to the defaults featuring in his logical system as 'possible hypotheses they [the "users", H.R.] are prepared to accept as part of an explanation to predict the expected observations'.

It is tempting but I think incorrect to conceive of expectations which are (considerably) weaker than full beliefs as *subdoxastic states* in the sense of Stich (1978). One of the traits of these states that distinguishes them from beliefs is that they are not consciously accessible. It is quite likely that even normal, co-operative, adult subjects lack the ability to become aware of, or to be conscious of, the contents of many of their expectations. In that respect expectations meet Stich's criteria of subdoxastic states. They fail, however, on his second criterion according to which subdoxastic states are 'largely inferentially isolated from the large body of inferentially integrated beliefs to which a subject has access', or simply 'inferentially impoverished' (Stich 1978, p. 507). We have good reasons for saying that beliefs form (or at least should form) an inferentially integrated cognitive subsystem. But exactly the same is true of expectations on every conceivable level of firmness. The very structure that sets up the relations of coherence between beliefs also connects expectations. For instance, if we were to postulate that *beliefs* are just those propositions whose degree of certainty or firmness of acceptance exceeds some specified threshold value, the same may just as well be claimed for *expectations*. Just the threshold value will be somewhat lower. Anyway, to accept a proposition then is to assign it a degree of certainty or firmness which is equal to or higher than a contextually fixed number r.[37] Expectations are entities of the same kind as beliefs, and there is no psychologically interesting distinction here as Stich claims it for the distinction between beliefs and subdoxastic states.

We have not found a sharp boundary between beliefs and expectations. Any potential standard for separating beliefs from expectations may be contextually

---

[36] For all this, see the excellent discussion of assent and 'being under an impression that', as part of the traditional occurrence analysis of belief, in Price (1969, Series I, Lecture 9).

[37] It is debatable whether the set of sentences exceeding some such threshold value r should meet any constraints of logical coherence. If so, then degrees of certainty or firmness may be called *degrees of expectation* or of *entrenchment* in the technical sense of Gärdenfors and Makinson (1988, 1994). Compare Section 1.2.6 and Chapter 8 below.

shifted, according to the situation one is facing. It is very doubtful whether an agent makes use of the same set of propositions in different situations—even if his doxastic state does not change at all. A loose conjecture may count as a full belief in a party chat, but in the courtroom one ought to be firmly convinced of the truth of a proposition in order to affirm that one believes it to be true. Pragmatic considerations are needed to determine what qualifies as a belief. A similar view is expressed by Gärdenfors and Makinson (1994, pp. 223–224):

> Epistemologically, the difference between belief sets and expectations lies only in our attitude to them, i.e., what we are willing to do with them. For so long as we are *using* a belief set K, its elements function as full beliefs. But as soon as we seek to *revise* K, thus putting its elements into question, they lose the status of full belief and become merely expectations, some of which may have to go in order to make consistent place for beliefs introduced in the revision process.

It is worth repeating that not only expectations, but also beliefs may turn out to be wrong. A fallibilist or corrigibilist (Levi's term) attitude is appropriate towards both expectations and beliefs. In fact, there may be logical conflicts between beliefs, between expectations, and between beliefs and expectations taken together. We shall also talk of *information bases*, with the understanding that not all items of information need to be correct, and some may eventually have to go. The 'beliefs' in the base should rather be called *prima facie beliefs*. They are somehow 'given' and represented in the base, but they are not really embraced, the agent is not committed to them. Only after a reflective equilibrium has been reached can we say what the 'real beliefs' of the agent are. Some of the prima facie beliefs may have been gone. Only what results after all inferences from the information base (and the expectations) have been drawn is actually accepted.

So far we have found many similarities between expectations and beliefs, but no substantial differences. I shall take account of that state of affairs by aiming at formal representations of beliefs and expectations that have the same data structure. If beliefs are represented in a belief base, then expectations should likewise be collected in a belief base; if beliefs are represented by a theory, then expectations should also form a theory.

Although there is no principal difference in the structure of beliefs and expectations, we may still want to distinguish between them, since they may be supposed to play different roles in the formation of the set of accepted sentences, i.e., in the formation of 'real' beliefs or opinions. In what follows I will accordingly distinguish between the 'harder' and the 'softer' beliefs that an agent may have, and I identify these sets with his 'explicit information' and his 'implicit expectations'. I will further assume that both kinds of belief come in varying *degrees* of firmness, plausibility or importance (see Section 1.2.6). The agent's epistemic state is thus encoded by two structures of the same format.

The *prima facie beliefs* (or *prima facie opinions*) of the agent are contained in the union of the information set H with the expectation set E. In most cases, this union will be inconsistent, since in real life explicit information very often contradicts what the agent had originally expected.

Now, what is *accepted* by an agent, what are his *opinions* or, more sloppily, his *beliefs*? In case of conflict, we want the explicit information to override the expectations. If we stick to the philosophy of theory bases, every belief set K is supposed to be obtained from some information base H as a result of drawing inferences from H. What is new in this treatment is that I want to make provision for the possibility that the inferences regulating the statics and dynamics of belief (a) are ampliative (content-increasing) and expectation-based[38] and (b) know how to deal with inconsistencies in information and expectation. Before elaborating on this topic, we have to provide some information about more standard forms of logic which serve as a foundation of more sophisticated forms of inference operations.

1.2.4 *Logics I: Consequence operations*

The process of taking the closure of belief bases in order to reach theories reflecting what an agent 'really' believes may be modelled by two quite different sorts of operations. I shall distinguish, by way of terminological decision, between standard *consequence operations* which transmit the truth from the premises to the conclusion and non-standard *inference operations* which are important for various kinds of (everyday) reasoning.[39]

We want to avail ourselves of a (more or less) classical consequence relation operative in the background and a (non-classical) inference relation which will be a principal object of our considerations.[40] We shall see that it is a nontrivial problem to formulate conditions describing how exactly consequences and inferences are, or should be, related.

The term *'logic'* will be used as a generic term covering both inference and consequence operations. In the notation, I will distinguish the two kinds of logic by using Cn for consequence operations and Inf for inference operations. As already indicated, Cn is supposed to preserve the truth of beliefs, Inf to produce coherence among beliefs. When I speak of 'standard closures' of belief bases later on, I mean closures under the consequence operation Cn. The concept of inference may be regarded as a liberalization of the standard concept of consequence since Inf is not required to satisfy the principles of Monotonicity and *Ex falso quodlibet*. The task of working out the consequences of, or drawing inferences from, a given set of premises, belongs to the realm of theoretical reason. It is not obvious how considerations of practical reason should enter this domain.

In this section we identify logic with some *consequence Operation Cn*. The next section will then start the discussion of logics as captured by sophisticated

---

[38] Compare Levi's (1983, Section 5) and (1991, Section 3.1) 'inductive' or 'inferential' or 'deliberate expansion'.

[39] A terminological distinction between logic and inference which is parallel to this distinction between consequence and inference operations was made by David Israel (1980) and by Sten Lindström (1991). Compare footnote 55.

[40] More precisely, inference operations are placed in the forefront until the end of Chapter 5. Chapter 6 is not about logic. In Chapters 7 and 8 consequence relations will play a central role, inference relations are only indirectly dealt with, via the bridge described in Chapter 4.

inference operations Inf. Both operations take sets H of premises as arguments and yield sets $Cn(H)$ and $Inf(H)$ as conclusions, the sets of consequences of H and of sentences which may be inferred from H, respectively. For inferences at least, this is a serious limitation, since human reasoners are able to draw inferences not just from sets of sentences, but also from pictures, other people's actions, from all kinds of clues, hints and evidence. We shall have a very modest reflection of additional generality in Chapter 5 where inferences are drawn on the basis of *structured* expectation and information bases.

Most of the time, I assume that Cn obeys certain structural rules, and require that it be Tarskian, that is, *reflexive* (or '*inclusive*'), *monotonic, transitive* (or '*idempotent*'), and *compact*. I refer to such a logic either as a consequence operator Cn or as a consequence relation $\vdash$, with the understanding that $\phi \in Cn(H)$ if and only if $H \vdash \phi$. Using the first notation our four requirements become

(Ref) $H \subseteq Cn(H)$

(Mon)   If $H \subseteq H'$ then $Cn(H) \subseteq Cn(H')$

(Tran) $Cn(Cn(H)) \subseteq Cn(H)$

(Comp)   If $\phi \in Cn(H)$ then $\phi \in Cn(H')$ for some finite subset H' of H

It is easy to verify that, given Reflexivity and Transitivity, Monotonicity implies the (following version of the) Cut rule: If $\phi \in Cn(H)$ and $\psi \in Cn(H \cup \{\phi\})$ then $\psi \in Cn(H)$. Conversely, given Monotonicity and Compactness, the Cut rule implies Transitivity.[41]

In the context of the present study, the monotonicity condition (Mon) is the most interesting one. Logics (consequence or inference operations Cn or Inf) that satisfy (Mon) are called *monotonic*, otherwise they are called *nonmonotonic*.

The background logic Cn to be used later on is further supposed to be *supraclassical*, i.e., what follows from a given premise set by classical propositional logic Cn_0 should also follow from it by Cn. We finally assume that Cn satisfies the *Deduction Theorem* (or '*Conditionalization*'):

(SupraClass) $Cn_0(H) \subseteq Cn(H)$

(DedThm)   If $\psi \in Cn(H \cup \{\phi\})$ then $\phi \supset \psi \in Cn(H)$

1.2.5   *Logics II: Inference operations*

If the body of our beliefs is incoherent in some sense, this creates *internal problems* that call for a kind of reflective resolution. Given certain basic pieces of information, the task is to find what can be derived from these pieces, what are their consequences or what may legitimately be inferred from them.

---

[41] *Proof:* Let $H \vdash \phi$ and $H \cup \{\phi\} \vdash \psi$. Then by (Ref) and (Mon), $Cn(H) \vdash \psi$, so by (Tran) $H \vdash \psi$ which proves (Cut).

Conversely, let $Cn(H) \vdash \psi$. By (Comp), there are $\phi_1, \ldots, \phi_n \in Cn(H)$ such that $\phi_1, \ldots, \phi_n \vdash \psi$. Applying (Mon) twice, we get that $H \cup \{\phi_1, \ldots, \phi_{n-1}\} \vdash \phi_n$ and $H \cup \{\phi_1, \ldots, \phi_n\} \vdash \psi$. Hence, by (Cut), $H \cup \{\phi_1, \ldots, \phi_{n-1}\} \vdash \psi$. Repeating the same argument $n - 1$ more times, we get that $H \vdash \psi$, so we have proved (Tran).

Everyday thinking requires nonmonotonic reasoning since real agents typically have to act on knowledge that must be considered incomplete and uncertain. What agents really *know* is usually very little, but they avail themselves
of a huge array of background assumptions, presumptions, defaults, prejudices,
and so forth, that help them jump to the conclusions they need in order to
take any action.[42] Everyday reasoning (practical reasoning, but probably scientific reasoning as well) needs to be ampliative and adventurous, to make risky
inferences in its attempt to both complete the meagre supply of explicit information and reach a cognitive equilibrium in a rich and well-balanced theory. The
background assumptions function as a kind of extralogical parameter that helps
making the appropriate inference operation stronger than classical logic.[43] On
the other hand, it may, and it in fact will, frequently happen that conclusions
which were drawn on the basis of earlier, insufficient information are defeated by
explicit information provided at a later stage. That is why everyday reasoning is
nonmonotonic.

Furthermore, it may and in fact does frequently happen that the reasoner
finds himself forced to draw inferences from contradictory pieces of information,
without lapsing into doxastic ruin. That is why everyday reasoning is paraconsistent. So while everyday reasoning is bold where boldness is needed, it also is
cautious where caution is needed, and consolidates 'overcomplete' belief bases.
Good formal accounts of everyday reasoning, that is, good inference operations
Inf, should be apt to reflect both the features of nonmonotonicity and paraconsistency.

If the basic pieces of information are inconsistent with one another, then the
process of deriving information that is implicit in the explicitly given premises is
particularly difficult. This task is addressed by *paraconsistent logics*, i.e., logics
which do not trivialize classically inconsistent sets of sentences. Classical logic
derives any arbitrary sentence ψ from a set H that includes both ϕ and $\neg\phi$.
Paraconsistent logics are designed to avoid just that.[44] Even if I am not going
to employ paraconsistent logics in the usual sense, it is clear that the cognitive
equilibration procedures to be applied to belief or information bases must be
suitable for handling inconsistencies. I shall address this feature in some more
detail presently, in Section 1.2.7.2 below.

---

[42]Take any of the hundreds of examples that have been discussed in the literature on nonmonotonic reasoning, e.g., Gärdenfors and Makinson (1994, pp. 199, 214–216).

[43]All ampliative inference operations need to draw on some extralogical information structure
that determines how to increase the content of the premises.

[44]More precisely, a logic or inference operation is *paraconsistent* if and only if it is not
explosive, i.e., if it does not satisfy the classical *Ex falso quodlibet*, i.e., if not everything is
derivable from an inconsistent set of premises. Without further systematic pretensions, we say
that a set H of sentences is inconsistent if there is a sentence ϕ such that either both ϕ and
$\neg\phi$ are contained in H or they are both derivable from H (with the help of some Tarksian
consequence operation). A comprehensive survey of paraconsistent logics is furnished in Priest,
Routley and Norman (eds.) (1989); for an interesting attempt to bring them to bear on the
study of belief revision, see Restall and Slaney (1995).

The need for nonmonotonic reasoning, reasoning that violates the above-mentioned rule of monotonicity, has been recognized by researchers in knowledge representation for quite some time and has been addressed in various formal approaches since 1980.[45]

Fortunately, it has turned out that the breakdown of monotonicity does not mean that common sense reasoning is completely irregular. On the contrary, many studies have been conducted concerning the question which general properties may be expected to hold for nonmonotonic everyday inferences. Usually one would think that $Inf(H)$ should contain all of H (i.e., that the inference relation Inf is reflexive), but if the premises themselves are inconsistent, we have reason to reject this conception of inference as too narrow.[46] I shall return to a discussion of the abstract properties that nonmonotonic logics should have in Section 4.4.

Following Makinson (1989, p. 3), one could argue that we then construe inference as revision rather than inference as logic: 'Procedures that do not satisfy the condition of *inclusion* [= Reflexivity, H.R.] ... are best thought of not as processes of "inference" but rather of theory or knowledge change.'[47] But is this really as clear a contrast as it seems? Perhaps belief formation *is* belief revision, perhaps processes of inference just *are* processes of changing one's acceptance system, of forming and reforming one's theory about the world. Stalnaker (1984) considers both deduction and inquiry in general to consist in the processes of changing acceptance states in response to 'new' information—new, that is, to some single acceptance state or 'compartment of mind'.[48] Similarly, Poole (1988, p. 28) considers 'reasoning, not as deduction, but as theory formation'.[49] The difference between Poole and Stalnaker here is mainly terminological; they agree about the pivotal role of theory change in processes of reasoning.

What is the relation between Inf and Cn? Two things may be said in answer to this question. First, standard logic Cn may be considered to be a particularly well-behaved (and perhaps particularly dull) form or limiting case of the more general concept of everyday inference which is to be captured by Inf.

---

[45] The year of birth of systematic treatments of nonmonotonic reasoning is 1980. For detailed information about the amazing variety of formalisms for nonmonotonic reasoning that have been developed since then, see the collections edited by Ginsberg (1987) and Gabbay, Hogger and Robinson (1994).

[46] We do this in Chapter 5.

[47] This sentiment is echoed, for instance, by Kraus, Lehmann, and Magidor (1990, p. 177): 'Relations that do not satisfy it [Reflexivity, H.R.], probably express some notion of theory change.'

[48] "Inquiry in general is a matter of adjusting one's beliefs in response to new information ... " (Stalnaker 1984, p. 87) and "Inquiry is the process of changing ... acceptance states, either by interaction with the world or by interaction between different acceptance states." (p. 99) See also pp. 85–87 for the application of this account to particular problem of 'purely deductive inquiry'.

[49] Poole's important point is that nonmonotonic reasoning can be seen as theory formation on the basis of an ordinary monotonic consequence operation. See Section 1.2.7.1 below.

Second, some standard logic Cn can be taken as a monotonic background logic for every nonmonotonic inference Inf, in the sense that Inf has somehow to be 'in harmony' with Cn. After all, the concept of logical consequence is related to the concept of truth, and beliefs in general aim at truth. Thus a good inference operation should somehow build on some truth-transmitting consequence operation. The interplay between standard monotonic and nonstandard nonmonotonic logic will be encoded in a set of rationality postulates for reasonable inference relations. These postulates are also presented in Section 4.4.

As mentioned before, I formally indicate the use of nonmonotonic reasoning in the process of transforming a belief base H into a coherent belief set K by writing $K = Inf(H)$ rather than $K = Cn(H)$. When we want to be more general, then we can think of the 'information base' on which some reasoning process can be based as any kind of informational structure. Then Inf takes such a structure $\mathcal{H}$ as an argument and yields the set $Inf(\mathcal{H})$ of sentences that are inferable from, or acceptable on the basis of, this general sort of information base.

I have also mentioned that there are two equivalent ways of referring consequences and inferences if symbolic notation is used in the discussion of logics. First, we may use, as I have done up to now, the *operational* point of view, with the operators Cn and Inf. Alternatively but equivalently, one may adopt a *relational* point of view, and use the symbols $\vdash$ and $\vdash\!\!\!\sim$. For two data structures Φ and Ψ, the formulae

$$\Phi \vdash \Psi \qquad \text{and} \qquad \Phi \vdash\!\!\!\sim \Psi$$

mean that Ψ is a logical consequence of Φ, and that Ψ can be legitimately inferred from Φ, respectively. This is a very general way of expressing logical relations. In this book, the data structure Φ is usually a set H of sentences (possibly equipped with some ordering), while Ψ is a single sentence ψ. The relations $\vdash$ and $\vdash\!\!\!\sim$ are related to the above operators as follows:

$$H \vdash \psi \quad \text{iff} \quad \psi \in Cn(H)$$
$$H \vdash\!\!\!\sim \psi \quad \text{iff} \quad \psi \in Inf(H)$$

1.2.6 *Orderings of beliefs*

We are now going to have a look at more refined representations of belief and opinion than unembellished belief bases and theories. Not all items accepted in a certain doxastic state have the same status. Some of them are more firmly believed, more certain, more plausible or important than others. In other words, beliefs and expectations may be *ordered* or *ranked* in some way.

Suppose we have a set of sentences ('opinions') which are accepted by the reasoner, and that these sentences are ordered by some relation of firmness, certainty, plausibility, importance, or the like. One can think of sentences with high positions or ranks in the ordering as *beliefs*, while those with low positions or ranks in the ordering are merely regarded as *expectations*. There is no sharp boundary line between beliefs and expectations, and if we should need one, we have to fix it by context-dependent, pragmatic factors.

I have argued above that expectations should be modelled by theories which have the same formal structure as the theories representing beliefs. But which sort of structure should orderings of opinions have? Should the orderings show any respect for the logical or inferential relations between the opinions? Must the orderings be 'coherent' in some sense? In answering these questions, there are two principal possibilities, just as the beliefs of someone can either be represented by a belief base or by a belief set (theory). By way of terminological decision, I distinguish *priorities* of beliefs from *entrenchments* of beliefs. The former are captured by a weak ordering[50] which does not have to meet any logical constraints. The latter reflect different degrees of retractability and do have to respect the inferential relationships between sentences, since, for instance, you cannot 'really', effectively retract $\phi \vee \psi$ without retracting ϕ. That means, ϕ is at most as entrenched in a doxastic state as $\phi \vee \psi$ is. This idea is motivated by the same kind of ideas that led us to representing beliefs by logically closed theories. I discuss these two kinds of orderings in turn.

1.2.6.1 *Priorities in belief bases*

A *prioritized belief base* is a finite sequence $\mathcal{H} = \langle H_1, \ldots, H_n \rangle$ of finite sets of sentences of L. The sentences in such a base have varying priorities, with the understanding that the elements of a lower-indexed set H_i are less 'important' or 'valuable' or 'certain' than the elements of a higher-indexed set H_j with $i < j$. This is a straightforward idea that has been popular among philosophers and computer scientists for quite some time. The idea of prioritized belief bases may be found in Rescher's (1964, 1973, 1976) 'modal categories' and 'plausibility indexings', in Fagin, Ullman and Vardi's (1983) 'database priorities', in Brewka's (1991) 'levels of reliability' and in Nebel's (1992) 'epistemic relevance orderings'.

The representation of a prioritized belief base in terms of a sequence of sets is somewhat more general than in terms of a set H equipped with a weak ordering $\prec$, since the H_i's in $\mathcal{H}$ need not be disjoint. We may have an occurrence of ϕ in H_i and another occurrence of ϕ in H_j, with i different from j—a situation which is not possible in the alternative formalization using $\prec$ over H. For the belief change methods considered in this book, however, one can drop all non-maximal occurrences of sentences without any change in the resulting set of beliefs. So we may suppose for the following that the H_is in $\mathcal{H}$ are actually disjoint, and we can equivalently speak of the set H equipped with a priority ordering $\prec$, with the understanding that for any two elements ϕ and ψ in H, we have $\phi \prec \psi$ if and only if $\phi \in H_i$ and $\psi \in H_j$ with $i < j$. Clearly, the flat belief base $\mathcal{H} = \langle H \rangle$ is associated with the empty ordering $\prec$.

Let me emphasize that each prioritized belief base conveys two kinds of information. First, there is the syntactic structure of the sentences expressing in-

---

[50]That is, a transitive and connected relation $\preceq$ (usually allowing ties), or, in the strict version, an asymmetric and modular relation $\prec$. Compare Section 6.3, in particular page 157.

formation and expectations (cf. Maxim B of Section 1.2.1). It distinguishes, e.g., the joint belief $\phi \wedge \psi$ from two simultaneous but separate beliefs ϕ and ψ. In the former case, ϕ and ψ will stand and fall together while in the latter case they are going to be treated as independent. Second, of course, the prioritized belief base acknowledges varying degrees of belief, with the idea that beliefs in higher layers should in case of conflict somehow be given precedence over beliefs in lower layers of the belief base $\mathcal{H}$.

One can abstract from the specific structure of an information base in two steps. First, the prioritized base can be telescoped into a simple set of sentences without priorities by taking the union $H = |\mathcal{H}| = \bigcup \{H_i : i = 1, \ldots, n\}$. The unstructured set H can be called the *flattened form* or the *propositional projection* of the prioritized base $\mathcal{H}$. (In the limiting case when there is no prioritization of the information base $\mathcal{H} = \langle H_1 \rangle$ in the first place, I shall sloppily identify $\mathcal{H}$ with $H = H_1$.) Second, one can abstract from the syntactic information that is still present in H and form the theory $K = Cn(H)$. The priority structure is lost in the first step, the syntactic structure is lost in the second step. At the end, only the logical content of the information base remains identifiable.[51] But although the belief set K *itself* does not record its origin in $\mathcal{H}$, one may suppose that potential *changes* of K are dependent on the 'provenance' or 'history' of its elements. The theory K may thus be supposed to be equipped with some kind of 'memory'. An interesting question is how the relevant structure of the prioritized belief base $\mathcal{H}$ can guide the changes made to the theory K generated from $\mathcal{H}$, and how it can be reflected on the theory level. It has been argued already by Gärdenfors (1990) that much of the conceptual resources of belief base revisions can be reflected on the belief set level, if the notion of epistemic entrenchment of beliefs is employed. In Rott (1992c, 2000a) I have used relations of epistemic entrenchment as a tool for a detailed (re)construction of a particular form of base-oriented theory revisions.

I have mentioned that for some purposes it is necessary to keep apart explicitly given information and implicitly cherished expectations. In line with the similarities in their nature, we aim at a similar data structure for items of information and expectation.

As before, the *information base* of an agent will be represented by prioritized belief base $\mathcal{H} = \langle H_1, \ldots, H_n \rangle$ with the understanding that the elements in $H = \bigcup_i H_i$ are regarded pieces of explicit or 'hard' information (Maxim B).

The *expectations* (default assumptions, presumptions, 'soft information') of an agent are likewise collected in a prioritized base $\mathcal{E} = \langle E_1, \ldots, E_m \rangle$ which has the same structure as $\mathcal{H}$ (except that we do not presume $E = |\mathcal{E}| = \bigcup_i E_i$ to be finite). In much of the following I am going to make the simplifying assumption that $\mathcal{E} = \langle E_1 \rangle$ is unprioritized—in which case I shall sloppily identify $\mathcal{E}$ with

---

[51] Another way to get the effects of abstracting from the syntactic structure is to take the conjunction $h = \bigwedge H$, when H is finite. This of course preserves the syntactic structure. But in most approaches of belief change, logically equivalent single sentences (as opposed to sets of sentences that are logically equivalent) get treated in identical ways.

$E = E_1$—and that this unprioritized E is both consistent and closed with respect to a standard consequence operation. In this case we call E an *expectation set*. Thus we have a background *theory* E which is supposed to be equipped with some selection structure for governing potential revisions. In contrast to the information base H which may be rather small, one should think of the set E as a huge collection of expectations. In almost every situation, we have lots of ideas about what will typically or normally be the case or happen. We have a great many expectations, but at the same time we are well aware that a large portion of them are going to be frustrated.

Having distinguished information and expectation bases, we can merge them again. Let $\mathcal{E} \circ \mathcal{H}$ be the result of putting the prioritized belief base $\mathcal{H}$ on top of the prioritized belief base $\mathcal{E}$.[52] Formally, for $\mathcal{E} = \langle E_1, \ldots, E_m \rangle$ and $\mathcal{H} = \langle H_1, \ldots, H_n \rangle$, the concatenation $\mathcal{E} \circ \mathcal{H}$ is defined as $\langle E_1, \ldots, E_m, H_1, \ldots, H_n \rangle$. Note that $\circ$ is just the operation of *appending* one list to another, not a genuine process of *revising* expectations or beliefs, and that $\circ$ is associative. We abbreviate $\mathcal{G} \circ \langle \{\phi\} \rangle$ by $\mathcal{G} \circ \langle \phi \rangle$.

1.2.6.2 *Entrenchments in belief sets* I have mentioned the straightforward idea that the sentences really accepted by the agent are those that are, in some ordering of opinions, above a certain level of firmness, certainty, plausibility or importance. If we add to this idea the arguments that speak in favour of the requirement that the set of accepted sentences be a theory with respect to some background logic, then it is clear that the ordering of beliefs and expectations has to meet some constraints that relate it to this logic: There must be a guarantee that the sets of sentences that are 'better' than some sentence, or 'at least as good' as some sentence, do indeed form a theory. Perhaps the most well-known orderings that satisfy the appropriate logical constraints are the 'epistemic entrenchment relations' and the related 'expectation orderings' of Gärdenfors and Makinson (1988, 1994).[53] In Chapter 8, these relations will turn out to be symptoms of a more basic choice structure in the agents' minds.

Priorities and entrenchments are quite different kinds of orderings, more different than appears at first sight. The point is not alone that the former naturally apply to bases while the latter naturally apply to theories. They both may be

---

[52] This is possible because there are no logical constraints to be obeyed in the composition of prioritized belief bases. If we had to meet logical constraints, for example the ones involved in 'epistemic entrenchment' or 'expectation orderings' (see below), the 'putting on top' operation would need to be followed by a subsequent stratifying construction. For two possibilities of doing this, see Rott (2000a).

[53] Rescher (1964, 1973, 1976) and Williams (1994) require belief bases to be ordered by some logically constrained relation ('conjunction-closed modal categories', 'plausibility indexings', 'ensconcements'). It is not very principled to allow logical irregularity (non-closure) pertaining to the *sets* of beliefs, and to ban logical irregularity from the *orderings* of these beliefs. However, there is nothing objectionable to using belief bases with logically well-behaved orderings as finite representations of theories with such orderings (Rott 1991b, 'E-bases') or as partial specifications of them (Williams 1995, 'partial entrenchment rankings').

used to guide the changes of doxastic states, but they are characteristically applied in these processes in completely different ways. It is possible to generate entrenchments in a theory $K = Cn(H)$ from the priorities in the associated base H in such a way that the entrenchments are, in a certain sense, 'approximately equivalent' to the base priorities. But then the fact that ϕ has priority over ψ in the base does not imply that ϕ is more entrenched than ψ in the theory, nor does the converse implication hold.[54]

1.2.7   *More on nonmonotonic and paraconsistent inference operations*

1.2.7.1   *Nonmonotonic reasoning by expectation revision*   The idea that non-monotonic reasoning can be modelled by the revision of theories or expectations was first formulated by Israel (1980) and Poole (1988), and was later turned into a fully fledged theory by Gärdenfors and Makinson (Makinson and Gärdenfors 1991 and Gärdenfors and Makinson 1994). The actual acceptance set, i.e., what can legitimately be inferred from the explicitly available information on the basis of the expectations, is identified with the revision of the expectations by the information.[55]

Let us assume as given a prioritized set of expectations, a finite prioritized set of items of information, and a primitive revision operation $\underset{*}{}$ for the expectation set. The basic idea for modelling beliefs is that the *acceptance set* (belief set or theory) of an agent is exactly that what can be inferred from the information base, and that in turn is *the expectations revised by the information*. We should like to represent this expectation revision by

$$K = Inf(\mathcal{H}) = \mathcal{E} \underset{*}{} \mathcal{H}$$

The set K is interpreted as containing everything that an epistemic agent with information $\mathcal{H}$ and expectations $\mathcal{E}$ implicitly accepts—after perfect deliberation.

If, however, we want to employ the formal tools of the paradigm for belief revision by Alchourrón, Gärdenfors and Makinson,[56] we have to abstract from much of the structural information given by the information and expectation

---

[54] For methods of employing priorities in belief change, see Sections 1.2.7.2, 5.2 and 5.3.3; for methods of employing entrenchments, see Section 8.7. For the relation between priorities and entrenchments, and how they can be seen as approximately equivalent, see Rott (1992c) and (2000a). For an example showing how priorities and entrenchments which are (approximately) equivalent may well order the same sentences differently, see Rott (1992c, p. 44).

[55] 'Real rules of inference are rules (better: policies) guiding belief fixation and revision. . . . Inference (reasoning) is non-monotonic: New information (evidence) and further reasoning on old beliefs . . . can and does lead to the revision of our theories and, of course, to revision by "subtraction" as well as by "addition".' (Israel 1980, p. 55) The 'idea of theory formation from a fixed set of possible hypotheses is a natural and simple characterization of default reasoning. Rather than defining a new logic for default reasoning, we would rather say that it is the natural outcome of considering reasoning, not as deduction, but as theory formation.' (Poole 1988, p. 29). Also compare footnote 39.

[56] See for instance Alchourrón, Gärdenfors and Makinson (1985), Makinson (1985), Gärdenfors (1988), and Gärdenfors and Rott (1995). More about this paradigm, the so-called *AGM approach* to belief revision, later.

bases, and use $E = Cn(|\mathcal{E}|)$ and $h = \bigwedge(|\mathcal{H}|)$ instead. In symbols, the suggestion is captured by what I call the *Makinson-Gärdenfors identity*

$$K = Inf(H) = E * h$$

The loss of structural information involved in the transition from $\mathcal{E}$ to E and from $\mathcal{H}$ to h is severe, but we have to put up with it in the standard model of belief revision, since it allows only for revisions of belief sets by single beliefs. We take the conjunction h rather than the base $\mathcal{H}$ or H as the second argument of $*$, since revisions by sets of sentences (so-called *multiple revisions*) have not been studied very much within the AGM paradigm, and in so far as it has been studied, the multiple revision $E * H$ is usually identified with $E * h$.[57] Furthermore, we have to presume that E is a logically closed set.[58] Given all this, we can take $*$ to be a revision function in the sense of Alchourrón, Gärdenfors and Makinson. As a minimal requirement for expectation revision, it is then assumed that $E * h$ is a logically closed set containing h.[59] In case of conflict, 'hard' information thus takes priority over 'soft' expectations. Expectations are prone to be defeated by explicit information.

Now let us have a look at some limiting cases. Notice that even an empty information base will leave an agent with enough beliefs to act upon. If the expectations are consistent and the information is *vacuous* in the sense that it contains at most logical truths, then no elements in the expectation set will be overturned, so one ends up with $K = E * h = Cn(E)$. More generally, if H is consistent with E—which, as we have noted, will be a rather exceptional case—then we have $K = Cn(E \cup H)$.[60] On the other hand, Inf reduces to the standard logic Cn if (and only if) the set of expectations is vacuous, since $K = Cn(\emptyset) * h = Cn(H)$.[61]

If two belief bases $\mathcal{H}$ and $\mathcal{H}'$ are logically equivalent under a standard consequence relation, should we expect them to be equivalent under nonmonotonic inferences? In other words, should $Cn(H) = Cn(H')$ entail $Inf(H) = Inf(H')$? According to the Gärdenfors-Makinson modelling of the basic idea, we have to check whether $E * h$ equals $E * h'$, which is in fact true.[62] But this reasoning ceases to be valid, or so I shall argue, in more elaborate versions of 'revisions' of

[57] Compare Fuhrmann and Hansson (1994, Section 12).

[58] If E is itself a non-theory, then an AGM revision operation $*$ formally violates the *Principle of Categorial Matching* which I discuss in the next section: It takes a base and yields back a theory. This is no defect here, however. First, there is no obstacle to thinking that really $Cn(E)$ is the background theory and letting the revisions of this theory be guided by the structure of the 'axiomatization' E. And second, more importantly, there is no reason to require that the expectation revision function $*$ must satisfy the principle of categorial matching since it isn't a model for the *dynamics* of doxastic systems at all. See Section 1.2.8 below.

[59] For more properties of AGM revision functions, see Section 4.3.

[60] This follows formally from conditions ($*3$) and ($*4$) as labelled in Section 4.3.

[61] For consistent H, this follows from ($*3$) and ($*4$), for inconsistent H from ($*1$) and ($*2$).

[62] By the AGM postulate ($*6$).

$\mathcal{E}$ by $\mathcal{H}$ that do not sacrifice the inner structure of the (possibly prioritized) belief bases. Later, in the course of Chapter 5, we introduce an alternative account which comes closer to the basic idea and proposes a non-AGM model for $\mathcal{E} * \mathcal{H}$, namely by exploiting the full structure and defining $Inf(\mathcal{H}) = \mathcal{E} \underline{*} \mathcal{H}$ to be the result of a 'consolidation' of $\mathcal{E} \circ \mathcal{H}$.

1.2.7.2 *Nonmonotonic and paraconsistent reasoning by consolidation*

Items of an information base may stand in conflict with one another. As a further *desideratum*, therefore, we would like our non-standard inference mechanism to support *paraconsistent reasoning*. For inconsistent information bases, where classical logic would have us derive all sentences of the language, a more useful reasoning facility should accommodate or resolve the inconsistency and yield only conclusions that may be considered 'reliable' or 'dependable'. I should warn the reader in advance, however, that we will not end up with a paraconsistent *logic* in the usual sense, because we are going to employ extralogical information for resolving inconsistencies: namely the syntactic structure and the prioritization of belief bases. In reasoning, agents are modelled as looking for pieces of information that may be held responsible for an inconsistency and basing their inferences only on premises that are more reliable or dependable than these.

In order to avoid doxastic ruin in the face of a classically inconsistent base, we have to look for an inference operation that is in some respects *weaker than classical logic*. We shall in fact overshoot this mark. Not only will our new inference relation fail to be supraclassical (it does not hold in general that $Cn_0(H) \subseteq Inf(H)$), but it will not even reflexive: Not all elements of the information base will come out accepted in the belief state that results from drawing inferences from the information base (it does not hold in general that $H \subseteq Inf(H)$). As already pointed out, this may be taken as a second reason to deny that we shall be dealing with a logic at all.

For the development of the following idea, we can make use of any revision or contraction operation that is applicable to prioritized belief bases.[63] Any such method that is antecedently defined as applying to combined sequences of information and expectations will do. For the sake of concreteness, however, let us pick a particular change operation that has been proposed by quite a number of researchers.[64]

Henceforth I shall use the label *Prioritized Base Contraction* as a technical term. The operation denoted by it is a variant of what I shall later call foundationalist belief change, and it proceeds in two steps. Suppose we are given a prioritized belief base $\mathcal{G} = \langle G_1, \ldots, G_k \rangle$ such that $G = \bigcup_i G_i$ entails ϕ, and we want to withdraw our doxastic commitment to ϕ. Thinking of our intended inter-

---

[63] Cayrol and Lagasquie-Schiex (1995) survey a wide range of such operations in the context of nonmonotonic reasoning.

[64] Rescher (1964, 1973, 1976), Fagin *et al.* (1983), Brewka (1991), Nebel (1992), Benferhat *et al.* (1993). A logical characterization has been obtained independently by Rott (1993) and del Val (1994)).

pretation $\mathcal{G} = \mathcal{E} \circ \mathcal{H}$, this means that we want to eliminate ϕ from $K = Inf(\mathcal{H})$.[65] As a first step, we look for the 'best' maximal subsets of G that do not imply ϕ. This is done by maximizing the sentences accepted at each level of priority, subject to the constraint that ϕ be not implied. We begin from the top G_k and advance step by step down to the bottom G_1, keeping what we we have got at the earlier stages.

In order to make this more precise, let me introduce a bit of notation. Let $\langle G_1, \ldots, G_k \rangle$ be identical with $\langle E_1, \ldots, E_m, H_1, \ldots, H_n \rangle$, of course with the understanding that $k = m + n$. For any two subsets F and F' of $G = G_1 \cup \cdots \cup G_k$, we say that F is *better than* F' (with respect to $\mathcal{G}$), in symbols $F' \prec F$, if and only if there is an i such that $F \cap G_i$ is a proper superset of $F' \cap G_i$, and for all j greater than i, it holds that $F \cap G_j$ and $F' \cap G_j$ are identical. The process of maximizing from the top to the bottom informally described above can now be recognized as the process of finding a $\prec$-preferred subset of G that does not imply ϕ.

This process of cascading maximization is not uniquely determined, for at each level, there will in general be several 'best' maximal non-implying subsets. For any prioritized belief base $\mathcal{G}$, the set $\mathcal{G} \parallel \phi$ is defined to be the set of all subsets F of G with the following property: $F \nvdash \phi$, but for all subsets F' of G such that $F \prec F'$ it holds that $F' \vdash \phi$. Intuitively, $\mathcal{G} \parallel \phi$ is the set of all best subsets of G (with respect to $\mathcal{G}$) that do not entail ϕ.[66]

A unique outcome of the withdrawal procedure is achieved in the second step. Here we adopt the 'cautious' or 'indifferent' rule to the effect that only those sentences should be in the contraction of $\mathcal{G}$ with respect to ϕ which are entailed by *all* of these best subsets.[67] The result of this construction gives us a subset of K which is both consistent and closed with respect to the standard logic Cn. That is, it is an inferentially coherent belief set.

Next we need the notion of *consolidation*. The operation of consolidation aims at making belief structures consistent. Given some prioritized base $\mathcal{G}$ and a contraction method for it, consolidation may be construed quite generally as a particular form of belief contraction, namely the contraction of $\mathcal{G}$ with respect to the falsity constant $\perp$. When I speak of consolidation in this book, however, I have in mind a particular method of achieving this, viz., the prioritized base contraction with respect to $\perp$. I refer to the result by $Consol(\mathcal{G})$.[68]

---

[65] We shall see that it may happen that $\phi \notin Inf(\mathcal{H})$ although G entails ϕ. This case is covered by the treatment that follows.

[66] If $\mathcal{G} = \langle G \rangle$ is not prioritized, $\mathcal{G} \parallel \phi$ coincides with the AGM set $G \perp \phi$ of all set-theoretically maximal subsets of G that do not imply ϕ.

[67] Technically speaking, a prioritized base contraction may be seen as a form of partial meet contraction of the theory $K = Cn(G)$ in the sense of Alchourrón, Gärdenfors and Makinson (1985). See Nebel (1992), Rott (1993, Section 6) and Gärdenfors and Rott (1995, Sections 5.1–5.2).

[68] The concept of consolidation was introduced, in a form that is slightly different (no closure of best non-implying subsets under Cn) and slightly more general (partial meet contractions rather than prioritized base contractions) than mine, by Sven Ove Hansson (1991, Section 2).

Now we return to our initial question. What can be inferred from a prioritized information base $\mathcal{H}$, given some fixed background expectation base $\mathcal{E}$? What may be accepted on that basis? Let us first assume, like Gärdenfors and Makinson, that the information base $\mathcal{H}$ is 'flat' (not prioritized), so we can identify it with a plain set H. Then we can define the belief set K generated from $\mathcal{E}$ and H in the following way.

$$Inf(H) = Consol(\mathcal{E} \circ \langle H \rangle)$$

In so doing we deviate from the Gärdenfors-Makinson idea $Inf(H) = E \underline{*} (\bigwedge H)$. The most important difference is that this operation Inf is paraconsistent, but not reflexive: The acceptance set $Inf(H)$ is consistent under the consequence operation Cn, even if H itself is inconsistent under Cn. Indeed, H fails to be a subset of $Inf(H)$ if and only if H is inconsistent under Cn. Explicit information is not necessarily believed.

Now the more general case when $\mathcal{H}$ is a genuinely prioritized information base is straightforward:

$$Inf(\mathcal{H}) = Consol(\mathcal{E} \circ \mathcal{H})$$

Using the above notation, we have $Inf(\mathcal{H}) = Consol(\mathcal{G}) = \bigcap\{Cn(G') : G' \in \mathcal{G} \parallel \bot\}$. Notice that here Inf is now an operation different in kind from the Gärdenfors-Makinson account, since it takes as arguments more complicated data structures than finite sets of sentences. The flat case is covered as a special case. Of course it is not possible to define $Inf(H)$ by identifying it with $Inf(\mathcal{H})$ for *some* $\mathcal{H}$ such that $H = |\mathcal{H}|$, since clearly there may be many such $\mathcal{H}$s which generate different acceptance sets. The canonical way to deal with non-prioritized bases H is to put $Inf(H) = Inf(\langle H \rangle)$.

The present approach is both a generalization (in one respect) and a specialization (in another respect) of the Gärdenfors-Makinson approach. If we define a two-place revision operation that takes prioritized bases as arguments by putting

$$\mathcal{E} \underline{*} \mathcal{H} = Consol(\mathcal{E} \circ \mathcal{H})$$

then we see that $Inf(\mathcal{H}) = \mathcal{E} \underline{*} \mathcal{H}$. (Notice that the result of this revision of a base by a base is a theory.) The idea is similar to that of the Makinson-Gärdenfors identity, $Inf(h) = E \underline{*} h$. It widens the latter since the data structures for both arguments of $\underline{*}$ that we allow are *much more general* than those of Gärdenfors and Makinson. As I have already noted, the representation of prioritized belief bases $\mathcal{E}$ and $\mathcal{H}$ with the help of $E = Cn(|\mathcal{E}|)$ and $h = \bigwedge(|\mathcal{H}|)$ is only a very rough expedient that loses lots of valuable structural information embodied in prioritized belief bases. In the converse direction, we can naturally and without any loss of information represent E by $\mathcal{E} = \langle E \rangle$, and h by $\mathcal{H} = \langle \{h\} \rangle$. We can thus capture $E \underline{*} h$ by $\langle E \rangle \underline{*} \langle \{h\} \rangle = Consol(\langle E, \{h\} \rangle)$. If we do this, however,

For a further generalization, see Olsson (1998b).

we need to realize that our proposal is *much less general* than the Gärdenfors-Makinson proposal in another respect. We have committed ourselves to a particular revision method $*$, namely the one using consolidation by prioritized base contraction, while Gärdenfors and Makinson leave it entirely open which model of theory change to use.[69] More detailed comments on the difference between the approach to nonmonotonic reasoning taken by Gärdenfors and Makinson and the one sketched here will be given in Chapter 5.

1.2.8 *Belief revision strategies*

In the last sections, I have been dealing with the *static dimension* of belief formation. The topic has been thinking, reasoning, deliberating, etc., or more exactly: the process of deriving beliefs from an information base, on the basis of a fixed set of expectations that the agent possesses.[70] Intuitively, however, we often find reason to think that mechanisms for *belief changes* are sensitive to the form and content of our *belief states*. Our static picture determines, to some extent at least, our dynamic picture of the field. Let us now start reflecting on how and why this may in fact be the case.

In the last section we have for the first time come across the dynamic perspective on doxastic systems. I have defined the process of drawing inferences from a given belief base by the process of consolidation which in turn was characterized as a special form of belief contraction. How the contraction is to be performed has been determined by two factors. First, by the syntactic structure and prioritization of the base, and second by the prescriptions of the method of prioritized base contraction. The latter are, so to speak, *laws* of belief dynamics, while the former are the *initial conditions*. Once we have set up the relevant laws, belief changes are completely determined by the structuring of the base. For other kinds of laws of belief dynamics, which will be discussed later in this book, the specification of a certain kind of choice function or an entrenchment relation has the same determinant effect. In other words, once we have agreed on 'the right' method for changing our mind, the dynamics of our belief are already contained in the comprehensive description of the statics. A doxastic state, so it seems, contains a complete description of belief revision strategies.

This may be compared to the interplay between statics and dynamics in physical theories. In Newtonian mechanics for example, a comprehensive specification of the momentary state of a systems must include values for masses and forces. Given these, we can, in principle, use Newton's laws to calculate the further development of the system (even though in practice the calculations often are far too complicated and cumbersome to be successfully carried through). The

---

[69]For consistent h, this is the same as the so-called full meet revision $E *_0 h$ which only makes good sense if E is a non-theory. If E is a theory, then $E *_0 h = Cn(h)$ whenever h is inconsistent with E. See Alchourrón and Makinson (1982).

[70]Needless to say, the drawing of inferences is really a *process* that takes time when performed by real agents or computer systems. This aspect is entirely disregarded in this book. I shall stick to the idealization involved in the use of abstract inference *operations* and keep on speaking of the 'static dimension' here.

theoretical terms of physical theories are part of the static picture of physical systems, but by entering into the basic laws of motion, they account for actual and potential changes of such systems. In precisely the same way, orderings of belief bases, for instance, may be regarded as part of the static picture of doxastic systems, but by entering into the basic laws of motion for beliefs, they do account for actual and potential revisions of such systems. This is the reason why *dynamic* are superior to merely *kinematic* theories.[71]

Like Newtonian forces and masses, parts of the specification of a momentary *state* of a system enter into the basic laws governing the changes of that system. The system's potential behaviour, as a response to possible external influences ('perturbations'), is implicitly, but completely fixed by a full specification of its constitution at a single given time instant.

There are two things to consider in belief change: a static and a dynamic aspect; belief formation and belief revision; the making up and the changing of one's mind. Now let us ask a very general question: Does, or should, my belief state include the structures guiding its own potential changes, or should other, external factors determine my changes of mind? Should belief revision strategies be part of my belief state (the dynamic option), or should they come from somewhere else and thus be independent of my beliefs (the static option)?[72]

It is difficult to imagine other sources for the structures guiding belief revisions than mental states or dispositions of the believing agent himself. If beliefs are subjective, how could possibly belief changes be more objective? Who is going to decide how beliefs should be changed, if not the agent himself? This is why I favour the view that any selection mechanism involved in belief changes should be interpreted as a part of the agent's doxastic state—and as subject to revisions just as the beliefs themselves are revisable.

However our representation of belief states will look like, any modelling of a process of belief change should obey to the following *Principle of Categorial Matching*:

(PCM)   The representation of a belief state after a belief change has taken place should be of the same format as the representation of the belief state before the change.[73]

---

[71] *Kinematics* is that branch of physics which describes the motion of objects without being concerned with the forces that produce the motion. *Dynamics* in contrast involves an examination of the forces which produce the motion. Interesting comparisons between the nature of physical theories and that of theories about rational belief systems are made by Ellis (1979, Chapter 1). Ellis argues that the (static) 'laws of logic ... are about belief systems in the sense in which, say, Newton's First-Law of Motion is a statement about how bodies move'. (1979, p. 101). His main point is that both kinds of laws determine *ideals* of behaviour rather than actual behaviour.

[72] Compare Hansson (1992, pp. 243–244) on selection mechanisms as background factors vs. as state-specific; and Freund and Lehmann (1994, Section 5.1) on what they call the 'static' vs. the 'dynamic viewpoint'. I shall give a more formal analysis of this problem in Section 3.6.

[73] This principle is too obvious that I could claim to have invented it. It is formulated, for instance, in Dalal (1988) and called there the principle of 'adequacy of representation'. The

If this principle is satisfied—but only if it is satisfied—then our modelling is able to provide for iterations of belief change. If there is information encoded into the initial belief state that helps the agent change his beliefs, then the same kind of information should be available after the change to give him guidance in further changes. If revision-guiding information is external to the initial belief state, then that information need not be available as part of the posterior belief state either. Given the intuitive plausibility of (PCM), it is strange that most of the suggestions in the literature on belief changes fail to meet this requirement. We shall be able to meet it both in the foundationalist modelling of Chapter 5 and in the coherentist modelling of Chapters 7 and 8. In the foundationalist case, orderings are part of the doxastic states, in the coherentist case, however, we need to stipulate state-independent choice functions that are not 'faithful' to the agent's current belief set.

It must be stressed, on the other hand, that there is no reason to require that the same kind of matching hold between a belief or information base and the set of beliefs generated from this base (the 'derived' acceptance state in full reflective equilibrium). The theory generated from a prioritized belief base through the application of some inference relation has a different structure from the belief base: It is flat and logically closed, neither of which is true of the base. And there is nothing wrong with the categorial mismatch. It is part of our very concept of inference operations as generating states in reflective equilibrium that no need for iteration arises in the generation of theories from bases (see Section 3.2.1). Even if we conceive of the inference relation as being modelled by revisions or contractions of expectation bases (as I did in the Section 1.2.7), the principle of categorial matching is not relevant at all.[74]

I have pointed out that the Principle of Categorial Matching is particularly relevant when iterations of belief change are at issue. Although this is an important topic it will not be treated extensively in this book. A few sections, however, are interspersed that briefly indicate the picture afforded by different approaches to belief change. Readers interested in this topic are referred to Sections 3.6, 5.2.3, and 7.10.

---

present version has first been formulated in Gärdenfors and Rott (1995, p. 37).

[74] In the terminology of distinctions drawn later, in Chapter 3, categorial matching must be aspired to along the horizontal dimension, not along the vertical dimension.

EPISTEMOLOGY AND BELIEF CHANGE

From a philosophical point of view, AI and computer science are rather sloppy when speaking of knowledge. What they mean by 'knowledge base', 'knowledge system', and 'knowledge representation' does not refer to what is commonly meant by 'knowledge'. It is widely agreed that if we *know* rather than just *believe that* ϕ, then ϕ must be true and our belief that ϕ is true must have some sort of justification.[75] Neither truth nor justification is presupposed or guaranteed when computer scientists use the above-mentioned terms. I do not mean to suggest that this mistake has serious consequences in practice. For the purposes of the information sciences, it seems perfectly harmless to commit this kind of inaccuracy. But philosophers, and in particular epistemologists, must not be allowed to confuse epistemic and doxastic concepts. It is rather their duty to clarify the subtle interconnections between knowledge and belief. Evidently, this is a formidable task and cannot be achieved in the present book; I shall not develop an epistemological theory of my own. I take as given a particular theory of knowledge that has gained much respect recently, namely the one of Keith Lehrer as elaborated in his 1990 book *Theory of Knowledge*.[76] Based mainly on Lehrer's work, the purpose of the present chapter is to show that a proper understanding of knowledge does not only require an understanding of belief *simpliciter*, but in addition a thorough understanding *of the dynamics of belief*.

The main point of this chapter is to show that a good theory of belief revision is necessary for a proper development of a theory of knowledge. I shall later also argue that the theory of belief revision can profit from a study of the concepts

---

[75] *Externalist* accounts of knowledge argue that people or animals need not be able to justify their knowledge, but rather hold that there must be some *reliable* (causal, counterfactual, nomological) connection between the knower and the things known. I acknowledge that it makes good sense to say that the dog, for instance, knows where in the garden its favourite bone is buried, but I want to focus on what Lehrer (1990, pp. 4, 36) calls 'characteristically human sort of knowledge'. I think that the verb 'to know' is ambiguous and that its variant meanings reflect different and conflicting intuitions—a fact that has caused much unnecessary dispute in recent epistemology. I do not claim to cover all uses of 'to know'.

[76] This book is the successor of, and shows considerable overlap with, a book with the title *Knowledge* published by the same author in 1974. Given that Lehrer is one of the most prominent contemporary epistemologists and has made important contributions to the field for more than three decades, it seems legitimate to use his theory as the point of reference for the following reflections. The basic structure of the 1990 definition of 'knowledge' is retained in Chapter 2 of Lehrer (1997) (and duplicated in an analogous definition of 'wisdom'). For the reasons mentioned in footnote 101, the second edition of the *Theory of Knowledge* (Lehrer 2000) presents a concept of knowledge that is quite different from the one of the first edition; it is less relevant to the things I am interested in in this book.

that have evolved in recent epistemology. We ought to have a basic picture of what knowledge is, and of how knowledge can be obtained, when we search for principled solutions of the problems of belief change. In Chapter 3 I exploit the fundamental distinction between foundationalist and coherentist accounts of knowledge that has played a central role in the epistemological literature of this century.[77] I want to argue that this distinction applies more properly to theories of *doxastic* states than to theories of *epistemic* states. It is unfortunate that philosophers have tended to focus on knowledge without attending equally closely to the—seemingly less problematic—notion of belief. I shall not try to compensate for this deficiency in this chapter, but I hope to profit from the light that is shed on belief revision by the dichotomy of foundationalist and coherentist approaches to the theory of knowledge. In fact this fundamental distinction will help us find two basic, as it were orthogonal perspectives on belief change that will structure the whole of this book.

2.1 Epistemology, knowledge representation and revision

What is knowledge? What can we know, and how do we know? These are the questions addressed by *epistemology*, also known as the *theory of knowledge*.[78] It is well-known that the problems of epistemology are far from being solved.

There are related questions which may be raised even if one considers the issues addressed by epistemology proper to be settled. How is knowledge represented in natural knowers? How is knowledge be represented in 'artificial intelligences', or how should it be represented—if such a representation is possible at all? The first problem is studied in psychology and the philosophy of mind, the second one in computer science. There is considerable debate as to whether one should attempt to create artificial knowers by trying to mimic natural, human knowers or by starting the project afresh, unconstrained by any android preconceptions. Arguably, the representation of pieces of knowledge should not differ from the representation of beliefs. From the agent's point of view, human or artificial, there is little if anything that allows him to distinguish between mere belief and real knowledge. From a first-person perspective, there is little if any difference, and it is doubtful whether we should expect the representation of belief and knowledge to represent more than what is accessible to the reasoner or reasoning system itself.

Now let us suppose, counterfactually, that the questions concerning the representation of knowledge and belief have been answered to everybody's satisfaction. Then many interesting questions are still left open. Solutions regarding the concept and the representation of knowledge do not automatically give answers to questions concerning the dynamics of belief and knowledge. How is 'knowledge', or better: alleged knowledge, revised in the light of new evidence? How should

---

[77]See, for instance, Lehrer (1990), Plantinga (1990) or Sosa (1980).

[78]The German term *Erkenntnistheorie* originates with the Kantian tradition and carries different connotations.

it be revised? Agents are fallible, and what they consider to be their knowledge quite often turns out to be false—and hence not to be proper knowledge at all. At this stage we reach a turning point. The focus of attention gets shifted from knowledge to belief, and normative questions are seen to become increasingly important. Epistemology, the philosophy of mind and psychology analyse and describe what knowledge is and how it is obtained and represented in human beings. Knowledge representation has to do with normative issues in so far as there are many different approaches to representing information, some of which are 'better' and some of which are 'worse'—with respect to what we demand from the relevant 'knowledge systems', e.g. computational tractability, efficiency, comprehensiveness, reliability, transparency. The change of alleged knowledge, that is, the change of belief and acceptance systems, is intrinsically beset by normative problems as well. It surely is a reasonable question to ask about an 'ethics' of belief and acceptance. But it seems that this question can be set aside when talking about knowledge. In so far as 'knowledge' is perceived to be in need of revision, however, it is perceived to be less than real knowledge, but only belief, opinion, prejudice or some similar sort of doxastic (rather than epistemic) state. Problems of the ethics of belief arise insofar—and perhaps only insofar—as problems of a genuinely dynamic sort arise. The ethics of belief in the traditional understanding[79] is primarily concerned with the question when it is rational or justified to *adopt* some new belief, thus involving an act of belief *acquisition*. The additional problem of belief change is that we have to face the question when to *eliminate* or *replace* which of the previously held beliefs, thus involving acts of belief *dislodgement*.

Prima facie, it appears that the questions posed by epistemology, knowledge representation and 'knowledge' revision are, though related, clearly separable. In any case, there does not seem to be a close connection between the theory of knowledge and the theory of belief revision. It is the thesis of the present chapter that precisely this is illusory. Even if we do not contemplate any problems of the 'intermediate' field of knowledge representation, we can find very close interdependencies of epistemology and belief revision. More precisely, I want to illustrate that

(i) the analysis of knowledge requires a proper solution of the problem of belief revision

(ii) the analysis of belief revision should not be conducted without a proper understanding of the notion and representation of knowledge

I must actually restrict claim (i) considerably because I base my considerations on the particular epistemological theory of Keith Lehrer. I take Lehrer to be following in great detail a trail that was started by Plato, in a beautiful passage in one of his earlier dialogues:

---

[79] For the exciting debate that took place in the nineteenth century, see the anthology edited by McCarthy (1986).

True opinions too are a fine thing and altogether good in their effects so long as they stay with one, but they won't willingly stay long and instead run away from a person's soul, so they're not worth much until one ties them down by reasoning out the explanation. ... And when they've been tied down, then for one thing they become items of knowledge, and for another, permanent. And that's what makes knowledge more valuable than right opinion, and the way knowledge differs from right opinion is by being tied down. (*Meno* 97e–98a, Plato 1994, p. 69)

My considerations will not depend on the details of Lehrer's theory; in fact I shall offer a few non-trivial improvements on some of his definitions. But I do base my arguments on the overall architecture of Lehrer's undertaking. If Lehrer were completely misguided, then what I say about the relation between epistemology and the theory of belief revision might equally be mistaken.

Claim (ii) needs to be qualified as well. The analysis of belief revision is not dependent on features that distinguish genuine knowledge from mere belief. It is rather dependent on the structure and the formation of beliefs as they are relevant in the theory of knowledge. What I have in mind above all is the fundamental distinction between *foundations* and *coherence* theories of knowledge. This distinction happens to have come to the fore in the theory of knowledge, but it may just as well be placed in a theory of belief.[80] It is primarily concerned with the inferential relations between various beliefs, that is to say, with the internal structure of our belief systems. The contrast lies in the answer to the question whether there is such a thing as a *belief base*, that is, beliefs that are not in need of an inferential justification by other beliefs and that taken together inferentially justify all the remaining ('derived') beliefs. All of this can be dealt with in a general theory of belief; nothing requires to refer this topic to the theory of knowledge. Claim (ii) can still be upheld, given the fact that many relevant aspects of the structure of beliefs have as a matter of fact come out most clearly in epistemological discussions.

2.2   Lehrer's theory of knowledge: Central definitions

In this section we unroll the central parts of Lehrer's theory of knowledge and show how they are rooted in problems and questions which belong to the theory of belief change. The following considerations are based on the author's summary in Lehrer (1990, pp. 147–149). I remove the time parameter t which does not play any interesting role in Lehrer's theory.[81]

For a long time it has been thought in philosophy that knowledge is justified true belief. A very short and very famous article by Gettier published in 1963 has made it clear that this analysis is inadequate. Let us have a look at one of Gettier's counterexamples.

Suppose that Smith has very strong evidence for

---

[80] Wolfgang Spohn has pointed out to me that this point is fully explicit in BonJour (1985).

[81] This is a general maxim for the responsible use of (semi-)formalizations: Never introduce variables or indices into your theory if they are not doing any real work! Never complicate your notation beyond necessity!

ϕ : Jones owns a Ford.

Suppose further that Smith is totally ignorant of Brown's whereabouts. Still he can (and does) correctly infer from ϕ that

ψ : Either Jones owns a Ford, or Brown is in Boston.

χ : Either Jones owns a Ford, or Brown is in Barcelona.

ξ : Either Jones owns a Ford, or Brown is in Brest-Litovsk.

By pure coincidence, and entirely unknown to Smith, Barcelona happens to be the place where Brown actually is. However, Jones does *not* own a Ford. Now, χ is a *true justified belief*, since ϕ is a justified belief and and χ may be logically derived from ϕ, and the second disjunct of χ is true. However, it would be utterly counterintuitive to say that Smith *knows* that χ is true, because he believes that χ is true 'for the wrong reasons'. He has just been lucky that the right belief occurred to him.[82]

Gettier's example has led many epistemologists to the conclusion that knowledge is *more* than justified true belief. They do not discard this venerable definition, but supplement it by a fourth clause. Lehrer's suggestion is the following:

DK   *S knows* that ϕ if and only if

  (i) S accepts that ϕ,

  (ii) it is true that ϕ,

  (iii) S is completely justified in accepting that ϕ, and

  (iv) S is completely justified in accepting that ϕ in a way that does not depend on any false statement.

Lehrer's distinction between belief and acceptance was commented upon in Chapter 1; I do not think that it would make a big difference if we substituted belief for acceptance in clause (i). Similarly, clause (ii) is not very controversial in the theory of knowledge. The essential parts of the definition are the last two clauses which appeal to the problematic notions of justification and dependence. I have substituted in clause (iv) a formulation taken from Lehrer (1990, p. 18) for the formulation in his official summary in Lehrer (1990, p. 147). In the latter he requires S to be completely justified in accepting that ϕ 'in a way that is not defeated by any false statement'. This statement seems to be slightly screwed up. What Lehrer means, I think, is that S is completely justified in accepting that ϕ in a way that cannot be defeated by pointing out that the justification relies on a false statement. The above phrasing of clause (iv) expresses this more accurately than Lehrer's official formulation.[83]

---

[82] Actually luck is not even needed on the account of belief given in Chapter 1. If we assume that the beliefs of an agent are logically closed, any justified false belief ϕ gives rise to an infinite set of justified true beliefs—all disjunctions $\phi \vee \phi'$ where ϕ' is any arbitrary truth. Since we do not require that a belief manifests itself in an occurrent event in the believer's mind, but may reside hidden in an implicit theory of his or hers, Smith is not just lucky, but his false belief that Jones owns a Ford will automatically generate infinitely many justified true beliefs like χ.

[83] Lehrer (1990, p. 138) apparently thinks that the two formulations are 'equivalent'.

Lehrer further characterizes the concept of knowledge in a two-layered strategy.[84] First he develops the notion of personal justification which is based on an agent's subjective acceptance system. In a second step the purely subjective standpoint gets transcended by several operations on the agent's current belief set that could help him to approximate the whole truth.[85]

2.2.1 Personal justification and the comparative reasonableness of acceptance

All of Lehrer's considerations are based on the notion of an acceptance system which is defined as follows.[86] We slightly adapt the notation.

D1   A system X is an *acceptance system* of S if and only if X contains just statements of the form, S accepts that ϕ, attributing to S just those things that S accepts with the objective of accepting that ϕ if and only if ϕ.

This definition could be amended a little by taking into account that on Lehrer's account, there is only one acceptance system for each agent at a certain time. So it seems that D1 should better start like that: 'A system X is *the* acceptance system of S if and only if ... '

Lehrer is one of the main advocates of a coherence theory of knowledge. According to this approach, all justification comes from coherence with a given acceptance system X. There is no justification *simpliciter*, only justification on the basis of some X.

D2 S is *justified in accepting* ϕ on the basis of system X of S if and only if ϕ coheres with X of S.

What is needed now is of course an elucidation of 'coherence-with-a-system'. Lehrer defines it in terms of the relations of competing, beating and neutralizing of propositions.

D3 ϕ *coheres* with X of S if and only if all competitors of ϕ are beaten or neutralized for S on X.[87]

Interestingly, Lehrer does not seem to require that ϕ belongs to X for D2 and D3. The function that takes an acceptance system X and yields back the set of all sentences cohering with X, or equivalently, of all sentences S is justified in accepting on the basis of X, may be interpreted as an inference operation in the sense of Chapter 1.

---

[84] Compare in particular Lehrer (1990, pp. 141–152) and (1997, 28–45).

[85] These operations are carried out only hypothetically, 'for the sake of argument' in fictitious test dialogues with an omniscient sceptic. In reality, the agent's state of mind remains unchanged.

[86] This is the term used in Lehrer (1990); in Lehrer (1997, pp. 25–29) the leading part is taken by the 'evaluation system' which includes not only a person's accepted propositions (geared to truth) but also her preferred propositions (geared to merit).

[87] Definition D3 on p. 148 of Lehrer (1990) actually starts as follows: 'S is *justified in accepting* ϕ on the basis of system X of S if and only if ... ' From definition D2 and the surrounding text, however, it is obvious that this is a misprint and the formulation given above is the intended one.

D4 ψ *competes* with ϕ for S on X if and only if it is more reasonable for S to accept that ϕ on the assumption that ψ is false than on the assumption that ψ is true, on the basis of X.

D5 ϕ *beats* ψ for S on X if and only if ψ competes with ϕ for S on X, and it is more reasonable for S to accept ϕ than to accept ψ on X.

D6 χ *neutralizes* ψ as a competitor of ϕ for S on X if and only if ψ competes with ϕ for S on X, but $\psi \wedge \chi$ does not compete with ϕ for S on X, and it is as reasonable for S to accept $\psi \wedge \chi$ as to accept ψ alone on X.

Intuitively, definitions D4 and D5 are not beyond reproach. Let us assume, for instance, that it is extremely reasonable to accept that ϕ, and that ψ weakens the reasonableness of accepting ϕ just a little bit, by a more or less negligible degree. Should we say that ψ 'competes with' ϕ? In what sense would we speak of a competition? ψ may be about quite another subject matter than ϕ, so there cannot be much rivalry or conflict between the two. And again, if it is still a little more reasonable to accept ϕ than to accept ψ, should we say that ϕ 'beats' ψ? Surely in the situation just described it would make good sense to accept both ϕ and ψ (assuming that it is reasonable to accept ψ in the first place), even if ϕ beats ψ in Lehrer's sense. We are ready to accept sentences that weaken each other a little, and we tend to connect such sentences by 'although'. If we say, 'They get along together very well, although they have no interests in common', in symbols 'ϕ although ψ', we do accept both ϕ and ψ. It does not matter that ψ 'competes with' ϕ in Lehrer's sense, and it does not matter whether ϕ 'beats' ψ or not.

The last three definitions, D4 – D6, all lead us to a comparative concept 'reasonable-to-accept', relativized to a given acceptance system X. Before turning to a brief discussion of that concept, we finish off the first, subjective part of Lehrer's analysis of knowledge. For personal justification, it is just the agent's current acceptance system which is the basis for judgements of coherence.

D7 S is *personally justified in accepting* that ϕ if and only if S is justified in accepting that ϕ on the basis of the acceptance system of S.

Now of course everything hinges on what is meant by 'reasonable-to-accept'— again, relativized to a given acceptance system. Surprisingly, Lehrer does not say very much about this, and he considers it as an advantage that his primitive term of reasonableness is open to many different interpretations. In the few pages he devotes to the topic (Lehrer 1990, pp. 127–131), however, he recommends to employ *cognitive decision theory*.[88] Lehrer suggests to identify the degree of reasonableness of accepting a hypothesis ϕ with its *expected epistemic utility*:

---

[88] The most eminent advocate of cognitive decision theory is Isaac Levi (1967, 1984). For a critical voice, see Weintraub (1990). In Lehrer's recent book (1997, in particular pp. 30–35), the contribution of cognitive decision theory to personal justification has vanished and its role is taken over by the principle of trustworthiness, a kind of rationality principle that is geared solely to the acquisition of truth. Cognitive decision theory is still present, however, in the second edition of the *Theory of Knowledge* (Lehrer 2000, pp. 144–148).

$$r(\phi) \;=\; p(\phi) \cdot Ut(\phi) + p(\neg\phi) \cdot Uf(\phi)$$

where $p(\phi)$ and $p(\neg\phi) = 1 - p(\phi)$ are the probabilities of ϕ being true or false respectively, and $Ut(\phi)$ and $Uf(\phi)$ are the positive or negative utilities of accepting the hypothesis ϕ, when ϕ is true or false. The utility $Ut(\phi)$ is supposed to reflect the informativeness of ϕ, and possibly other virtues such as ϕs explanatory power, simplicity, or pragmatic value, and the advantage of conserving existing beliefs (Lehrer 1990, p. 131).

It should be noted that this absolute degree of reasonableness of accepting is not quite sufficient for what we need in order to understand the foregoing definitions. In the definition of competition, Lehrer appeals to the reasonableness of accepting ϕ *on the assumption that ψ is true or false*. What we need, then, is something like expected *conditional* epistemic utilities $r(\phi|\psi)$ and $r(\phi|\neg\psi)$, and it is left unspecified how we can get them. There is no problem with the well-known concept of conditional probabilities,[89] but it is not quite clear whether the utilities of accepting a hypothesis ϕ *conditional on accepting ψ or $\neg\psi$* should be thought of as different from the plain, unconditional utility of accepting ϕ. But if the utility of accepting a sentence is dependent on accepting some other sentence, should not the utility of accepting a sentence be sensitive to the context of acceptance, that is, to the acceptance system X as a whole?

This question leads us to a problem that is both more general and more important. As indicated in definitions D4 – D6, the reasonableness to accept a statement may be—and probably should be—relative to the acceptance system of the agent. But the above definition of $r(\phi)$ does not reflect this. It is rather an absolute measure of reasonableness. This is quite contrary to the coherentist's aim of evaluating *systems* of hypotheses rather than *single* hypotheses taken in isolation. Lehrer can counter this objection by saying that the utility functions Ut and Uf depend on the current acceptance system X. But then one may ask whether it is illuminating to base an analysis of 'coherence of ϕ with a system X' on an unexplained notion of 'utility of accepting ϕ on the basis of system X'.[90]

Appealing to cognitive decision theory suggests that the acceptance of a proposition is a matter of decision. We have seen in Chapter 1 that such an assumption is at least controversial. But in this respect Lehrer's replacement of 'belief' by 'acceptance' is a prudent move. It is certainly much more plausible to say that an agent decides to *accept* something than that he decides to *believe* something. Another potential point of criticism is that it need not be (objective?, subjective?) *probability* that is taken into account when the reasonableness of

---

[89] Let us assume that $p(\psi)$ and $p(\neg\psi)$ are positive, so that conditionalizing by either ψ or $\neg\psi$ is not beset with the problem of an ill-defined division by zero.

[90] In his brief discussion of expected epistemic utilities, Lehrer shows little awareness of the fact that Ut and Uf may or should depend on X, if reasonableness of acceptance is to be relative to the current acceptance system. Lehrer contrasts $Ut(\phi)$ with ϕs (objective or subjective) probability and links it to ϕs truth, but not to the acceptance of other sentences. He does not say anything about Uf.

accepting certain hypotheses gets assessed. Perhaps *plausibility*, a notion with different formal characteristics, is an equally suitable candidate. This objection again could be countered, by pointing out that decision theory simply is a theory based on probabilities.[91]

2.2.2 Varying imperatives for coherence

The most serious questions for Lehrer's account seem to be the following. How can we be sure that coherence in his sense, explained by means of a complicated mechanism of competing, beating and neutralizing, which in turn is based on a decision-theoretic criterion, will give us a coherent acceptance set?[92] If every element of an acceptance set coheres with that very set (as it presumably should), can we be sure that the acceptance set is consistent?[93] Will the acceptance set be closed under logical consequences? Both consistency and closure are themselves requirements of—inferential—coherence. If all of Lehrer's central definitions are finally based on a numerical degree of reasonableness, why not take that very same degree as the sole arbiter of acceptance, without the taxing detour via definitions D2 – D7?

All these questions call for a fresh and systematic look at the principles that are involved in Lehrer's concept of coherence.

First and foremost, we have coherence according to Lehrer's own theory. The corresponding imperative is this: *Accept precisely those sentences the competitors of which are beaten or neutralized!*[94]

Second, we have seen that Lehrer's theory is grafted on top of cognitive decision theory which of course brings along its own standards of coherence. Lehrer does not address this point. An attempt to make it explicit reveals that there are several ways to go. The following idea seems initially plausible: *Accept those sentences that promise the greatest expected cognitive utility!* A moment's reflection, however, shows that this idea is premature. Why should we reject sentences of positive (or at least non-negative) expected utility for the sole reason that there are still more useful ones? Don't all sentences with positive *r*-values contribute to the overall expected utility? Thus the right imperative seems to be this: *Accept all sentences with positive (or non-negative) expected cognitive utility!*[95] But

---

[91] Theories of plausibility of the kind I have in mind are offered, amongst others, by Grove (1988), Rescher (1976), Shackle (1961), and Spohn (1988). They would have to be supplemented by a qualitative decision theory.

[92] The distinction between relational coherence (coherence as a relation) and systemic coherence (coherence as a property of a system) is discussed in connection with Lehrer's theory by Olsson (1999).

[93] This problem for Lehrer's theory has also been treated by Erik Olsson (1998a).

[94] There is some problem in reformulating the *definition* of justified acceptance into an *imperative* or *rule* of acceptance. The former may be applied to all beliefs that have somehow come to be accepted and does not have any normative force, which the latter obviously has. I think, however, that for the purposes of the present discussion the difference can be neglected.

[95] This precept is plausible only if one assumes, as Lehrer apparently does, that the expected utility of *rejecting* a hypothesis is zero. An alternative idea would be to accept just those

perhaps this is again mistaken. It may be wrong to suppose that the utilities of individual sentences simply add up to yield the utility of the whole body of beliefs. This would mean that we cannot rely on the above imperatives, because conditional utilities (see above) may be very different from the unconditional ones. For instance, I may entertain two alternative hypotheses, both with high expected epistemic utility, which contradict each other. Accepting either one of them seems reasonable, but accepting both of them would lead to inconsistency which is not particularly useful. Thus one should take into account interactions between the individual beliefs, and regard beliefs as constituting a system that may be assessed *only* holistically. This is very much in the spirit of coherentists anyway, who argue that only the corpus of belief or knowledge taken as a whole is a proper unit of epistemic appraisal. The degree r of reasonableness of acceptance then must not be applied to individual sentences, but to *sets* of sentences, and corresponding imperative reads thus: *Accept those sets of sentences that promise the greatest expected cognitive utility!*

A third concept of coherence is the logical or inferential one. I shall discuss its implications in detail later, but it is expedient to anticipate the main points already here. A set of sentences is inferentially coherent if it is consistent and closed under consequences. The corresponding imperative is: *Accept all the logical consequences of what you accept, but avoid accepting contradictions!*

I will show in Chapters 7 and 8 that the second and the third concepts of coherence are compatible. But even if one is ready to grant Lehrer's mechanics of justification in the sense of his definitions D3 – D6 some plausibility in itself, it is doubtful whether it can be made to cohere with other concepts of coherence. One has to face the fact that different coherence criteria may conflict with one another, and decide which of these criteria are the most justified ones. As it stands, Lehrer's theory is a hybrid of at least three different intuitions.

2.2.3 *Undefeated justification and justification games*

In order to arrive at knowledge one has to go beyond the actual acceptance system of an individual agent. In a sense very close to the passage of Plato's *Meno* quoted on page 48, knowledge must be stable under criticism. In Lehrer's 'justification games' the part of the critic is taken over by an omniscient 'sceptic'. We now have a look at the formal definitions that take Lehrer from merely personal justification to complete and indeed indefeasible justification.

D8   A system V is a *verific system* of S if and only if V is a subsystem of the acceptance system of S resulting from eliminating all statements of the form, S accepts that ϕ, when ϕ is false.

As in the case of definition D1, it would be preferable to say that the V mentioned in D8 is *the* verific system of S, since it results from the unique

---

propositions ϕ which have a degree of reasonableness that exceeds a contextually fixed threshold value.

acceptance system of S (at a given time) by just cutting out the false beliefs. The verific system is the basis for verific and complete justification:

D9 S is *verifically justified in accepting* that ϕ if and only if S is justified in accepting that ϕ on the basis of the verific system of S.

D10 S is *completely justified in accepting* that ϕ if and only if S is personally justified in accepting ϕ and S is verifically justified in accepting ϕ.

Complete justification in this sense, however, does not solve the Gettier problem. Smith's belief that Jones owns a Ford may not depend on any false belief. Let us suppose that Jones told Smith that he has a Ford, showed him papers stating that he, Jones, owns a Ford, and always drives a Ford on his way from his home to his office. All this is believed and known (in some pre-theoretical sense) by Smith, and justifies his belief that Smith does in fact own a Ford. So the reason for Smith's not knowing that sentence χ above is true is not that he accepts false sentences, but rather that he is not aware of all true sentences that are relevant to the case. Notice that it is not enough to know *some* relevant facts since a biased selection of true facts may be utterly misleading and turn the agent away from some other truths. Lehrer suggests to solve this difficulty be looking at what he calls the 'ultrasystem' of an agent. Here are his definitions.

D11 S is *justified in accepting* that ϕ *in a way that is undefeated* if and only if S is justified in accepting ϕ on the basis of every system that is a member of the ultrasystem of S.

D12   A system M is a member of the *ultrasystem* of S if and only if either M is the acceptance system of S or results from

- eliminating one or more statements of the form, 'S accepts that ψ', when ψ is false,
- replacing one or more statements of the form, 'S accepts that ψ', with a statement of the form 'S accepts that not ψ', when ψ is false,
- or any combination of such eliminations and replacements in the acceptance system of S

with the constraint that if ψ logically entails χ which is false and also accepted, then 'S accepts that χ' must also be eliminated or replaced just as 'S accepts that ψ' was.

Before moving on to the criticism of Lehrer's, we should mention his key result: 'Knowledge reduces to undefeated justification, a just reward for our arduous analytical efforts.' (Lehrer 1990, p. 149) Clearly, undefeated justification implies personal justification: the actual acceptance system of S is a member of the ultrasystem; it also implies verific justification: the sceptic can make S eliminate all his false beliefs, thus effecting a transition to the agent's verific system;[96] finally, it also implies truth: if ϕ were false, the sceptic could make S eliminate ϕ.

---

[96] We neglect the possibility that the removal of a false belief may tear along some true beliefs.

One may wonder why Lehrer requires for the undefeated justification of ϕ that *every* member of the ultrasystem must support ϕ. It would seem sufficient that S is *ultimately justified*. By this we mean that for every member M of the ultrasystem of S there is another member M' of the ultrasystem of S which improves on M and on the basis of which S is justified in accepting ϕ. That M' *improves on* M means, of course, that M' can be reached from M by some combination of 'truth-conducive' eliminations and replacements of the kind specified in definition D12. This concept seems more adequate since even if S knows that ϕ, pre-theoretically understood, a mischievous sceptic may well advance an impressive battery of true facts speaking *against* the truth of ϕ, so that S looses his confidence that ϕ is true. Only later in his conversation with the omniscient sceptic, when S comes to know mpre about the truth, will he regain his old true and justified belief. Although the correct belief would be dropped on the receipt of true but misleading information, we may consider it to constitute knowledge, since one can later learn that this information has in fact been misleading. Temporary doubts about ϕ should not count, so it seems, as long as all potential paths of the ultra justification game finally lead to ϕ's acceptance.[97] Lehrer himself seems to agree with that when discussing his Grabit example:[98]

Suppose I see a man, Tom Grabit, with whom I am acquainted and have seen often before, standing a few yards from me in the library. I observe him take a book off the shelf and leave the library. I am justified in accepting that Tom Grabit took a book, and, assuming he did take it, I know that he did. Imagine, however, that Tom Grabit's father has, quite unknown to me, told someone that Tom was not in town today, but his identical twin brother, John, who he himself often confuses with Tom, is in town at the library getting a book. Had I known that Tom's father said this, I would not have been justified in accepting that I saw Tom Grabit take the book, for if Mr. Grabit confuses Tom for John, as he says, then I might surely have done so, too. (Lehrer 1990, p. 139)

Lehrer summarizes the lesson to be drawn from this example as follows: 'I may be said to know that Tom Grabit took the book despite the fact that, had I known what his father said without knowing about his madness, I would not know whether it was Tom who took it.'

In contrast to Lehrer's undefeated justification, ultimate justification no longer implies that S is justified on the basis of his current personal or verific acceptance systems. But what we have been looking for is an *objectified* notion of justificationjustification!objectified, which can be conjunctively added to the (at least partially) subjective notions or personal and verific justification. In particular, we cannot dispense with verific justificationjustification!verific if we want

---

[97] Could there be an eternal wiggling of the acceptance value of ϕ in response to the sceptic's challenges? Not in the present context. Since we want to if we set aside questions of infinity in this chapter and we decided (in footnote 96) to neglect the possibility that a truth-conducive change can make S drop truths, the sceptic has the means to make S accept the true and complete theory about the world which, of course, cannot be further improved.

[98] A similar example about barns and papier-mâché facsimiles originally due to Carl Ginet is discussed in Nozick, (1981, pp. 174–175) and Bach (1984, pp. 40–41).

to end up with the right analysis of Gettier type examples. In the example discussed on page 49, Smith is both personally and ultimately justified in believing
that either Jones owns a Ford or Brown is in Barcelona (ξ). What prevents him
from knowing this is that his belief that Jones owns a Ford is false, so he is not
verifically justified to believe that ξ.[99] Ultimate justification is not sufficient for
knowledge.

Since every theory can be improved by a transition to *the one true and
complete theory about the world*,[100] ultimate justification reduces to justification
on the basis of that theory. On the one hand, it does not seem objectionable to
call for that theory as the final arbiter of knowledge. On the other hand, I cannot
see why the whole truth must be coherent, in Lehrer's or in any other but the
purely logical sense. Shouldn't we try to avoid stipulating that the one true and
complete theory is coherent, because that would mean basing epistemology on a
questionable metaphysics? Similarly, I do not see any intuitive reason why every
truth should be justified, on the basis of the true and complete theory. If this is
right, then it is *impossible*, by Lehrer's own definition of knowledge as well as
by the definition using ultimate justification, that an agent will know the whole
truth. Shouldn't we try to avoid this conclusion? This suggests that undefeated
justification may not be necessary for knowledge, and even ultimate justification
may not be.

I shall stop these cosmic speculations now and return to more definite matters
again.

2.3   Lehrer's logical constraints for eliminations and replacements, and how to improve them

We have seen that *eliminations* and *replacements* of accepted propositions are
of paramount importance for Lehrer's approach to the theory of knowledge.[101]
These operations are very close to the operations of contraction and revision as

---

[99] This argument depends on the assumption that if ξ is not part of an acceptance system,
then it cannot be justified on the basis of that system. Strictly speaking, Lehrer does not make
that assumption; compare Lehrer's definitions D2 – D6 and especially my comment on D3
above.

[100] Saying this actually steps beyond Lehrer's account, who does not provide for the possibility
of *knowledge expansion* through the sceptic—which marks an important difference between
Lehrer's concept of replacements and the usual understanding of the concept of revision. The
uniqueness involved in talking about 'the one true and complete theory about the world' is of
course relative to the language used, which I assume as given.

[101] Eliminations and replacements are respectively called 'weak corrections' and 'strong corrections' in Lehrer (1997, pp. 45–49). In Lehrer (2000, especially pp. 153–154, 168–169), there
are no replacements any more, and there is no talk of strong corrections. In this new account,
the ultrasystem is closer in essence to what was called the verific system earlier, and undefeated
justifiction is similar to verific justification. I cannot discuss here the question as to whether
there is an epistemologically significant difference between the sceptic's removing errors and
his supplying new information.

they are known in the theory of belief revision.[102] As Lehrer places no constraints on the structure of acceptance systems (see Definition D1), there seems to be no need for him to apply non-trivial change operations to acceptance systems.[103] May we not just eliminate a false sentence ψ by simply dropping the statement 'S accepts that ψ' from the system X, and similarly, may we not simply substitute the statement 'S accepts that $\neg\psi$' for the the statement 'S accepts that ψ' in X, when a false sentence ψ is to be replaced by its negation?

The answer is, 'No'. Effortless eliminations and replacements are excluded by the *logical constraint* stated in Lehrer's definition D12: 'if ψ logically entails χ which is false and also accepted, then "S accepts that χ" must also be eliminated or replaced *just as* "S accepts that ψ" was.'[104]

These are the only constraints Lehrer enters into his official definitions, but in the running text he acknowledges more constraints of a similar kind. They are in a sense complementary to the ones we just mentioned. Let ψ again be the false sentence to be eliminated or replaced by its negation $\neg\psi$. While the first group of constraints concerns (false) sentences *implied by* ψ, the second group deals with (necessarily false) sentences *implying* ψ. The first group of constraints is forward-looking, the second group is backward-looking. Here is the quotation from Lehrer (1990, p. 141):

The sceptic ... may require the claimant to eliminate anything the claimant accepts that is false, and the claimant must eliminate the specified item from his acceptance system *and at the same time* eliminate anything he accepts that logically implies the eliminated item. Or the sceptic may require the claimant to replace anything the claimant accepts that is false with the acceptance of its denial *and at the same time* replace anything that logically implies the replaced item with acceptance of its denial. (My italics)

In order to make the discussion of *Lehrer's logical constraints* more easily surveyable, I now give shorter, semi-formalized formulations.

(FE)   *Forward Elimination.* If ψ is to be eliminated and $\psi \vdash \chi$, then item χ, if false, must be eliminated.

(FR)   *Forward Replacement.* If ψ is to be replaced by $\neg\psi$ and $\psi \vdash \chi$, then item χ, if false, must be replaced by $\neg\chi$.

(BE)   *Backward Elimination.* If ψ is to be eliminated and $\chi \vdash \psi$, then item χ (which is false) must be eliminated.

---

[102] I have just mentioned one important difference between 'replacements' and 'revisions' in footnote 100. See page 72 below.

[103] If this were true, then Lehrer would turn out to be a foundationalist! Compare Chapter 3.

[104] My italics. Compare footnote 11 on p. 194 of Lehrer (1990): ' ... with the constraint that if ψ logically entails χ, which is false and also accepted, then "S accepts that χ" must also be eliminated or replaced *in the same way as* "S accepts that ψ" was.' (Again, my italics) I am reading the phrases "in the same way as" (in Lehrer's footnote 11) and "just as" (in D12) as indicating that a replacement of ψ by its negation should enforce a replacement of χ by its negation.

(BE)   *Backward Replacement.* If ψ is to be replaced by $\neg\psi$ and $\chi \vdash \psi$, then item χ (which is false) must be replaced by $\neg\chi$.

I have not mentioned here that only χs that are accepted should possibly be eliminated or replaced. This should be self-evident.

I am not going to discuss the falsity conditions in (FE) and (FR) which presume an impartial, objective, omniscient supervisor for the eliminations and replacements in the 'ultra justification game'. I want to reflect on the logical constraints only in so far as they are accessible to the agent himself. In doing so, I want to avoid the assumption that the agent is or should be omniscient; however, I do embrace the idealizing assumption that he or she is capable of drawing all and only valid logical inferences that can be drawn on the basis of their acceptance systems.

The point I want to make is that Lehrer's logical constraints are too simplistic. Being based on consequence relations between single sentences, they in effect consider the accepted items *in isolation* rather than as items in an acceptance *system*. The constraints do not really address the logical coherence of a belief with all its surrounding beliefs. It is important to take into account the *context of the remaining accepted items* when formulating logical constraints for eliminations and replacements. Without any claim that these are 'the right' constraints, the following ones are certainly more adequate in that they show some sensitivity to the context in which beliefs are situated. We keep on using the variable 'ψ' for the false sentence that is to be eliminated or replaced by its negation, and give both a formulation that is close to Lehrer's own statements and a slightly more formalized version.

(FE$^-$)/(FR$^-$) *Forward Elimination/Forward Replacement.*

If ψ is an *essential* premise for the logical derivation of χ and χ is false, then '*S* accepts that χ' must also be eliminated. In symbols:

If ψ is to be eliminated or replaced by $\neg\psi$, and $F \cup \{\psi\} \vdash \chi$ for some set F of accepted items which are not eliminated, but $F \nvdash \chi$ for any such F, then item χ (if false) must be eliminated.

(FR$^+$)   *Forward Replacement.*

If the addition of $\neg\psi$ creates a *new implication* of ξ (and ξ is true) then '*S* accepts that ξ' must be added to the acceptance system. In symbols:

If ψ is to be replaced by $\neg\psi$, and $F \cup \{\neg\psi\} \vdash \xi$ for some set F of accepted items which are not eliminated, then item ξ (if true) must be added, if it is not already accepted.

(BE$^-$)/(BR$^-$) *Backward Elimination/Backward Replacement.*

If F is a set of accepted premises that logically implies the eliminated or replaced item ψ, then for at least one (false) member χ of F, '*S* accepts that χ' must be eliminated. In symbols:

If ψ is to be eliminated or replaced by $\neg\psi$, and $F \vdash \psi$ for some set F of items, then at least one member of F (which is false) must be eliminated.

These conditions correct a number of counterintuitive features of Lehrer's constraints. In the case of a replacement of ψ by $\neg\psi$, there is no reason to *replace* all the χs that are either critically implying ψ or essentially implied by ψ, by their negations. It is enough that such χs are *eliminated*. Let us temporarily employ the following concepts of critical and essential implication. A sentence χ *critically implies* another sentence ψ in the context of a set F, if F implies ψ, $F - \{\chi\}$ does not imply ψ and χ is one of 'the weakest' or 'most vulnerable' elements of F. A sentence χ is *essentially implied* by another sentence ψ in the context of a set F, if F implies χ, but $F - \{\psi\}$ does not imply χ. Critically implying and essentially implied χs are to be eliminated when ψ is just eliminated without being replaced by its negation. And in this respect, there is no reason at all to think that replacements present problems any different from those presented by contractions.

However, a replacement occasions one kind of adjustment of the acceptance system which cannot be initiated by an elimination. Since there is an addition of $\neg\psi$ to the old acceptance system, it is reasonable to require that any new derivations that are made possible by this new item should in fact be made, and the results be added to the acceptance system. This is condition (FR^+) which has no counterpart in Lehrer's definitions.

The most important respect in which the above constraints improve upon Lehrer's constraints is that they pay attention to the fact that ψ or any of the χs mentioned are part of a system. The Fs mentioned in the constraints represent the contexts in which the respective tests for implications have to be made. The contexts are subsystems of the original acceptance system of the agent. By making the contexts explicit, we raise important new questions: Which items may be included in those Fs that figure in the forward-looking constraints? Which items should be excluded from those Fs that figure in the backward-looking constraint? More generally, how do we know which items in an acceptance system survive the process of elimination or replacement? Lehrer should give answers to these questions if his theory of knowledge is to be considered complete, but he fails to give them. We shall see that it is precisely the theory of belief change that addresses these questions and offers a variety of ways to answer them.

The constraints (BE^-) and (BR^-), which concern sets F that imply the false belief ψ that has to be given up or replaced, leave much room for choices. It requires the agent to give up at least one (false) belief in F, but it does not tell us *which* belief or beliefs ought to be given up. The constraint does not fully determine what to do, but leaves us the freedom to choose. But how are the choices which members to eliminate from the acceptance set to be made? I will give various answers to this question in later chapters. This is not to say that I want to enumerate concretely the cognitive values that govern our decisions what to give up. But I shall inquire into the 'logic' behind the relevant choices that will serve to mark out 'coherent' or 'rational' changes of belief.

A choice-theoretic perspective has already come up in Lehrer's suggestion to explicate the notion of personal justification. We saw that he ends up with a decision-theoretic explication of the comparative notion of reasonableness-of-acceptance. According to him, it is more reasonable to accept ϕ than to accept ψ just in case ϕ has greater expected epistemic utility than ψ. But this is a comparison between two single items of belief only, and it is not clear how it can be extended to a criterion for assessing whole systems of belief. In fact it is not at all evident that the *logical* constraints that Lehrer mentions on his way from personal to undefeated justification mix well with the *decision-theoretic* advice given in the case of personal justifications. It will be one of my tasks in this book to show that a combination of logical and choice-theoretic constraints is indeed possible, and I will show this independently of Lehrer's particular idea that go for maximal or at least non-negative expected epistemic utility. Using choice-theoretic methods, we can maintain the idea that acceptance systems resulting after some change ought be inferentially coherent, that is, consistent and closed with respect to a given logic. Choice and logic can indeed coexist in harmony.[105] It is unclear whether Lehrer's coherence mechanics in terms of competing, beating and neutralizing is compatible with such an account. Since both logic and the theories of choice and decision seem to be better motivated and more principled than Lehrer's own account, I shall not try to fit his theory together with the former ones, but rather study closely the interaction between logic and choice or decision theory. Chapter 4 develops the logical, theoretical theory; Chapter 6 turns to the practical theory of choice; Chapters 7 and 8 apply the latter theory, on a semantical and, alternatively, on a syntactical level, in order to deal with the problems of belief formation and revision. On both levels we shall be able to derive logical postulates from choice-theoretic ones.

2.4   Epistemology and belief change – a symbiotic relationship

So far I have been using concepts and ideas borrowed from belief revision theory to structure and correct an important contemporary account in the theory of knowledge. But the symbiosis between the two research areas can equally well be viewed from the opposite perspective. I now want to give a first indication of how concepts and ideas developed in epistemology can help to structure and interpret much of the work that has been in belief revision theory over the last two decades.[106]

One of the most relevant distinctions for belief revision is that between *foundationalist* and *coherentist* approaches in epistemology. Lehrer (1990, p. 13) characterizes the fundamental difference between foundationalist and coherentist views of knowledge as follows.

---

[105] Another fundamental idea which is not mentioned in Lehrer's account is that doxastic changes should be *conservative*, should incur only *minimal* changes to the previous acceptance system. It will turn out that we are not committed to this idea of diachronic coherence, but we have room for it. The finer, the more discriminating the agent's preferences are, the more conservative the changes will be.

[106] There has been an ongoing controversy over the coherentism-vs.-foundationalism issue in the belief revision literature for more than a decade, see Gärdenfors (1990), Doyle (1991), Nayak (1994a), del Val (1994, 1997), Hansson and Olsson (1999) and Bochman (2000b, 2000c).

According to *foundationalists*, knowledge and justification are based on some sort of foundation, the first premises of justification. These premises provide us with basic beliefs that are justified in themselves, or self-justified beliefs, upon which the justification for all other beliefs rests.

Coherentists argue that justification must be distinguished from argumentation and reasoning. For them, there need not be any basic beliefs because all beliefs may be justified by their relation to others by mutual support. [My italics, HR]

Ernest Sosa (1980, pp. 23–24) makes essentially the same point in more metaphorical terms:

For the *foundationalist*, every piece of knowledge stands at the apex of a pyramid that rests on stable and secure foundations whose stability and security does not derive from the upper stories or sections.

For the *coherentist* a body of knowledge is a free-floating raft every plank of which helps directly or indirectly to keep all the others in place, and no plank of which would retain its status with no help from the others. [My italics]

My thesis now is that the categorical distinction between foundationalism and coherentism is more properly applied to theories of belief than to theories of knowledge, since no reference is made in the drawing of this distinction to the question of whether the beliefs in question are actually true. (As Lehrer realizes very clearly, this objective question has to be linked to the subjective question of coherence by a separate step.) There is also a second, more 'dynamic' sense in which the distinction becomes relevant, and that is when it comes to modelling the dynamics of belief. So there are two questions that separate foundationalists from coherentists.

(1) Can a distinction between basic beliefs and derived beliefs be validly drawn?
(2) And if so, are changes of beliefs made primarily on the base level or on the level of 'coherent theories'?

It is clear what the foundationalist's and the coherentist's answers will look like. The former affirms while the latter denies the first question. In response the second question, the former would say 'on the base level', while the latter, lacking a distinguished base level, must opt for the level of coherent theories. It is necessary to work out in greater detail the concepts and distinctions on which these answers are based. In the next chapter I am going to provide a framework that will help us to understand the issues involved and to characterize two fundamentally different perspectives on the process of belief revision. These perspectives will turn out to be related to, but not to be identical with the dichotomy between foundationalism and coherentism. I will not go as far, however, as Hansson and Olsson (1999) who argue that the coherence theory in the epistemologist's sense is trivialized in the context of the coherence perspective on belief revision.

When transferring the idea of 'foundations' of knowledge to the area of belief change, it is not particularly important whether the basic beliefs are true, let alone infallibly true. Neither is it important that they are justified to such a high degree that they may be seen as certain. In the following, belief bases are not supposed to carry that connotation. Basic beliefs are distinguished from derived beliefs just by the fact that they are somehow 'given', either explicitly or as things that are taken for granted. Givenness is not at all supposed to imply indefeasibility here.

Now I will try to show how categories coming from epistemology can be exploited for the analysis of the dynamics of belief.

3

CHANGING DOXASTIC STATES: TWO COMPLEMENTARY PERSPECTIVES

In the last chapter we have seen how the analysis of knowledge may lead to the problem of belief revision. What one knows at a specific time instant does not only relate to what one believes at that instant, but also to a great variety of potential changes of the agent's current beliefs. The relevance of the revision of beliefs to problems of epistemology provides an important reason for studying it.

In Chapter 1 we attended to what I called internal problems of belief. This has meant the task of drawing the right inferences on the basis of given information and expectations. If the task is successfully carried out, then the agent reaches a state of cognitive or reflective equilibrium. The present chapter is about problems caused by external interference. The influences from outside the doxastic system that we are going to consider come in two basic types. Agents may either learn new things or become doubtful about things that they used to think are true. Interaction with the world or with other agents may induce them to accept new items of belief, or to cease to accept old items of belief. In accordance with established terminology, I will say that doxastic states, as far as they are represented by belief bases or belief sets, are to be *revised* and/or *contracted* in response to new information. External influences are represented by discrete pieces of input to the doxastic system, which will in turn be represented by single input sentences of a given object language.[107] Whether truth-conducive or not, such input first of all means a perturbation to a system that has previously been coherent. The addition of new beliefs frequently leads to a violation of the ideal of consistency, the subtraction of new beliefs violates the ideal of inferential closure.

The problem to be solved by doxastic agents is to restore the reflective equilibrium, to restore coherence, in response to some such external perturbation. I shall argue in this chapter that there are three essentially different ways of doing this, depending on the philosophical perspective one takes. Two of these approaches on belief revision are then followed up and developed in detail in Chapter 5 and Chapter 7 of the present book. Chapter 6 will set the stage for,

---

[107]The input may also come in sets, in which case we speak of *multiple* revisions and contractions. If we are required to accept or reject at least one element of a set G of sentences, we have to perform a *pick* revision or contraction. If we are required to accept or reject simultaneously all elements of a set G of sentences, we have to perform a *bunch* revision or contraction. See Fuhrmann and Hansson (1994).

and Chapter 8 will deal with a preference-based reformulation of, the account presented in Chapter 7. As a preparation for all this, Chapter 4 discusses rationality postulates for belief change and inference operations, thereby providing a gauge by which various constructions for belief change can be measured. While Chapter 4 represents an abstract, functional input-output approach to changes of doxastic states, the later chapters provide more constructive modellings based on certain structural features of the doxastic states in question.[108]

This chapter offers a general philosophical framework which will be filled in concretely by the formal constructions performed later on.

3.1   Two dimensions of belief change

It is a major tenet of this book that we can, and should, carefully distinguish two dimensions in the change of beliefs. I shall indicate these dimensions by vertical and horizontal arrows in the diagrams that will follow. The vertical lines will represent the *static dimension*.

This dimension is concerned with belief states (acceptance systems, bodies of opinions) and with operations of reasoning, thinking, or drawing inferences from a given information base, without any external interaction of the doxastic system. Although in reality these inference operations are processes that take time we will abstract from the temporal aspect along this dimension. An inference operation will be conceived of as an abstract function that takes some set of premises (the information base) and yields a set of beliefs which are taken to be in some kind of 'reflective equilibrium'. It is essential that in the process of drawing inferences all premises are given right from the start, that there is no disturbance of the reasoning by any influence from external sources.

The other dimension, orthogonal to the static dimension as it were, is the *dynamic dimension*. It is represented by horizontal lines in our diagrams.

The dynamic dimension is concerned with the adaptation to new pieces of information: with learning, changing one's beliefs, making adjustments in response to

---

[108]Compare Section 4.1 for further remarks on this topic.

some perturbation of the doxastic system.[109] Here we have some measure of time, if only a very crude one. Time is discrete on this account, with the transition from point n to point $n + 1$ being effected by some operation of belief change. In contrast to the vertical dimension which by its nature consists of at most two storeys,[110] the horizontal dimension may have any finite number of steps. Since human agents differ from Lehrer's 'sceptic', there is no pre-determined endpoint to learning new things or calling old things into question. For the sake of iterations, we want the Principle of Categorial Matching to hold in the horizontal dimension, but there is no reason whatsoever why we should like to impose it in the vertical dimension.

In the vertical perspective we are dealing with the internal problem of making one's belief coherent. In the horizontal perspective we are dealing with the reaction to some disturbing input to the doxastic system. This may be achieved directly, by drawing inferences from the altered, possibly inconsistent belief base, or indirectly, by rearranging the coherent initial state in such a way that the input is accommodated, without recourse to the original (or any other) belief base. With a grain of salt, one can say that the vertical line is about making up one's mind, while the horizontal line is about changing one's mind. One has to bear in mind two provisos, though. First, these terms are used for purely cognitive decisions that are not not necessarily accompanied by decisions pertaining to any visible action. Second, the changes of mind referred to are not made spontaneously by the agent himself, but induced by external sources of information or criticism. This is another difference between what is represented by the horizontal and vertical arrows.

3.2 Three types of maxims of coherence for belief representation and revision

We shall distinguish three senses of coherence, with different types of entities to which the labels 'coherent' or 'incoherent' can be attached. The first reading applies the term coherence to single belief states, the second reading to sequences of belief states (with a length of at least two), the third one to the agent's dispositions to choose across varying choice situations. None of these concepts is identical with the concept of coherence that 'coherentists' in epistemology like Lehrer have in mind.

These concepts of coherence give rise to certain prescriptions to doxastic subjects. They can all be subsumed under the principal maxim 'Be coherent'. More concrete maxims that fall under this general rule may be obtained if one

---

[109]These processes of adaptation may actually be performed, or else they may be made hypothetically, just for the sake of argument. Learning is understood as getting more information about the same state of affairs. Learning as a result of a changing world is quite different, even as regards the logical principles underlying it. See Katsuno and Mendelzon (1992).

[110]Assuming, as we do, that Inf is idempotent, i.e., that $Inf(Inf(\mathcal{H})) = Inf(\mathcal{H})$ for every information base $\mathcal{H}$.

fleshes out the concept of coherence in one way or other.[111] These submaxims are fundamental guiding principles for belief change; they are assumed to be operative, somehow, in the minds of the rational agents. One may also call the maxims 'integrity constraints' (Gärdenfors and Rott 1995) since they all aim, in some sense or other, at the integrity of belief states and changes.

3.2.1   Inferential coherence, static constraints

An agent's beliefs should be well-balanced, they should be in a reflective equilibrium. Here 'reflection' is understood as reasoning according to some set of logical rules. Two kinds of dissatisfying states of mind are to be avoided by the reasoner: states that are not consistent and states that are not closed. This is the import of the two important maxims of inferential coherence that I consider in this section.

The first maxim says that the full set of beliefs should be *consistent* with respect to the given background inference relation.

(i) The beliefs of an agent should be consistent.[112]

This is a minimal synchronic concept of coherence. Maxim (i) is essential for any account of the concept of an equilibrium state. In my view, inconsistency is the best motive one can have for further deliberation. Maxim (i) can be seen as distinguishing the enterprise of belief revision from that of paraconsistent logic which tolerates sets of beliefs that are *classically* inconsistent. Paraconsistent logic shares with belief revision the aim of explicating rational deliberation in the face of contradictory information. While belief revision is inconsistency-phobic, paraconsistent logic is not. However, there is no sharp separating line between the two. Advocates of paraconsistent logic argue against an outright elimination of inconsistencies from the information base, and claim that one can well live with them. This alternative view can indeed be reconciled with the belief revision approach advocated in this book if one properly distinguishes 'prima facie beliefs' that are somehow *represented* or *entertained* in the doxastic state of the agent and the 'real' beliefs actually *held* or *accepted* by the agent in that doxastic state.

One motive for embracing maxim (i) is that the beliefs of an agent with an inconsistent belief set cannot all constitute knowledge. This is because it is impossible that the elements of an inconsistent information base are all true. We are all striving to have beliefs that are true, we are all aiming at knowledge and avoidance of error, so it cannot be proper to believe both ϕ and $\neg\phi$.

More demanding than the first maxim is the second requirement that the set of beliefs be *closed* under the given inference relation. That is, further reflection

---

[111] This idea of splitting coherence into several aspects or respects is comparable to Grice's (1975) famous conversational maxims of Quantity, Quality, Relation and Manner which are, in a way, all special cases of the more general 'Cooperative Principle'.

[112] We need not decide at this juncture which of the various concepts of consistency is the right one for our purposes. Given an inference operation Inf, H can be called *inconsistent* if $\bot \in Inf(H)$, or if $\phi \wedge \neg\phi \in Inf(H)$, or if $\{\phi, \neg\phi\} \subseteq Inf(H)$ or if $Inf(H) = L$, etc.

should not provide the reasoner with new beliefs that are not already included in his belief set. No drawing of inferences from what her or she believes would lead him or her to believing (or disbelieving) anything new. The maxim concerns *inferential closure*:

(ii) If a sentence ϕ may be inferred from the beliefs of an agent, then ϕ should be believed by the agent.

Condition (ii) captures the idea that an agent with an 'open' set of beliefs (one that is not logically closed) actually knows more than is explicitly represented in his information base. If a sentence follows logically from the the sentences included in the information base and the base itself is fully accepted, then that sentence should be accepted as well. An agent who believes that ϕ and believes that $\phi \supset \psi$ 'implicitly' or 'really' believes that ψ. True, it may well be that the agent is not conscious or aware that he believes ψ. Upon persistent reflection, however, the agent should—insofar as he is rational—acknowledge that he does believe that ψ, and include ψ in his body of belief (which we then call 'belief set', see Section 1.2.2).

It must be seen, though, that from a psychological or computational point of view, inferential closure involves an intolerable amount of idealization. Perhaps real-life agents are able to remain consistent, but belief sets satisfying (ii) clearly do not give a realistic description of mental or computational states. A more adequate interpretation of the maxims is that they concern what we *ascribe* to rational agents rather than what they would assent to if asked.[113]

Maxims (i) and (ii) are pertinent to the *static* picture of a doxastic agent, that is, to the problems of reasoning, or thinking or drawing inferences, when there is no new item of information to be accommodated and no old item of information to be questioned. (We have said a few things about this topic before, in Chapter 1 and Section 3.1 above.) The first two maxims may jointly be taken to codify a minimal sense of *coherence* in belief states. The notions of coherence and equilibrium that are appropriate here are dependent on the underlying logic (inference operation or consequence operation) that governs the language in question. I feel even tempted to suggest that it is the *task of logic*, to transfer any body of premises into a coherent and well-balanced theory. What we have to keep in mind, though, is that this concept of the synchronic coherence of beliefs is much simpler and less interesting than the concept of coherence that coherentist epistemologists have had in mind. More elaborate versions of coherence between beliefs in a framework congenial to the one presented here are discussed by Olsson (1998b).

Let us try to get a clearer picture of the problems involved here by formalizing the static constraints (i) and (ii). Assume that a doxastic state is represented by a set K of sentences without prioritization. Our maxims say that K should be coherent, i.e., consistent and closed with respect to an underlying inference

---

[113]For a little more discussion of this topic, compare page 12 and footnote 31 of Chapter 1.

operation Inf. Formally, the first maxim which says that drawing inferences from a coherent belief state should not result in doxastic ruin may be formulated as

$$Inf(K) \neq L$$

The second maxim says that drawing inferences from a coherent belief state, a belief state in reflective equilibrium, should not change the belief state at all. It is formally expressible as

$$Inf(K) = K$$

We can put down a combined *maxim of inferential coherence*

$$K = Inf(K) \neq L$$

Notice that the set K appearing here in the argument position of Inf is infinite.[114] The maxim says that a belief set K should be a non-trivial fixed-point in the process of drawing inferences. Consistently closing one's beliefs under Inf is a form of making them coherent. If the belief state doesn't change when we do this, it has already been coherent.

Let us now change the perspective and look at conditions for the inference operation Inf rather than for belief sets K. We can add to the previous considerations the above idea that applying an inference operation means producing coherence. This means that for any (possibly prioritized) belief base $\mathcal{H}$, $Inf(\mathcal{H})$ must be coherent, which in turn means, by the above clauses, that both $Inf(Inf(\mathcal{H})) \neq L$ and $Inf(Inf(\mathcal{H})) = Inf(\mathcal{H})$. From that it follows that $Inf(\mathcal{H}) \neq L$. The iterated application of the inference operation Inf presumes that it can be applied both to belief bases $\mathcal{H}$ and to theories K—an assumption that is not unproblematic. I have presented an important example of such an inference operation in Section 1.2.7.2, putting $Inf(\mathcal{H}) = Consol(\mathcal{E} \circ \mathcal{H})$.[115]

Let us say that a belief set K is *base-generated* if $K = Inf(\mathcal{H})$ for some belief base $\mathcal{H}$. If a belief set is base-generated and the second maxim is to hold for K, then we get $Inf(\mathcal{H}) = Inf(Inf(\mathcal{H}))$ (for all $\mathcal{H}$). In fact the interpretation of inferences as coherence-generating operations implies that the latter equation holds for arbitrary $\mathcal{H}$. If it is to follow from the concept of *inference* that the set of all sentences that can be inferred from some basic information (by whatever structure this information might be represented) is coherent, then we must have

$$Inf(\mathcal{H}) = Inf(Inf(\mathcal{H})) \neq L$$

for arbitrary belief base structures $\mathcal{H}$.

---

[114]It is an interesting question to ask whether the maxim of inferential coherence is a constraint for K or a constraint for Inf. Perhaps it is best to say that is a simultaneous constraint for both.

[115]It is easy to verify that this inference operation is coherence-producing: $Inf(Inf(\mathcal{H})) = Consol(\mathcal{E} \circ (Consol(\mathcal{E} \circ \mathcal{H}))) = Consol(\mathcal{E} \circ \mathcal{H}) = Inf(\mathcal{H}) \neq L$.

3.2.2   An example: The Polish swan

The problems for belief *change* created by the 'static' maxims (i) and (ii) are illustrated by the following example.example!Polish swan

'Henry' is the name I have given to a bird that has been caught in a trap near Wrocław, the capital of Silesia in Poland.[116] Henry looks exactly like a swan. My friend, the hobby ornithologist, tells me that all European swans are white, which is interesting for us only insofar as it concerns Henry: If Henry is a European swan, then he is white. Moreover, as everybody knows, Poland is a country in east central Europe. We use the obvious abbreviations

s : Henry is a swan

p : Henry is Polish

e : Henry is European

w : Henry is white

Now let our little theory about Henry be just what can be logically inferred from these pieces of information. Clearly, w is a logical consequence of the sentences in our information base, i.e., it is contained in the belief set generated by it. However, having reasoned carefully, in perfect accordance with the sound rules of our preferred logic, we finally turn empirical and venture a closer look at the bird in the trap. To our great surprise, it turns out that Henry is black. We have got to revise our beliefs!

Let us, for the sake of simplicity, suppose that we have just found out that w is in fact false (i.e., that Henry is not black). The minimal change, set-theoretically speaking, required to accommodate this observation is clearly just to put $\neg w$ into the information base. If we want to keep the information base consistent— maxim (i)—, however, you need to *revise* it. This implies that some of the beliefs in the original information base must be *retracted*. The problem of belief revision thus involves a problem of belief retraction. Usually, we do not want to give up all of our beliefs since this would mean an unnecessary loss of information, and information is, after all, valuable. So we have to choose between retracting s, retracting p, retracting $p \supset e$, and retracting $s \wedge e \supset w$. When giving up one belief we need to give up more beliefs, and to make a decision which of the reasons for this belief to retain and which to retract.

In trying to accommodate $\neg w$, there is *no purely logical reason* for making one choice rather than the other among the sentences to be retracted. There are several ways of specifying the revision of our belief set by $\neg w$. We have to draw on additional information about the respective merits and demerits of our beliefs. What is needed here is a well-defined method of constructing the revision that respects maxims (i) and (ii) (and the other maxims to come) as far as possible.

The contraction process faces parallel problems. To illustrate this, we just continue with a variant of the same example. Suppose we saw Henry only from a distance and at nightfall, and we are not sure that Henry is in fact black, but we

---

[116]Basically the same example is discussed in Gärdenfors and Rott (1995, pp. 36–37). I shall later give it a different twist.

seriously doubt now that he is white. That is, we want to contract our belief set with respect to w. Of course, w itself must be deleted from the belief set. But at least one of the sentences s, p, $p \supset e$, and $s \wedge p \supset w$ must be given up as well in order to be able to maintain deductive closure—maxim (ii)—while successfully withdrawing w. Again there is *no purely logical reason* for making one choice rather than the other. We see that coherently deleting w from our information base poses quite the same problem as coherently adding $\neg w$.

Abstracting from this particular example, we can distinguish two basic ways of modifying belief sets.[117]

In a *belief revision*, a new sentence that is typically inconsistent with a belief set K is added, but in order that the resulting belief system be consistent—i.e., satisfy integrity constraint (i)—some of the old sentences in K are deleted. The result of revising K by a sentence ϕ will be denoted $K * \phi$.

In a *belief contraction*, some sentence in K is retracted without adding any new beliefs. In order that the resulting system be inferentially closed—i.e., satisfies maxim (ii)—some other sentences from K must be given up. The result of contracting K with respect to the sentence ϕ will be denoted $K \dot{-} \phi$.[118]

The consistency constraint (i) creates problems in revisions, the closure constraint (ii) does so in contractions. But we can draw a more general and more important moral. *Any* kind of static coherence requirement for belief sets makes belief change operations difficult. Unconstrained belief changes (or belief changes according to the maxim of minimum mutilation which we are going to consider presently) could be obeyed by plain set-theoretic add-and-delete operations. This is no longer possible if a belief change is required to result in a coherent belief state.

3.2.3 *Diachronic coherence, economic constraints*

Many epistemologists and philosophers of science have held that we should cherish our beliefs and should not give them up beyond necessity. This is the doctrine of conservatism in epistemology, with Willard V. O. Quine's *Maxim of Minimum Mutilation* as its most prominent version.[119] Gärdenfors has used the maxim of minimum mutilation as a powerful principle that motivates many formal postulates for belief change. He calls it the criterion of *Informational Economy*:

---

[117]These ways are studied in the seminal work of Alchourrón, Gärdenfors and Makinson. As mentioned in Chapter 2, their notions of 'revision' and 'contraction' correspond to Lehrer's notions of 'replacement' and 'elimination'.

[118]One may have to eliminate basic beliefs even if ϕ itself is not in the belief base H. On this, and two essentially ways of understanding 'discarding information', see Section 3.4.1.

[119]The maxim pervades most of Quine's work. See for instance Quine and Ullian (1978, pp. 67–68): 'Conservatism is ... sound strategy ... , since at each step it sacrifices as little as possible of the evidential support ... that our overall system of belief has hitherto been enjoying. ... Conservatism holds out the advantages of limited liability and a maximum of live options for each next move.' It should be pointed out that in Quine's philosophy *simplicity* is a forceful competitor against conservatism. It stands in a similar tension with minimal change principles as inferential coherence does in the present account (see below).

The key idea is that, when we change our beliefs, we want to retain as much as possible of our old beliefs—information is in general not gratuitous, and unnecessary losses of information are therefore to be avoided (Gärdenfors 1988, p. 49)

It is possible to think of the idea of minimal change as a diachronic concept of coherence that comes out most clearly if the evolution of doxastic states is watched over a longer stretch of time. If adjacent doxastic states were not, as a rule, very similar to one another, we could not discern any pattern of coherence of a reasoner's belief in sequences of repeated changes of belief. Like the individual photographs of a motion picture, single states of belief would lose to make sense if they were not embedded in a neighbourhood of closely resembling states. Or, to change the metaphor, a person's mind could not be understood as *one and the same mind* in motion, were it not for a principle of inertia that accounted for the resistance of beliefs against changes. It is an aim of the present book to elucidate further the laws of motion that govern ideal or rational belief change.

By what method can we measure how big a change from one belief state to another one is? This is dependent on *how much* is changed and *how important* the things that are changed are. The following two maxims apply to the problem of belief contraction.[120]

(iii) The amount of information lost in a belief change should be kept minimal.

We can take this to refer to a set-theoretic comparison of the original and the contracted belief set.[121] An alternative, more sophisticated comparison takes into account not the quantity, but also the quality of the lost beliefs:

(iv) In so far as some beliefs are considered more important or entrenched than others, one should retract the least important or entrenched ones.

While maxims (i) and (ii) are logical constraints pertinent to the static picture, the maxims (iii) and (iv) are economic constraints specific to the dynamics of belief. They are requiring the changes of belief to be *minimal* in a certain sense. Maxim (iii) suggests looking at minimal losses with respect to set-theoretic inclusion, i.e., at the maximal sets of beliefs that do not imply the sentence to be withdrawn, so it leads most naturally to the classic theory of partial meet contractions of (Alchourrón, Gärdenfors and Makinson 1985) which I will briefly touch on in Section 7.2.

Maxim (iv) requires one to respect minimality with respect to an ordering of priority or entrenchment between beliefs (as discussed in Section 1.2.6); it may serve as an argument to recommend, for instance, the prioritized base contractions as already sketched in Section 1.2.7.2, or the contractions guided by

---

[120]Corresponding maxims for belief revision are not hard to come by, but they vary from context to context. In most of Chapter 5, the operation of belief base revision is simple insertion and thus is maximally conservative; in the rest of this book we use the so-called Levi identity (see Section 4.3).

[121]For finite sets of beliefs we might interpret this maxim as referring to the numbers of accepted sentences. Reference to numbers, however, does not make sense if the belief state is represented by a theory, since all theories in countably infinite languages are infinite.

relations of 'epistemic entrenchment' which will be studied in detail in Chapter 8. While (i) and (ii) are purely logical maxims and (iii) is a purely set-theoretic maxim, (iv) has recourse to some extra-logical and extra-mathematical structuring of belief sets. Priorities and entrenchments are not a priori fixed, but are structures that are either empirically given or ascribed to an agent by rationalizing observers. It is extremely important to realize that processes of belief change are not a purely logical matter, but draw heavily on extra-logical factors, if they are to yield intuitively plausible results. This fact is reflected in the formal models that have gained currency in the past two decades, and it is true of the models I am going to present in the course of this book. The most important kinds of structures that represent the resources guiding our changes of belief are syntactic structures of belief bases, as well as various sorts of preference relations and choice functions.[122] Although the two dynamic maxims look rather different in nature, it is possible to exhibit close connections between some well-known ways of implementing them.[123]

There is a fundamental tension between the static and the dynamic Constraints for belief change. Dynamic constraints tell you that you should change your belief as little as possible, but static constraints urge you to change your beliefs at least so much as to ensure the coherence of the posterior belief state. And of course, the closest *coherent* belief state assimilating some input information will in general deviate more from the original belief state than the closest belief state *simpliciter* (whether coherent or not). The minimal change of a (logically) coherent or well-balanced belief state will in general *not* result in a state that is (logically) coherent or well-balanced. Foundationalists and coherentists tend to respond to this tension in different ways. Foundationalists give priority to the idea of minimal change, allow that belief bases are changed in a minimal way, at the possible expense of importing incoherencies, and rely on the idea that a reflective equilibrium may be reached by some process that comes only after the change. For the coherentist, there is no way of compromising the idea of coherence. The change operation itself—and not a subsequent process of straightening out one's beliefs—has to make sure that the posterior belief state is coherent.

Conservatism in epistemology, in particular the sort of ideology standing behind maxim (iii), has been harshly criticized by Christensen (1994). Pagnucco and Rott (1999) cast doubts on conservatism in formal models of belief revision, appealing to 'Principles of Preference and Indifference'. These principles clash with the idea of minimum mutilation and may be considered to be more important than the latter at least in some contexts. The maxims of minimum

---

[122] Notice also that in order to apply (iv) to the choice of the most appropriate posterior belief states, one needs a method of combining the importance or entrenchment assessments of single sentences into some quality assessment of sets of beliefs. Some such methods are suggested in Rott (1991a, 1992a, 1992c, 2000a).

[123] For the relationship between partial meet contractions and prioritizations, see Nebel (1992) and Rott (1993), for that between partial meet contractions and entrenchments, see Rott (1991a).

mutilation will not play a pivotal role in this book.[124]

3.2.4 *Coherence of choice, dynamic constraints*

Let us now introduce a fifth maxim for the dynamics of belief. In contrast to the first four ones, this maxim has not yet been put forward as a general principle that should be operative in the fields we are interested in.

(v) In so far as choices are to be made when performing a belief change, these choices should be coherent.

I propose to take seriously the idea that the choices involved in the potential revisions of a certain state of mind should be coherent (or, as some writers say, 'consistent') with each other. Since I will explore the notion and the logic of coherent choice in detail in Chapter 6, a few lines may suffice here. Coherent choice is possible if—and perhaps only if—there is something stable in the reasoner's mind that underlies his dispositions to choose. In many cases, the reasoner can be ascribed some context-independent preference orderings that determine his choices, decisions and actions. The existence of underlying preferences is exactly what guarantees that his choice behaviour is coherent.[125] For example, if you choose bananas when you could as well have apples, and if you choose bananas when you could as well have pears, then you should also choose bananas when you could as well have either apples or pears (unless you change your mind). This is so because you are perceived as preferring bananas both to apples and to pears, and, in so far as you are a rational agent, your choices should not depend on the contingencies of the sample of options (the 'menu') that happens to be offered to you. I shall explore the 'logic' of coherent choices in detail in Chapter 6. Since a choice between different ways to go stands at the beginning of every action, I take the logic of coherent choice to be an important part of the general theory of practical reason.

3.3   The vertical and the horizontal perspective: Direct and coherence-constrained belief change

We distinguish two basic modes of belief revision. The *direct* or *immediate mode* operates on the level of belief bases. No constraints of coherence are placed on the structure of the prior and the posterior bases H and H'. Therefore change operations may be simple and in full accordance with the idea of minimal change. In contrast, in the *coherence-constrained mode*[126], both K and K' are supposed to be inferentially coherent systems of beliefs. That is, they are supposed to be

---

[124]I argue in Rott (2000b) that the maxims have not even played a pivotal role in the work of Alchourrón, Gärdenfors and Makinson. Also compare Melmoth (1994).

[125]The preferences mentioned here will in the end (in Chapter 8 turn out to be identical with the ones referred to in maxim (iv). The way they are employed, however, does not guarantee diachronic coherence or minimum mutilation. Compare again Pagnucco and Rott (1999).

[126]This term is supposed to be a generalization of the 'logic-constrained' mode of Gärdenfors and Rott (1995).

consistent and closed under an inference operation *Inf*. Constraints of minimal change are relevant in this mode, but they are subordinated to constraints of inferential coherence and, to some extent, of the coherence of choices. Whenever the constraints come into conflict, it is the latter that take priority over minimal change.

We can now characterize the two fundamentally different perspectives on belief change that will be studied in this book. They are distinguished by how much care they devote to the static and dynamic dimension of belief formation. The *vertical perspective* is marked by sophisticated statics and plain dynamics. The *horizontal perspective*, on the other hand, is distinguished by plain statics and sophisticated dynamics. This is but a pictorial way of representing the dichotomy of direct and coherence-constrained modes of belief change.

3.4   Foundationalist and coherentist belief change

I have discussed the distinction between foundationalism and coherentism in epistemology in Chapter 2. In the context of theories of belief revision, the distinction takes a somewhat different shape. For the coherentist, the problem of belief revision is non-trivial because not every belief state qualifies as a 'good' one. We want and need a coherent set of beliefs to act upon. The belief state after a change should be as well-balanced as the initial belief state. This is a special application of the more general Principle of Categorial Matching stated in Section 1.2.8. There are no such aspirations in the foundationalist direct mode of belief change.

In order to illustrate the difference between the coherentist and the foundationalist modellings, we take up our example of Henry, the Polish swan (see Section 3.2.2). In a real-life database about Henry, the only information that we would enter explicitly is probably that Henry is a swan and that he is Polish (since he was caught in Poland). We would not write down in his individual file that all European swans are white, nor would we take down what everybody knows anyway, namely that Poland is a part of Europe. These data, and many more things, should rather be supposed to be part of a general and comprehensive background theory that describes our expectations about the 'normal' state and development of the world.

Clearly, it would be very hard, if not impossible, to spell out in detail what the contents of this background theory should look like. But we would certainly want to use our implicit expectations when drawing conclusions about Henry (such as conclusions about his colour). Continuing with the obvious meaning of the propositional letters, s and p would be explicit items in the database, while both $p \supset e$ and $s \wedge e \supset w$ are pieces of information remaining in the background. Notice that background information is not generally less reliable than explicit information in a database. In the present example it is plausible to assume that $p \supset e$ is in fact the most certain piece of information, followed by s, p and $s \wedge e \supset w$ with decreasing certainty. In our story, it is pretty obvious that Henry is a swan, but we cannot exclude the possibility that we have missed the

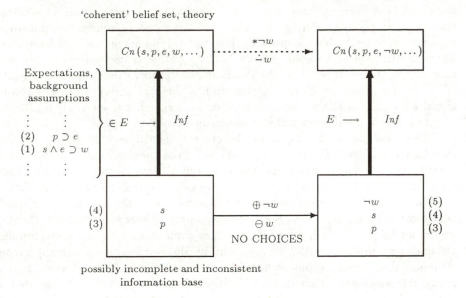

FIG. 3.1. Example: Foundationalist modelling (vertical perspective)

discovery of a new subspecies. It is not so clear that Henry is in fact Polish, for he might have come over from Czechia or from Germany. And finally, our friend, the hobby ornithologist, may not be fully competent. But we had no reason to distrust him before discovered that Henry is in fact black. In any case it seems safe to take our initial theory about Henry, phrased in the minimal language that we have introduced, to be $K = Inf(\langle p; s \rangle) = Cn(s, p, e, w)$.

The revision of our beliefs in response to our surprising observation that Henry is actually black can be effected in at least two different ways. First, we can simply add this new observation, $\neg w$ say, to the entries in the information base. Since we trust our senses, this new entry is going to receive the highest priority; see Fig. 3.1.[127]

But of course the total information provided by the base and the background is now inconsistent. So we have to apply some process of reasoning that is capable of either dealing with or removing inconsistencies. If we stick to the above-mentioned indication concerning the relative vulnerability of the beliefs, it seems advisable to discard our belief in our friend's statement about European swans. I have suggested a general and principled solution of how to remove inconsistencies in Section 1.2.7.2. At the present stage it is important only to notice two general things. First, the change we made on the base level

---

[127]The numbers in parentheses in the figure indicate the relative priorities of the represented items. The expectation base $E = |\mathcal{E}|$ should be understood as being very large.

is extremely simple; we have just added the new observation, with a specification of its (topmost) priority. Second, a sophisticated inference mechanism is required in order to handle possibly incomplete and inconsistent information bases. Putting these two things together, we obtain a derived change operation on the theory level. The old theory K concerning Henry is changed into $K' = Inf(\langle p; s; \neg w \rangle) = Cn(s, p, p \supset e, \neg w) = Cn(s, p, e, \neg w)$. Since the conditional $s \wedge e \supset w$ is the weakest element of the expectation set E, it must give way to the new item $\neg w$ in order to have consistency restored.[128]

The second way of effecting a revision of our beliefs in response to our noticing that Henry is black is fundamentally different from the first. One can deny that what happens in a belief change is that we simply add $\neg w$ to our information base about Henry. In the alternative picture, the new observation instead confronts the coherent theory K holistically, and it is parts of that theory *taken as a whole* that have to be given up (see Fig. 3.2). According to this view of belief change, we do not keep track of how the theory once came into being. At a given point of time, we simply have the theory, usually plus some kind of accompanying structure that we can exploit when forced to perform a theory change. It is a reasonable assumption that this structure will in some way reflect the origins or the historical development of the theory in question, and that the initial degrees of certainty or plausibility of the items in the expectation and information bases are somehow transferred to the theory level.[129] Clearly it is not sufficient to just remove w from our original theory, since K includes, for instance, both $s \vee w$ and $\neg s \vee w$ that jointly entail w. (At least) one of these disjunctions must be given up together with w in order to have a consistent new *theory K'* about Henry that assimilates the surprising observation. Which of $s \vee w$ and $\neg s \vee w$ should be given up is not a matter of logic, but a matter of choice. And there are of course many more sets of beliefs entailing w that need to be considered. (Remember the critical discussion of Lehrer's conditions in Section 2.3.) The choices should be made in a principled and well-considered way. A major part of this book (Chapters 6 to 8) is devoted to working out some general ideas of how the requisite choices are to be organized so that they can possibly qualify as rational.

[128] In Chapter 5 I say more about the fact that my preferred model of foundational belief change ranks explicit beliefs invariably higher than implicit beliefs. In the present example, the case seems to be harmless, since it is hard to imagine explicit information that runs counter the indubitable background information that $p \supset e$. But this is not always so. For instance, had we transferred the item $s \wedge e \supset w$ from the expectation base to the information base (our friend now makes an official statement as a world expert on waterfowl), the proposed model of prioritized base change would have us end up with discarding $p \supset e$ in response to the contradiction caused by $\neg w$. So we must indeed be anxious to enter the most certain pieces of information to the stock of explicit beliefs that make up the information base—clearly an annoying feature of the idea of identifying $Inf(\mathcal{H})$ with $Consol(\mathcal{E} \circ \mathcal{H})$. Explicitness is not the same as priority or certainty.

[129] Transferring rankings of the expectation and information bases to the coherent level of theories is not a trivial matter. For special ways of encoding the structure of belief bases in terms of 'epistemic entrenchment', see Rott (1992c, 2000a).

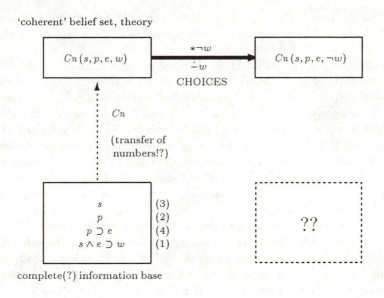

FIG. 3.2. Example: Coherentist modelling (horizontal perspective)

Again it is important to notice just two things here. First, the changes performed on the theory level require quite sophisticated processes of choice in order
to maintain inferential coherence (i.e., comply with the maxims of consistency
and closure). Second, no sophisticated inference mechanism is needed to account
for the generation of coherent theories from incoherent belief bases. In fact, once
the attention is shifted from the base level to the theory level, this problem vanishes as it were. Perhaps the agent can remember how his initial theory K had
been constructed as an inferential product of his initial information base. After
the change, however, there need not be a characterization of an information base
any more. On the coherentist account, it may well be that K is changed into
a successor theory K' in such a way that no plausible axiomatization in terms
of explicit beliefs can be given. The simplest example to illustrate this point
is this. Consider the information base $\{p, q\}$ and retract $p \wedge q$ from the theory
$K = Cn(\{p, q\})$ (or revise K with $\neg p \vee \neg q$). Without additional information, we
would intuitively expect $K \dot{-} (p \wedge q)$ to be $Cn(\{p \vee q\}$ (and $K * (\neg p \vee \neg q)$ to be
$Cn(\{p \leftrightarrow \neg q\})$. Even after conceding that one of p and q may be (or: is) false, we
would probably still cling to the belief that the other one is true. But there is no
base which can be constructed naturally from our initial belief base that reflects
this. In particular, the obvious solution $\{p \vee q\}$ certainly does not record any
piece of *explicit* belief, since in this example only p and q have an 'independent
standing'. We are faced with a deep-seated dilemma now. *Either* we must give
up the idea that the elements of a belief base record explicit beliefs, *or* else we

violate the methodological Principle of Categorial Matching (see Section 1.2.8) as applied to belief base revision, which requires that the result of a belief base change should be a belief base again.[130]

Foundationalists argue that belief changes take effect, and should take effect, only on the level of basic beliefs. *Coherentists* argue that changes take effect on the level of beliefs-in-a-system. According to them, beliefs cannot be sorted out into basic and derived beliefs, but rather have to fit together in a way that they form a coherent, well-balanced, equilibrated whole. Since they do not acknowledge a process of deriving beliefs from a set of independently warranted premises, coherentists have no interest in the static dimension. What they do, and must do, is develop systematic models of change that explain how a coherent system of beliefs can be transformed into another coherent system in response to some externally caused perturbation.

I should say clearly how the distinction between foundationalism and coherentism is not drawn in this book. For instance, I do not agree with the philosophical outlook of the otherwise excellent paper of del Val (1994). When introducing 'the foundations theory', he writes: 'the syntax based approach [del Val's main representative of the foundational theory, H.R.] ... has a serious drawback, namely, the result of revision becomes extremely dependent on the syntactic form of the database.' (1994, p. 911) By taking this to be a drawback, del Val shows that he is ready to relinquish precisely what I consider to be the central idea of the foundationalist philosophy of belief change. His aim is to dissociate the 'database', which may be any arbitrary logically equivalent transform of the original belief base, from the notion of 'basic beliefs'. This could be just a terminological disagreement, if he did not stipulate, in his definition of a 'basic beliefs function', that all databases with the same logical content are assigned the same set of basic beliefs. Since his theory revision operations are determined by what the basic beliefs are, each theory can have only a single revision operation associated with it. This is unacceptable for a foundationalist, as I understand the term. Agents may well hold the same theory on the grounds of two completely different sets of basic beliefs. Maxim B (Section 1.2.1) now tells us that elements in the belief base have a distinguished doxastic status, since they and only they are the beliefs that have an independent standing. Different belief bases for the same theory, therefore, lead to different dispositions to revise that theory. Maxim B, which concerns the statics of belief in the first instance, turns out to have important dynamic implications. The structure of the belief base carries the kind of extra-logical information that is required to guides rational changes of belief.

A similar disagreement about the role and function of belief bases is already visible in an earlier dispute between Belnap and Rescher. In his criticism of Rescher's (1964) theory of hypothetical reasoning, Belnap (1979, p. 23) declares the difference between a single item $p \wedge q$ and two separate items p and q in a

---

[130] An alternative would be to work with flocks of belief bases in parallel. Compare Fagin, Ullman, Kuper and Vardi (1986).

belief base to be just 'notational bondage' that ought to be straightened out by some process of 'articulation'. I agree that agents should seek an articulation of their beliefs in the process of drawing inferences from their beliefs, but this does not sanction forgetting about one's base after that process has been completed. Each element of a belief base must be taken as a singular piece of information, irrespective of the syntactical complexity of the sentences expressing it. I think that Rescher (1979, p. 31) is entirely correct in pointing out that if we take belief bases seriously—whether for the purposes of hypothetical reasoning or for the purposes of belief revision—then there is indeed a crucial difference between 'juxtaposing commas' and 'conjoining ampersands'.[131] If they are separate items juxtaposed by a comma, the fates of p and q are independent of one another, unless there is additional information to the contrary (e.g., another basic belief $p \supset q$); if they are conjoined by an ampersand, they stand and fall together. Against del Val and Belnap, I do not think that it is legitimate to subject the belief base to any manipulations. While closing a base under conjunctions or disjunctions or under some non-classical logic, say, may have desirable effects from a formal point of view, it runs counter to Maxim B. The structure of the original belief base provides us with information about how to change our beliefs, and that structure should not be changed without compelling reasons.

3.4.1 *Two variants of foundationalist belief change*

In this section I take the foundationalist's distinction between basic and derived beliefs for granted. The distinction is in fact required to build up the framework for the two perspectives developed in this chapter.

Foundationalist belief change is characterized by changes taking place at the level of belief bases. We may distinguish two altogether different modes of belief base change. According to the first, which we call the *direct mode*, there are no coherence constraints whatsoever for prima facie beliefs,[132] so the process of revising the belief base can be construed as absolutely trivial. According to the second mode, which we call *coherence-constrained*, belief bases have to be consistent. Hence the process of revising the base is non-trivial.

There is another sense of non-triviality of belief changes that does not arise in connection with belief *revisions* but is inherent in the notion of belief *contraction*. 'Discarding information' may mean two different things. First, it can be done in a purely surface-oriented manner such that only syntactically specifiable items of belief are erased. Second, it can be done in a way respecting the contents of our pieces of information. If the task is to discard the belief that ϕ in this second sense, then it does not suffice to discard the token representing ϕ. It has to be ensured that ϕ does not only disappear from the surface of the belief base, but also that it is not derivable from the remaining items of belief.

---

[131] Belnap keeps on denying the difference in the amended version of his critique published in Anderson, Belnap and Dunn (1992, pp. 541–553).

[132] Remember from Section 1.2.3 that prima facie beliefs are the elements of a possibly inconsistent belief base.

Coherentists cannot help focusing on the horizontal (dynamic) dimension—they simply lack the vertical (static) one. Foundationalists, however, are not committed to restricting themselves to the vertical (static) dimension. A foundationalist may hold that even on the level of basic beliefs we need to obey some minimal constraints of coherence. Most importantly, he may say, we have to see to it that our basic beliefs are consistent. It has long been noted in epistemology that basic beliefs are not infallible.[133] Some of them may be due to illusion or to misinformation. We may be deceived, end up with contradictory sets of basic beliefs, and thus find it necessary to purify the base so that consistency is restored. This can be done either as part and parcel of the very process of adopting new basic beliefs; changing bases then becomes an interesting and difficult task. Or alternatively, the purifying process can be initiated only after some new input has been received and accepted, in a subsequent process of drawing inferences from the new, possibly inconsistent information base. In the first case, the foundationalist focuses on the dynamic (horizontal) dimension in quite the same way as coherentists do, with the crucial difference that foundationalists acknowledge only consistency—but not closure—as the requisite form of coherence. In the second case, the foundationalist does not care much about the process of acquiring information but rather focuses on the static (vertical) dimension of straightening out the beliefs. This difference defines, within the foundationalist variety, the fundamental distinction between taking the horizontal and taking the vertical perspective. Thus, the distinction of the two perspectives is not identical with the distinction between foundationalism and coherentism in the theory of belief and belief change. We have just found that one may view foundational belief change (changes operating on bases) from a horizontal as well as a vertical perspective. I start my further considerations with the latter one.

3.4.1.1 *The direct mode of revising information bases* The direct mode of revising information bases sticks to purely foundationalist ideas and does not require any coherence on the level of the belief base. There is no constraint filtering what can count as a prima facie beliefs. As a consequence, the idea of *minimal change* is very easily conformed to. Adding ϕ set-theoretically (or putting ϕ on top of a list) just *is* a minimal change effecting the acceptance of ϕ. Taking ϕ away set-theoretically (or taking ϕ out of a list) just is a minimal change that removes ϕ. The use of these change operations is what makes this mode 'direct' or 'immediate'.[134]

As a compensation to these trivial dynamics, however, the static dimension must be quite elaborated. A useful inference relation taking the agent from his information base to what he 'really' believes or accepts is typically required to be both paraconsistent and nonmonotonic for the direct mode. Paraconsistent, because the information base may well be inconsistent—and that should not

---

[133] See for instance BonJour (1985, Chapter 2) and Lehrer (1990, Chapter 4).

[134] This is also terminologically continuous with Gärdenfors and Rott (1995).

prevent the agent from holding a reasonable set of beliefs. Nonmonotonic, because information bases tend to be both incomplete and uncertain, so he should
be ready to jump to more informative conclusions—conclusions, however, which
may easily be given up in the light of further, more precise information.

The direct or immediate mode of revising information bases thus has three
essential ingredient (compare Fig. 3.3):

(I) In line with the foundationalist philosophy of base change, derived beliefs
change as a result—and *only* as a result—of changes on the base level.
The new belief set is obtained in two steps. First, the base is changed, and
second, inferences are drawn from the new base. Put as a slogan: Theory
revision, $*$, is trivial base revision, $\oplus$, plus sophisticated inference, Inf. Similarly, theory contraction, $\dot{-}$, is trivial base change, $\ominus$, plus sophisticated
inference, Inf.

(II) Concerning $\oplus$ and $\ominus$: The changes leading to the new base are made in a
rather trivial (plain, artless, undemanding, uncomplicated, ...) fashion,
that is: There are *no choices* to be made.

(III) Concerning Inf: In line with the philosophy of nonmonotonic reasoning,
we assume that we usually have only very few explicit beliefs and depend
essentially on a huge bulk of implicit, defeasible background assumptions.
In line with the philosophy of paraconsistent reasoning, we assume that
usually the union of explicit beliefs and implicit, defeasible background
assumptions is contradictory.[135]

Let us draw attention to an important point made in (II). That a belief
change is trivial (plain, artless, undemanding, uncomplicated, ...) just *means*
that there are not any problems of choice to be solved when effecting the belief
change. Problems of choice do not arise, because all perturbations cause the belief
base to be accommodated in a most straightforward way. This in turn is due to
the fact that there are no coherence constraints whatsoever that are placed on
the structure of information bases.

External disturbances caused by some input to an doxastic system may be
interpreted as just instigating a new inferential closure process. We have distinguished two principal kinds of changes: revisions and contractions. In the simple
model that identifies states of belief with sets of sentences (we neglect prioritizations here for the sake of simplicity), we can correspondingly think of two
primitive kinds of input. Set-theoretic addition of ϕ to a set H and set-theoretic
subtraction of ϕ from a set H, in symbols, $H \oplus \phi = H \cup \{\phi\}$ and $H \ominus \phi = H - \{\phi\}$.
The direct mode in effect reduces the problem of external perturbations to the
problem of finding inferential closures. The straightforward ideas are that $K * \phi$
be identified $Inf(H \oplus \phi)$ and that $K \dot{-} \phi$ be identified $Inf(H \ominus \phi)$. (We shall study
the merits and demerits of these suggestions at some length in Chapter 5.) Well-
balanced sets K, reached by applying the inferential closure operation to some

---

[135] Contradictions may arise even within the set of explicit beliefs, and/or within the set of
implicit assumptions.

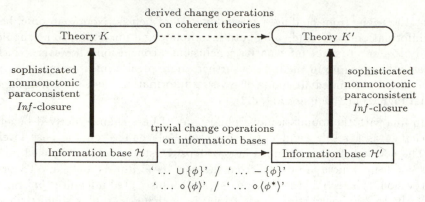

FIG. 3.3. Foundationalist belief change in the vertical perspective:
The direct or immediate mode of base revision

basic information set H, are changed not by adding or subtracting ϕ to/from K but rather by adding or subtracting ϕ to/from the belief base H and then applying the process of closure anew. Notice that since a given theory K can in general be construed as the inferential closure of many different bases H_i, the operation of belief change in response to some input is not well-defined, if it is applied to K *simpliciter*. 'Revising a theory' does not really make sense on the above-mentioned account, we must always understand it as 'revising a theory-as-generated-by-a-certain-belief-base'. The process of revising well-balanced belief sets is indirect in the sense that it involves a retreat to some generating belief base H_i, a set-theoretic operation of addition or subtraction.

While the process of simple set-theoretic subtraction from a belief base is unproblematic for the direct mode of base revision, set-theoretic addition tends to generate problems. If ϕ is inconsistent with H, the immediate mode can be applied sensibly *only* if we endorse a paraconsistent logic. As usually understood, paraconsistent logics are *logics* in the minimal sense that they are reflexive, i.e., the set of conclusions always includes the set of premises. But it is not obvious that this should be so. Consider a set H that includes both ϕ and $\neg\phi$. Paraconsistent logicians argue that the set of consequences of H should include ϕ and $\neg\phi$, but not everything. Hence, paraconsistent logics must be weaker than classical logic in a some respects, namely, in that they must not include the classical principle of *ex falso quodlibet*. The set of paraconsistent consequences of H is not in general closed under classical logic. Instead of arguing that H should entail *both* ϕ and $\neg\phi$ (but not everything), one could equally well argue that H should entail *neither* ϕ *nor* $\neg\phi$ (but not nothing)—at least as long as nothing in H (or more precisely, nothing in the agent's mind) favours one over the other. In this way one can allow the set of conclusions to be closed under classical logic. However, giving up the principle that the premises must be contained in the set of consequences may be regarded as disqualifying the process of drawing inferences as an application of some *logic*.[136] It will become clear in the course of this book that *extra-logical* information is indeed required in all the attempts at making this alternative account of handling inconsistent premise sets precise. I have already suggested a variant of Sven Ove Hansson's operation of 'consolidation' as a way of reducing internal problems of inference to external problems of belief change. (More on this in Section 3.5 below.)

3.4.1.2 *The coherence-constrained mode of revising information bases* Paraconsistent logicians tolerate and sometimes even welcome inconsistencies, but most people prefer to resolve them. If Harry tells us that ϕ is the case and Sally tells us that $\neg\phi$ is the case, then we usually think that one of them must be wrong.[137] One of these bits of information, either ϕ or $\neg\phi$, cannot be true, and therefore should not be accepted. This is the naive or common-sense reaction, and it is the reaction modelled by the standard accounts of belief revision. Instead of acquiescing in inconsistencies, agents aim, when striving to reach a well-balanced state of mind, at avoiding inconsistencies and restoring consistency. So when ϕ is inconsistent with H, one had better refrain from using mere addition to the belief base in the first place. Instead of applying trivial set-theoretic operations and thereby committing oneself to some nonstandard logic (in any case, to the repudiation of classical logic), one may prefer to apply a more sophisticated change operation and then use standard logic when taking the inferential closure. This is the same logic as the one which tells us when to apply the non-trivial change operation: when ϕ is inconsistent with H.

Remembering that consistency is a minimal form of coherence, we call this mode of foundational belief change *coherence-constrained*, or *logic-constrained* (since consistency is a form of inferential coherence). Let us introduce a new terminology for this approach. We call $H * \phi$ the *revision* of H by ϕ, with the idea that revision is identical with simple addition $H \cup \{\phi\}$ only if ϕ is consistent with H. If ϕ is inconsistent with H, some elements of the latter have to be given up in order to make consistent place for ϕ. (For the sake of simplicity, classical approaches to belief revision assume that the 'new' information ϕ has priority over the 'old' beliefs H.) We also have an operation of *contraction*, $H \dot- \phi$, of belief bases which corresponds to the operation of subtraction.[138] The idea of contraction, as opposed to mere set-theoretic elimination, is that ϕ should be 'really' removed from the set of beliefs, that is, ϕ should not be reintroduced by a subsequent step of closing the new base under (standard) logical consequences. If ϕ is an element of H, $H \dot- \phi$ will in general not be identical with $H - \{\phi\}$. And even if ϕ is not included in the old base H, we should not expect that $H \dot- \phi$ is always identical with H. Although a subtraction is unproblematic in so far as it cannot introduce an inconsistency, it is problematic in that it may fail to have

---

[136] Compare the discussion in Section 1.2.5.

[137] Assuming, somewhat optimistically, that they speak the same language.

[138] For the sake of notational economy, I use the same operators $*$ and $\dot-$ for base changes as I use for theory changes.

achieved its goal after the final 'internal' balancing out of the belief state has been carried out.[139]

We now give a characterization of the coherence-constrained mode of base changes that parallels the one given for the direct mode (compare Fig. 3.4).

(I) In line with the foundationalist philosophy of base change, derived beliefs change as a result—and *only* as a result—of changes on the base level. The new belief set is obtained in two steps. First, the base is changed, and second, inferences are drawn from the new base. Theory changes are derived from sophisticated change operations, $*$ and $\dot{-}$, that transform one consistent belief base into another consistent belief base, subject to certain constraints.

(II) Concerning $*$ and $\dot{-}$: The changes leading to the new base are made in a (more or less) sophisticated, skilful, demanding, complicated fashion, that is: There are *choices* to be made.

(III) Concerning Cn: In contrast to the philosophy of paraconsistent reasoning, we maintain the logical consistency of the explicit beliefs. Due to the employment of sophisticated change operations, standard consequence operations may be taken as the logic on which belief change is based.[140]

The constraints mentioned at the end of (I) are of at least three kinds. The first are constraints of *success* implied by the nature of the changes to be made. A revision by ϕ should lead to a belief base that contains ϕ, while a contraction with respect to ϕ should lead to a belief base that does not imply ϕ. The second kind of constraints requires information bases to be *consistent*. A base must exhibit this minimal sort of coherence, but it need not be in full reflective equilibrium which would be reached only by a subsequent process of reasoning or drawing inferences from the items in the base. The third kind of constraints concerns questions of *minimal Change*. Usually there are numerous belief bases satisfying the first two sorts of constraints. Many philosophers argue that we should then choose, among the multiplicity of 'admissible' belief bases, those that are closest to the original belief base. In the case of belief contractions, this means we should choose those that incur as small a loss of information as possible. This is the doctrine of minimum mutilation or doxastic conservatism sketched in Section 3.2.3. In most of the present book, I presume that the constraints of success arising from the very conceptions of revision and contraction are met.[141] I also focus on the logical

---

[139]Here there is an asymmetry between revisions and contractions. When considering revisions, I argued that the maintaining of consistency (i.e., the non-implication of a contradiction) can be conceived as a task of logic. But it seems obvious that we cannot commission *logic* to take care of the non-implication of the *contingent* sentences to be withdrawn in a contraction.

[140]Since there is no need for a sophisticated inference relation, we do not adopt here the philosophy of nonmonotonic reasoning, according to which we usually have only very few explicit beliefs and depend essentially on a huge bulk of implicit assumptions.

[141]I allow, however, for the possibility that agents refuse to contract or revise.

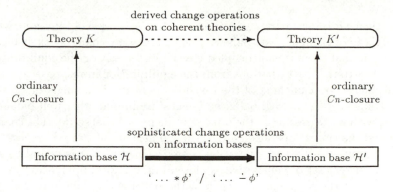

FIG. 3.4. Foundationalist belief change in the horizontal perspective:
The coherence-constrained mode of base revision

consistency constraint, which is a coherence constraint applicable to belief bases
as well as to theories. I will not, however, insist on minimum mutilation.

In order to keep information bases consistent, one often has to call into question an old base, or some of its elements, when new information comes in. This
will involve choices concerning which elements of the base to discard, according
to their supposed reliability and certainty. These are the choices mentioned in
(II), and one can arguably insist that these choices should be coherent, in a sense
to be explained later (in Chapter 6). This requirement represents another type
of constraint for base revision operations.

We have seen that this mode combines foundational and coherentist features
on the base level. Unlike the direct mode, the logic-constrained mode of revising
information bases is interested mainly in the dynamics of belief change, more
precisely in the ways of maintaining consistency on the base level while processing new items of information. The static dimension of reasoning and drawing
inferences can be kept at a fairly low level of sophistication. Usually closure under standard classical logic is applied in the process of generating of coherent
theories from information bases.

This mode of belief revision, as well as many concrete ways of constructing relevant contractions and revisions, have been championed by Sven Oven
Hansson.[142] The present book makes no contribution to the discussion of this
'middle way' between foundationalism in the vertical perspective and coherentism in the horizontal perspective.

3.4.2 *Coherentist belief change: The logic-constrained mode of revising theories*

Once we have started shifting the attention in processes of belief change from the
inference operation (the vertical dimension in the above figures) to the change

---

[142]See Hansson (1989, 1992, 1993a, 1993b) and the textbook presentation provided in Hansson
(1999a).

operation (their horizontal part), we can now go further and avoid the somewhat roundabout way of changing well-balanced belief states by changing their bases. One may argue that there is no principled reason for separating the equilibration in response to external perturbations from the equilibration in response to internal insufficiencies. We can think of the revision operation as a single integrated process of transforming one well-balanced belief state into another well-balanced belief state, without regressing to the origins of the prior belief set.[143] The change operation and the closure operation are combined into an integrated process, or, to put it differently, the change operation is so designed that it is bound to end up with a consistent set of beliefs that is closed under logical inferences.

This is in fact the paradigm approach of the belief revision literature set by Alchourrón, Gärdenfors and Makinson in the 1980ies—an approach now commonly branded with their initials 'AGM'. But what exactly is the famous *AGM paradigm*? To me the most appropriate answer seems to say that it is the study of belief contraction and belief revision operations by means of both abstract 'rationality postulates'[144] and explicit constructions.[145] Alchourrón, Gärdenfors and Makinson themselves investigated the methods of partial meet contraction, safe contraction and epistemic entrenchment contraction. A fourth approach, suggested by Grove (1988), shows that partial meet contractions are virtually identical with contractions in terms of systems of spheres à la Lewis (1973),[146] and may naturally be counted as forming part of the AGM paradigm. The final characteristic mark of the work of AGM consists in a number of *representation theorems*[147] that show that all their constructions satisfy the rationality postulates, and conversely, that any change function satisfying these postulates can be construed as one arising from any of the AGM methods.[148] I advocate a broad understanding of the AGM paradigm that does not exclude any kind of liberalizations of the AGM postulates, extensions of the construction methods or application to representations of doxastic states with a richer structure than belief sets (such as belief bases or belief sets plus selection mechanisms).

The appropriate change operations are again coherence-constrained, but this time in the full sense of inferential coherence (see Section 3.2.1 above). Like the change operations in the last section, they aim at consistency ('falsity' $\bot$ is not

---

[143]This does not exclude that the selection mechanisms employed in coherentist modellings of belief change show traces of the doxastic history (previous belief changes) of the prior belief base.

[144]See Chapter 4.

[145]Of the style of those studied in Chapter 7.

[146]There is one important difference. While Lewis uses possible worlds, Grove uses models of the language as *Ersatz* possible worlds. Therefore, Grovean possible 'worlds' models are always injective in the sense of Freund (1993); no two distinct worlds satisfy the same set of sentences. Compare Section 7.2 below.

[147]Of the style of those studied in Chapter 7.

[148]Figure 3.5 in addition indicates the direct links between the three paradigmatic AGM constructions that have been established by Alchourrón and Makinson (1985), Rott (1991a) and Rott (1992a).

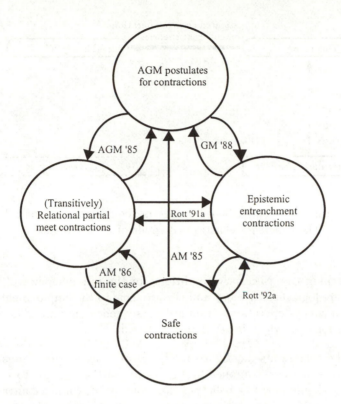

FIG. 3.5. The core of the AGM paradigm

entailed—the relevant criterion for revisions) and at successful removals (ϕ is not entailed—the relevant criterion for contractions). In addition, they need to ensure that the final belief system is internally well-balanced; everything that may be legitimately inferred must already already by accepted in the posterior belief state. The dynamics already take into account all the desiderata one can advance from the static point of view. This is the picture of ideally rational belief change operations.

Again we briefly list the main ingredients of coherentist belief change (compare Fig. 3.6).

(I) In line with the coherentist philosophy of theory change, there is no distinction between basic and derived beliefs. The new belief set is obtained in a single step, there is no recourse to belief bases. Theory change, $*$ or $\dot{-}$, requires a sophisticated change operation transforming one coherent belief system into another coherent belief system, subject to certain constraints.

(II) Concerning $*$ or $\dot{-}$: The changes leading to the new theory are made in a (more or less) sophisticated, skilful, demanding, complicated fashion, that is: There are *choices* to be made.

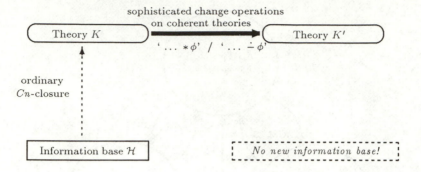

FIG. 3.6. Coherentist belief change in the horizontal perspective:
The coherence-constrained mode of theory revision

(III) Concerning Cn: Necessary conditions for the full coherence of belief systems are logical consistency and closure. Due to the employment of sophisticated change operations, standard consequence operations may be taken
as the logic on which belief change is based.[149]

Again the constraints mentioned in (I) are of at least three kinds. The first
are constraints of *success* implied by the nature of the changes to be made. A
revision by ϕ should lead to a belief set that contains ϕ, while a contraction with
respect to ϕ should lead to a belief set that does not contain ϕ. The second kind of
constraints requires belief sets to be both *consistent* and *closed*. A belief set must
exhibit these features of coherence, which are necessary for a full reflective equilibrium reached only by a completed process of reasoning or drawing inferences
from the items in the set. The third kind of constraints concerns questions of
minimal change. Since there are usually many belief sets satisfying the first two
sorts of constraints, philosophers have argued that we should then choose, among
the multiplicity of 'admissible' belief sets, those that are closest to the original
belief set. In the case of belief contractions, this means we should choose those
that incur the smallest loss of information possible. Once more, this is the doxastic conservatism sketched in Section 3.2.3. We generally obey the constraints
of success arising from the very conceptions of revision and contraction,[150] as
well as the logical coherence constraints. As already mentioned, we do not insist
on minimum mutilation.[151]

[149]See footnote 140 above.

[150]I shall allow, however, for agents that refuse to contract or revise, which is most natural
in the case where ϕ is a logical theorem or a contradiction. See conditions ($\dot{-}4$) and ($*2^-$) on
pp. 103 and 111 below.

[151]Weighty arguments against the use of minimum mutilation in a coherentist dynamics
of belief can be based on 'Principles of Preference and Indifference'. They are discussed in
Pagnucco and Rott (1999).

Only the upper part of Fig. 3.6 is relevant for this type of belief change. The initial information base $\mathcal{H}$ may leave traces in the structure that guides the revision on the level of coherent theories, but there is no new base resulting from revision. Any subsequent revision must be based on a structure of a different kind.

I have left open a crucial question. As we have seen, there are foundationalist approaches in the vertical and in the horizontal perspective. I have also described a coherentist approach in the vertical perspective. Is there, finally, a coherentist approach in the horizontal perspective? No. This is a conceptual impossibility. Since coherentists simply do not allow to distinguish a base for the full body of beliefs, they cannot lay stress on the process of deriving belief systems from belief bases. Hence all coherentist belief change must be viewed from the horizontal perspective. The following combinations are possible:

	foundationalist	coherentist
vertical perspective direct mode	+	−
horizontal perspective coherence-constrained mode	+	+

3.5 Relating belief change and nonmonotonic reasoning

It should have become clear by now—and will become clearer later[152]—that belief change and nonmonotonic reasoning are intimately connected. The direction of the connection is dependent on the perspective taken. The vertical perspective constructs belief change through reasoning that must be understood as nonmonotonic. If we take the horizontal perspective, on the other hand, we can explain nonmonotonic reasoning through belief change (or rather: expectation change), where beliefs (or expectations) may be represented either in the form of a base or in the form of a theory. So the two different perspectives have complementary views on the conceptual priority of belief revision and nonmonotonic reasoning.

To make this point as perspicuous as possible, I now sum up a few things that have essentially been said before.

If we take the vertical perspective of the direct mode of belief revision (which is necessarily foundationalist in character), belief change is essentially a secondary effect of the drawing of sophisticated inferences from anarchically evolving belief bases. We start from a belief base most generally represented by a structure $\mathcal{H}$. Assume that we have given a powerful, and that must mean: nonmonotonic and paraconsistent, inference relation *Inf* and trivial change operations $\oplus$ (adding ϕ set-theoretically, or appending ϕ to a list etc.) and $\ominus$ (subtracting ϕ set-theoretically, or cancelling ϕ from a list etc.). The initial coherent

---

[152] In Chapters 4 and 5.

belief state (belief set) that we ascribe to the agent then is $K = Inf(\mathcal{H})$. As we know, revisions and contractions of this initial system can then easily be defined:

$$K * \phi = Inf(\mathcal{H} \oplus \phi)$$

$$K \dot{-} \phi = Inf(\mathcal{H} \ominus \phi)$$

Given the above interpretation of inferences as generating (inferential) coherence, both $K * \phi$ and $K \dot{-} \phi$ must be consistent and closed. Do they have to be successful, that is, do we have a guarantee that ϕ is accepted in $K * \phi$, but is not accepted in $K \dot{-} \phi$? No, not necessarily. The problem is already visible in revisions, but by a judicious choice of $\oplus$ and Inf we can see to it that ϕ gets a privileged status and is in fact included in $K * \phi$. The problem is much more severe for contractions. There is no way of guaranteeing that the 'superficial' removing of (occurrences or tokens of) ϕ will not be annulled when the inference operation is subsequently applied to the new belief base. So it becomes evident that a totally different way of achieving the effect of contractions in the direct mode is required. Concrete suggestions of how to give sense to the general idea of deriving theory change operations from trivial base changes plus sophisticated inferences will be presented in Chapter 4.

Let us now consider the horizontal perspective with its coherence-constrained mode of belief revision. I have pointed out that there is no need for a sophisticated inference relation in this approach. But if we wish, for independent reasons, to avail ourselves of ampliative forms of everyday reasoning, we need a nonmonotonic logic after all. And we can use the very models developed for rational belief change as devices for the construction of nonmonotonic inferences. Suppose that we are given some powerful revision and contraction operations $\underline{*}$ and $\dot{-}$ for expectations. Poole (1988) and Gärdenfors and Makinson (1994) have held the view that nonmonotonic reasoning consists essentially in the revision of expectations by information (see Section 1.2.7.1). In order to be able to use revision operations as they are common in the standard paradigm of belief change, the expectations must be given in the form of a logically closed theory E, and the information must be put into a single sentence h. Then we can define

$$Inf_E(h) = E \underline{*} h$$

This identity has been suggested by Gärdenfors and Makinson (Makinson and Gärdenfors 1991, Gärdenfors and Makinson 1994).

As an alternative to the Gärdenfors-Makinson method, one can adopt a more general one that is applicable to belief and expectation bases that may be infinite and prioritized. This method is no longer part of the coherentist ideology of belief change, but it retains the idea of understanding nonmonotonic reasoning in terms of a non-trivial form of belief revision. We start from an expectation base in the most general representation format $\mathcal{E}$, and we let $\mathcal{H}$ denote the initial information base that we ascribe to the agent. In Section 1.2.7.2, I have defined the set of sentences inferable from the information base $\mathcal{H}$, given our expectations have the structure $\mathcal{E}$, to be $Consol(\mathcal{E} \circ \mathcal{H})$. Now remember that consolidation really is a special form of belief change, viz., a belief contraction with respect to the contradiction. Thus this approach to nonmonotonic (and, by the way, paraconsistent) reasoning is indeed based on belief change, too.

$$Inf_{\mathcal{E}}(\mathcal{H}) = (\mathcal{E} \circ \mathcal{H}) \dot{-} \perp$$

Notice that while the first, more special approach of Gärdenfors and Makinson is based on belief revision, this second one is based on belief contraction. The expectation base effective in the background would usually be left implicit, and the parameter $\mathcal{E}$ attached to Inf indicating the dependence of the inference operation on that base remains suppressed. There is no problem in doing this, so long as the structure and content of the expectations do not change.

One more point should be clarified here. On the accounts described in Section 1.2.7, 'hard' information is combined with 'soft' expectations in the process of drawing inferences from the information base. This happens along the vertical dimension. But it is obvious that an agent's expectations may also change in time, in the process of his interactions with the world, giving rise to an evolution of inference operations. I will not deal with the important problem of the dynamics of expectations in this book. For the sake of simplicity, I shall abstract from any expectation change in the horizontal dimension and assume that the expectations remain always the same. I simply take it for granted that we *have* expectations representable by a formal structure (a prioritized base), and do not inquire whether there is a rational process underlying the *formation* of expectations. Causal mechanisms may be more effective in this domain than reasoning in accordance with approved logical canons, but I shall refrain from speculations about this matter.

3.6 The dialectics of the statics and the dynamics of doxastic states

We now have a further look at the interplay between static and dynamic aspects of belief, continuing a topic that was first addressed in Section 1.2.8 above. We ask whether belief revision strategies determine beliefs, whether beliefs determine belief revision strategies, or whether beliefs and belief revision strategies are rather two altogether independent components of cognitive states.

We start with a definition. A *belief revision model* is a pair $\langle \mathbb{K}, * \rangle$, where $\mathbb{K}$ is a non-empty set of potential or 'admissible' belief sets and $*$ is a function assigning to every pair consisting of a belief set K in $\mathbb{K}$ and a sentence ϕ of a language L a unique revised belief set $K * \phi$ in $\mathbb{K}$. Belief revision models $\langle \mathbb{K}, * \rangle$ are discussed by Gärdenfors (1986, 1988) and others in the context of doxastic interpretations of conditionals by means of the Ramsey test. I am going to restrict the following discussion to the case where doxastic states are represented by belief sets and changes are initiated by inputs in the form of single sentences. But similar reflections would apply to arbitrary formats of states and inputs.

There are two possible ways of abstractly capturing the processes of revision and contraction. The first one, which is the one employed in belief revision models, is based on the idea that a change operation should tell us *for every potential belief set* K (from a reference class $\mathbb{K}$) and every sentence ϕ (from a language L) how K should be revised by ϕ, or contracted with respect to ϕ. Formally, we are presented with two-place change operations

$$* : \mathbb{K} \times L \longrightarrow \mathbb{K}$$
$$\dot{-} : \mathbb{K} \times L \longrightarrow \mathbb{K}$$

The alternative idea is to think of each belief change operation as being pertinent to a *single belief state* only. Since that belief set is the only one that is revised by a given operation, there is no need to mention it explicitly, thus input sentences figure as the sole arguments of the revision and contraction functions

$$* : L \longrightarrow \mathbb{K}$$
$$\dot{-} : L \longrightarrow \mathbb{K}$$

As far as I can see, this is the idea behind almost all of the seminal work of Alchourrón, Gärdenfors and Makinson.[153]

Surprisingly, these barren and abstract considerations about belief revision models and formal concepts of revision functions have considerable philosophical implications. Prima facie, it seems advisable always to use two-place functions, since they appear to be the more general and powerful tools for the of study belief change. In particular, they allow us to capture iterated changes of belief, simply because the value $K * \phi$ (or $K \dot{-} \phi$) of a two-place operation can again be taken as the first argument of $*$ (or $\dot{-}$) in a subsequent step of belief change. Still the alternative view that conceives of change operations as unary functions has an advantage. Two-place revision functions have a consequence that seems hardly acceptable—'exactly one revision strategy for every belief set'. Surely two persons may differ in their doxastic states while accepting exactly the same set of sentences. They may differ in their dispositions to chance their beliefs. The less powerful notion of a unary revision-function, that in itself offers no framework for iterated belief revision, allows us to interpret a doxastic state as being identical with a belief revision strategy, which can in turn be formalized either by $*$ itself or by some selection structure on which the construction of $*$ can be based.

The main argument we have to consider runs as follows. Revision and contraction functions are formalizations of belief change strategies. If a single revision function, $*$ say, may take different belief sets as arguments, this in effect means

---

[153]Perhaps it would be clearer to use the notation $*_K$ and $\dot{-}_K$ here, to indicate that these functions are understood to apply to a single given K only. Segerberg (1995, p. 550) disagrees with my interpretation of Alchourrón, Gärdenfors and Makinson. I base my judgment on the fact that almost nothing in their work is concerned with revisions of varying belief sets. Exceptions can be found in Alchourrón and Makinson (1985, Section 7) and Gärdenfors (1986).

that ∗ is applicable to varying belief sets, and thus a *belief revision strategy does not determine a belief set*. Intuitively, this seems strange. For a belief revision strategy is a belief revision strategy *of an agent*, and it is the set of the agent's own actual beliefs that he wants to change in the first place. In any case, an agent's belief revision strategy should be faithful to his beliefs, *faithful* in the sense that ϕ is a belief if and only if a contraction of the agent's doxastic state with respect to ϕ does actually change something.[154] So the decision of how to revise one's beliefs should be dependent on which beliefs to revise, and conversely, the way how someone revises his beliefs tells us everything about his beliefs.

This intuitive objection notwithstanding, technical models realizing the idea of two-place revision functions have been investigated by Schlechta (1991a, 1991b) and Rott (1992a, 1992b) amongst many others. In these papers a fixed background ordering between sentences is assumed to exist which is not—or at least need not be—dependent on the agent's current stock of beliefs and can thus be applied to arbitrary belief sets. The technical problem of 'faithfulness' just mentioned is easily solved. In defining rules of application for belief revision strategies, one can make special provision for the current belief set K in such a way that the resulting change operation is faithful to K.[155] A more substantial philosophical problem, however, is how to justify such a 'belief-independent' way of revising beliefs. Where does such an ordering come from, if it does not spring from the agent's mind?

The correct answer, I think, is this: The ordering springs from the agent's mind indeed, but not from his set of beliefs. An interpretation of belief revision models and two-place revision functions that makes good sense is this. A belief revision model represents an agent's disposition to change his *potential* belief sets in response to *potential* input. The representation of revision functions as taking varying belief sets as arguments indicates that the dynamic dispositions are construed as independent from, and presumably less contingent and more stable than, the current set K of the agent's beliefs. Belief revision strategies somehow reflect the (informational) values the agent attaches to propositions— irrespective of whether he does or does not accept these propositions as beliefs. The question then is whether it is a sound strategy to postulate something in the agent's mind that is represented by a belief revision model. In the larger part of this book, I will in fact hypothesize that there are preferences or choice functions in some compartment of the agent's mind (be it located in the cognitive or in the volitional faculty), preferences or choice functions that are independent of

---

[154] I am neglecting here the possibility that the agent may refuse outright to withdraw a (very strong) belief ϕ, with the consequence that the contraction with respect to ϕ is not successful.— Beliefs can also be defined in alternative ways, viz., as those sentences the revision with which does not change anything, or those sentences that are contained in the contraction of K with respect to $\perp$ or the revision of K by $\top$ (provided K is consistent).

[155] In the central Definitions 6 and 8 of Chapter 7, the semantic and syntactic choice functions are independent of K, yet only elements of K can be found in the contraction $K \dot{-} \phi$. Compare in particular Section 7.10 on iterated belief change in the coherentist perspective.

his propositional beliefs, may take various bodies of beliefs as arguments and will most of the time stay stable when the beliefs change. Except for the foundationalist modelling presented in Chapter 5, I shall not engage in a more detailed discussion of the changes of belief change strategies.[156]

I have just argued that belief revision strategies do not and should not determine beliefs. It is equally important to emphasize that the reverse holds as well. *Belief sets do not determine belief revision strategies.* The sentences accepted by an agent are just one aspect—the propositional aspect as it were—of his mental state. And of course the identity of one aspect does not entail the identity of other aspects. Belief revision strategies do not only depend on *which beliefs* are endorsed but also, e.g., on how *valuable* these beliefs are considered to be or how they are (thought to be) *connected*. Prima facie, this simple insight seems to be at odds with the formal conception of revisions and contractions as two-place functions which take a belief set and an input sentence and return a new belief set. If the first argument of such a function is fixed as K, say, then it *is* thereby fully determined how K is going to be revised.[157] In reality, however, an agent may have exactly the same set K of beliefs on Monday as he had on Saturday, and yet he may have changed his disposition to change K in response to incoming information.[158] But then this simply means that the agent has undergone a change of his belief revision strategy while his beliefs remain constant—a phenomenon that is perhaps not as common as a change of beliefs (with the belief revision strategy remaining constant), but not so uncommon either. Both are changes in the mind, in the doxastic state of the agent. I will posit background structures governing revisions of varying belief sets (i.e., two-place revision functions), but I think of revision-guiding structures not as immutable and objectively given, but as integral parts of the cognitive states of an agent. At this point, then, it seems best to represent a *doxastic state* by a pair consisting of two independent components: a belief set representing the current set of beliefs, and a two-place revision function representing the current belief change strategy.[159]

However, there is yet another argument to the effect that belief sets are

---

[156]In Chapter 5, a change of beliefs (more precisely, a change of a belief base) is *ipso facto* a change of the strategy to revise one's current beliefs. In particular, it may well happen that a change of the belief base does not at all affect the identity of the inferential closure K (i.e., $K = Inf(B) = Inf(B')$), but does alter the dispositions to change K. See in particular Section 5.2.3 on iterated belief change in the vertical perspective.

[157]Formally, a two-place revision function is like a family of indexed unary revision function, with the prior belief states as indices. For each K, there is only one unary revision function $*_K$ indexed by K.

[158]Such things typically happen on weekends: No new information, but new attitudes instead.

[159]In the foundationalist approach presented in Chapter 5, the coordinates are more like $\langle \mathcal{H}, \oplus \rangle$ from which the first component of $\langle K, * \rangle$ can be constructed with the help of Inf: $K = Inf(\mathcal{H})$. But there is no *theory* revision strategy that can be determined by $K' * \phi = Inf(\mathcal{H}' \oplus \phi)$, since $\mathcal{H}'$ is not unique for K'. The belief revision strategy just is the simple base revision operation $\oplus$.

determined by belief revision strategies after all, an argument that may even be regarded as a trivial consequence of the general fact that a model for the dynamics of some system must always presuppose and include a model of its statics. More specifically, the argument says that if the set of current beliefs is inferentially well-balanced (i.e., consistent and closed under logical inferences), then this set may be identified with the result of a (vacuous) revision of the belief state by the truth constant $\top$ or a (vacuous) contraction with respect to the falsity constant $\perp$.[160] If we are given a unary belief change operation $*$ or $\dot{-}$, then it is possible to retrieve from it a coherent belief set by putting $K = *(\top)$ or $K = \dot{-}(\perp)$. In that case there is no need for a separate registering of the belief set, and one may say that the doxastic state just *is* a belief revision strategy, formally captured by a *unary* function taking an input sentence and returning a new belief set. This idea is pursued, for instance, in Rott (1991b, Section 2) and Nayak (1994b).

From the point of view taken here, however, the problem with this argument is that unary belief change operations are never just 'given', but are rather obtained as a result of filling in the current belief set (the propositional component of the doxastic state) into the first argument place of the current two-place belief revision function (the dispositional, conative component). And anyway, we recall from Section 1.1 that it is implausible to suppose that the set of an agent's beliefs is solely determined by the internal structure of the cognitive state (represented, e.g., by some doxastic preference ordering). What can count as a belief is also dependent on pragmatic factors such as institutional context, rigidity of standards, linguistic conventions, etc. The set of beliefs may thus be considered vague or context-dependent. This is another reason for denying the suggestion that we know the set of an agent's beliefs if we know his belief change strategy.

Let us stop with these more philosophical reflections here, even though no definitive conclusion has been reached.[161] We return to the topic in Sections 5.2.3 and 7.10 and see what kinds of pictures emerge from the foundationalist and coherentist perspectives taken in Chapters 5 and 7 respectively.

3.7 Justifications in models of belief change

Another question that has to be answered when representing a state of belief is whether *justifications* on which the beliefs are based should be part of the representation or not.[162] With respect to this question there are two main approaches. The first approach holds that one should keep track of the justifications for one's beliefs, because propositions that lose all of their justification should no longer be accepted as beliefs. The other approach holds that one need not consider the

---

[160] Incidentally, in the standard theory of theory change, a revision by $\top$ and a contraction by $\perp$ lead to identical results. This follows from the Levi identity and properties of standard consequence operations: $K * \top = Cn((K \dot{-} (\neg\top)) \cup \{\top\}) = K \dot{-} \perp$.

[161] Such a conclusion is offered in Rott (1999, 2000c).

[162] The term 'justification' is used here in a pre-theoretical sense and not to be identified with any of Lehrer's concepts of justification that we came across in Chapter 2.

pedigree of one's beliefs. The focus should only be on the inferential structure of the beliefs—what matters is how a belief coheres with the other beliefs that are accepted in the present state.

The following general principle formulated by Harman (1986, p. 39) is an important criterion.

Principle of Positive Undermining: One should stop believing ϕ whenever one positively believes one's reasons for believing ϕ are no good.

A coherentist reading is obtained if the latter part of the principle is interpreted as 'having positive reasons for not believing ϕ'. Interpreted in this way, the principle expresses a certain kind of *inertia* of a reasoner's beliefs. One does not cease to accept a proposition just for the fact that the justifications are lost. An extra impulse is needed that actively undermines the belief.

On the other hand, according to the justifications theory, belief revision should consist, first, in giving up all beliefs that do no longer have a *satisfactory justification* and, second, in adding new beliefs that have become justified. Here a different principle of undermining is operative:

Principle of Negative Undermining: One should stop believing ϕ whenever one does not associate one's belief in ϕ with an adequate justification (either intrinsic or extrinsic). (Harman 1986, p. 39)

A consequence of this principle is that if a state of belief containing ϕ is revised so that the negation of ϕ becomes accepted, not only should ϕ be given up in the revised state, but also all beliefs that depend on ϕ for their justification.[163] If one believes that ψ, and all the justifications for believing ψ are given up, continued belief in ψ is no longer justified, so it should be rejected.

Justifications come in two kinds. The standard kind of justification is that a belief is *justified by one or several other beliefs*, but not justified by itself. However, since all beliefs should be justified and since infinite regresses are disallowed, some beliefs must be *justified by themselves*.[164] Another requirement on the set of justifications is that it be *grounded* or *well-founded*. In particular, justifications are supposed to be *non-circular* so that we do not find a situation where a belief in ϕ justifies ψ, a belief in ψ justifies χ, while χ in turn justifies ϕ. We have seen in Section 2.4 that Sosa (1980, pp. 23–24) uses a pyramid metaphor to describe the foundationalist view of the justificational structure of a doxastic state.

For a more detailed account of the distinction between justifications and coherence models, see Harman (1986).

If changes of belief have to respect justificatory dependencies between items of belief, then they become non-trivial. Aiming at the preservation of the structure of justifications introduces a similar tension with the maxim of minimal

---

[163] This is the 'Simple Filtering' principle (Rott 2000a), a descendant of Fuhrmann's (1991) 'Filtering' principle.

[164] This is the famous regress argument. See Sosa (1980, Section 5) and BonJour (1985, Section 2.1).

change as the maxim of inferential coherence or Quine's maxim of simplicity. The models I am going to consider in this book are not justification-based. No track-keeping of reasons of beliefs is provided. For some remarks on the tenuous links of justification-based models with the models investigated here, see Gärdenfors and Rott (1995, Sections 2.4 and 7).

4

CONCEPTS OF THEORETICAL RATIONALITY: POSTULATES FOR BELIEF CHANGE AND NONMONOTONIC REASONING

In the previous chapters we have given some introductory elucidations of the concepts of revision and contraction in belief change, as well as of the notion of nonmonotonic or defeasible inference. We now turn to a more detailed and systematic characterization of theoretical rationality by means of abstract postulates. Later, in Chapter 6, we shall discuss equally abstract postulates of rational choice. In so far as choices belong to the field of practical reasoning, that chapter is concerned with practical philosophy. The postulates we are going to consider in this chapter, however, clearly belong to the field of theoretical reasoning, and thus to theoretical philosophy. We shall see that within the latter field three topics are very closely related: two kinds of changes of opinions, namely, contractions and revisions, and nonmonotonic reasoning (nonmonotonic inference relations or logics).

4.1 Black boxes, structures and rules of application

There are two kinds of approaches to the study of processes of any kind. The first approach may be called a *black box approach*. In this the processes are modelled by means of input-output dependencies. These dependencies need not be functional, it is enough to specify equations or conditions that describe the manifest behaviour of an individual or a system. It is not that the output can be determined by the input with the help of these equations or conditions, a relational dependency is enough. It is also worth pointing out that many things in the black box approach are not actually observable. We shall see that the relevant conditions frequently describe *potential* behaviour rather than *actual* behaviour. Once a system has acted or changed, it may well be that it never returns to its initial state, so very little can be said about actions or changes that actually took place. However, there is ample room for a systematic account of the dispositions to change in response to potential inputs.

If the processes in question are at least partially under the voluntary control of an agent, then the equations or conditions of the black-box-approach are not natural laws, but an expression of normative requirements of rationality. We shall talk of *rationality postulates* for belief change and nonmonotonic reasoning in this section. For ease of terminological distinction, the term *coherence constraints* will be used in Chapter 6 for conditions of essentially the same kind (but applying to practical rather than theoretical reasoning).

What is not mentioned in the black box approach are *internal states* mediating and accounting for the observable behaviour. The second kind of approach that we shall distinguish does specify internal states. Instead of input-output relations, there are transitions from internal states in response to some input, resulting in new internal states and (possibly) outputs. This model accords well with that variant of the functionalist picture in the philosophy of mind according to which human cognition can be thought of as representable by the machine table of a Turing machine. Such a table encodes effectively functional transitions from pairs consisting of an internal state and an input to a posterior internal state and an output.[165]

Usually the internal states and their effects can be represented by some kind of structure and some rule as to how the structure gets applied. For that reason we call the second approach the *structure-plus-rule-of-application approach.* The structure is normally not directly accessible. It determines or governs behaviour through the rule of application, and it may be revealed by the behaviour it determines. The black box approach tells us *what* may or will happen, the structure cum rule of application approach tells us *why* it happens. Usually we can think of a one-many relation between processes and internal structures. Since internal states help to disambiguate what is indiscernible in the black box approach, it is much more plausible in the second approach to conceive of the transitions as functional in the sense that the posterior state is uniquely determined by the prior state and the input. Conversely, however, one and the same manifest transition behaviour may be accounted for by many different kinds of structures and rules of application. In this sense, behaviour constrains, but underdetermines its underlying structure, and different structures (plus rules of application) that effect the same behaviour may rightly be called *equivalent*.

In the terms of a distinction made in Section 1.2.8, the black box approach may be called a *kinematic theory* of belief change, while the structure-plus-rule-of-application approach is a genuinely *dynamic theory* of belief change.

The distinction between the two kinds of approaches has become folklore in the belief revision literature for quite some time; see in particular Makinson (1985). For a closely related distinction between an *external description* (in input-output terms, by means of a choice function) and an *internal description* (by means of mechanisms or procedures) in the theory of choice, see Aizerman and Malishevski (1981, p. 1031) and Aizerman (1985, pp. 235–239).

As mentioned before, we are going to present rationality postulates for belief change and nonmonotonic reasoning in this chapter, postulates that flesh out in detail a certain type of black box approach. Corresponding structure-plus-rule-of-application approaches in a foundationalist and coherentist framework will be presented in Chapters 5 and 7, respectively.

---

[165]See the classical paper by Putnam (1960, Section 1) where the output consists in a compound write-and-read action.

Constructive approaches of the latter kind are clearly very important. Without them, the setting up of rationality postulates would perhaps have 'the advantages of theft over honest toil'.[166]

4.2 Postulates for rational contractions

Some of the following postulates have gained considerable prominence in the literature on belief change as studied in the research program of Alchourrón, Gärdenfors and Makinson (marked by acronym 'AGM'—see Section 3.4.2). We begin with the problem of belief contraction. There are six *basic AGM postulates* are commonly denoted by $(\dot{-}1) - (\dot{-}6)$, and two *supplementary* ones, denoted by $(\dot{-}7)$ and $(\dot{-}8)$. These postulates are formulated below; for their motivation, we refer the reader to the standard reference paper Alchourrón, Gärdenfors and Makinson (1985) and to Gärdenfors (1988, Chapter 3). The remaining postulates, recognizable by labels with a letter appended to the number, have either been introduced by other authors in belief revision or are new. Our motivation to discuss the novel postulates here is that they will surface in later developments of more constructive models for belief change. Nevertheless it should be borne in mind that they have an independent standing in the literature.

We base our considerations on a slightly different grouping of the AGM postulates and relegate the fourth postulate, $(\dot{-}4)$, to the group of supplementary postulates. The set of basic postulates for contractions then is small.

$(\dot{-}1)$	$K \dot{-} \phi = Cn(K \dot{-} \phi)$ whenever $K = Cn(K)$	*(Closure)*
$(\dot{-}2)$	$K \dot{-} \phi \subseteq K$	*(Inclusion)*
$(\dot{-}3)$	If $\phi \notin K$, then $K \subseteq K \dot{-} \phi$	*(Vacuity)*
$(\dot{-}5)$	$K \subseteq Cn((K \dot{-} \phi) \cup \{\phi\})$	*(Recovery)*
$(\dot{-}6)$	If $Cn(\phi) = Cn(\psi)$, then $K \dot{-} \phi = K \dot{-} \psi$	*(Extensionality)*

Notation: While most postulates make sense without this restriction, the reader may interpret the letter 'K' as generally standing for a theory, i.e., $K = Cn(K)$. The above postulates, and all postulates that follow, are understood as applying to any arbitrary theory K and as quantified over all sentences ϕ and ψ. The conditions $(\dot{-}2)$ and $(\dot{-}3)$ or (Pres) establish a connection between the statics and the dynamics of belief. In a certain sense, they express the idea that $\dot{-}$ is *faithful* to the belief set K. If K is consistent, its content can be read off from the contraction behaviour, since then $K = K \dot{-} \bot$ by $(\dot{-}2)$ and $(\dot{-}3)$. Postulate $(\dot{-}3)$ is given in a slightly weaker form than usual (Gärdenfors 1988, p. 61). The more common consequent is $K \dot{-} \phi = K$; it is easily recovered with the help

---

[166]This is Russell's (1919, p. 71) famous bon mot about the method of 'postulating' (it is mentioned in Makinson 1985, p. 350). Regarding similar problems in the theory of choice, compare Sen (1982, p. 190): 'Arrow was undoubtedly right in saying that "one of the great advantages of abstract postulational methods is the fact that the same system may be given several different interpretations, permitting a considerable saving of time" (Arrow 1951/63, p. 87). But *that* probably is *also* one of the great disadvantages of these methods.'

of ($\dot{-}$2). Slight weakenings and strengthenings of ($\dot{-}$1), ($\dot{-}$4) and ($\dot{-}$5) will be discussed below (on page 111).

It should be noted that the most controversial AGM axiom, namely the *Recovery condition* ($\dot{-}$5) is included in this first and most basic set. The literature is abundant with arguments for and against Recovery; see in particular, Makinson (1987) and Makinson (1997), Levi (1991) and Pagnucco and Rott (1999).

The remaining conditions are called *supplementary postulates*.

($\dot{-}$4)   If $\phi \in K \dot{-} \phi$, then $\phi \in Cn(\emptyset)$ *(Success)*

This postulate states that the agent may refuse to withdraw a belief only if it is a logical truth; we will comment on the postulate presently. The remaining postulates deal with the contraction of conjunctions. We shall give a very detailed account of such contractions in Chapters 7 and 8. For a first understanding, one should note that if the beliefs of a person form a theory, then the presence of a conjunction $\phi \wedge \psi$ is equivalent to the presence of its conjuncts ϕ and ψ. Moreover, in the idealized context of theories, there is no clear intuitive difference between accepting $\phi \wedge \psi$ and simultaneously accepting both ϕ and ψ. By the same token, it is hard to see any difference between disbelieving $\phi \wedge \psi$ and disbelieving either ϕ or ψ, or between withdrawing $\phi \wedge \psi$ and withdrawing either ϕ or ψ.[167]

The next group of postulates, all variations on the AGM postulate ($\dot{-}$7), specifies conditions under which the contraction K with respect to a sentence ϕ is a subset of a contraction with respect the conjunction of ϕ with another sentence ψ.

($\dot{-}$7c)   If $\psi \in K \dot{-} (\phi \wedge \psi)$, then $K \dot{-} \phi \subseteq K \dot{-} (\phi \wedge \psi)$

($\dot{-}$7) $K \dot{-} \phi \cap K \dot{-} \psi \subseteq K \dot{-} (\phi \wedge \psi)$

($\dot{-}$7P) $(K \dot{-} \phi) \cap Cn(\phi) \subseteq K \dot{-} (\phi \wedge \psi)$

($\dot{-}$7p)   If $\phi \in K \dot{-} (\phi \wedge \psi)$, then $\phi \in K \dot{-} (\phi \wedge \psi \wedge \chi)$

($\dot{-}$7g) $K \dot{-} (\phi \vee \psi) \subseteq Cn((K \dot{-} \phi) \cup \{\neg\psi\})$

($\dot{-}$7a)   If $\phi \notin K \dot{-} \phi$, then $K \dot{-} \phi \subseteq K \dot{-} (\phi \wedge \psi)$

The letters appended to the usual AGM numbers of the postulates will become clear in Section 4.4. Postulate ($\dot{-}$7c) was found to be relevant in Rott (1992b), where it has the same name ('c' for 'cut' or 'cumulative'); given ($\dot{-}$1) and ($\dot{-}$5), it is a weakening of ($\dot{-}$7). In the context of the basic postulates including ($\dot{-}$4) and ($\dot{-}$5), both ($\dot{-}$7P) and ($\dot{-}$7p) are equivalent to ($\dot{-}$7) (see the *Partial Antitony* property in Alchourrón, Gärdenfors and Makinson 1985, Observation 3.3, and Rott 1992b, Lemma 2). Postulate ($\dot{-}$7g) is likewise equivalent to ($\dot{-}$7) (see Gärdenfors 1982, p. 98, 'g' for 'Gärdenfors'). Finally, ($\dot{-}$7a) is an extraordinarily strong condition of *Antitony* which will play a crucial role only in Section 8.7 below. It is argued in Pagnucco and Rott (1999) that ($\dot{-}$7a) is *not* unduly strong.

---

[167]These occurrences of 'either–or' are non-exclusive. In the absence of any good reason to prefer one over the other, one should arguably disbelieve or withdraw both ϕ and ψ. Compare the discussion of a 'Principle of Indifference' in Pagnucco and Rott (1999).

The last group of postulates consists of variations on the AGM postulate ($\dot{-}8$). They specify conditions under which the contraction K with respect to a conjunction $\phi \wedge \psi$ is a subset of the contraction with respect a conjunct ϕ.

($\dot{-}$8c)   If $\psi \in K \dot{-} (\phi \wedge \psi)$, then $K \dot{-} (\phi \wedge \psi) \subseteq K \dot{-} \phi$

($\dot{-}$8sd) $K \dot{-} (\phi \wedge \psi) \subseteq K \dot{-} \phi$ or $K \dot{-} (\phi \wedge \psi) \subseteq K \dot{-} \psi$

($\dot{-}$8d) $K \dot{-} (\phi \wedge \psi) \subseteq K \dot{-} \phi \cup K \dot{-} \psi$

($\dot{-}$8wd) $K \dot{-} (\phi \wedge \psi) \subseteq Cn(K \dot{-} \phi \cup \{\neg \psi\}) \cup Cn(K \dot{-} \psi \cup \{\neg \phi\})$

($\dot{-}$8vwd) $K \dot{-} (\phi \wedge \psi) \subseteq Cn(K \dot{-} \phi \cup K \dot{-} \psi)$

($\dot{-}$8n) $K \dot{-} \phi \subseteq K \dot{-} (\phi \vee \psi) \cup K \dot{-} (\phi \vee \neg \psi)$

($\dot{-}$8)   If $\phi \notin K \dot{-} (\phi \wedge \psi)$, then $K \dot{-} (\phi \wedge \psi) \subseteq K \dot{-} \phi$

($\dot{-}$8g)   If $\phi \vee \psi \notin K \dot{-} \phi$, then $(K \dot{-} \phi) \subseteq Cn((K \dot{-} (\phi \vee \psi) \cup \{\neg \psi\})$

($\dot{-}$8m) $K \dot{-} (\phi \wedge \psi) \subseteq K \dot{-} \phi$

($\dot{-}$8wm)   If $\phi \wedge \psi \in K$, then $K \dot{-} (\phi \wedge \psi) \subseteq K \dot{-} \phi$

Postulate ($\dot{-}$8c) was found to be relevant in Rott (1992b), where it has the same name ('c' for 'cautious' or 'cumulative'); given ($\dot{-}$4) and ($\dot{-}$5), it is a weakening of ($\dot{-}$8).[168]

The 'd' in the next group of postulates stands for 'disjunctive' and derives from corresponding conditions to be discussed later in this chapter. We have a strong ('sd'), a plain ('d'), a weak ('wd') and a very weak ('vwd') variant. Condition ($\dot{-}$8sd) was called the *Covering condition* in Alchourrón, Gärdenfors and Makinson (1985, p. 317).[169] It clearly implies ($\dot{-}$8d). But given ($\dot{-}$1), the seemingly weaker condition ($\dot{-}$8d) also implies ($\dot{-}$8sd).[170] It is trivial that ($\dot{-}$8d) implies ($\dot{-}$8wd). And in the context of ($\dot{-}$2) and the Recovery postulate ($\dot{-}$5), ($\dot{-}$8wd) implies ($\dot{-}$8vwd). The latter postulate was called ($\dot{-}$8r) ('r' for 'relational') in Rott (1993); given ($\dot{-}$2), ($\dot{-}$4) and ($\dot{-}$5), ($\dot{-}$8vwd) is a weakening of ($\dot{-}$8).[171]

It is easy to see that given ($\dot{-}$4) and ($\dot{-}$6), ($\dot{-}$8sd) implies ($\dot{-}$8c). However, there is no logical relationship between ($\dot{-}$8c) and ($\dot{-}$8vwd), not even in the finite case and in the presence of ($\dot{-}$1) – ($\dot{-}$7) (for examples satisfying the one of them but not the other see Rott 1993, p. 1438).

I want to emphasize the strong intuitive appeal of the postulates ($\dot{-}$7c) and ($\dot{-}$8c). They are much weaker than the original AGM postulates ($\dot{-}$7) and ($\dot{-}$8), and they are very plausible. This can be seen if we remember that giving up

[168] A sharper observation is this: Given ($\dot{-}\emptyset$1) and ($\dot{-}5^0$) introduced below and ($\dot{-}$2), ($\dot{-}$8c) is a weakening of ($\dot{-}$8).

[169] The Covering condition corresponds to a condition for the choice functions on maximal non-implying subsets stating that $\gamma(K \perp \phi) \subseteq \gamma(K \perp (\phi \wedge \psi))$ or $\gamma(K \perp \psi) \subseteq \gamma(K \perp (\phi \wedge \psi))$ (mentioned in Alchourrón, Gärdenfors and Makinson 1985, p. 527), or more generally, $\gamma(S) \subseteq \gamma(S \cup S')$ or $\gamma(S') \subseteq \gamma(S \cup S')$.

[170] Thanks to Sven Ove Hansson for pointing this out to me.

[171] Again, a sharper observation is this: Given ($\dot{-}\emptyset$1), ($\dot{-}5^0$) and ($\dot{-}$2), ($\dot{-}$8vwd) is a weakening of ($\dot{-}$8).

$\phi \wedge \psi$ means giving up at least one of ϕ and ψ. Taken together, ($\dot{-}$7c) and ($\dot{-}$8c) say that if ψ is still included in the contraction with respect to $\phi \wedge \psi$ then the agent just withdrew ϕ in his effort to remove either ϕ or ψ.

The 'n' of ($\dot{-}$8n) stands for 'negation', and the presence of this postulate is only explained by its analogue in a later part of this chapter.

Postulate ($\dot{-}$8) is the original AGM postulate. In the context of the basic postulates, ($\dot{-}$8g) is equivalent to ($\dot{-}$8) (see Gärdenfors 1982, p. 98, 'g' for 'Gärdenfors').

Given ($\dot{-}$4), ($\dot{-}$8) implies ($\dot{-}$8sd) (Alchourrón, Gärdenfors and Makinson 1985, Observation 3.4) and hence ($\dot{-}$8d).[172]

Postulate ($\dot{-}$8m) is an extremely, and indeed and unduly, strong condition. One can give intuitive arguments against the acceptability of ($\dot{-}$8m), adapting a similar argument given by Gärdenfors (1988, p. 60). The following lemma shows the inadequacies of ($\dot{-}$8m) on more formal grounds.

Lemma 1 *(i) A contraction function $\dot{-}$ over a consistent theory K which satisfies ($\dot{-}$8m) in combination with ($\dot{-}$1), ($\dot{-}$2), ($\dot{-}$3) and ($\dot{-}$5) is trivial in the sense that it puts $K \dot{-} \phi = K$ for all sentences ϕ;*

(ii) a contraction function $\dot{-}$ over K satisfies postulates ($\dot{-}$7) and ($\dot{-}$8m) if and only if there is a theory H such that for every ϕ, $K \dot{-} \phi = H$.

(iii) a contraction function $\dot{-}$ over K satisfies postulates ($\dot{-}$2), ($\dot{-}$3), ($\dot{-}$7) and ($\dot{-}$8wm) if and only if there is a subtheory H of K such that for every ϕ,

$$K \dot{-} \phi = \begin{cases} H \text{ if } \phi \in K \\ K \text{ otherwise} \end{cases}$$

Condition ($\dot{-}$8wm) avoids the most blatant inadequacies of ($\dot{-}$8m), but part (iii) of the lemma shows that it also makes the contraction operation largely uninteresting: The agent has only one fallback position from his belief set.[173] ($\dot{-}$8wm) is called *Monotony condition* in Alchourrón, Gärdenfors and Makinson (1985, p. 526).[174]

We close this section with some considerations about situations in which the agent refuses to contract his belief set. It follows from the basic AGM postulates ($\dot{-}$1) and ($\dot{-}$5) that if ϕ is in $Cn(\emptyset)$, then ϕ is in $K \dot{-} \phi$, and in fact $K \dot{-} \phi = K$. On the other hand, AGM endorse a postulate that states that the converse should be true as well. We have already mentioned postulate

($\dot{-}$4)   If $\phi \in K \dot{-} \phi$, then $\phi \in Cn(\emptyset)$

---

[172]It is proved in Alchourrón, Gärdenfors and Makinson (1985, Observation 6.5) that the conjunction of ($\dot{-}$7) and ($\dot{-}$8) is equivalent to the strong *Ventilation condition*

$$K \dot{-}(\phi \wedge \psi) = K \dot{-} \phi \text{ or } K \dot{-}(\phi \wedge \psi) = K \dot{-} \psi \text{ or } K \dot{-}(\phi \wedge \psi) = K \dot{-} \phi \cap K \dot{-} \psi.$$

[173]The felicitous term 'fallback' is due to Lindström and Rabinowicz (1991).

[174]Given the other AGM postulates, ($\dot{-}$8wm) is shown to be relatively equivalent to the condition $K \dot{-} \phi \wedge \psi = K \dot{-} \phi \cap K \dot{-} \psi$, and to characterize so-called full meet contractions.

It seems that it is a scarcely motivated loss of generality to require that rational agents should invariably stick to one of his or her beliefs only if it is a logical truth. There may well be other, extralogical reasons that lead an agent to think that a belief is irrevocable. Any kind of statement ϕ which is considered to be a priori, or analytically or necessarily true by the agent might qualify for such a sacrosanct status. In such cases, worlds in which ϕ is false are taken to be not even remotely possible, and giving up ϕ is taken to be doxastically unacceptable. Let us call a sentence ϕ *taboo* with respect to a given contraction function $\dot{-}$ if and only if ϕ is in any potential contraction $K \dot{-} \psi$ of the agent's beliefs. More formally, the *taboo set of a contraction function* $\dot{-}$ is the set $K_{\overline{R}} = \{\phi : \phi \in K \dot{-} \psi$ for every $\psi\}$. By postulate $(\dot{-}1)$, it follows immediately from the definition of $K_{\overline{R}}$ that the taboo set is closed under Cn. We now consider some alternative ways of characterizing the taboo set.

Lemma 2 *Let the contraction function* $\dot{-}$ *satisfy the basic postulates* $(\dot{-}1)$, $(\dot{-}2)$, $(\dot{-}5)$ *and* $(\dot{-}6)$. *Then each of the following conditions implies its successor. (i) For all* ψ, $\phi \in K \dot{-} \psi$ *(ii)* $\phi \in K$ *and* $K \dot{-} \phi = K$ *(iii)* $\phi \in K \dot{-} \phi$ *(iv) There is a* ψ *such that* $\phi \wedge \psi \in K \dot{-}(\phi \wedge \psi)$ *If* $\dot{-}$ *satisfies in addition* $(\dot{-}7)$ *and* $(\dot{-}8c)$, *then (iv) implies (i). Hence, in that case all four conditions are equivalent.*

There is one part mentioned in Lemma 2 that plays a role as the antecedent of a condition that is worth displaying as a postulate in its own right.

$(\dot{-}5^0)$ If $\phi \in K \dot{-} \phi$, then $K \subseteq K \dot{-} \phi$.

This is a very plausible condition. It says that if an agent refuses to withdraw a belief when instructed to withdraw it, he should not withdraw any other beliefs instead. This is an immediate consequence of the Recovery postulate $(\dot{-}5)$, but it is much weaker and much more impeccable than the latter. It is surely acceptable in most contexts in which Recovery is doubtful.[175] If we accept postulates $(\dot{-}1)$ and $(\dot{-}4)$, then $\phi \in K \dot{-} \phi$ is equivalent with $\phi \in Cn(\emptyset)$, so $(\dot{-}5^0)$ essentially adds that $K \subseteq K \dot{-} \top$. Alternatively, if we accept postulates $(\dot{-}6)$ and $(\dot{-}7c)$, then we get that $\phi \in K \dot{-} \phi$ implies $K \dot{-} \top \subseteq K \dot{-} \phi$, so what $(\dot{-}5^0)$ requires is again only $K \subseteq K \dot{-} \top$.

Let us finally look at two weaker versions of $(\dot{-}7P)$ and $(\dot{-}8c)$ that are also related to refusals to withdraw certain beliefs:

$(\dot{-}7r)$ If $\phi \in K \dot{-} \phi$, then $\phi \in K \dot{-} \phi \wedge \psi$

$(\dot{-}8rc)$ If $\phi \wedge \psi \in K \dot{-} \phi \wedge \psi$, then $\phi \in K \dot{-} \phi$

Postulate $(\dot{-}7p)$ directly implies $(\dot{-}7r)$, and so does $(\dot{-}7P)$ in the context of $(\dot{-}6)$. Postulate $(\dot{-}8c)$ implies $(\dot{-}8rc)$, provided that the basic condition $(\dot{-}1)$ is satisfied.[176] The 'r' in the labels $(\dot{-}7r)$ and $(\dot{-}8rc)$ stands for 'refuse'. In later sections (6.1.2 and 8.2.5) we will explain in what sense postulates $(\dot{-}7r)$ and

---

[175]This is why postulate $(\dot{-}5^0)$ will be needed in Section 8.7.

[176]Postulates $(\dot{-}7r)$ and $(\dot{-}8rc)$ also follow from $(\dot{-}1)$ and $(\dot{-}4)$, but we want to avoid a commitment to $(\dot{-}4)$ whenever possible. In the absence of $(\dot{-}4)$, $(\dot{-}7)$ does not imply $(\dot{-}7r)$.

$(\dot{-}8rc)$ correspond to conditions for refusals to make a choice in certain types of situations.[177]

4.3   Postulates for rational revisions

It is not quite clear how frequently pure contractions occur in real life. Contractions are usually not made for their own sake. In most cases, they are prompted by some piece of information telling us that such-and-such is the case, and we had thought that this was not the case. So probably, most contractions are made as part of a construction of a belief revision where the eliminated item is replaced by its negation.

There is another problem with belief contractions. Many writers with foundationalist intuitions[178] have argued that it is not reasonable to assume that the Recovery principle hold or should hold. There are too many situations, so these writers, in which an attempt to 'undo' a contraction by a subsequent addition of the sentence just withdrawn will not fully recover the old belief set. Other researchers of a more coherentist creed, on the other hand, are not convinced by the arguments against Recovery. They point out that Recovery is satisfied by a number of plausible constructions, and anyway they want to build upon Recovery because it is useful for the derivation of interesting theorems. The discussion can be avoided if one focuses on revisions in the first place. For if revisions are based on contractions, the question whether these contractions satisfy Recovery is without any consequence. The problem situation about this most controversial postulate for contractions is lucidly summarized by David Makinson (1987, 1997).

Revisions can of course be investigated as objects in their own interest, independently of any study of contractions. In Chapters 5 and 7 we shall introduce and discuss five direct ways of constructing revisions.

Alternatively, one can adopt a two-pronged simultaneous study of contractions and revisions, with close attention to their interrelation.[179] Two principal ways of relating revisions to contractions have been suggested. First, a revision by ϕ, with the aim of accepting ϕ while maintaining consistency, may be construed as a contraction with respect to $\neg\phi$ followed by a consistent addition of ϕ. Second, and conversely, a contraction is something like a 'disjunctive belief state': Withdrawing a belief ϕ means keeping only those of the original beliefs that one would continue holding even if one came to know that ϕ is in fact false. We consider the two proposals in turn.

Revisions are obtained from contractions by means of the so-called *Levi identity*:[180]

---

[177] Alternative names for $(\dot{-}8rc)$ and $(\dot{-}7r)$ will be $(\dot{-}\emptyset1)$ and $(\dot{-}\emptyset2)$.

[178] Including in particular A. Fuhrmann, I. Levi, S.O. Hansson, S. Lindström and W. Rabinowicz

[179] The first systematic treatment is Gärdenfors (1982).

[180] It is not easy to pin down the exact origin of this name. Levi (1977, Section 2) argues that the only legitimate forms of changes of a corpus of knowledge are 'contraction' and 'expansion'

$$K * \phi = Cn((K \dot{-} \neg\phi) \cup \{\phi\})$$

Gärdenfors (1988, pp. 63, 71, 213) showed that the Levi identity, together with the basic axioms of theory contraction, allows to derive a kind of converse to the Levi identity, which has been called the *Harper identity* or the *Gärdenfors identity*:

$$K \dot{-} \phi = K \cap (K * \neg\phi)$$

The Harper identity, though perhaps intuitively slightly less appealing than the Levi identity, can thus be derived from the latter with the help of the basic conditions for contractions—with the controversial condition of Recovery being an essential ingredient of the proof. Conversely, Recovery is an immediate consequence if contractions are derived from revisions by means of the Harper identity. (*Proof.* By ($\dot{-}1$), if ψ is in K then $\phi \supset \psi$ is both in K and in $K * \neg\phi$, since $\neg\phi$ is in $K * \neg\phi$, by (*2).) Anyway, the Harper identity has received much less support (and criticims) in the literature than the Levi identity.

Most researchers in belief change have followed Levi's thesis that contraction is the more basic change operations in belief change. In Chapter 5, however, we shall argue that in the context of foundational belief change it are *revisions* that should be given primacy over contractions. This is because the most natural way of constructing a successful contraction with respect to ϕ in the most attractive foundationalist mode of belief change proceeds by a revision by what I call a 'phantom belief' $\neg\phi^*$. Though this idea bears some similarity to the idea of the Harper identity, one should not confuse a specific construction recipe with a general postulate; in fact the construction in the framework of that chapter violates either direction of the Harper identity.

Together with the basic properties of contractions, the Levi identity immediately gives us:

(*1) $K * \phi = Cn(K * \phi)$ whenever $K = Cn(K)$ *(Closure)*

(*2) $\phi \in K * \phi$ *(Success)*

(*3) $K * \phi \subseteq Cn(K \cup \{\phi\})$ *(Expansion 1)*

(*4) If $\neg\phi \notin K$, then $Cn(K \cup \{\phi\}) \subseteq K * \phi$ *(Expansion 2)*

(*6) If $Cn(\phi) = Cn(\psi)$, then $K * \phi = K * \psi$ *(Extensionality)*

Some comments are in order. First note that revisions generated by the Levi identity are logically closed, even if the original set of beliefs is not. A variant of the Levi identity for belief bases which are not assumed to be logically closed

(these terms being used here as later in the AGM tradition). Levi (1983, Section 9)—observe that this article was finished in 1975—states that 'replacement [AGM's revisions, H.R.] can be regarded, for purposes of analysis, as a contraction followed by an expansion'. Anyway, Levi puts forth a substantial philosophical thesis, not just a formal artifice. On the intended meaning of Levi's identity and its relation to his 'Commensuration Requirement', see also Levi (1991, Sections 2.9 and 4.6, including footnotes). — For an interesting reversal of Levi's idea, see Hansson (1993a).

would be to take just the set $(K \dot{-} \neg\phi) \cup \{\phi\}$ rather than its closure.[181] Postulates (∗1) and (∗2) follow from the Levi identity alone, while the other conditions can be obtained only with the help of the basic conditions for contractions. We have taken over the standard AGM numbering of these postulates, although it somehow obscures the correspondences. It is easy to see that (∗3) and (∗4) basically correspond to ($\dot{-}$2) and ($\dot{-}$3), respectively. It would mean no difference to the logical force of (∗3) if it were qualified by the same antecedent condition as (∗4). Instead of (∗4), the weaker *Preservation Principle*

(Pres) If $\neg\phi \notin K$, then $K \subseteq K * \phi$ *(Preservation)*

would be even better suited for the matching with ($\dot{-}$3). In addition, (Pres) is a somewhat purer condition than (∗4) since the additional strength of the latter is already covered by conditions (∗1) and (∗2). The conditions (∗3) and (∗4) or (Pres) establish a connection between the statics and the dynamics of belief. If K is consistent, its content can be read off from the revision behaviour, since then $K = K * \top$ by (∗3) and (∗4).

It is also clear that (∗6) is the counterpart of ($\dot{-}$6). Although (∗1) does not need ($\dot{-}$1) for its derivation, the reverse derivation of ($\dot{-}$1) via the Harper identity requires (∗1), so these two postulates can also be regarded as a pair. For reasons that will become clear in Section 4.4, we split up (∗1) into two conditions. For the sake of simplicity, we assume that K always denotes a theory from now on.

(∗1a) If $\psi \in K * \phi$ and $\chi \in Cn(\psi)$, then $\chi \in K * \phi$

(∗1b) If $\psi \in K * \phi$ and $\chi \in K * \phi$, then $\psi \wedge \chi \in K * \phi$

Similar to the case of ($\dot{-}$1) and (∗1), the Recovery condition ($\dot{-}$5) and the *Success postulate for revisions*, (∗2), can be paired by means of the Harper (rather than the Levi) identity. Notice that unlike (∗2), the *Success condition for contractions* ($\dot{-}$4), is not included in the basic set for contractions, but has been treated as an optional extra in the preceding section. Its pendant for revisions, the condition of *Consistency Preservation*[182] has the AGM label (∗5) and will likewise be considered optional.

(∗5) If $Cn(\phi) \neq L$, then $K * \phi \neq L$ *(Consistency Preservation)*

Apart from this special case of number shuffling, the following supplementary postulates for revisions correspond, via the Levi identity, to the similarly labelled properties of contractions.

The remaining postulates specify various relations between revisions by stronger and revisions by weaker sentences. In the postulates of the 7 series the contractions of stronger sentences appear on the left-hand side of the subset relation.

(∗7c) If $\psi \in K * \phi$, then $K * (\phi \wedge \psi) \subseteq K * \phi$

---

[181] The Harper identity can be applied to belief bases without adaptation.

[182] A more appropriate name would be just *Consistency*, since $K * \phi$ is required to be consistent even when K is not. We stick to the more well-established usage.

(∗7) $K * (\phi \wedge \psi) \subseteq Cn((K * \phi) \cup \{\psi\})$

(∗7g) $K * \phi \cap K * \psi \subseteq K * (\phi \vee \psi)$

(∗7) is the standard AGM axiom. Postulate (∗7g) ('g' for 'Gärdenfors') is first mentioned by Gärdenfors in (1979, p. 393), and later proved to be equivalent with (∗7) in Gärdenfors (1988, pp. 57, 211–212).

In the postulates of the 8 series the contractions of weaker sentences appear on the left-hand side of the subset relation.

(∗8c)   If $\psi \in K * \phi$, then $K * \phi \subseteq K * (\phi \wedge \psi)$

(∗8sd) $K * (\phi \vee \psi) \subseteq K * \phi$ or $K * (\phi \vee \psi) \subseteq K * \psi$

(∗8d) $K * (\phi \vee \psi) \subseteq K * \phi \cup K * \psi$

(∗8wd) $K * (\phi \vee \psi) \subseteq Cn(K * \phi \cup \{\psi\}) \cup Cn(K * \psi \cup \{\phi\})$

(∗8vwd) $K * (\phi \vee \psi) \subseteq Cn(K * \phi \cup K * \psi)$

(∗8n) $K * \phi \subseteq K * (\phi \wedge \psi) \cup K * (\phi \wedge \neg\psi)$

(∗8)   If $\neg\psi \notin K * \phi$, then $K * \phi \subseteq K * (\phi \wedge \psi)$

(∗8m) $K * \phi \subseteq K * (\phi \wedge \psi)$

(∗8wm)   If $\neg\phi \in K$, then $K * \phi \subseteq K * (\phi \wedge \psi)$

Like ($\dot{-}$7c) and ($\dot{-}$8c), their cousins (∗7c) and (∗8c) are eminently plausible. If we find that the sentence ψ is accepted after a revision by ϕ, then we may identify $K * \phi$ with $K * (\phi \wedge \psi)$. Taken together, (∗7c) and (∗8c) are equivalent with the *Reciprocity condition*:[183]

(Rec)   If $\phi \in K * \psi$ and $\psi \in K * \phi$, then $K * \phi = K * \psi$

Hansson (1999a, p. 216) calls the AGM postulates (∗7) and (∗8) *Superexpansion* and *Subexpansion*, respectively. Gärdenfors (1988, p. 60) gives arguments against (∗8m), which in effect show that this very strong postulate is incompatible with the combination of postulates (∗1), (∗2), (∗4), (∗5). In the case where $\neg\phi$ is not, but $\neg\psi$ is, contained in K, we want to see that $\neg\psi$ is still in $K * \phi$ but no longer in $K * (\phi \wedge \psi)$. Postulate (∗8wm) does not admit this straightforward rebuttal. However, like ($\dot{-}$8wm), it commits the agent to a very inflexible method of changing his beliefs:[184]

Lemma 3 *A revision function* ∗ *over K satisfies the basic revision postulates plus* (∗7) *and* (∗8wm) *if and only if there is a subtheory H of K such that for every ϕ,*

$$K * \phi = \begin{cases} Cn(H \cup \{\phi\}) \text{ if } \neg\phi \in K \\ Cn(K \cup \{\phi\}) \text{ otherwise} \end{cases}$$

---

[183]Makinson (1985, p. 354) calls the condition the *Stalnaker property*. For the term 'reciprocity' and a short discussion, see Makinson and Gärdenfors (1991, p. 198). For the equivalence, see Gärdenfors and Rott (1995, p. 54). A related observation for contractions can be found in Rott (1992b, p. 54), and one for nonmonotonic logics in Makinson (1989, p. 6).

[184]I skip the proof because of the analogy of this result with Lemma 1(iii) and Observation 5(ii) which I prove both.

When dealing with limiting cases, settling for some particular solution is often a matter of taste rather than of stringent necessity. Such a limiting case is given when an agent is to revise by some sentence ϕ that is inconsistent. In the AGM theory the agent is obliged to actually accept this sentence, in order to obey the universal postulate of Success ($*2$). This, however, makes it necessary to restrict the regulative ideal of consistency to those cases where the input is itself consistent; see the Consistency condition ($*5$). Alternatively one might decide the matter, with at least as much justification, in such a way that the regulative ideal of Consistency takes precedence over the regulative ideal of Success, and accordingly weaken ($*2$) while strengthening ($*5$):

($*2^-$)   If $Cn(\phi) \neq L$, then $\phi \in K * \phi$ *(Weak Success)*

($*5^+$) $K * \phi \neq L$ *(Strong Consistency)*

For the case where ϕ is inconsistent, we can plausibly set $K * \phi = K$. In this modelling, rational people just *refuse to accept* inconsistent information. We shall advocate a model for belief change satisfying ($*2^-$) and ($*5^+$) in the next chapter.

What are the corresponding ideas for the contraction postulates? Here a limiting case is given when an agent is to contract with respect to some sentence ϕ that is a logical truth. In the AGM theory the agent is obliged to refuse to *retract* this sentence, in order to obey the universally valid postulates ($\dot{-}1$) and ($\dot{-}5$). This, however, makes it necessary to restrict the regulative ideal of success to those cases where the 'input' is no logical truth; see the Success condition ($\dot{-}4$). One might argue in the reverse that Success is more important than both Closure and Recovery, and accordingly weaken ($\dot{-}1$) and ($\dot{-}5$) while strengthening ($\dot{-}4$):

($\dot{-}1^-$)   If $\phi \notin Cn(\emptyset)$, then $K \dot{-} \phi = Cn(K \dot{-} \phi)$ *(Weak Closure)*

($\dot{-}5^-$)   If $\phi \notin Cn(\emptyset)$, then $K \subseteq Cn((K \dot{-} \phi) \cup \{\phi\})$ *(Weak Recovery)*

($\dot{-}4^+$) $\phi \notin K \dot{-} \phi$ *(Strong Success)*

It does not appear to be too high a cost to restrict the Recovery condition ($\dot{-}5$) in this way. However, the restriction of the closure condition ($\dot{-}1$) is rather unappealing.[185] So if we want to implement ($*2^-$) and ($*5^+$) in the context of AGM style belief revision, it seems better to turn to an alternative idea, stick to the standard basic AGM postulates and instead tamper with the Levi identity:

$$K * \phi = \begin{cases} Cn((K \dot{-} \neg\phi) \cup \{\phi\}) & \text{if } Cn(\phi) \neq L \\ K & \text{if } Cn(\phi) = L \end{cases}$$

This will give us the desired properties.

4.4   Postulates for rational nonmonotonic reasoning

In Chapter 1 we discussed the use of logics that do not satisfy the principle of Monotony. A liberalization of the standard concept of a logics was seen to be a

---

[185] For the case where ϕ is in $Cn(\emptyset)$, we would then probably need to set $K \dot{-} \phi = \emptyset$.

desideratum. However, logics should still have *some* nice features that logicians liked about classical logics. As pointed out in Sections 1.2.4–1.2.5, we make a terminological distinction between *inference operations* which are important for various kinds of reasoning and *consequence operations* which preserve truth values. In this section we shall give a list of important properties that an inference operation may or should have even if it violates Monotony. The inference operations considered in this section are all standard in so far as the 'theorems' include the 'premises', and they are not paraconsistent. The list below is structured in such a way that the relations to the above list of postulates for belief revision is easily recognized.

We are now going to study finitary nonmonotonic inference operations *Inf*, that is, inference operations that can take only finite sets of premises. Instead of a finite set of premises, one can just as well take the conjunction of these premises and then work with an inference operation *Inf* which only takes single sentences. This is the way we shall follow in the present section.[186] A nonmonotonic inference operation *Inf* will be regarded as being built upon a monotonic consequence operation *Cn* which is supposed to be Tarskian (i.e., satisfy Reflexivity, Idempotence, Monotonicity and Compactness) and to satisfy the Deduction Theorem. That inferences are 'built upon' consequences means that there are adequacy conditions for *Inf* that connect *Inf* and *Cn*—in just the same way as *Cn* has figured in a number of conditions for belief contractions and revisions. In the following we frequently alternate between the use of the consequence and inference *operations Cn* and *Inf* and the use of the corresponding consequence and inference *relations* $\vdash$ and $\hspace{0.2em}\vdash\hspace{-0.6em}\sim$. We usually write $Cn(\phi)$ and $Inf(\phi)$ for $Cn(\{\phi\})$ and $Inf(\{\phi\})$, respectively.

Here now is the relevant catalogue of conditions for finitary nonmonotonic logics.

(Ref)	$\phi \in Inf(\phi)$	*(Reflexivity)*
(RW)	If $\psi \in Inf(\phi)$ and $\chi \in Cn(\psi)$, then $\chi \in Inf(\phi)$	
		(Right Weakening)
(And)	If $\psi \in Inf(\phi)$ and $\chi \in Inf(\phi)$, then $\psi \wedge \chi \in Inf(\phi)$	*(And)*
(LLE)	If $Cn(\phi) = Cn(\psi)$, then $Inf(\phi) = Inf(\psi)$	
		(Left Logical Equivalence)
(CP)	If $Cn(\phi) \neq L$, then $Inf(\phi) \neq L$	*(Consistency Preservation)*
(Cut)	If $\psi \in Inf(\phi)$, then $Inf(\phi \wedge \psi) \subseteq Inf(\phi)$	*(Cut)*
(Cond)	$Inf(\phi \wedge \psi) \subseteq Cn(Inf(\phi) \cup \{\psi\})$	*(Conditionalization)*
(Or)	$Inf(\phi) \cap Inf(\psi) \subseteq Inf(\phi \vee \psi)$	*(Or)*
(CMon)	If $\psi \in Inf(\phi)$, then $Inf(\phi) \subseteq Inf(\phi \wedge \psi)$	
		(Cumulative Monotony)

---

[186] It is possible but difficult to extend finitary inference operations in a conservative way such that they can take infinite sets and their structural properties are preserved. Freund and Lehmann (1993) carefully develop the most successful ways of doing this.

(SDRat) $Inf(\phi \vee \psi) \subseteq Inf(\phi)$ or $Inf(\phi \vee \psi) \subseteq Inf(\psi)$

(DRat) $Inf(\phi \vee \psi) \subseteq Inf(\phi) \cup Inf(\psi)$ *(Disjunctive Rationality)*

(WDRat) $Inf(\phi \vee \psi) \subseteq Cn\,(Inf(\phi) \cup \{\psi\}) \cup Cn\,(Inf(\psi) \cup \{\phi\})$
                                                    (Weak Disjunctive Rationality)

(VWDRat) $Inf(\phi \vee \psi) \subseteq Cn\,(Inf(\phi) \cup Inf(\psi))$
                                                    (Very Weak Disjunctive Rationality)

(NRat) $Inf(\phi) \subseteq Inf(\phi \wedge \psi) \cup Inf(\phi \wedge \neg\psi)$ *(Negation Rationality)*

(RMon)   If $\neg\psi \notin Inf(\phi)$, then $Inf(\phi) \subseteq Inf(\phi \wedge \psi)$ *(Rational Monotony)*

(Mon) $Inf(\phi) \subseteq Inf(\phi \wedge \psi)$ *(Monotony)*

(WMon)   If $\neg\phi \in Inf(\top)$, then $Inf(\phi) \subseteq Inf(\phi \wedge \psi)$ *(Weak Monotony)*

With the exception of (SRat), (WDRat), (VWDRat) and (WMon), all of these conditions are well-known from the literature.[187] For a detailed discussion of the postulates, the reader is referred to Kraus, Lehmann and Magidor (1990), Makinson and Gärdenfors (1991), Lehmann and Magidor (1992), Makinson (1994) and Freund (1993). (WMon) seems much more harmless than (Mon) but we shall see presently that it is pretty strong indeed.

The concept of *basic reasoning* is characterized by the properties (Ref), (LLE), (RW), (And). Notice that (Ref) and (RW) entail that for every sentence ϕ

(Supra-Cn) $Cn\,(\phi) \subseteq Inf(\phi)$

Since Cn is assumed to be compact and well-behaved with respect to $\wedge$,[188] (And) and (RW) entail even more, namely that for every sentence ϕ

(Cn-closure) $Cn\,(Inf(\phi)) \subseteq Inf(\phi)$

An operation Inf satisfying this condition is called *left-absorbing* (cf. Freund and Lehmann 1993, p. 27, Makinson 1994, p. 45).

The concept of *cumulative reasoning* is defined as basic reasoning plus (Cut) and (CMon).[189] *Preferential reasoning* is cumulative reasoning plus (Or), or equivalently, cumulative reasoning plus (Conditionalization). *Rational reasoning* is preferential reasoning plus (RMon). *Disjunctive reasoning* is basic reasoning plus (Or). Finally, the concepts of *(weak, very weak) disjunctively rational reasoning* are defined as preferential reasoning plus (DRat) (respectively, (WDRat), (VWDRat)).

---

[187]The conditions (Ref) and (Mon) are different from the ones with the same labels used in the characterization of consequence operations Cn in Section 1.2.4. This is due to the fact that inference operations Inf as defined here take only single sentences as arguments, and I hope it does not seem to cause any confusion. Conditions (Tran) and (DedThm) of Section 1.2.4 correspond to (Cut) and (Cond) in the present list which has no counterparts of (Comp) and (SupraClass).

[188]In some sentence is in $Cn\,(G)$ for some set G, it is already in $Cn\,(G_0)$ and in $Cn\,(\bigwedge G_0)$ for some finite subset G_0 of G.

[189]Given, (Ref), (RW) and (And), (Cut) and (CMon) taken together are equivalent to the following *Reciprocity condition*: If $\phi \in Inf(\psi)$ and $\psi \in Inf(\phi)$, then $Inf(\phi) = Inf(\psi)$.

As suggested by the above names, the postulates for nonmonotonic reasoning form a hierarchy as regards their plausibility. Basic reasoning is often considered to be the unnegotiable hard core. Cumulative reasoning is something people aim at when implementing real-life reasoning systems. The conditions (Or) and (Cond) are somewhat more dubious, but still more widely accepted than the non-horn conditions labelled as 'rationality'.[190] The strongest of them, Rational Monotony, is very often violated when concrete recipes how to go about constructing nonmonotonic inferences are considered.

The following facts should be noted. Every basic inference operation satisfying (Or) also satisfies (Cut). However, (Cut) is a weaker property than (Or), because the latter does not follow from the former. Given the postulates for basic reasoning, (Or) is relatively equivalent with the condition of Conditionalization, which is sometimes referred to as the *Hard half of the deduction theorem* (Kraus, Lehmann and Magidor 1990, p. 191). In the context of the basic postulates, (RMon) implies (DRat) which in turn implies (NRat); the converse directions are not valid, as is shown in Kraus, Lehmann and Magidor (1990, pp. 196–198) and Lehmann and Magidor (1992, pp. 16–20).

Not so common are the non-horn conditions of *Strong Disjunctive Rationality* (WDRat), *Weak Disjunctive Rationality* (WDRat) and *Very Weak Disjunctive Rationality* (VWDRat) which are variations of the better-known condition of *Disjunctive Rationality* (DRat). (DRat) has been studied very carefully from a semantic point of view by Freund (1993). (SDRat), although looking considerably stronger than (DRat) will soon turn out to be essentially equivalent. (VWDRat) appears in Freund (1993, p. 243) and, in an infinitary form, in Lindström (1991, condition (Gamma)). In forms transformed for (finitary) belief revision and contraction, it is found in Katsuno and Mendelzon (1991, p. 273, condition (R8)) and Rott (1993, p. 1437, condition ($\dot{-}$8r)). The strange condition (WDRat) will appear in Chapter 7 below.

The significance of the unusual variants of Disjunctive Rationality will become clear later. We first want to locate them in the field of related conditions. The following lemma gives us some clues.

Lemma 4 *(i) (DRat) implies (WDRat).*
(ii) (WDRat), taken together with (Ref), implies (VWDRat).
(iii) (DRat), taken together with (RW) and (And), implies (SDRat).
(iv) (WDRat), taken together with (RW) and (And), implies

(WDRat$^+$) $Inf(\phi \vee \psi) \subseteq Cn(Inf(\phi) \cup \{\psi\})$ or $Inf(\phi \vee \psi) \subseteq Cn(Inf(\psi) \cup \{\phi\})$

(v) For any basic inference relation $\mid\!\sim$, (WDRat) is equivalent with

(Triplet)   If $\phi \vee \psi \vee \chi \mid\!\sim \neg\phi$, then $\phi \vee \psi \mid\!\sim \neg\phi$ or $\phi \vee \chi \mid\!\sim \neg\phi$

---

[190] A conditional assertion is called *horn*, if its antecedent is the conjunction of zero or more unnegated atomic sentences and its consequent is a single unnegated atomic sentence or the falsity constant, $\perp$.

(vi) For any basic inference relation $\succ$ *, (WDRat) and (CMon) taken together imply (DRat).*

Parts (i) and (ii) of this Lemma justify the labels (WDRat) and (VWDRat). Like (NRat), both (WDRat) and (VWDRat) are weakenings of (DRat) which add to the power of preferential reasoning (cf. Lehmann and Magidor 1992, Section 3, Makinson 1994, Section 4.1). Nevertheless there does not seem to be any logical relationship between (NRat) and either of (WDRat) and (VWDRat).

In parts (iii) and (iv) of the Lemma the conditions (DRat) and (WRat) are shown to be equivalent with conditions which prima facie seem to be substantially stronger.[191] Part (v) takes a first step to relate (WDRat) to a postulate that is amenable to a direct choice-theoretic interpretation.[192] Part (vi) finally shows that for logics satisfying Cumulative Monotony, Weak Disjunctive Rationality coincides with Disjunctive Rationality.

We now have a look at two weaker versions of Cumulative Monotony and (Or):

(∅Cond)   If $\perp \in \mathit{Inf}(\phi \wedge \psi)$, then $\neg\psi \in \mathit{Inf}(\phi)$

(∅CMon)   If $\perp \in \mathit{Inf}(\phi)$, then $\perp \in \mathit{Inf}(\phi \wedge \psi)$

It is easy to verify that indeed (Cond) implies (∅Cond),[193] and (CMon) implies (∅CMon) provided the basic conditions for nonmonotonic logics are satisfied. (∅Cond) and (∅CMon) correspond to the conditions ($\dot-$7r) and ($\dot-$8rc) for refusals to withdraw beliefs, respectively. They will be linked to two conditions, (∅1) and (∅2), that constrain coherent refusals of choice (see Section 6.1.2 below).

It is surprising, but not difficult to perceive that the postulates for nonmonotonic reasoning are counterparts of the postulates for belief revisions. The key idea, due to Makinson and Gärdenfors (1991), is to identify the inferences $\mathit{Inf}(\phi)$ drawn from a sentence ϕ with the revision $E * \phi$ of a fixed background theory E by ϕ. The theory E is supposed to consist of the agent's expectations. More pre-

---

[191] A condition for contractions that corresponds to (DRat⁺) is discussed in Alchourrón, Gärdenfors and Makinson (1985, p. 317) under the name *Covering condition*; see page 104 above.

[192] As will become clear in Chapters 6–8. Notice that (Triplet) must be carefully distinguished from Lehmann and Magidor's (1992, p. 22) Horn conditions (16) and (17):

(16)   If $\phi \vee \psi \vee \chi \succ \neg\phi \wedge \neg\psi$, then $\psi \vee \chi \succ \neg\psi$

(17)   If $\phi \vee \psi \succ \neg\phi$, then $\phi \vee \psi \vee \chi \succ \neg\phi$

(16) is validated in cumulative reasoning, (17) in preferential reasoning. In terms of the syntactical choice-theoretical analysis presented later in this book, (Triplet) corresponds to the horn condition (III), while (16) corresponds to the non-horn condition

If $\psi \in \delta(\psi, \chi)$, then either $\phi \in \delta(\phi, \psi, \chi)$ or $\psi \in \delta(\phi, \psi, \chi)$

which follows from (∅1) and (III), and (17) corresponds to the horn condition (I).

[193] A condition equivalent with (∅Cond) is

(∅Or)   If $\neg\phi \in \mathit{Inf}(\phi)$, then $\neg\phi \in \mathit{Inf}(\phi \vee \psi)$

cisely, nonmonotonic inference operations are obtained from revision by means of what may be called *Makinson-Gärdenfors identity*:[194]

$$\phi \mathrel{\vert\!\sim} \psi \ \text{ iff } \ \psi \in E * \phi \text{ for some fixed expectation set } E$$

Furthermore, there are two weakenings of Conditionalization and Rational Monotony which correspond to the AGM revision postulates (*3) and (*4).

(WCond) $Inf(\phi) \subseteq Cn(Inf(\top) \cup \{\phi\})$

(WRMon) If $\neg\phi \notin Inf(\top)$, then $Inf(\top) \subseteq Inf(\phi)$

Essentially the same conditions were introduced by Makinson and Gärdenfors (1991, p. 193) and Gärdenfors and Makinson (1994, p. 201). In my view, they do not have any independent motivation as requirements on nonmonotonic reasoning. All their interest derives from their correspondence with the AGM postulates (*3) and (*4). If $Inf(\top)$ is inconsistent, then (WCond) and (WRMon) are vacuously satisfied. It is worth mentioning another formulation of (WCond) and (WRMon). Taken together, they say that there is a consistent theory E such that for all sentences ϕ

$$Inf(\phi) \subseteq Cn(E \cup \{\phi\}) \qquad \text{and}$$
$$\text{if } \neg\phi \notin E, \text{ then } E \subseteq Inf(\phi)$$

The theory E is the set of the agent's expectations.[195] Evidently, (WCond) and (WRMon) imply the latter formulation, if $Inf(\top)$ is consistent (what it normally should be). Interestingly enough, the latter also imply the former. If we substitute $\top$ for ϕ, we immediately recognize that E is identical with $Inf(\top)$. In the previous sections, we said that $\dot-$ and $*$ are contraction and revisions functions *over the belief set* K, just because K was the point of reference in $(\dot-2)$ and $(\dot-3)$, as well as in (*3) and (*4). In a similar vein, if Inf satisfies (WCond) and (WRMon), we can now say that it is an inference relation *with respect to the set* $E = Inf(\top)$

The condition of Consistency Preservation corresponds to (*5). Although it is important for certain purposes,[196] it is now placed quite at the bottom of our above list and will be treated in parallel with $(\dot-4)$ and (*5).

Our last observation in this chapter characterizes the situation in which a basic nonmonotonic inference operation Inf (built on a monotonic background logic Cn) is itself disjunctive and monotonic, or at least disjunctive and weakly monotonic. In those cases the inference operation can be (re-)constructed in a straightforward manner with the help of the monotonic operation Cn.

---

[194]For the motivation of this identity, compare Sections 1.2.7 and 3.5 above.

[195]Notice that both of the conditions just displayed are in the scope of a single quantifier phrase involving E. There will be more examples of reformulations of purely 'operational' postulates (like (WCond) and (WRMon)) into postulates referring to the existence of a fixed set (like E in the case we are now considering) in Chapter 6.

[196]See, for instance, the remark in Rott (1993, p. 1449).

Observation 5 *(i) A basic inference operation Inf satisfies (Or) and (Mon) if and only if there is a fixed set of sentences F such that for every ϕ, $Inf(\phi) = Cn(F \cup \{\phi\})$;*

(ii) a basic inference operation Inf satisfies (WRMon), (Or) and (WMon) if and only if there are two fixed sets of sentences F and G such that $F \subseteq G$ and for every ϕ,

$$Inf(\phi) = \begin{cases} Cn(G \cup \{\phi\}) \text{ if } G \cup \{\phi\} \not\vdash \bot \\ Cn(F \cup \{\phi\}) \text{ otherwise} \end{cases}$$

In this observation, the set F of sentences takes the role of the 'laws' that are 'necessarily true', while the set G is the set of sentences that are 'normally true'.

The Levi and the Makinson-Gärdenfors identities can be combined into the following direct definition of nonmonotonic inferences in terms of contractions.

$$\phi \mathbin{\vdash\mkern-7mu\sim} \psi \text{ iff } \phi \supset \psi \in E \dot{-} \neg\phi \text{ for some fixed expectation set } E$$

For an agent to be licensed to defeasibly infer ψ from ϕ, it is not sufficient to have the material conditional $\phi \supset \psi$ among his expectations. This conditional must be retained even if the agent were to withdraw his belief or expectation that ϕ is false, i.e., his belief that the conditional $\phi \supset \psi$ really is 'contrary-to-fact' or 'counterfactual'.

Since both revisions of beliefs and nonmonotonic reasoning can be construed as ultimately depending on the primitive operation of contraction (of beliefs or expectations), many authors have chosen to exclusively concentrate on the problem of contraction. Corresponding results for $*$ and $\mathbin{\vdash\mkern-7mu\sim}$ can then easily be derived as corollaries. There are at least (four) reasons, however, that we shall devote equal attention to both belief revision and nonmonotonic reasoning. First, many people think that in real life there are only few—if any—occurrences of contractions of belief, as compared to the multitude of situations in which one is induced to perform a revision or draw a nonmonotonic inference. Second, there is much room for doubts about the validity or reasonableness of the Recovery condition ($\dot{-}5$)—and just this condition is an indispensable prerequisite for many of the most interesting results about contractions.[197] Third, it is desirable to see at work certain structures that are applied in more constructive approaches to revision and defeasible inference not only indirectly, via the Levi and Makinson-Gärdenfors identities, but also directly without any detours. Fourth, there are models which are much more natural when applied to revisions than when applied to contractions.[198]

The correspondences exhibited in this chapter are displayed graphically in the table on page 118. In general the table lists postulates which are equivalent via the Levi and Makinson-Gärdenfors identities (linking contractions with

---

[197]We shall make heavy use of Recovery in Chapter 7. Once the the Levi identity has been applied, however, the effects of Recovery vanish.

[198]The best examples of such models will soon be seen in Chapter 5.

Contractions	Revisions	Nonmonotonic Inferences
($\dot{-}$1) If $K = Cn(K)$ then $K \dot{-} \phi = Cn(K \dot{-} \phi)$	($*$1) If $K = Cn(K)$ then $K * \phi = Cn(K * \phi)$	(Cn-closure) $Inf(\phi) = Cn(Inf(\phi))$
($\dot{-}$1a) If $\psi \in K \dot{-} \phi$ and $\chi \in Cn(\psi)$, then $\chi \in K \dot{-} \phi$	($*$1a) If $\psi \in K * \phi$ and $\chi \in Cn(\psi)$, then $\chi \in K * \phi$	(RW) If $\psi \in Inf(\phi)$ and $\chi \in Cn(\psi)$, then $\chi \in Inf(\phi)$
($\dot{-}$1b) If $\psi \in K \dot{-} \phi$ and $\chi \in K \dot{-} \phi$, then $\psi \wedge \chi \in K \dot{-} \phi$	($*$1b) If $\psi \in K * \phi$ and $\chi \in K * \phi$, then $\psi \wedge \chi \in K * \phi$	(And) If $\psi \in Inf(\phi)$ and $\chi \in Inf(\phi)$ then $\psi \wedge \chi \in Inf(\phi)$
($\dot{-}$5) $K \subseteq Cn((K \dot{-} \phi) \cup \{\phi\})$	($*$2) $\phi \in K * \phi$	(Ref) $\phi \in Inf(\phi)$
($\dot{-}$2) $K \dot{-} \phi \subseteq K$	($*$3) $K * \phi \subseteq Cn(K \cup \{\phi\})$	(WCond) $Inf(\phi) \subseteq Cn(Inf(\top) \cup \{\phi\})$
($\dot{-}$3) If $\phi \notin K$, then $K \subseteq K \dot{-} \phi$	(Preservation) / ($*$4) If $\neg\phi \notin K$, then $K \subseteq K * \phi$ If $\neg\phi \notin K$, then $Cn(K \cup \{\phi\}) \subseteq K * \phi$	(WRMon) If $\neg\phi \notin Inf(\top)$, then $Inf(\top) \subseteq Inf(\phi)$
($\dot{-}$6) If $Cn(\phi) = Cn(\psi)$, then $K \dot{-} \phi = K \dot{-} \psi$	($*$6) If $Cn(\phi) = Cn(\psi)$, then $K * \phi = K * \psi$	(LLE) If $Cn(\phi) = Cn(\psi)$, then $Inf(\phi) = Inf(\psi)$
($\dot{-}$4) If $\phi \notin Cn(\emptyset)$, then $\phi \notin K \dot{-} \phi$	($*$5) If $Cn(\phi) \neq L$, then $K * \phi \neq L$	(CP) If $Cn(\phi) \neq L$, then $Inf(\phi) \neq L$
($\dot{-}$7c) If $\psi \in K \dot{-} (\phi \wedge \psi)$, then $K \dot{-} \phi \subseteq K \dot{-} (\phi \wedge \psi)$	($*$7c) If $\psi \in K * \phi$, then $K * (\phi \wedge \psi) \subseteq K * \phi$	(Cut) If $\psi \in Inf(\phi)$, then $Inf(\phi \wedge \psi) \subseteq Inf(\phi)$
($\dot{-}\emptyset$2) / ($\dot{-}$7r) If $\phi \in K \dot{-} \phi$, then $\phi \in K \dot{-} (\phi \wedge \psi)$	($*\emptyset$2) If $\perp \in K * (\phi \wedge \psi)$, then $\neg\psi \in K * (\phi)$	($\emptyset$Cond) If $\perp \in Inf(\phi \wedge \psi)$, then $\neg\psi \in Inf(\phi)$
($\dot{-}$7) / ($\dot{-}$7g) $K \dot{-} \phi \cap K \dot{-} \psi \subseteq K \dot{-} (\phi \wedge \psi)$ $K \dot{-} (\phi \wedge \psi) \subseteq Cn((K \dot{-} \phi) \cup \{\neg\psi\})$	($*$7) / ($*$7g) $K * \phi \cap K * \psi \subseteq K * (\phi \vee \psi)$ $K * (\phi \wedge \psi) \subseteq Cn((K * \phi) \cup \{\psi\})$	(Or) / (Cond) $Inf(\phi) \cap Inf(\psi) \subseteq Inf(\phi \vee \psi)$ $Inf(\phi \wedge \psi) \subseteq Cn(Inf(\phi) \cup \{\psi\})$
($\dot{-}\emptyset$1) / ($\dot{-}$8rc) If $\phi \wedge \psi \in K \dot{-} (\phi \wedge \psi)$, then $\phi \in K \dot{-} \phi$	($*\emptyset$1) If $\perp \in K * \phi$, then $\perp \in K * (\phi \wedge \psi)$	($\emptyset$CMon) If $\perp \in Inf(\phi)$, then $\perp \in Inf(\phi \wedge \psi)$
($\dot{-}$8c) If $\psi \in K \dot{-} (\phi \wedge \psi)$, then $K \dot{-} (\phi \wedge \psi) \subseteq K \dot{-} \phi$	($*$8c) If $\psi \in K * \phi$, then $K * \phi \subseteq K * (\phi \wedge \psi)$	(CMon) If $\psi \in Inf(\phi)$, then $Inf(\phi) \subseteq Inf(\phi \wedge \psi)$
($\dot{-}$8d) $K \dot{-} (\phi \wedge \psi) \subseteq K \dot{-} \phi \cup K \dot{-} \psi$	($*$8d) $K * (\phi \vee \psi) \subseteq K * \phi \cup K * \psi$	(DRat) $Inf(\phi \vee \psi) \subseteq Inf(\phi) \cup Inf(\psi)$ (DRat$^+$) $Cn(\phi \vee \psi) \subseteq Inf(\phi)$ or $Cn(\phi \vee \psi) \subseteq Inf(\psi)$
($\dot{-}$8wd) $K \dot{-} (\phi \wedge \psi) \subseteq Cn(K \dot{-} \phi \cup \{\neg\psi\})$ $\cup Cn(K \dot{-} \psi \cup \{\neg\phi\})$	($*$8wd) $K * (\phi \vee \psi) \subseteq Cn(K * \phi \cup \{\psi\})$ $\cup Cn(K * \psi \cup \{\phi\})$	(WDRat) $Inf(\phi \vee \psi) \subseteq Cn(Inf(\phi) \cup \{\psi\})$ $\cup Cn(Inf(\psi) \cup \{\phi\})$ (WDRat$^+$) $Inf(\phi \vee \psi) \subseteq Cn(Inf(\phi) \cup \{\psi\})$ or $Inf(\phi \vee \psi) \subseteq Cn(Inf(\psi) \cup \{\phi\})$
($\dot{-}$8vwd) $K \dot{-} (\phi \wedge \psi) \subseteq Cn(K \dot{-} \phi \cup K \dot{-} \psi)$	($*$8vwd) $K * (\phi \vee \psi) \subseteq Cn(K * \phi \cup K * \psi)$	(VWDRat) $Inf(\phi \vee \psi) \subseteq Cn(Inf(\phi) \cup Inf(\psi))$
($\dot{-}$8n) $K \dot{-} \phi \subseteq$ $K \dot{-} (\phi \vee \psi) \cup K \dot{-} (\phi \vee \neg\psi)$	($*$8n) $K * \phi \subseteq$ $K * (\phi \wedge \psi) \cup K * (\phi \wedge \neg\psi)$	(NRat) $Inf(\phi) \subseteq$ $Inf(\phi \wedge \psi) \cup Inf(\phi \wedge \neg\psi)$
($\dot{-}$8) If $\phi \notin K \dot{-} (\phi \wedge \psi)$, then $K \dot{-} (\phi \wedge \psi) \subseteq K \dot{-} \phi$	($*$8) If $\neg\psi \notin K * \phi$, then $K * \phi \subseteq K * (\phi \wedge \psi)$	(RMon) If $\neg\psi \notin Inf(\phi)$, then $Inf(\phi) \subseteq Inf(\phi \wedge \psi)$
($\dot{-}$8m) $K \dot{-} (\phi \wedge \psi) \subseteq K \dot{-} \phi$	($*$8m) $K * \phi \subseteq K * (\phi \wedge \psi)$	(Mon) $Inf(\phi) \subseteq Inf(\phi \wedge \psi)$

revisions, and, respectively, revisions with nonmonotonic inferences). The first four rows, however, are exceptional. (∗1), (∗1a), (∗1b), and (∗2) follow from the Levi identity alone, without presuming anything about contractions. However, as remarked above, a connection can be established via the Harper identity.[199] Given this identity, (∸1) follows from (∗1), the Recovery postulate (∸5) follows from (∗2) and so on. Sometimes the equivalences of postulates further down in the table presume as background some of the basic postulates which are the ones listed in the top seven rows of the table. We do not give proofs of the correspondences here. Many of them are discussed in Alchourrón, Gärdenfors and Makinson (1985), Gärdenfors (1988, Section 3.6), Makinson and Gärdenfors (1991), Lindström (1991), Rott (1992b), and Gärdenfors and Rott (1995, Sections 3.3 and 6.2–6.3).

---

[199]The Harper identity could also be used for the lower rows. We prefer the Levi identity in these cases because it has more intuitive appeal than the Harper identity.

5

FOUNDATIONAL BELIEF CHANGE USING NONMONOTONIC INFERENCE

In this chapter we study ways of making the direct mode of belief revision (Section 3.4.1) more concrete. We start with two quite straightforward ideas. One may either treat the items of an information base as infallible, come what may, or one may see to it that the information base is always cleared of contradictions. Assume that the theory K has been generated from a (flat, unprioritized) belief base H with the help of an inference relation Inf. Then the first method puts $K * \phi$ to be $Inf(H \cup \{\phi\})$, while the second method puts it to be $Inf(H *_0 \phi)$ where $*_0$ signifies full meet base revision (a method of consistently 'adding' ϕ to H that will be explained in due course). While logically well-behaved, the first idea fails with the crucial condition of Consistency Preservation. The second idea turns out to be logically rather ill-behaved. A more decisive objection against using it, however, will be found in the fact that this method allows defaults to override explicit beliefs.

In Section 1.2.7, we have reviewed an idea of Israel's (1980) and Poole's (1988) according to which nonmonotonic inferences can be conceived of as revisions of one's default theory or expectations. Elaborating on this insight, Makinson and Gärdenfors (1991) studied (finitary) inference relations that results from putting $Inf(\phi) = E * \phi$ for some fixed body of expectations E, and they showed how logical properties of nonmonotonic reasoning are thus inherited from logical properties of revision operations. Here E is supposed to be a logically closed theory, and H is a finite information base, so that an AGM revision can be performed with respect to the conjunction $\bigwedge H$. Expectation and information receive an asymmetric treatment since (even in the case of conflicts within H) the latter is invariably accepted. We shall see, however, that Makinson and Gärdenfors's expectation-based reconstruction of nonmonotonic reasoning is of no help if we want to overcome the serious shortcomings of our first two attempts to make sense of the direct mode of belief revision.

For that reason I shall propose to keep the idea of working with both expectation and information, but to modify the Makinson-Gärdenfors account in two important respects. First, I argue that both the expectations and the information base should be endowed with additional structure, and secondly, I modify the way in which inferences are drawn on the basis of expectation and information. I treat both expectation and information as prioritized bases $\mathcal{E}$ and $\mathcal{H}$, that is, as sequences of sets of sentences with increasing priorities. I finally propose a third method of making sense of the direct mode of belief change. This method integrates expectation and information—both old and new information—into a single prioritized belief base, which will usually be inconsistent. The inference operation then proceeds by considering all 'optimal' ways of restoring consistency, by 'consolidating' the base through the method of prioritized base contraction. More formally, we investigate the idea of putting $Inf(\mathcal{H}) = Consol(\mathcal{E} \circ \mathcal{H})$ and of putting $K * \phi = Consol(\mathcal{E} \circ \mathcal{H} \circ \langle\{\phi\}\rangle))$, where $Consol$ is a special method of eliminating inconsistencies from prioritized belief bases.[200] The method supports not only nonmonotonic, but also paraconsistent reasoning, in the sense that from a base H or $\mathcal{H}$ which is classically inconsistent no inconsistent conclusion is derived. The new inference operation is no longer supraclassical, and in fact it is not even *reflexive*.[201] It does not generally hold that H is a subset of $Inf(H)$; this inclusion will actually fail just in case H is classically inconsistent. We shall argue that although our third and final method fails to validate the Preservation Principle which states that after a 'consistent revision' one should not have lost any of one's previous beliefs, it is still the most reasonable way to make sense of the direct mode.

For reasons that will become apparent later, we shall mainly focus on the operation of revision in this chapter. Indeed, contrary to the influential thesis of Isaac Levi, revisions will emerge as more fundamental than contractions in the context of belief change as modelled along the lines suggested in this chapter.

The foundationalist explorations of this chapter with a formal analysis of the various methods in the direct or immediate mode of belief change, trying to figure out in particular the extent to which logical properties of revision operations are inherited from logical properties of nonmonotonic inference operations.

5.1 Modelling the direct mode: Drawing inferences from information bases

Let $\mathcal{H}$ be a finite prioritized information base, $H = |\mathcal{H}|$ be the finite collection of all data in $\mathcal{H}$ without prioritization, and let h be the conjunction of the elements of H. Assume that $\mathcal{H}$ gives rise to a belief set $K = Inf(\mathcal{H})$, using some inference operation Inf.[202]

For belief *change*, the method applied in this chapter is to change K into a new belief set K' by *first* changing the base in a rather plain (artless, undemanding, uncomplicated) fashion from $\mathcal{H}$ or H to a new base $\mathcal{H}'$ or H', and *then* taking the sophisticated inferential closure of the new base, that is, putting $K' = Inf(\mathcal{H}')$ or $K' = Inf(H')$.

---

[200] As will be recalled from Section 1.2.7, the result of the $Consol$ operation is not a prioritized belief base, but a flat theory.

[201] It would not be quite right to say that Inf violates (Supra-Cn) and (Ref). The point is that in our final method, Inf will usually take prioritized information bases as arguments, rather than single sentences. The most general version of reflexivity that we will find violated is this: $|\mathcal{H}| \subseteq Inf(\mathcal{H})$.

[202] Of course, if our inference operation can only take single sentences, as those considered in the last chapter, then the inferences to be drawn from $\mathcal{H}$ are bound to disregard both priorities and syntactic structure, and '$Inf(\mathcal{H})$' can mean nothing else than $Inf(h)$.

There is an important question which will find varying answers in the course of this chapter. Given that the propositional contents of two bases $\mathcal{H}$ and $\mathcal{H}'$ are logically equivalent with respect to standard (monotonic) logic, $Cn(H) = Cn(H')$, do we want them invariably to be equivalent with respect to sophisticated (nonmonotonic) inference, $Inf(\mathcal{H}) = Inf(\mathcal{H}')$? Now *if* we telescope the finite bases $\mathcal{H}$ and $\mathcal{H}'$ into the conjunctions h and h' of their propositional parts, then this should hold, according to the widely accepted condition of Left Logical Equivalence (LLE).[203] But this ceases to be true, or so we shall argue, when we work with the more finely structured bases H and $\mathcal{H}$ themselves, rather than with the conjunctions of their elements.

5.1.1 *Infallible foundations*

The first method of dealing with the information base is to treat its members as completely certain or *infallible*.[204] Once an information base is given, its items are always taken at face value and never called in question. When drawing inferences from the information base, no operations of 'doubting', 'filtering' or 'correcting' are performed. The only things that happen are *set-theoretic addition* and *subtraction* induced by external factors. Revisions and contractions based on this idea, however, have the drawback of frequently running into inconsistencies or, respectively, into bases which allow to recover the eliminated element from other items. If we put

$$K * \phi \;=\; Inf(H \cup \{\phi\})$$

and *Inf* is irreflexive, then this theory revision function does not satisfy the condition of Consistency Preservation. Similarly, if we put

$$K \dot{-} \phi \;=\; Inf(H - \{\phi\})$$

then this theory contraction function fails to satisfy the Success postulate that requires that ϕ should be *effectively* discarded from K, in the sense that ϕ cannot be inferred from the new belief base. The 'contraction' function just defined has a very special interpretation: It just means that ϕ is removed as an *explicit* item from the information base. Changes are made here on the surface only.

Depending on the inference relation used, this method of contraction, as well as all the other contraction methods considered in this chapter, does not in general satisfy the basic Inclusion postulate $(\dot{-}2)$ that states that $K \dot{-} \phi$ should be a subset of K. The failure of $(\dot{-}2)$ is, I submit, a realistic feature of ordinary belief contraction, as soon as we are ready to allow for processes of nonmonotonic

---

[203]Remember that in the preceding chapter (LLE) has been introduced as part of the most *basic* concept of (finitary) nonmonotonic reasoning.

[204]Recent advocates of infallibility for 'observations' include Friedman and Halpern (1999). I agree with Segerberg (1998), however, that infallibility may best be taken as a mark of hypothetical reasoning.

reasoning. Calling in question one's explicit information will often reinstate one's previous expectations. Still the fact that Inclusion is violated may be regarded as one of the reasons why we shall focus on belief *revisions* rather than on belief contractions in this chapter.

The present method is most suitable and enlightening in situations where a new incoming piece of information ϕ contradicts the overall corpus of beliefs and opinions of the subject, but does not contradict his or her explicit information. In this case we have the great advantage of being able to reduce a belief-contravening *revision* $K * \phi$ on the level of coherent belief sets to a simple and consistent *addition* $H \cup \{\phi\}$ on the base level and a subsequent derivation of the new beliefs with the help of a sophisticated inference operation *Inf*.

The method is not suitable, however, when the explicit information collected in a belief base H is itself inconsistent. In that case we are invariably led into the inconsistent belief set $K = L$. Thus the main idea of belief revisions which are supposed to maintain coherence gets lost here, and the postulate of Consistency Preservation, (*5), is badly violated. What would be needed here is a resolution of informational conflicts in H before the inference relation *Inf* starts operating. If *Inf* is assumed to be supraclassical, one must effectively cancel some of the items in the information base, and one must find a principled strategy how to do this. If we choose a different inference operation which is not supraclassical, in particular one of the paraconsistent variety, then we need not cancel items from the information base. The problem with subclassical logics, however, is that besides the cautious avoidance of contradictions we at the same time want a bold inference relation supplying us richly with defeasible conclusions, and these two goals seem difficult to reconcile in a single inference relation. In this book we aim at strong and powerful inference operations.

As we have seen in Chapter 4, there is a very close correspondence between postulates for belief revision and postulates for nonmonotonic reasoning. The fundamental idea of Makinson and Gärdenfors (1991), further elaborated in Gärdenfors and Makinson (1994), was, first, to construe the inference process (condensed in an operator *Inf*) as a kind of expectation revision process, and, second, to capture this process by an AGM style operator. Like Makinson and Gärdenfors, we assume that the function $\underline{*}$ satisfies the postulates (*1) − (*8) of Section 4.3.[205] This corresponds exactly to *Inf* being a rational inference relation. In this way the nonmonotonic inference operation inherits nice properties from corresponding nice properties of the belief revision operation (or more exactly, the expectation revision operation). In our foundationalist approach, the inheritance goes just the other way round. Nonmonotonic reasoning is used for deriving coherent belief sets from information bases. In the case of infallible information, this sophisticated inference process is supplemented by an extremely simple process of the addition of information. Except for some minor cases, the

---

[205] I use the symbol $\underline{*}$ to distinguish expectation revision from the revision of an information base. In the context of expectation revision, I will label the AGM postulates $(\underline{*}1) - (\underline{*}8)$.

correspondences turn out to be the same.[206] However, we must emphasize once more that the crucial condition of Consistency Preservation, ($*5$), is violated by the construction we have considered in this section. Since the maintenance of coherence and consistency is a principal *raison d'être* of the very idea of theory revision, this is certainly a devastating defect. There is no analogous defect in the Gärdenfors-Makinson derivation of nonmonotonic inference operations from expectation revision operations.

5.1.2 *Fallible foundations: Straight clearing of information conflicts*

Set-theoretic addition may lead to inconsistency, and set-theoretic subtraction may not really eliminate a belief because it is recoverable through inferences drawn from the remainder of the belief base. One may take this as a decisive argument against applying purely set-theoretic manipulations of belief bases. These operations are just *too* trivial. In order to ensure that revisions are more coherent and contractions are more successful, we are now going to change the information base in a more appropriate way, treat pieces of information as fallible and make them subject to revisions. The idea of fallible foundations is well-known in contemporary epistemology. See BonJour (1985, Section 2.2) and Lehrer (1990, Chapter 4). But basic beliefs in our sense need not be basic in the epistemologist's sense. As explained in Section 1.2.1, they are just given *somehow*, and it is irrelevant for the purposes of our discussion to ask about their sources or qualities.

Let us replace set-theoretic addition and subtraction by two simple but effective change operations on the information bases (to instantiate the general operations $\oplus$ and $\ominus$ figuring in Section 3.4.1). This is done in order to avoid running into an inconsistency in a revision and, respectively, to avoid ending up with a base that allows us to recover the eliminated element after the contraction. The most straightforward way to achieve this aim is to apply the method of full meet contraction and revision of the information base H. The method is plain enough to comply with the philosophy of the vertical perspective. It works as follows. When performing a revision by ϕ (or a contraction with respect to ϕ), we keep only those sentences which are contained in *all* maximal subsets of H that are consistent with ϕ (respectively, that do no imply ϕ). More precisely, $H \dot{-}_0 \phi$ is the intersection of all the maximal subsets of H that do not imply ϕ, and $H *_0 \phi$ is the set $H \dot{-} \neg\phi$ enlarged by ϕ.[207]

---

[206] See Section 5.3 below. The exceptions are: ($*3$) and ($*4$) require the full force of (Cond) and (RMon). In Gärdenfors and Makinson's converse undertaking, ($*3$) and ($*4$) generate only the weak forms (WCond) and (WRMon).

[207] Formally, $H \dot{-}_0 \phi = \bigcap \{G \subseteq H : G \not\vdash \phi$ and for all $G' \subseteq H$ such that $G \subset G'$ it holds that $G' \vdash \phi\}$ and $H *_0 \phi = (H \dot{-}_0 \neg\phi) \cup \{\phi\}$. In the limiting cases, one can put $H \dot{-}_0 \phi = H$ if $\phi \in Cn(\emptyset)$, and $H *_0 \phi = H$ if $\neg\phi \in Cn(\emptyset)$. See Alchourrón and Makinson (1982) for the deficiency of this method when it is applied to logically closed theories. For its application to belief and expectation bases, see Fagin, Ullman and Vardi (1983), Poole (1988), Nebel (1989) and Hansson (1993b, Theorem T3).

Full meet contraction and revision are still fairly trivial, and in fact very crude, change operations.[208] They sacrifice a very large part of the information base. Yet, thanks to our nonmonotonic reasoning background, this will not result in grossly impoverished belief sets after such a change has been performed. Fewer items of explicit information leave intact more default assumptions. Even an empty information base will leave us with enough beliefs to act upon.

Let $*_0$ be the operation of full meet base revision, and $\dot{-}_0$ be the operation of full meet base contraction. Then our improved method of theory revision is this:

$$K * \phi \;=\; Inf(H *_0 \phi)$$

In contrast to the earlier suggestion, this revision function satisfies the condition of Consistency Preservation, $(*5)$. Our improved method of theory contraction puts

$$K \dot{-} \phi \;=\; Inf(H \dot{-}_0 \phi)$$

Unfortunately, this contraction function still does not in general satisfy the condition of Success, $(\dot{-}4)$. Although we can now be sure that ϕ is not implied by $H \dot{-}_0 \phi$, ϕ may still be regained from those of the agent's expectations or default assumptions that get accepted as a result of the nonmonotonic reasoning process. Moreover, this method does not satisfy the Inclusion condition $(\dot{-}2)$ either. The contracted belief base $H \dot{-}_0 \phi$ may allow more defaults to be activated then the original belief base H.

The present method has the advantage of providing for a paraconsistent resolution of inconsistent information. Although the revised base has to meet a minimal form of coherence, viz., consistency, the method fully complies with the philosophy of the direct mode as expounded in Section 3.4.1, since no extralogical information for the change of the base is necessary, and thus requirement (II) formulated in that section is satisfied. The method does not lead to a drastic impoverishment of the agent's beliefs since his expectations will usually make up for lost pieces of information.

On the other hand, the method of straight clearing of information conflicts is not satisfactory. The operations of full meet contraction and revision cancel too much information, thus effectively allowing implicit expectations to override explicit data. Here is an example illustrating the problem. Let the information base H be the set $\{p, q, r\}$ so that $K = Inf(H)$ includes p, q and r. Let us further assume that each of these items is surprising, in the sense that the set

---

[208] A *much* gentler method would be to take *the logical closures* of the maximal non-implying sets and put $H \dot{-}_0 \phi = \bigcap \{Cn(G) : G \subseteq H,\, G \nvdash \phi$ and for all $G' \subseteq H$ such that $G \subset G'$ it holds that $G' \vdash \phi\}$ and $H *_0 \phi = Cn((H \dot{-}_0 \neg\phi) \cup \{\phi\})$. However, then we end up with a *theory* after revising the information *base*. There simply is no new information base in this case, and thus the philosophy of the direct mode of belief revision is violated. For the connection between this sort of base change operation and and the corresponding operations on the associated theories, see Nebel (1989) and Rott (1992c, in particular pp. 27–28).

of defaults or expectations includes $\neg p$, $\neg q$ and $\neg r$. Now suppose that the agent receives from a reliable source evidence for the fact that $\neg p \vee \neg q \vee \neg r$ is true. We consider $K \dot{-} (p \wedge q \wedge r)$. Since $H \dot{-}_0 (p \wedge q \wedge r)$ is empty, straight clearing yields the result that $K \dot{-} (p \wedge q \wedge r)$ is $Inf(\emptyset)$, and this set contains $\neg p$, $\neg q$, and $\neg r$. Similarly, $K *_0 \neg p \vee \neg q \vee \neg r$ contains all three of $\neg p$, $\neg q$, and $\neg r$. So although the agent initially has the explicit information that each of p, q and r is true, *all* the expectations to the contrary are enforced when he makes room for the assumption that (at least) *one* of p, q and r is false. This surely is undesirable.[209]

If we follow Gärdenfors and Makinson in construing *Inf* as involving a transition from the set of expectations to its revision occasioned by explicit information, another possible objection is that the method presented is a hybrid method that does not allow for an integrated treatment of two entirely unrelated revision processes: that of the initial information base (taking us from H to $H' = H *_0 \phi$ or $H' = H \dot{-}_0 \phi$) and that of the background theory of expectations (taking us from E to $E \underline{*} h'$). We are now going to further develop the vertical perspective in such a way as to achieve greater methodological unity.

5.1.3 *Modelling the direct mode by expectation revision*

For the changes of the coherent belief sets, the central method of the immediate mode is to change $K = Inf(\mathcal{H})$ into a new belief set K' by *first* changing the base in a rather plain (artless, undemanding, uncomplicated) fashion from $\mathcal{H}$ to a new base $\mathcal{H}'$, and *then* drawing inferences from $\mathcal{H}'$: that is, taking $K' = Inf(\mathcal{H}')$ as the new belief state. We now have a closer look at the possibility of applying the finitary inference relations characterized in the last chapter and their Gärdenfors-Makinson style reconstruction in terms of revisions of expectation sets. Let H again be the flattened form (the 'propositional projection') of a prioritized information base $\mathcal{H}$, and let h be the conjunction of the elements of H.

We now restate the definitions introduced in the previous sections in terms of expectation revision. First we have the method of infallible information:

$$K * \phi \;=\; Inf(H \cup \{\phi\}) \;=\; E \underline{*} (\textstyle\bigwedge (H \cup \{\phi\}))$$

and

$$K \dot{-} \phi \;=\; Inf(H - \{\phi\}) \;=\; E \underline{*} (\textstyle\bigwedge (H - \{\phi\}))$$

As mentioned above, this contraction function fails to satisfy the condition of Success for contractions, $(\dot{-}4)$. There are two reasons why $(\dot{-}4)$ may get violated. First, it may happen that $H - \{\phi\}$ still entails ϕ, and since $\bigwedge(H - \{\phi\})$ must be in the theory $K \dot{-} \phi$, by the Success and Closure conditions for $\underline{*}$, ϕ will still be in $K \dot{-} \phi$. But even if it is not entailed by $H - \{\phi\}$, ϕ may reappear in a second way, due to its being entailed by some of the expectations, possibly taken

---

[209]Note that the result would not be counterintuitive if H were $\{p \wedge q \wedge r\}$. The agent's actual belief base $\{p, q, r\}$ represents three independenty aupported pieces of information.

together with some pieces of information. Neither does this method satisfy the basic AGM postulate of Inclusion, $(\dot{-}2)$. For instance, it may happen that ϕ is in a consistent set K, while $\neg\phi$ is an expectation contained in the set E wich can also be found in $K\dot{-}\phi$.

Next we have straight clearing of information conflicts.

$$K * \phi \;=\; Inf(H *_0 \phi) \;=\; E\underline{*}(\bigwedge(H *_0 \phi))$$

This revision function satisfies the condition of Consistency Preservation, $(*5)$.

$$K\dot{-}\phi \;=\; Inf(H\dot{-}_0\phi) \;=\; E\underline{*}(\bigwedge(H\dot{-}_0\phi))$$

This contraction function still does not generally satisfy the condition of Success, $(\dot{-}4)$. Although the first of the above-mentioned reasons is blocked (since by definition $H\dot{-}_0\phi$ does not include ϕ), the second reason remains.

Due to the presence of a great many expectations in E, even an empty information base $H = \emptyset$ would leave us with enough beliefs to act upon. We then just get $K = E\underline{*}\top$ which is identical with the full, unrevised expectation set E.[210]

5.2 Modelling the direct mode: Combining information and expectation

5.2.1 *Information bases without priorities*

In Section 5.1 we improved upon the method infallible information by seeing to it that the information base is always kept consistent. The method that resulted, however, was not really satisfactory. So perhaps we chose the wrong remedy. Instead of locating the mistake in the process of base change (along the horizontal axis), we may look for it in the inference process (along the vertical axis). Perhaps it was wrong to insist that $H = |\mathcal{H}|$ and $H' = |\mathcal{H}'|$ must always be contained in $Inf(\mathcal{H}) = E\underline{*}h$ and $Inf(\mathcal{H}') = E\underline{*}h'$ respectively—an effect that resulted essentially from the melting together of all the information represented in a prioritized base into a single sentence. We had to do this in order to be able to perform an AGM revision on the expectation set. But we need not necessarily endorse that way of reconstructing nonmonotonic reasoning. In this section, I propose to leave the structure of the (prioritized) information base $\mathcal{H}$ intact and take advantage of this structure in the process of revising the expectations through explicit information.

If we want to make sure that the expectation set E is revisable by an ordinary AGM revision operation $\underline{*}$, we must also assume that E is a theory. But now I no longer want to commit myself to the Makinson-Gärdenfors model of nonmonotonic reasoning and aim at an improved interpretation of the direct mode of belief revision. For this purpose we let $\mathcal{E}$ retain its structure of a (prioritized)

---

[210]It is supposed here that E is consistent and closed under Cn and that $\underline{*}$ is an AGM revision operation, so that we can apply $(*3)$ and $(*4)$.

belief base and do not telescope it into its propositional projection $E = |\mathcal{E}|$. We are ready to allow for varying priorities or 'degrees of expectation'.

The basic idea now is first to put the prioritized information base $\mathcal{H}$ 'on top of' the prioritized expectation base $\mathcal{E}$, and then to perform our favourite mechanism for (prioritized) base change or consistency restoration. The functional roles of $\mathcal{E}$ and $\mathcal{H}$ in the model to be presented now are very similar. In particular, $\mathcal{E}$ and $\mathcal{H}$ have the same format: They are not required to be logically closed, they may be infinite, and they may be prioritized. There need not even be a sharp dividing line between 'hard' information and 'soft' expectations. Once expectation and information are glued together, one cannot perceive the difference between them any more. The difference is just that expectations are only implicit, while information is always explicit.[211] Anyway, K is now defined to be

$$K \;=\; Inf(\mathcal{H}) \;=\; Consol(\mathcal{E} \circ \mathcal{H})$$

for some fixed prioritized expectation base $\mathcal{E}$. This is different from, and not reducible to, the Makinson-Gärdenfors identity.

For a start, let us assume that the information base $\mathcal{H}$ is 'flat', i.e., unprioritized. So $\mathcal{H} = \langle H \rangle$. We now look how the consolidation construction which has been designed for the static (vertical) dimension helps us to generate an interesting operation for the dynamics of coherent theories. The first revision operation we consider is suitable for flat bases only:

$$K * \phi \;=\; Consol(\mathcal{E} \circ \langle H \cup \{\phi\}\rangle)$$

The operations $\circ$ and $Consol$ of list concatenation and consolidation have been introduced in Sections 1.2.6 and 1.2.7 above. Although the operation of consolidation may in general be just any way of contracting a belief set with respect to $\bot$, in this chapter we always have in mind the special case of a consolidation by means of a prioritzed base contraction (with respect to $\bot$).

This new revision function satisfies the condition of Consistency Preservation, and in fact even the condition of Strong Consistency ($*5^+$) that entails that even the revision by an inconsistent sentence must not lead to an inconsistent belief set. However, the new revision function does not in general satisfy the Success condition ($*2$). Success is only guaranteed for consistent input ϕ, since every consolidation ends up with a consistent body of beliefs. The method does satisfy Weak Success, ($*2^-$).

For the problem of contraction, the following recipe suggests itself.

$$K \dot{-} \phi \;=\; Consol(\mathcal{E} \circ \langle H - \{\phi\}\rangle)$$

This surface-oriented contraction function is similar to the corresponding revision function. Like the previous suggestions in this chapter, however, it fails

---

[211]I admit of course that the method of invariably giving explicit opinions priority over implicit opinions is problematic, but it is the best method available in our framework. I have briefly commented on this problem of in Section 1.2.3.

to satisfy the condition of Success for contractions, ($\dot{-}$4). An alternative method is to use *phantom beliefs* and put

$$K \dot{-} \phi \;=\; Consol(\mathcal{E} \circ H \circ \langle \neg \phi^* \rangle)$$

This notation is meant to denote the following somewhat more general process of consolidation. The phantom belief $\neg\phi^*$ acts just like an ordinary belief $\neg\phi$ at the top level of priority in the process of determining the 'best' maximal consistent subsets of $E \cup H \cup \{\phi\}$ (the first step of prioritized base contraction with respect to $\bot$). However, when the Cn-closures of these best subsets are formed (the second step), $\neg\phi^*$ is no longer counted. It is simply omitted from the maximal consistent sets. Still, as a result of the provisional appearance of $\neg\phi$, the contraction function thus constructed satisfies the condition of Success: If ϕ is not a logical truth, it is certainly omitted from $Consol(\mathcal{E} \circ \mathcal{H} \circ \langle \neg \phi^* \rangle)$. A phantom belief $\neg\phi^*$ is exactly what Poole (1988) calls a *constraint* for nonmonotonic reasoning.[212] To my knowledge, the idea of using Poolean constraints for the modelling of belief contractions was first put forward by Brewka (1991). But the idea of using processes of consolidation for the construction of revisions and contractions can be applied quite generally, using any kind of consolidation function $Consol$ one favours.

It is an advantage of the present suggestion that it supports an integrated treatment of (the revisions of) 'hard' information and 'soft' expectations. Moreover, it is very flexible as regards the choice of a method for contraction used in the consolidation process. Finally, items of information that would have been radically cancelled by the straight clearing method of the last section now take effect. Recall the example discussed on page 125, with $H = \{p, q, r\}$ and $K = Inf(H)$ including p, q and r, but $E = |\mathcal{E}|$ including $\neg p$, $\neg q$ and $\neg r$. Let us assume $\mathcal{E} = \langle \{\neg p, \neg q, \neg r\} \rangle$ for the sake of simplicity. Then the revision $K * (\neg p \vee \neg q \vee \neg r)$ obtained by the new method contains $(p \wedge q) \vee (p \wedge r) \vee (q \wedge r)$. When told that at least one out of three items of information may be false (but we are not told which), we would still stick to two of them (although we do not know which). This is a much better result than the one we got earlier, and in fact it seems to be what we intuitively expect.

One may consider it as a problem, however, that in the process of a revision no absolute priority is given to new information ϕ. It may well drop out when the new expectation-plus-information base is consolidated. The present proposal is thus closer to the concepts of *autonomous* (Galliers 1992) or *selective fact-finding belief revision* (Cross and Thomason 1992) which does not commit itself to assigning absolute priority to new information.[213] Recency of learning does

---

[212]'The idea is to define a set of constraints used to prune the set of scenarios. They are just used to reject scenarios and cannot be used to explain anything. Constraints are formulae with which scenarios must be consistent.' (Poole 1988, p. 39)

[213]Sven Ove Hansson (1994b) uses the term 'non-prioritized belief revision' for the revision concept that does not in general assign priority to new information over old information. In

not indicate evidential superiority. In this model input sentences do not have a privileged status and may well be neutralized by earlier beliefs.

5.2.2 *Prioritized information bases*

Let us be more consistent in taking the original structure of the information base $\mathcal{H}$ seriously. Rather than telescoping it into the flat set H, we shall now leave the prioritization of $\mathcal{H}$ totally untouched.

In order to fit ϕ somehow into a prioritized base $\mathcal{H}$ when performing a belief revision, we would need an integer i to come along with the new belief ϕ that specifies its priority position in the changed information base. (There is no analogous problem for the contraction of (irredundant) belief bases.[214]) The problem is that information does not usually come with a certainty parameter, and if it does, the meaning of the parameter is scarcely clear. In order to avoid integers as additional input parameters, and for the sake of simplicity, we assume in the following (quite in the spirit of the AGM approach) that the input sentence always gets assigned top priority. Rather than fit ϕ into $\mathcal{H}$, we place it on top of $\mathcal{H}$.

Definition 1 Let $\mathcal{E}$ be a prioritized expectation base, $\mathcal{H}$ be a prioritized information base, and let the belief set K be generated by $Inf(\mathcal{H}) = Consol(\mathcal{E} \circ \mathcal{H})$. Then the *consolidation revision function* $*$ over K, in symbols $* = \mathcal{R}(\mathcal{E}, \mathcal{H})$, is defined by

$$K * \phi \;=\; Consol(\mathcal{E} \circ \mathcal{H} \circ \langle \phi \rangle))$$

This revision function satisfies the conditions of Strong Consistency ($*5^+$) and Weak Success ($*2^-$). It is striking, however, that this revision method fails to satisfy the Preservation Principle (Pres), according to which new pieces of information that are consistent with the old theory K should never lead to the removal of any old beliefs. We now give a simple counterexample illustrating why (Pres) fails.

Assume that we have no expectations whatsoever, i.e., $\mathcal{E} = \langle \emptyset \rangle$, and that we have an unprioritized (single-layered) information base $\mathcal{H} = \langle H \rangle = \langle \{q, \neg q, p \leftrightarrow q\} \rangle$. Using for 'best subsets' of H the notation explained in Section 1.2.7.2 (page 41), we have

---

the present context, however, this term would be misleading, since we are always ready to use priorities for resolving inconsistencies. See also Hansson (1999b).

[214] The problem shows up for the first of the above-mentioned contraction methods, however, if H is *redundant* in the sense that one and the same belief ϕ is in H_i and H_j for different i and j. In that case one has to say *which* of the occurrences of ϕ is to be retracted.

$$\mathcal{H} \parallel \bot \qquad = \{\{q, p \leftrightarrow q\}, \{\neg q, p \leftrightarrow q\}\}$$

$$K \quad = Inf(\mathcal{H}) \qquad = \bigcap \{Cn(F) : F \in \mathcal{H} \parallel \bot\} \qquad = Cn(p \leftrightarrow q)$$

$$(\mathcal{H} \circ \langle p \rangle) \parallel \bot = \{\{p, q, p \leftrightarrow q\}, \{p, \neg q\}\}$$

$$K * p = Inf(\mathcal{H} \circ \langle p \rangle) = \bigcap \{Cn(F') : F' \in (\mathcal{H} \circ \langle p \rangle) \parallel \bot\} = Cn(p)$$

We see that $\neg p$ is not in K. Nevertheless, $p \leftrightarrow q$ which is in K is lost in $K * p$—a blatant violation of the Preservation Principle.

It might seem annoying to find the Preservation Principle violated. As soon as one admits nonmonotonic reasoning, however, many phenomena that had been considered unacceptable from the point of view of traditional logics, become quite natural. It is just to be expected that new information can lead to a reexamination of old information. In our simple example, the input p makes it clear that $p \leftrightarrow q$ is actually more vulnerable than it was in the initial belief state.

One problem the present model has with such examples is to justify why new incoming information is unfailingly given top priority. I grant that this is a very crude method that could—and perhaps should—be refined by having each piece of new information accompanied by some priority tag indicating the position in the prioritized bases at which this item is to be placed. This idea almost invariably calls for numerical assignments which we want to avoid for the reasons given above.[215] But *if* one is ready to accept the absolute priority of new information, it is not strange that the new sentence p in the above example in fact leads to a revaluation, and finally to a removal, of the old item $p \leftrightarrow q$ at the lower level of the new base $\mathcal{H} \circ \langle p \rangle$.

For the contraction function, we again have a surface-oriented removal of explicit information

$$K \dot{-} \phi \quad = \quad Consol(\mathcal{E} \circ (\mathcal{H} \backslash \phi))$$

where '$\backslash \phi$' removes all occurrences of ϕ in $\mathcal{H}$.[216] It is obvious that this contraction function in general fails to satisfy the condition of Success.

As an alternative and I think preferable method of belief contraction, we can again use phantom beliefs.

Definition 2 Let $\mathcal{E}$ be a prioritized expectation base, $\mathcal{H}$ be a prioritized information base, and let the belief set K be generated by $Inf(\mathcal{H}) = Consol(\mathcal{E} \circ \mathcal{H})$. Then the *consolidation contraction function* $*$ over K, in symbols $\dot{-} = \mathcal{C}(\mathcal{E}, \mathcal{H})$, is defined by

$$K \dot{-} \phi \quad = \quad Consol(\mathcal{E} \circ \mathcal{H} \circ \langle \neg \phi^* \rangle)$$

---

[215] This is done in Brewka (1991).

[216] Yet another idea would be to put $K \dot{-} \phi = Consol((\mathcal{E} \circ \mathcal{H}) \backslash \phi)$ where now '$\backslash \phi$' removes all occurrences of ϕ in $\mathcal{E} \circ \mathcal{H}$.

If $\neg\phi$ can be found in $|\mathcal{E} \circ \mathcal{H}|$, then $K \dot{-} \phi$ contains $\neg\phi$ and actually equals $K * \neg\phi$.[217]

This contraction function satisfies the condition of Success, ($\dot{-}4$). However, the condition of Inclusion, ($\dot{-}2$), is still invalid, as is seen from a simple example. Let $\mathcal{E} = \langle\emptyset\rangle$ and $\mathcal{H} = \langle\{\neg p\}, \{p\}\rangle$. Then $K = Inf(\mathcal{H}) = Cn(p)$ and $K \dot{-} p = Inf(\mathcal{H} \circ \langle\neg p^*\rangle) = Cn(\neg p)$. The new phantom belief $\neg p^*$ blocks the old belief p (as it should be) which had in turn suppressed the old and less firmly presented item of information $\neg p$. After the contraction has taken place, the latter comes to the fore again, and $K \dot{-} p$ is no subset of K.

As in foundational belief change in general, the Recovery condition ($\dot{-}5$) fails. Take for instance the empty expectation base $\mathcal{E} = \langle\emptyset\rangle$ and the information base $\mathcal{H} = \langle\{p \vee q, p \leftrightarrow q\}\rangle$. Then $K = Consol(\mathcal{E} \circ \mathcal{H}) = Cn(p \wedge q)$ and $K \dot{-} p = Consol(\mathcal{E} \circ \mathcal{H} \circ \langle\neg p^*\rangle) = Consol(\langle\{p \vee q, p \leftrightarrow q\}, \{\neg p^*\}\rangle) = Cn(p \vee q) \cap Cn(p \leftrightarrow q) = Cn(\emptyset)$. Since K contains q but $Cn(K \dot{-} p \cup \{p\})$ does not, the Recovery condition is violated.

Definitions 1 and 2 make up my final model of the direct mode of belief revision. They have all the essential characteristics (I) – (III) of the direct mode of belief revision that we mentioned in Section 3.4.1 above. First, derived beliefs change as a result, and only as a result, of changes on the base level, since the changes only operate on $\mathcal{H}$. Second, the changes leading to the new base are made in a rather plain fashion, viz. by list concatenation, so there are no non-trivial choices to be made. And third, I have provided a model of nonmonotonic reasoning on the basis implicit, defeasible background assumptions represented in the expectation base $\mathcal{E}$. For prioritized bases $\mathcal{H}$ and $\mathcal{H}'$ with $|\mathcal{H}| \subseteq |\mathcal{H}'|$ (or more specifically, with $\mathcal{H}' = \mathcal{H} \circ \mathcal{G}$ for some $\mathcal{G}$), it need not—and usually does not—hold that $Inf(\mathcal{H}) = Consol(\mathcal{E} \circ \mathcal{H})$ is a subset of $Inf(\mathcal{H}') = Consol(\mathcal{E} \circ \mathcal{H}')$.

I have deliberately deviated from the Makinson-Gärdenfors intuition that hard information revises soft expectations. On the present account, expectations are conjoined with information, with the latter taking priority over the former, and then an inference mechanism is applied to consolidate the integrated structure. Clearly, construing $Inf(\mathcal{H})$ as $Consol(\mathcal{E} \circ \mathcal{H})$ for some fixed prioritized base $\mathcal{E}$ is substantially different from construing $Inf(\mathcal{H})$ as $E * h$ for some fixed theory E. It is not possible to define an AGM style expectation revision operation $*$ by putting $E * h$ to be $Consol(\mathcal{E} \circ \mathcal{H})$ or $Consol(\mathcal{E} \circ \langle H\rangle)$. Apart from the fact that this would not give us a 'successful' AGM revision function, $*$ would not even be well-defined. Clearly, for every contingent theory E and every contingent sentence h, there are many prioritized bases $\mathcal{E}$ and $\mathcal{H}$ which are represented by E and h (in the sense that $E = Cn(|\mathcal{E}|)$ and $h = \bigwedge(|\mathcal{H}|)$) but differ decisively in the inferences they warrant after consolidation.[218]

---

[217]The contracted set $K \dot{-} \phi$ equals $\bigcap\{Cn(G - \{\neg\phi\}) : G \in (\mathcal{E} \circ \mathcal{H} \circ \langle\neg\phi\rangle) \Downarrow \perp\}$ in the case where $\neg\phi$ is not in $|\mathcal{E} \circ \mathcal{H}|$; and $\bigcap\{Cn(G) : G \in (\mathcal{E} \circ \mathcal{H} \circ \langle\neg\phi\rangle) \Downarrow \perp\}$ in the case where $\neg\phi$ is in $|\mathcal{E} \circ \mathcal{H}|$.

[218]Notice that what we are discussing here are operations transforming prioritized belief bases into theories and sentences. So the situation is now fundamentally different from that referred

Let me finally point out that for my favourite method of foundationalist belief change to make sense, a special representational format for the information and the expectations is needed, namely prioritized belief bases $\mathcal{H}$ and $\mathcal{E}$. The method of consolidation through prioritized contraction with respect to $\perp$ needs comparatively fine-grained priority distinctions in order to give good results. In the limiting case of flat bases, prioritized contraction becomes full meet contraction, which is in general very crude and eliminates too many items of belief and expectation. Compare, for instance, the indiscriminating bases $\mathcal{E} = \langle\{r, r \supset \neg p\}\rangle$ and $\mathcal{H} = \langle\{q, q \supset \neg p\}\rangle$ with the more fine-grained bases $\mathcal{E}' = \langle\{r\}, \{r \supset \neg p\}\rangle$ and $\mathcal{H}' = \langle\{q \supset \neg p\}, \{q\}\rangle$. In the first case, we lose all the information contained in the prior belief state and get $K * p = Cn(p)$, while in the second case, a lot of information is preserved, and $K * p$ equals $Cn(p, q, r \supset \neg p)$.

5.2.3   Iterated belief change in the direct mode

The direct mode offers a simple and perspicuous method for modelling iterated changes of belief. Some properties are immediate from the way we have described the processes of adding new information at a new top level of priority, and of consolidating the new information base by selecting maximal consistent sets from the top to the bottom. For arbitrary expectation and information bases $\mathcal{E}$ and $\mathcal{H}$, we thus get that $Consol((\mathcal{E} \circ \mathcal{H} \circ \langle\phi\rangle) \circ \langle\psi\rangle)$ is identical with $Consol(\mathcal{E} \circ \mathcal{H} \circ \langle\phi \wedge \psi\rangle)$ whenever ϕ and ψ are consistent with one another. If, however, ϕ and ψ are mutually incompatible, the item coming in later completely invalidates the earlier one, and $Consol((\mathcal{E} \circ \mathcal{H} \circ \langle\phi\rangle) \circ \langle\psi\rangle)$ equals $Consol(\mathcal{E} \circ \mathcal{H} \circ \langle\psi\rangle)$. So, we have already verified the following equation for iterated revisions of prioritized belief bases.

$$(K * \phi) * \psi = \begin{cases} K * (\phi \wedge \psi) & \text{if } \{\phi, \psi\} \not\vdash \perp \\ K * \psi & \text{otherwise} \end{cases} \tag{$\dagger$}$$

Equation ($\dagger$) is very strong, and we will shortly see that it must be taken with a grain of salt. Its upper line is essentially identical with the condition that Nayak *et al.* (1996) call *Conjunction*. They point out that it implies what the orthodox AGM theory says about iterated revisions, viz.,

(AGM-It) If $\neg\psi \notin K * \phi$, then $(K * \phi) * \psi = K * (\phi \wedge \psi)$

This condition follows from ($*3$), ($*4$), ($*7$) and ($*8$), provided the postulates are taken as applying to two-place revision functions. Moreover, Nayak *et al.* also show that Conjunction entails three of the four well-known postulates for iterated revision suggested by Darwiche and Pearl (1994, 1997):

(DP1)   If $\phi \in Cn(\psi)$, then $(K * \phi) * \psi = K * \psi$
(DP2)   If $\neg\phi \in Cn(\psi)$, then $(K * \phi) * \psi = K * \psi$
(DP3)   If $\phi \in K * \psi$, then $\phi \in (K * \phi) * \psi$

to in footnote 208 above.

(DP4) If $\neg\phi \notin K * \psi$, then $\neg\phi \notin (K * \phi) * \psi$

The lower line of equation (†) for iterated belief change is identical with (DP2). So clearly our mechanism for foundational prioritized belief revision satisfies all postulates of Darwiche and Pearl.

It is easy to check that in the presence of the basic AGM postulates (∗1) – (∗6), Conjunction implies (DP1), (DP3) and (DP4), as well as the supplementary AGM-postulates (∗7) and (∗8) (compare Nayak *et al.* 1996, Observation 2). For all this, it is again essential that the AGM postulates are understood to hold not only for some fixed belief set K (for a one-place revision function ∗) but for all potential belief sets at the same time (for a two-place revision function ∗). For the conceptual relevance of this question, see Section 3.6 above.

A problem pointed out by Nayak *et al.* is that *both* Conjunction *and* (DP2) are incompatible with the basic AGM postulates. We already know that some of the AGM postulates fail for our favourite kind of foundational belief change: We have got only a weakened form of (∗2) and nothing like (∗4) (which would imply the Preservation Principle). So the incompatibilities of Conjunction and (DP2) with the AGM postulates need not really worry us. After all, the set of postulates satisfied by the revision method introduced in Section 5.2.2 *must* be consistent, since they are all satisfied by a concrete revision recipe—just the one I was advocating. Still it is very instructive to follow the discussion of Nayak *et al.* (1996, pp. 121, 128) and see how their arguments relate to what happens in the present framework.

Example 1. The example Nayak *et al.* use to show that (DP2) is in conflict with the basic AGM postulates (taken as postulates for two-place revision functions ∗) is as follows. Let p and q be atoms and let $K = Cn(\emptyset)$. Then, by (∗1) – (∗6), we get that $(K * p) * \neg q = Cn(\{p, \neg q\}) = K * (p \wedge \neg q)$. Hence, since ∗ is a two-place revision function, we also get $((K * p) * \neg q) * q = (K * (p \wedge \neg q)) * q$. By two-fold application of (DP2), we get from this that $(K * p) * q = K * q$. But, by (∗1) – (∗6) again, $(K * p) * q = Cn(\{p, q\})$ and $K * q = Cn(\{q\})$, which are clearly different, and we have a contradiction.

Example 2. Now let us make clear that Conjunction is in conflict with the basic AGM postulates (again taken as postulates for two-place revision functions ∗).[219] Let p and q be atoms and let $K = Cn(\{p\})$. By (∗1), both $p \vee q$ and $p \vee \neg q$ are in K. But by (∗1), (∗2), and (∗5), at least one of $p \vee q$ and $p \vee \neg q$ is not in $K * \neg p$. Without loss of generality, let us suppose that $p \vee q$ is not in $K * \neg p$ (the other case is similar). Now consider $(K * p \vee q) * \neg p$. Since p and q are atoms, $p \vee q$ and $\neg p$ are consistent with one another, so Conjunction and (∗6) tell us that $(K * p \vee q) * \neg p = K * (\neg p \wedge q)$. On the other hand, $K * p \vee q = K$, by (∗3) and (∗4). Hence, since ∗ is a two-place revision function, we thus get

---

[219]The proof that Nayak *et al.* (1996, p. 128) offer for their Theorem 1 is not entirely correct since there the assumption that there is a revision function ∗ *with particular properties* is led to a contradiction, while the theorem requires them to show that there is *no* revision function ∗ which satisfies all the conditions mentioned.

$(K * p \lor q) * \neg p = K * \neg p$. Hence $K * (\neg p \land q) = K * \neg p$. But we said that $p \lor q$ is not in $K * \neg p$, so it is not in $K * (\neg p \land q)$ either, contradicting (*1) and (*2).

As I have pointed out, my favourite model of Section 5.2.2 satisfies both (a version of) Conjunction and (DP2). So one may be inclined to conclude that 'the reason' that the examples get a treatment different from the AGM approach is that the basic AGM postulates are violated. I am now going to recast the examples in the framework of the vertical perspective and show that this conclusion is not the right one.

Example 1, in the vertical perspective.

We start from the empty doxastic state $\mathcal{E} = \mathcal{H} = \langle \emptyset \rangle$. This gives us $K = Inf(\emptyset) = Cn(\emptyset)$. Concerning the relevant changes, we get $(K * p) * \neg q = Inf((\mathcal{H} \circ \langle p \rangle) \circ \langle \neg q \rangle) = Cn(\{p, \neg q\}) = Inf(\mathcal{H} \circ \langle p \land \neg q \rangle) = K * (p \land \neg q)$. So far we are in agreement with the earlier discussion. But the next step is blocked. We can easily calculate that $((K * p) * \neg q) * q = Cn(p \land q)$, but $(K * (p \land \neg q)) * q = Cn(q)$. So even though the theory $(K * p) * \neg q$ is identical with the theory $K * (p \land \neg q)$, the two theories are revised in a different way.

Example 2, in the vertical perspective. We start from the doxastic state without any expectations, $\mathcal{E} = \langle \emptyset \rangle$, and with the information base $\mathcal{H} = \langle \{p\} \rangle$. This gives us $K = Inf(\mathcal{H}) = Cn(p)$. Both $p \lor q$ and $p \lor \neg q$ are in K. Neither $p \lor q$ nor $p \lor \neg q$ is in $K * \neg p = Inf(\langle \{p\}, \{\neg p\} \rangle) = Cn(\neg p)$. Now consider $(K * p \lor q) * \neg p$. We find that it is $Inf(\langle \{p\}, \{p \lor q\}, \{\neg p\} \rangle) = Cn(\neg p \land q)$. And we also get that $K * p \lor q$ is $Inf(\langle \{p\}, \{p \lor q\} \rangle) = Cn(p) = K$. All this is in perfect agreement with our earlier discussion. But the next step is again blocked. We already saw that $(K * p \lor q) * \neg p = Cn(\neg p \land q)$, but that $K * \neg p = Cn(\neg p)$. So even though the theory $K * p \lor q$ is identical with the theory K, the two theories are revised in a different way.

The moral of the two examples is the same. In both cases we have theories that can be proven identical, and yet the revisions they undergo in response to the same input is different. We are not presented here with two-place revision *functions* operating on theories. In accordance with the foundationalist philosophy of belief change, what gets revised in the first place is *not theories*, but rather *prioritized belief bases*. And it becomes evident that two bases that generate the same theory will not in general continue to generate the same theory after a new sentence has been tagged to the top of them. Clearly our constructive model does not lead to a contradiction. (This is the general advantage of attending to constructive models rather than abstract postulates!) The reason that it is consistent, is not that it fails to satisfy the basic AGM postulates, but that it *per se* transcends the idea of a two-place revision function governing iterated changes of theories. The revision of the theories in question depends crucially on the bases they have been generated from. In other words, the change of a belief base does not only lead to a change of propositional content, but also to a change of the dispositions to change beliefs.[220] Propositional input changes the agent's dispositions to change other propositions. This does not fit together with the rationale of two-place revision functions as encoding attitudes of a single doxastic agent that may stay constant across varying informational states. We find ourselves in good concordance with Nayak *et al.* (1996, p. 122) who advocate a kind of meta-revisionist view: 'When we revise our current belief set, it is not only the belief set that changes; the accompanying revision operation also undergoes modification.' The difference is that Nayak *et al.* construct a complicated mechanism for the meta-revision of revisions,[221] while from our vertical perspective the whole work is being done easily and automatically by the trivial change operation that appends a new sentence on top of the old information base. The contrast between these two particular methods of revision is a nice illustration of the general fact that belief change operations are simple in the vertical perspective, but quite demanding in the horizontal perspective.

Finally, then, it may perhaps be wise to follow the advice of Nayak *et al.* to distinguish in equation (†) for iterated belief change two different revision functions, and write something like

$$(K *_1 \phi) *_2 \psi = \begin{cases} K *_1 (\phi \wedge \psi) & \text{if } \{\phi, \psi\} \vdash \perp \\ K *_1 \psi & \text{otherwise} \end{cases}$$

The point here is to indicate that $*_1$ und $*_2$, taken as one-place revision functions associated with K and $K *_1 \phi$ respectively, do *not* derive from a belief revision strategy as encoded by a two-place revision function. It seems, though, that in the present foundational framework the sloppier notation is not liable to create misunderstandings.

5.3 Logical properties of revision functions generated by the direct mode

Many forms of belief revision studied in the AGM paradigm (which we have subsumed under the general heading of coherentist belief revision) satisfy all of the characteristic AGM postulates $(*1) - (*8)$. It is a natural question to ask whether the different philosophy underlying the direct mode of foundationalist belief change makes itself felt by giving rise to different formal properties of the engendered revision operators. The answer is: Yes, it does.

We have been commenting much on the Success, Consistency Preservation, and Inclusion postulates in the previous sections of this chapter. Now we are going to have a look at the remaining rationality postulates for theory revision.

---

[220] As pointed out in Sections 1.2.8 and 3.6, one may even identify a doxastic *state* with a belief *revision* strategy.

[221] They do this by revising entrenchment relations using a method of Nayak (1994b) and Nayak, Nelson and Polansky (1996) . The method, by the way, has much to recommend itself; see Rott (2000c).

5.3.1 The case of infallible information

Observation 6 *Let H be a finite set of sentences and Inf be an arbitrary inference operation. Let $K = Inf(\bigwedge H)$, and define $K * \phi$ as $Inf(\bigwedge H \wedge \phi)$, for*

every sentence ϕ. Then the revision operation $*$ over K satisfies
$$\left.\begin{array}{c} (*1) \\ (*2) \\ (*3) \\ (*4) \\ (*6) \\ (*7) \\ (*7c) \\ (*8) \\ (*8c) \end{array}\right\} \ if$$

Inf satisfies
$$\left\{\begin{array}{c} Cn\text{-}Closure \\ (Ref) \text{ and } (RW) \\ (Cond) \\ (RMon) \\ (LLE) \\ (Cond) \\ (Cut) \\ (RMon) \\ (CMon) \end{array}\right\}.$$
However, the revision operation $*$ does not

satisfy Consistency Preservation, ($*5$).

5.3.2 The case of cleared information

Observation 7 *Let H be a finite set of sentences and Inf be an arbitrary inference operation. Let $K = Inf(\bigwedge H)$, and define $K*\phi$ as $Inf(\bigwedge(H *_0 \phi))$, for every sentence ϕ, where $*_0$ is the operation of full meet base revision. Then the revi-*

sion operation $*$ over K satisfies
$$\left.\begin{array}{c} (*1) \\ (*2) \\ (*3) \\ (*4) \\ (*5) \\ (*6) \end{array}\right\} \ if \ Inf \ satisfies \left\{\begin{array}{c} Cn\text{-}Closure \\ (Ref) \text{ and } (RW) \\ (Cond) \\ (RMon) \\ (CP) \\ (LLE) \end{array}\right\}.$$

However, the revision operation $*$ satisfies neither ($*7$) nor ($*8$), nor ($*8r$), even when Inf $= Cn$. It satisfies both ($*7c$) and ($*8c$) when Inf $= Cn$, but fails to satisfy either of the two when Inf is different from Cn.

Several writers in belief revision have linked coherentist and foundationalist approaches. An axiomatic characterizations of *theory* contraction and revision generated, amongst other methods, by full meet *base* contraction and revision are given by Hansson (1993b), and in another form by Rott (1993) and del Val (1994).[222] These theory change operations correspond to those discussed in this section for the special case when H is consistent and $Inf = Cn$.

5.3.3   The case of integrated prioritized information: Non-preservative belief change

We disregard the case of flat information bases and turn immediately to the more principled case where both the expectations and the pieces of information are represented in the form of prioritized belief bases.

Observation 8 *Let $\mathcal{E}$ be a prioritized expectation base and let Inf be the inference operation on prioritized belief bases defined by $Inf(\mathcal{H}) = Consol(\mathcal{E} \circ \mathcal{H})$, for any prioritized information base $\mathcal{H}$. Take some such $\mathcal{H}$, let $K = Inf(\mathcal{H})$, and define $K * \phi$ as $Inf(\mathcal{H} \circ \langle \phi \rangle)$, for every sentence ϕ. Then the revision operation*

$$* \text{ over } K \text{ satisfies } \begin{Bmatrix} (*1) \\ (*2^-) \\ (*3) \\ (*5^+) \\ (*6) \\ (*7) \\ (*8c) \\ (*8r) \end{Bmatrix}. \text{ However, the revision operation } * \text{ does not satisfy}$$

*the Preservation Principle (Pres), and a fortiori, it satisfies neither (*4) nor (*8).*

The idea of using phantom beliefs for the construction of contractions bears considerable similarity with the idea of the Harper identity (see page 108).[223] Both ideas construct a contraction with respect to ϕ by somehow combining a revision by $\neg\phi$ with the original belief state. Surprisingly, however, the Harper identity fails badly for our favourite constructions of Section 5.2.2; in general there is no subset relation between $K \dot{-} \phi$ and $K \cap (K * \neg\phi)$. The Levi identity, on the other hand, remains intact for consistent input sentences ϕ.

Observation 9 *Let $\mathcal{E}$ and $\mathcal{H}$ be prioritized expectation and information bases, respectively. Let $K = Consol(\mathcal{E} \circ \mathcal{H})$, and define*

$$K * \phi = Consol(\mathcal{E} \circ \mathcal{H} \circ \langle \phi \rangle)$$
$$K \dot{-} \phi = Consol(\mathcal{E} \circ \mathcal{H} \circ \langle \neg\phi^* \rangle)$$

---

[222]Actually Rott (1993) characterizes a 'Nebelized' version of theory contraction that is generated by full meet base contraction (first studied in Nebel 1989). This version enforces satisfaction of the Recovery postulate in a somewhat artificial way. The difference vanishes when we pass from contractions to revisions with the help of the Levi identity.

[223]This has been brought to my attention by Aditya Ghose.

for all sentences ϕ. Then the revision and contraction operations $*$ and $\dot{-}$ over K satisfy the Levi identity with respect to consistent sentences ϕ. However, both parts of the Harper identity are violated.

In this chapter we have investigated some of the properties of belief change as seen from the vertical perspective. We took three concrete constructions in the direct mode of belief revision and showed that some selected rationality postulates are satisfied or violated. However, we have given *no representation theorems* to the effect that every revision function satisfying such-and-such postulates can be represented as generated by one or the other of our constructions in the direct mode.

Recently, representation theorems for finitary theory contractions and revisions generated by prioritized base contractions and revisions have been found by Rott (1993, Theorem 7) and Alvaro del Val (1994, Theorems 2–5; 1997, Theorems 2–5).

Del Val's presentation is in terms of revisions in the style of Katsuno and Mendelzon (1991) rather than in terms of contractions in the style of Alchourrón, Gärdenfors and Makinson. After appropriate translation, however, the results (that have been obtained independently) are very similar. Del Val bases his considerations on a more general way of ordering the subsets of the base H than Rott, with the effect that he has a somewhat more general soundness part of the representation theorem.[224] Both del Val and Rott effectively solve a problem left open by Proposition 12 of Benferhat *et al.* (1993) that only locates the logic of 'inference relations generated by inclusion-based preorderings' between preferential and rational reasoning.[225]

However, it must be emphasized that both del Val and Rott deal with belief changes satisfying the Preservation Principle, due to the fact that their original belief sets K are supposed to be generated from consistent belief bases by some monotonic, essentially classical, logic Cn. Instead of modelling the crucial idea of this chapter $K = Inf(\mathcal{H})$, they start from $K = Cn(|\mathcal{H}|)$.[226] They in effect

---

[224] It seems to me that del Val's (1997) discussion of Rott (1993) should be supplemented by a few remarks. Part '(iii)$\Rightarrow$(ii)' of my Theorem 7 is a precise correlate of del Val's Theorem 4, the only difference being that del Val works with Katsuno-Mendelzon style revision and I work with (Nebelized) AGM style contraction. So the 'completeness parts' of the representation theorems are virtually identical. Del Val does have additional generality, though, in the 'soundness part' of his result. His Theorem 2 is stronger than part '(ii)$\Rightarrow$(iii)' of my Theorem 7, because he considers arbitrary preorders that extend set containment (while I consider only the ones generated by prioritized base contraction). This additional strength is very nice, of course, but del Val's comparison of his result with mine does not go well together with his earlier statement that Theorem 2 is 'easy' while Theorem 4 is 'most novel'.

Finally, del Val's proofs are much simpler indeed, but that is mainly due to the fact that I connect prioritized base contraction with the theory of choice, not to the fact that I do not have a semantics (which is in fact obtained in Rott (1993) via the 'Grove connection').

[225] These inference relations can easily be seen to correspond to Rott's and del Val's contraction and revision constructions; for the translation and the terminology, see Section 4.4.

[226] It seems that no reduction of our favourite method to the case of flat belief bases is possible. This also contrasts with the results of Rott (1993, p. 1448) and del Val (1994, p. 913) who

keep the horizontal perspective of the AGM framework, and no new belief base is generated after a belief change. There does not seem to be a straightforward way of transferring their representation results to the preferred modelling of the vertical perspective that I have taken in this chapter.[227]

both point out the important and surprising fact that the priority ordering of base elements can *in principle* be dispensed with, since it may be encoded in the syntactic structure of a flat belief base.

[227] Additional problems are posed by the fact that here we deal only with consolidations, i.e., with contractions with respect to ⊥, and that our method satisfies only the Weak Success postulate for revisions. These are minor technical problems, however, which would not seem to present insurmountable difficulties.

6

A GENERAL CONCEPT OF PRACTICAL RATIONALITY: CONSTRAINTS FOR COHERENT CHOICE

This remainder of this book consists of a single extended argument. It is an argument concerning the unity of theoretical and practical reason. I am going to argue that practical reason is the more comprehensive concept. I am not contending that practical reason dominates or even enslaves theoretical reason. There will be no conflict or rivalry between the two in what follows. My main point is that a particular, abstract account of practical reason allows us to subsume a large part of what would usually be seen as belonging to purely theoretical reason. Roughly, the view that is being advocated here takes principles of reasoning and revising beliefs, that is, principles of logic in a broad sense, to be derivable from principles of rational choice.

That methods for reaching a cognitive equilibrium can be derived from principles of practical (or economic) rationality is a sweeping claim, and it might seem that it is too vague to be capable of being substantiated at all. In order to counter this suspicion, I want to argue for my claim very carefully. In particular I want my claims to be both intelligible and verifiable. It will be required to engage in some degree of technicality. The formalist style of presentation, however, is by no means an exercise for its own sake. It is rather necessary for seeing clearly and distinctly the philosophical point I set out to make. Every formalized proof could of course be turned into an argument in natural language. Its length, however, would increase dramatically, and its comprehensibility would decrease by about the same factor. There may be some philosophers who think that there can be no deep philosophical insight behind something that is expressible in a formal language. I would like to argue for the contrary: Because of the perspicuous representation it affords, a formalization enables one to *see* things clearly that would have gone unnoticed otherwise, and therefore it may decisively help to gain deep insights.[228] It can of course be very difficult to interpret formal re-

---

[228] Somewhat surprisingly, users of symbolic logic may draw encouragement and inspiration from the late Wittgenstein (1953, §122): 'A main source of our failure to understand is that we do not *command a clear view* of the use of our words.—Our grammar is lacking in this sort of perspicuity. A perspicuous representation [*übersichtliche Darstellung*] produces just that understanding which consists in "seeing connexions". Hence the importance of finding and inventing *intermediate cases*.

The concept of a perspicuous representation is of fundamental significance for us. It earmarks the form of account we give, the way we look at things. (Is this a "Weltanschauung"?)'

If we substitute 'arguments' for 'words' and 'logic' for 'grammar', this is an excellent description of a chief virtue of formalized representations.

sults. But they are valuable for the philosopher precisely to the extent that they instigate a reflection about the very meaning and significance of these results.

The link between practical and theoretical reason will be worked out by matching our three closely related lists of rationality postulates on the side of theoretical reason—the ones presented in Chapter 4—with a fourth list on the side of practical reason. Like their theoretical counterparts, the practical postulates have great intuitive appeal and a long standing in the relevant literature. To the best of my knowledge, however, almost no explicit connection has been made in previous work between the different sets of postulates.[229] From now on, we shall no longer emphasize the differences between belief contraction, revision and nonmonotonic reasoning. Considering the close parallels exhibited in Chapter 4, they can be treated as three aspects of essentially the same subject matter and they can be confronted as a package with the postulates for practical reason.

It is one of the surprising results of the present study that there are extremely close and pointwise correspondences between the two kinds of postulates. We shall see that one can in general associate exactly one prominent postulate for belief change (*alias* nonmonotonic reasoning) with one prominent postulate for rational choice. It is this one-to-one coupling which lends our attempt to subsume theoretical reason under practical reason its particular force. This book says next to nothing, however, about the nature and origin of the decisions and choices made by a rational agent.

Many theories of rational choice give fairly detailed accounts of *how* choices are actually made. They specify, first, underlying structures which are relevant and, second, how these structures are applied in particular problem situations. Bayesian decision theory, for instance, is based on utilities and probabilities, and it recommends maximizing the expected utility values. Game theory is based on utilities alone, and in some versions it recommends the strategy of minimizing the maximal losses. Other examples involving non-numerical preferences are given by Aizerman and Malishevski (1981) and Aizerman (1985). We refrain here from any such concrete recommendations based on structures and rules of application. The postulates approach of the present chapter only lists general constraints that should be satisfied by any choice function, whatever its nature and origin.[230] Structures that underlie the choices and rules of how to apply these structures only appear in a limited way. We suppose that a rational agent chooses best options, where what is 'best' is judged by some given binary preference relation over the elements of a certain domain. This very simple and crude idea of optimization will be all that there is to the models studied in the following. A more substantive filling in of details of how the agent's choices are motivated must be reserved to future work.

---

[229]Exceptions are Lindström (1991) and Rott (1993, 1994, 1998). For various related observations, especially in the field of conditional logic, see Lewis (1973, p. 58), Nute (1980, p. 22) and (1994, p. 370), Delgrande (1987, p. 114), Lamarre (1991, p. 362), and Schlechta (1992, p. 682) and (1996).

[230]Compare the discussion in Section 4.1.

The simplicity of our account is not for free. It involves a great deal of idealization, a fact that has led to much criticism of the classical model of rational choice (which is the model on which we will rely). The preferences on which the choices are based are thought to be totally context-insensitive. Indeed this is taken to be characteristic of the notion of coherent or rational choice. There are famous examples, however, that strongly suggest that even the most central condition of the classical theory is inadequate as a universal constraint on choices—due to the fact that realistic choices are not insensitive to the contingencies of the choice situations.[231] Still we stick to this classical theory. Our point will be that this theory is quite perfectly suitable for an analysis of the theoretical problems of reasoning and revising beliefs. While the classical theory of choice is highly *idealized* so that many concrete situation defy an analysis in its terms, the logics generated by its employment constitute a great *liberalization* (*de-idealization*) in comparison with classical logical systems (where 'classical' is construed very broadly here). What is too rigid for practical reasoners seems to be quite versatile when applied to the specific aims of theoretical reason.[232]

Another important point of this book is that the rationality criteria for choices can be applied both on a semantic level (model level) and on a syntactic level (propositional level) with almost the same consequences for the resulting belief change and nonmonotonic reasoning operations. We will see that the semantic and the syntactic application lead to the same postulates for cognitive dynamics and defeasible inference—except for one single case where syntactic choices meeting a certain constraint lead to a stronger postulate for theoretical reason than semantic choices meeting the same constraint.[233] We consider this parallelism to be a significant and surprising result which adds to the tightness of the link of practical and theoretical reason.

The correspondence between the semantic and the syntactic way of interpreting logical principles in terms of rational choice calls for an explanation. And in fact it will be shown that every choice function on models gives rise to an equivalent choice function on sentences and vice versa—where 'equivalence' just means that these functions lead to identical operations for belief change and nonmonotonic reasoning. But more interestingly and importantly, in almost all (even in infinite) cases the coherence properties of the choice function on one level will be preserved in the transition to the corresponding choice function on

---

[231] This condition is often called 'Independence of Irrelevant Alternatives' or 'Condition α' and says that optimal alternatives in a menu remain optimal alternatives if some non-optimal alternatives are cancelled from the menu. For some reasons why this condition might fail, see Luce and Raiffa (1957, pp. 288–289), Sen (1993, pp. 500–503; 1995, pp. 24–26) and Kalai, Rubinstein and Spiegler (2001).

[232] It would be a very interesting and challenging task to see if the problems of the classical theory of rational choice carry over to the specific domain of logic and belief change. For instance, can we find good examples where Condition α cannot plausibly be applied in the cognitive domain? Such questions, too, must be addressed in future work.

[233] This is the case of condition (II) below. There are two more tiny differences which are too unimportant to be mentioned here.

the other level. And, of course, the preservation of coherence principles fails just where the parallel mentioned in the last paragraph has been broken.[234]

This has been a short preview of what will be coming in the rest of this book. Before applying the theory of rational choice in logical domains, we must develop it so far as it will be necessary for carrying out our program. This is the present chapter's task.

6.1   General choice functions

The fundamental concept of our analysis is that of a *choice function* or *selection function* σ. In the typical situation an agent is presented with a set S of potential options or alternatives (for instance, commodities or ways to act). This set among the elements of which the agent is to choose is called the *menu* or the *issue*.[235] If the choice function is applied to S, it returns the *choice set* $\sigma(S)$ of all those elements of S that are 'best choices' in S. Choice functions give us choice sets not just for one such situation, but for a field of many different menus which are all subsets of a universal set X of *options* or *alternatives*. Since in reality the agent is only confronted with a single menu at a given time, but choice functions deal with a whole field of menus, it is evident that a choice function represents the agent's *disposition* to choose, that is, his choices in *potential* situations that he might encounter at that time.[236] It is important that the choice function is itself independent of the contingent choice situation in which the agent happens to find himself.

In these abstract respects, choice functions are comparable to the operations of inference and revision discussed earlier. An inference operation *Inf* returns the set of conclusions for a vast field of potential premise sets H, and it is itself independent of the particular premise set which an agent actually entertains at a certain time. Similarly, a belief revision operation $*$ returns the revised belief set for vast field of potential inputs, and is independent of the particular proposition that happens to come in as input. In Chapter 4 we have discussed coherence conditions that relate the conclusions and revisions occasioned by (logically) related premises and inputs. In the same way we are now going to study coherence conditions that relate the choices made in (set-theoretically) related menus.

Before doing that we briefly address a principal problem (which will turn out to transfer to the theoretical field of inferences and revisions). Choice sets often

---

[234] Condition (II) again.

[235] Or *admissible agenda*, in Blair *et al.* (1976, p. 365).

[236] It is therefore virtually impossible to put choice functions to any kind of empirical test. Sen (1982, p. 3) very nicely summarizes the methodological problem of identifying a identifying a choice functionreal person's choice function at some time: 'Over short periods people may seek variety (fish today and steak tomorrow is not inconsistent), but over longer periods tastes can easily change (apparent inconsistencies may then reflect instead a changing choice function).' So although the theory of rational choice is certainly contentful because it excludes certain mental dispositions, there is a problem in making it *empirically* contentful.

contain more than just one element.[237] Let us suppose, for example, that $\sigma(S)$ is the set $\{x, y\}$ (the case of sets with more than two elements is similar). For this to happen, x and y must be preferred to all the other elements in S; but since x and y are both in the choice set, they must be either equally good or incomparable with each other. *Ties* and *incomparabilities* with respect to some underlying preference relation pose a problem to the choosing agent, since then choice sets will not in general be singletons. In the case we are considering, it is not determined by the choice function alone what the agent is actually going to choose (or to decide, or to do). There are two principal possibilities.

(1) First, the agent may *pick one* element of $\sigma(S)$ at random. He may do so because there is only one way to go, no way of getting a compromise or combination of x and y (as when x and y stand for two different continuations at a road fork). Another reason for picking either x or y may be that both of them possess certain merits, but their combination—though possible—is not good. Meeting Harry is an excellent idea, meeting Sally is an equally excellent idea, but meeting both or neither of them might be a very frustrating way of spending the evening. If we make a random selection of some arbitrary element in $\sigma(S)$ (for instance, if we take the first best option that we happen to come across), then of course there is no inherent rationality or 'logic' in this act of picking. So no matter how rational our way of constructing or justifying our choice function σ may be, the pick function used to break the ties left unresolved by σ will destroy the appearance of rationality in our behaviour to some extent. Our choices are neither fully predictable nor fully explainable, nor does our behaviour conform to a discernible pattern of coherence.[238]

(2) Second, there may be the opportunity of going for a satisfactory compromise or combination or aggregation of the members of the choice set. The rationale is that if σ does not distinguish between x and y then they should be given equal rights. Obeying what may be called the Principle of Indifference, we can argue that like alternatives should receive like treatment.[239] Then agents proceed to choose some suitable combination $x \bullet y$ of all the optimal alternatives x and y. The problem is that there is no guarantee that the combination $x \bullet y$ will itself be as good or eligible for choice as the elements combined. Since a combination of best options—if actually it is available as an option—will not in general be a best option, the Principle of Indifference may incur an unnecessary loss of quality. So although this strategy is more principled and logically well-behaved than the first

---

[237] Single-valued choice functions are a special case for which many of the distinctions to be made in this chapter collapse.

[238] On the important distinction between picking and choosing, see Ullmann-Margalit and Morgenbesser (1977).

[239] Compare Pagnucco and Rott (1999).

strategy, it has the drawback of deviating from the recommendations of the choice functions σ.[240]

We may call the first strategy 'bold' and the second strategy 'cautious'.[241] We shall later see that the main two methods of coherentist belief change and nonmonotonic reasoning follow the second strategy. Although it is hard to motivate the potential loss of 'epistemic value', this strategy seems more tolerable than the invocation of chance processes in the selection of best alternatives. And of course, we are interested in substantive logical properties which would soon be lost if we were to rely on the throwing of dice.

These considerations furnish a second argument that a choice function that yields non-singleton choice sets is best viewed as a representation of *potential choices* rather than actual choices. It is inherent in the concept of an option that agent can in reality only go for one option. The choice set σ is basically a recommendation that *any* of its element *would be* a good choice, but the agent can *actually* only choose one option. We have just argued that a best choice is either an element of the set $\sigma(S)$ or some well-motivated combination of its elements.

We need a number of formal definitions for future reference.

Let X be a set and $\mathcal{X}$ be a non-empty set of subsets of X. A *choice function* (or *selection function*) over $\mathcal{X}$ is a function $\sigma : \mathcal{X} \rightarrow Pow(X)$ such that the choice set $\sigma(S)$ is a subset of the menu (or issue) S, for every $S \in \mathcal{X}$. Intuitively, choice functions are supposed to give us the best elements of each S in $\mathcal{X}$. We do not require the menu to be non-empty ($S \neq \emptyset$) or the choice to be unfailingly successful ($\sigma(S) \neq \emptyset$). Suzumura (1983, p. 18) calls X the set of conceivable states and S the set of available states.

For the following considerations, domain conditions for choice functions turn out to be important. A set $\mathcal{X}$ of subsets of X is called *n-covering* ($n = 1, 2, 3, \ldots$) if it contains all subsets of X with exactly n elements. $\mathcal{X}$ is called $n_1 n_2$-*covering* (or $n_1 n_2 n_3$-*covering*) if it is n_1-covering and n_2-covering (and n_3-covering). $\mathcal{X}$ is called ω-*covering* if it is n-covering for all natural numbers $n = 1, 2, 3, \ldots$; it is called *at most finitary* if it contains only finite subsets of X, and it is called *finitary* if it is both ω-covering and at most finitary. The set $\mathcal{X}$ is called *additive* if it is closed under arbitrary unions, and it is called *finitely additive* if it is closed under finite unions; it is called *subtractive* if for every S and S' in $\mathcal{X}$, $S - S'$ is also in $\mathcal{X}$, and it is called *full* if $\mathcal{X} = Pow(X)$. Of course, if $\mathcal{X}$ is 1-covering and finitely additive, then it is ω-covering. Finally, we say that $\mathcal{X}$ is *compact* if for every T and S_i, $i \in I$, from $\mathcal{X}$, if $T \subseteq \bigcup \{S_i : i \in I\}$ then $T \subseteq \bigcup \{S_i : i \in I_0\}$ for some finite $I_0 \subseteq I$.

We say that a *choice function* σ with domain $\mathcal{X}$ is n-covering, finitary, additive, subtractive etc. if its domain $\mathcal{X}$ is n-covering, finitary, additive, subtractive

---

[240] This fundamental problem has been emphasized pointedly by Isaac Levi on many occasions.

[241] Cf. Makinson (1994, pp. 38–39) on 'sceptical', 'liberal' and 'choice' perspectives on nonmonotonic reasoning.

etc. Finally, we call σ *iterative* if its range is included in its domain, that is, if $\sigma(S)$ is in $\mathcal{X}$ for all S in $\mathcal{X}$.

We are going to introduce constraints on choice functions that restrict the choices that an agent can or should make *across different situations*, that is, when confronted with *differing menus*.[242] For these inter-menu conditions to make sense, it must be understood that the agent does not change his mind during the comparison of the choices made in varying situations. There is no temporal succession involved in the comparisons. As mentioned above, one may rather think of choices in hypothetical situations.

6.1.1 A limiting case: Not having to choose

In this and the next subsection we will discuss limiting cases in which the agent has *no genuine need* or *no genuine possibility* to choose. Only after these limiting cases have been dealt with shall we present the core of the classical theory of rational choice.

Let R be a non-empty set of elements of X which are *absolutely satisfactory*. By absolute satisfaction we mean that it does not depend on the menu with which we are presented whether the elements of R are chosen or not. If a menu contains an element of R, it is included in the choice set. Following Herbert Simon (1959), we can call R the 'level of aspiration'. What it means for an option to be absolutely satisfactory can be interpreted in at least two ways, depending on the specific situation and the psychological temperament. For an optimist in a happy situation, absolutely satisfactory options are 'optimal', excellent, first-rate, unsurpassed. In the eyes of a pessimist in a less enviable situation, absolutely satisfactory options are only 'good enough', tolerable, not bad, just all right. Simon, an eminent critic of the Principle of Utility Maximization, advocates a model of 'bounded rationality' and recommends an alternative *Principle of Satisficing*. According to that principle a rational agent should not care to strive after best options, but rather take any option that meets some fixed criteria of acceptability—even if he knows that there are better options. Simon thus suggests to choose options that are in some sense just good enough.[243] In contrast to this 'just-good-enough interpretation', it will be seen that our application to doxastic concepts better fits into the optimistic worldview in so far as absolutely satisfactory options are ones that really cannot be surpassed. These differences of interpretation, though philosophically important, make no difference from a logical point of view.

We add the remark that on one interpretation of Simon's satisficing, there is no contradiction with the standard view that rational agents should maximize or optimize. On this interpretation, the point is rather seen in the fact that

---

[242]From now on, we talk of *constraints* for choice functions and reserve the term *postulate* for conditions for belief change and inference operations (compare Section 4.1).

[243]See also Rescher (1989) and Slote (1989). Rescher emphasizes cost effectiveness as a salient aspect of rationality. But his insistence on cost-benefit analyses does not really solve the problem of resource-boundedness in epistemological matters, since it turns itself on a rather costly maximizing process: the maximization of benefits minus expenditures.

satisficing agents willfully employ an extremely crude preference relation, namely one which only distinguishes 'good' from 'bad' options, without admitting any finer gradations or incomparabilities (which the agent's 'real' preferences may reflect). If maximization is guided by this more or less trivial, two-class preference relation, then it becomes identical with satisficing. It is then clear that satisficing agents will meet even the strongest requirements of 'rational' or 'coherent' choice we shall study in this section. Amartya Sen, however, argues that this description misses the point of Simon's satisficing and its interpretation in terms of bounded rationality:

> ... the gap between satisficing and maximizing may be, at least formally, reduced. However, the content of the claim of satisficing is that the person in question *can* tell between the different levels of achievement which are all beyond the target level required, and despite this discernibility, choice behaviour departs from relentless maximization of the level of achievement. In this version of the story, a substantial difference is indeed made by the notion of satisficing, and the implications of satisficing behaviour may, in this interpretation, be quite different from those of maximization.[244]

The first constraint we consider is very plausible. It says that whenever we are presented with a menu containing some absolutely satisfactory options, exactly these options should be designated as best choices by the choice function σ. Choosing in such menus is very easy or, to put it differently, one does not really have to choose at all. Only menus in which there are no absolutely satisfying options require genuine choices. The corresponding constraint is this:

(Faith) $\quad \exists R \neq \emptyset \ \forall S \in \mathcal{X} :$ if $S \cap R \neq \emptyset$ then $\sigma(S) = S \cap R$

Normally, there is only one set R suitable for Faith, where 'normally' more precisely means: when each alternative from X appears in some issue in $\mathcal{X}$. The role of R can usually be played by $\sigma(X)$. If $X \in \mathcal{X}$ then $R = \sigma(X)$ can even be derived from (Faith). Substituting $\sigma(X)$ for R gives us a *purely operational* formulation of Faith that do not use the existential quantifier:

(Faith′) $\quad \forall S \in \mathcal{X} :$ if $S \cap \sigma(X) \neq \emptyset$ then $\sigma(S) = S \cap \sigma(X)$

If R is the unique set of options that is referred to in Faith, then we shall say that σ satisfies Faith *with respect to R*, or that it is *faithful to R*. An agent whose choice function obeys this prescription is faithful to what he considers good enough *per se*, irrespective of the special context of choice. In Chapters 7 and 8, we shall study agents who are faithful, in a certain sense, to what they believe. In the semantic application, R will be the set of models or 'worlds' $[\![K]\!]$ considered possible by the agent, and in the syntactic application where the choice function picks the sentences which are best to give up, R will be the set $L - K$ of the agent's non-beliefs.

It will turn out to be instructive to split Faith into two parts. Keeping fixed the non-empty set R of absolutely satisfactory options, we distinguish

(Faith1) $\quad \forall S \in \mathcal{X} :$ if $S \cap R \neq \emptyset$ then $\sigma(S) \subseteq R$
(Faith2) $\quad \forall S \in \mathcal{X} : S \cap R \subseteq \sigma(S)$

---

[244] Sen (1987, pp. 70–71). A different diagnosis is given in Sen (1997, pp. 768–769).

Faith1 says that if the menu S contains 'good' elements, then *only* good elements are chosen; Faith2 says that *all* good elements in the menu are chosen. If R is a set suitable for Faith1 or Faith2, we shall again say that σ satisfies Faith1 or Faith2 *with respect to* R.[245]

A much more radical idea than Faith is to stick to the choice of absolutely satisfactory elements even when the menu does not contain any such elements. This is the *Indiscrimination Condition*

(Indis) $\exists R \ \forall S \in \mathcal{X} : \sigma(S) = S \cap R$

According to this definition, options from $X - R$, i.e., options that are not absolutely satisfactory, are *never* chosen. Clearly, Indiscrimination implies Faith. Like the latter, Indiscrimination may be thought of encoding the ideology of absolutely satisfactory choices (independent of the menu) instead of relatively best choices (in the menu). But Indiscrimination goes much farther than Faith. It partitions the set X of all options into good and bad options, without any finer distinctions that allow to discriminate within the set of good or within the set of bad options. Later (in Section 6.5) we shall find a condition which is essentially equivalent to Indiscrimination, but does not refer to the existence of a fixed set of absolutely satisfactory options.

The Indiscrimination Condition, which sets the standards of eligibility for choice independently of the menu at hand, determines choice functions which *prima facie* seem to be utterly uninteresting. Still Indiscrimination will be illuminating as a limiting case constraint in later parts of this essay. Notice that there are in general many menus for which undiscriminating choice functions yield empty choice sets—unless every option in X is considered to be absolutely satisfactory. An undiscriminating agent refuses to choose, if there are not absolutely satisfactory options available. We now discuss refusals of choice in a more general setting.

6.1.2   *Another limiting case: Refusing to choose*

A desideratum for choice functions is that $\sigma(S)$ be non-empty whenever S is non-empty. That is to say that a choice function should be *successful* or *decisive*: for every menu presented to an agent, he is expected to reach a decision as to which options in the menu are best. Although this is a guiding idea for the very concept of a choice function, we shall not impose it as an obligatory feature. We want to provide for the possibility that agents *refuse to choose*.[246]

There are at least two possible ways that an agent may end up with an empty choice set $\sigma(S)$ even when S is non-empty. The first way can only apply in infinite domains X. One can imagine that the agent is being paralysed by the fact that there is no best option for him to take, because there are better and better and

---

[245]Let us once more consider the case where $X \in \mathcal{X}$. Then (Faith1) implies $\sigma(X) \subseteq R$ and (Faith2) implies $R \subseteq \sigma(X)$. However, it does not hold that (Faith1) with respect to R, taken individually, implies (Faith1) with respect to $\sigma(X)$, and similarly for (Faith2).

[246]This is obviously a sense of 'refusing to choose' different from Reiter's (1980, p. 86, note 2) who uses the phrase to describe the cautious method of dealing with ties (see p. 146 above).

better ... alternatives without limit. This happens in the case of an underlying preference relation with infinite ascending or descending chains.[247] An rational agent is supposed to look for 'the best way' to take—but in such a situation there simply is no best way. We will disregard this problem in the present essay since it seems to present purely technical rather than substantive philosophical difficulties. Were the agent to encounter a situation like this in the real world, he would usually be well advised to coarsen his preference relation a bit, so as to allow for an appropriate range of best options.

There is, however, another problem that may cause an agent to end up with the empty choice set. It may happen that the agent refuses to choose, because every option in the menu appears entirely unacceptable to him. Not a single option is perceived to be a reasonable way to deal with the situation at hand. While in the above-mentioned type of situation, there is an over-abundance of better and better options, there is a total lack of reasonable options in the type of situation envisaged now. In contrast to many studies in the theory of rational choice—and in contrast to many corresponding doctrines in belief revision and nonmonotonic reasoning—we want to be capable of modelling a reasoner who is sufficiently autonomous to decide what he considers acceptable and what not.[248] We shall later call sets of options that an agent would never consider as eligible for choice *taboo sets*.

The behaviour of agents that never refuse to choose can be captured by what may be called a *successful* choice function.[249] A successful choice function meets the following constraint.[250]

(**Success**)   If $S \neq \emptyset$, then $\sigma(S) \neq \emptyset$

As mentioned above, many choice theorists exclude the empty set from the domain of any choice function σ.

It is our task now to come up with some rationality constraints for agents that may refuse to choose in some situations. We must think of rules for potentially 'unsuccessful' choice functions. If the reasoner is not always going to choose, he or she should at least be consistent in his or her refusals to choose. We suggest a basic coherence condition which we consider to be a minimal constraint for all kinds of rational choice functions.

---

[247]The impact of a 'limit assumption' has been widely discussed in conditional logic. The classic reference is Lewis (1973, pp. 19–21 and 120–122). Many writers solve the problem by stipulating the existence of an underlying preference relation which is (conversely) well-ordered.

[248]There are a few writers who have taken this case into account, most notably Aizerman and Malishevski (1981), Aizerman (1985) and Lindström (1991).

[249]This terminology is not meant to carry an implicit reproach of agents that refuse to choose. In order to avoid this unwanted connotation, the predicate 'resolute' might be preferable to 'successful'. However, we adopt the alternative name since we will find that the 'corresponding' postulate for belief contractions is ($\dot{-}4$) which has long been called the postulate of Success in the literature (see Chapters 4 and 7). The label 'Success' is also used in this way by Nayak (1994a).

[250]From now on, we leave the quantifiers implicit in the formulation of the conditions. The variables S and S' are always supposed to range over sets in $\mathcal{X}$.

($\emptyset$)   If $\sigma(S) = \emptyset$, then $\sigma(S') \cap S = \emptyset$

The converse implication is trivially true. Success entails ($\emptyset$), but is not entailed by it. Again it turns out to be helpful to distinguish two parts of condition ($\emptyset$).

($\emptyset 1$)   If $S \subseteq S'$ and $\sigma(S') = \emptyset$, then $\sigma(S) = \emptyset$

($\emptyset 2$)   If $S \subseteq S'$ and $\sigma(S) = \emptyset$, then $\sigma(S') \cap S = \emptyset$

Actually condition ($\emptyset$) is slightly stronger than the conjunction of ($\emptyset 1$) and ($\emptyset 2$) since it is derivable from the latter only if the domain of σ is closed under finite intersections. Like finite additivity, closure under finite intersections is a rather weak requirement, and both of them will be satisfied in the two main (semantic and syntactic) applications of the theory of choice to doxastic contexts in Chapter 7.

Condition ($\emptyset 1$) says that if one refuses to choose any one element from some large set of alternatives, one must not choose an element of a smaller submenu. This condition in effect rules out the above-mentioned possibility that a choice set is empty for technical reasons only. For instance, if we are presented with a menu consisting of elements which are better and better with respect to some preference relation, then one could imagine resolving the problem by just restricting the menu to a finite subset which is of course free from such infinitely ascending or descending chains. Such a 'solution', however, is prohibited by condition ($\emptyset 1$).[251] As we said above, we neglect this case of technical complication found in some infinite menus.

According to the condition ($\emptyset 2$), if one refuses to choose any one element from some small set of alternatives, these elements must not be chosen from any larger menu. The common idea of both ($\emptyset 1$) and ($\emptyset 2$) is, of course, that once an option is considered definitely ineligible for choice, it should be considered so in any other context. A somewhat stronger principle than ($\emptyset 2$) may also be mentioned.

($\emptyset 2^{+}$)   If $S \subseteq S'$ and $\sigma(S) = \emptyset$, then $\sigma(S') = \sigma(S' - S)$

This condition says that if an agent refuses to choose any one element from some small set, then this set may be completely disregarded in the process of choosing of any bigger set.

In my opinion, ($\emptyset 1$), ($\emptyset 2$) and also ($\emptyset 2^{+}$) are plausible conditions for choice functions. They are vacuously satisfied by the choice functions of the classical theory of choice which are all successful. We presuppose in the following that choice functions satisfy ($\emptyset 1$) and ($\emptyset 2$) without always explicitly saying so.

The new conditions give rise to a new concept. We shall say that a choice function σ possesses a *taboo set* $\overline{R}$ (or that $\overline{R}$ is a taboo set of σ) if the following two conditions hold for all S in $\mathcal{X}$:

(t1) $\sigma(S) \cap \overline{R} = \emptyset$

(t2)   If $\sigma(S) = \emptyset$, then $S \subseteq \overline{R}$

---

[251] Condition (3) in Lewis (1973, p. 58) corresponds or our condition ($\emptyset 1$).

In other words, $\overline{R}$ is a taboo set of σ if and only if

$$\bigcup\{S \in \mathcal{X} : \sigma(S) = \emptyset\} \subseteq \overline{R} \quad \text{and} \quad \overline{R} \subseteq X - \bigcup\{\sigma(S) : S \in \mathcal{X}\}$$

Now we want to find out what conditions are required for such a set $\overline{R}$ to exist and to be unique.

Lemma 10 *Let x be in X and consider the following conditions:*
(i) $x \in S$ *for some* $S \in \mathcal{X}$ *such that* $\sigma(S) = \emptyset$
(ii) $x \notin \sigma(S)$ *for all* $S \in \mathcal{X}$
If σ satisfies ($\emptyset$) and (i), then it satisfies (ii). If σ is 1-covering and satisfies (ii) then it satisfies (i) .

The implication from (i) to (ii) tells us that $\bigcup\{S \in \mathcal{X} : \sigma(S) = \emptyset\} \subseteq X - \bigcup\{\sigma(S) : S \in \mathcal{X}\}$, so that there is room for a set $\overline{R}$. The implication from (ii) to (i) makes clear that the converse inclusion—and thus the uniqueness of the set $\overline{R}$—is guaranteed to hold if σ is 1-covering. If σ is not 1-covering then it may well happen that some x is never chosen but it is nevertheless not contained in any set in which the agent refuses to choose. That x is never chosen may be due to the fact that in each menu containing x there happens to be an element which is preferable to x. If x is in a menu in which the agent refuses to choose, i.e., if x is in $\bigcup\{S \in \mathcal{X} : \sigma(S) = \emptyset\}$, we say that it is *necessarily taboo* (or *necessarily ineligible for choice*); if x is in $X - \bigcup\{\sigma(S) : S \in \mathcal{X}\}$ than we call it *contingently taboo* (or *contingently ineligible for choice*). Both the set of necessarily taboo options and the set of contingently taboo options are taboo sets of σ, actually the smallest and the greatest taboo sets possible. If the choice function admits singleton menus, the two sets coincide.

We can show that an agent refuses to choose just in case he is confronted with a menu consisting solely of taboo options.

Lemma 11 *If σ satisfies ($\emptyset 1$) and ($\emptyset 2$) and $\bigcup\{S \in \mathcal{X} : \sigma(S) = \emptyset\}$ is in the domain $\mathcal{X}$ of σ, then it holds for every S' in $\mathcal{X}$ that $\sigma(S') = \emptyset$ if and only if $S' \subseteq \bigcup\{S \in \mathcal{X} : \sigma(S) = \emptyset\}$.*

Finally, we are able to relate our operational approach to refusals of choice as encoded in ($\emptyset$) with the existence of a taboo set.

Lemma 12 *A choice function satisfies ($\emptyset$) if and only if it possesses a taboo set.*

Since we presuppose that ($\emptyset$) is generally satisfied,[252] we thereby know that every choice function σ possesses a (possibly empty) taboo set, namely $\overline{R} = X - \bigcup\{\sigma(S) : S \in \mathcal{X}\}$. Thus we associate with every choice function σ over $\mathcal{X}$ this set as *its* taboo set $\overline{R}_\sigma$ (knowing that σ may actually have more than one taboo set) and we stipulate that $\overline{R}_\sigma$ be in $\mathcal{X}$. When there is no danger of confusion, we continue to write $\overline{R}$ instead of $\overline{R}_\sigma$.

---

[252]Or that ($\emptyset 1$) and ($\emptyset 2$) are satisfied and domains are closed under finite intersections.

If σ has a taboo set $\overline{R}$ and is at the same time faithful to a set R, it follows that $\overline{R} \cap R = \emptyset$. Absolutely satisfactory options are always chosen, taboo options are never chosen. Undiscriminating choice functions (with respect to some R) always have a taboo set, namely $\overline{R} = X - R$.

6.2   The classical theory of rational choice

In this section we formulate the core of what may be called the *Classical Theory of Rational Choice*. The pioneering work for this theory was done not by philosophers, but by economists, with the most important contributions including Samuelson (1938, 1947), Houthakker (1950), Arrow (1951/63, 1959), Chernoff (1954), Uzawa (1956), Richter (1966, 1971), Sen (1970, 1971) and Herzberger (1973). An excellent systematic survey of the field is given by Suzumura (1983). Although the conditions we are going to present in this section have been discussed for about 50 years, their particular arrangement as the most natural way of capturing the theory was put forward in full clarity only in the 1980s by Aizerman and Malishevski (1981), Aizerman (1985) and Moulin (1985).

The following conditions impose *coherence constraints* or *consistency constraints* on choices across varying menus. They do not specify conditions for choices pertaining to a single situation, or more precisely, pertaining to a single issue or menu.[253] Rationality has in this context been explicated by an appeal to coherence considerations. Here are the four most central conditions:

(I)   If $S \subseteq S'$, then $S \cap \sigma(S') \subseteq \sigma(S)$

(II) $\sigma(S) \cap \sigma(S') \subseteq \sigma(S \cup S')$

(III)   If $S \subseteq S'$ and $\sigma(S') \subseteq S$, then $\sigma(S) \subseteq \sigma(S')$

(IV)   If $S \subseteq S'$ and $\sigma(S') \cap S \neq \emptyset$, then $\sigma(S) \subseteq \sigma(S')$

The reason why these conditions may be regarded as most central will become apparent in Section 6.4.

Condition (I) is Sen's *Property* α and sometimes called *Chernoff* or *Heritage*. For finitely additive and subtractive domains, it is equivalent to

(I') $\sigma(S \cup S') \subseteq \sigma(S) \cup \sigma(S')$

Condition ($\emptyset 2$) of the last section follows from (I). If σ is iterative, then (I) entails that $\sigma(S) \subseteq \sigma(\sigma(S))$ for all S in $\mathcal{X}$. Therefore, choice functions satisfying (I) are *idempotent*, i.e., they satisfy the condition $\sigma(\sigma(S)) = \sigma(S)$ which Aizerman (1985, p. 242) calls *Complete Rejection* and Suzumura (1983, p. 42) calls *Stability axiom*.

While (I) expresses some kind of 'contraction consistency' (Sen) in proceeding from larger menus to smaller ones, condition (II) proceeds from smaller menus to larger ones. (II) is a finitary version of Sen's *Property* γ which is also called the *Expansion Axiom* or *Concordance*.

---

[253] We shall do this only in Section 7.4, where intra-menu conditions are applied which are motivated by logical rather than choice-theoretic considerations.

(II$^\infty$) $\quad \bigcap \{\sigma(S_i) : i \in I\} \subseteq \sigma(\bigcup \{S_i : i \in I\})$

(III) is called *Aizerman's axiom* in Moulin (1985) and in Lindström (1991). If the subset relation in its consequent is replaced by an identity, then we get what is called *Independence of rejecting the outcast variants* in Aizerman and Malishevski (1981, p. 1033) and *Nash's axiom* in Suzumura (1983, p. 41).[254] (III) is stronger than the *Superset axiom*[255]

$$\text{If } S \subseteq S' \text{ and } \sigma(S') \subseteq \sigma(S), \text{ then } \sigma(S) \subseteq \sigma(S')$$

Taken together with (I) (actually, even with (I$^-$) introduced below), the superset axiom implies (III). Condition ($\emptyset 1$) of the last section follows from (III), and ($\emptyset 2^+$) follows from (I) and (III). Condition (III) is independent of condition (II), even for finite X and in the presence of condition (I).

(IV) is Sen's *Property $\beta+$*, also called *Dual Chernoff* (Suzumura 1983). Taken together with ($\emptyset 1$), it implies both (II) and (III).[256] As there are many functions σ satisfying (I) – (IV), this set of constraints is consistent. An excellent discussion of these properties and an explanation of why they play a central role in the theory of choice is given by Moulin (1985). Actually, instead of (IV) Moulin uses *Arrow's axiom*, also known as *Strict Heritage*:[257]

$$\text{If } S \subseteq S' \text{ and } \sigma(S') \cap S \neq \emptyset, \text{ then } \sigma(S') \cap S = \sigma(S)$$

Taken together with (I), (IV) is equivalent with Arrow's axiom.

6.3   Rationalizable choice

In terms of the terminology introduced in Section 4.1, the specification of co-herence constraints presents a black-box approach to rational choice. Plausible though these constraints may be, we do not know why exactly they are sup-posed to embody rationality and how exactly it should come that choice func-tions comply with such constraints. In order to elucidate this point, we need a structure-plus-rule-of-application approach.

The classical theory of choice and preference is characterized by an idea which can be phrased by the slogan *'Rational choice is relational choice'*. That

[254] Independence of rejecting the outcast variants is the conjunction of (III) with the condition (I$^-$) considered in Section 6.5. It reappears as condition (14) in Moulin (1985). A doxastic variant of it is formulated in Gärdenfors and Makinson (1994, p. 207).

[255] The Superset axiom is listed in Rott (1993) where also the infinitary version (II$^\infty$) of (II) is used. Later in that paper, however, both variants are transformed into conditions called (II$'$) and (III$'$). These conditions are identical with the present conditions (II) and (III) and do the real work in the paper.

[256] But ($\emptyset 1$) is indeed necessary. For instance, Delgrande's (1987, pp. 113–114) *NP*-models satisfy (I$'$), and (IV), but lacking ($\emptyset 1$), these models do not satisfy (III). Consequently, Del-grande's conditional logic *NP* fails to satisfy Cautious Monotony (cf. Section 7.7.3 below). Also compare Nute (1994, pp. 367–373).

[257] A special case of this condition is condition (4) in Lewis (1973, p. 58).

is, rational choice is choice that can be construed as based on an underlying preference relation. Herzberger (1973, p. 189) traces this idea back to Chernoff (1954):

The intended interpretation of the set $\sigma(S)$, called the *choice set* for S, is that its elements are regarded as equally adequate or satisfactory choices for an agent whose values are represented by the function σ, and who faces a decision problem represented by the set S. Following Chernoff ... , this relativistic concept of equiadequacy for a given decision problem bears sharp distinction from invariant concepts like preferential matching or indifference which for a given agent are not relativized to decision problems, and which may be subject to more stringent constraints, both for rational agents and for agents in general. (Notation adapted)

We have already mentioned that the underlying intuitive idea is that choices should be made in a 'context-insensitive way', that is, independently of the special size and content of the menu.[258] In particular, the menu itself is not supposed to carry any choice-relevant information, and inclusion or exclusion of alternatives which are considered 'irrelevant' should have no influence on the choices made.

Choice sets are taken to be sets of *best elements*. There are basically two ways of making this idea precise. In accordance with the predominance of *minimization principles* in formal semantics (see Makinson 1993)—and because of the rather special interpretation given the syntactic choice functions in Chapter 7 —we identify the best options in a menu with its *least* or *minimal ones*. But of course there is no substantial difference of minimization principles from maximization principles—one just has to revert the direction of the preference relation used, and the structure of argumentation does not change at all. Minimizers do not differ in principle from maximizers.

The first formalization of the best-elements idea is based on a non-strict (reflexive) preference relation $\leq$, and is defined for a choice function σ over $\mathcal{X}$ with taboo set $\overline{R}$ as follows:

$$\sigma(S) = \begin{cases} \text{smallest}_\leq (S) = \{x \in S : x \leq y \text{ for all } y \in S\} & \text{if } S \not\subseteq \overline{R} \\ \emptyset & \text{if } S \subseteq \overline{R} \end{cases}$$

The second formalization is based on a strict (asymmetric and therefore irreflexive) preference relation $<$, and puts for all $S \in \mathcal{X}$,

$$\sigma(S) = \begin{cases} \text{min}_< (S) = \{x \in S : y < x \text{ for no } y \in S\} & \text{if } S \not\subseteq \overline{R} \\ \emptyset & \text{if } S \subseteq \overline{R} \end{cases}$$

These suggestions are respectively referred to as *stringent* and *liberal minimization* by Herzberger (1973, p. 197), *G-rationality* and *M-rationality* by Suzumura (1983, p. 21), and *optimization* and *maximization* by Sen (1997, p. 763).

---

[258] For more on menu-independence, see Sen (1997).

Richter (1971) says that $\leq$ *rationalizes* σ in the case of stringent maximization; Kim and Richter (1986) say that $<$ *motivates* σ in the case of liberal maximization. (Stringent maximization is often attributed to Condorcet 1785.) If $<$ is understood as the converse complement of $\leq$, then stringent and liberal minimization coincide. However, if $<$ is understood as the asymmetric part of $\leq$, then every liberal minimizer with respect to $<$ is a stringent minimizer with respect to the *augmentation* $\leq^+$ of $\leq$ which is defined by $x \leq^+ y$ iff $x \leq y$ or $y \not\leq x$, i.e., iff not $y < x$.[259]

Clearly, $\leq^+$ is *connected*, i.e., it holds for every x and y that either $x \leq^+ y$ or $y \leq^+ x$. While $\leq$ allows us to keep apart indifferences (both $x \leq y$ and $y \leq x$) from incomparabilities (neither $x \leq y$ nor $y \leq x$), $\leq^+$ blurs just this distinction, for whenever we have neither $x \leq y$ nor $y \leq x$ we have both $x \leq^+ y$ and $y \leq^+ x$. Thus we may not expect $\leq^+$ to be transitive. If, on the other hand, $\leq$ is already connected, then $\leq^+ = \leq$, and liberal minimization with respect to the asymmetric part $<$ of $\leq$ coincides with stringent minimization with respect to $\leq$. For all this, cf. Herzberger (1973, Section 3) and Sen (1997, Section 3).

In contrast to the dominant approach in the theory of choice and preference as well as the classic paradigm in belief revision (Alchourrón, Gärdenfors and Makinson 1985), we shall focus on liberal minimization which appears to be preferable on intuitive grounds. Liberal minimization is based on strict relations which do not allow to distinguish between incomparabilities and indifferences.[260] Non-strict relations do make this distinction, but stringent minimization tends to require connected relations.[261] This is so because only options that are comparable with *all* the other options in a menu are eligible for choice by stringent minimization. But intuitively, we are very often confronted with some options that are 'isolated' in the sense that they are somehow incomparable with any other option and related by $\leq$ only to themselves (often the use of multiple criteria is responsible for the feeling that some thing cannot be compared to some other thing). For these cases stringent minimization will give us an empty choice set—even if there is only a single isolated option in S. If, for instance, $\leq$ is a preference relation on $X = \{x, y, z\}$, and we only have $x \leq y$, but not the converse and no preferential comparability with z whatsoever, then the set of all smallest elements under $\leq$ in S is empty, while the set of all minimal elements in S under the asymmetric part $<$ of $\leq$ is $\{x, z\}$. Since the latter solution seems much more plausible, we conclude that stringent minimization is too stringent and shall therefore base our further investigation on liberal minimization based

---

[259] If $<$ is the asymmetric part of $\leq$, then smallest$_\leq$ (S) is a subset of min$_<$ (S), but not in general the other way round.

[260] But see page 157.

[261] Connected relations can often be obtained only at the cost of turning incomparabilities into indifferences—i.e., of using augmentations. The interpretation of non-strict relations as the converse complements of strict relations explains the slightly awkward role of *negative transitivity* and *negative well-foundedness* in Rott (1993). More reasons for taking strict relations as primitive are given in Rott (1992b, Section 4).

on a strict preference relation $<$:

Definition 3 A choice function σ (with taboo set $\overline{R}$) is called *relational with respect to a strict preference relation* $<$ *over* X, in symbols $\sigma = \mathcal{S}(<)$, if and only if for every $S \in \mathcal{X} - Pow(\overline{R})$

$$\sigma(S) = \min_{<}(S)$$

That is, σ always selects the minimal elements of S under the preference relation $<$, unless the menu S contains only elements that are taboo. When S is a subset of the taboo set $\overline{R}$, then we of course put $\sigma(S) = \emptyset$.[262] By saying that σ is *relational* or *rationalizable*, we mean that there is a preference relation $<$ over X with respect to which σ is relational. A preference relation $<$ over X is called *modular* (or *negatively transitive, virtually connected, ranked*) if $x < y$ implies that either $x < z$ or $z < y$, for all x, y and z in X.[263] Modularity and asymmetry jointly entail transitivity. A choice function σ is called *transitively (modularly) relational* or *transitively (modularly) rationalizable* iff there is a transitive (modular) preference relation $<$ over X with respect to which σ is relational.

A strict binary relation $<$ over X is called *n-acyclic*, if no n objects $x_1, x_2, \ldots,$ x_n in X form a cycle under $<$, i.e. if there is no chain $x_1 < x_2 < \cdots < x_n < x_1$ in X. 1-acyclicity is irreflexivity, 2-acyclicity is asymmetry. $<$ is called *acyclic*, if it is n-acyclic for every $n = 1, 2, 3, \ldots$. $<$ is called *well-founded* if there is no infinite descending chain under $<$, i.e., if never $\cdots < x_3 < x_2 < x_1$. Obviously, if $<$ is well-founded then it is acyclic. $<$ is *smooth* with respect to $\mathcal{X}$ if there are no infinite descending $<$-chains in S, for every S in $\mathcal{X}$. Smoothness is a restricted form of well-foundedness.

Having decided to base the following considerations on strict rather than non-strict relations, we may ask whether there is a way to differentiate between ties and incomparabilities using strict relations only. With non-strict relations, it is very easy to say when two items x and y are comparable: Either $x \leq y$ or $y \leq x$. There is no problem of distinguishing items that are tied or equally good (both $x \leq y$ and $y \leq x$) and those that are incomparable (neither $x \leq y$ nor $y \leq x$). But it is not a trivial matter to express the notion of comparability when only a strict relation $<$ is available. It is clear that x and y are comparable if either $x < y$ or $y < x$ holds. But if this is not the case, how can we distinguish two items which are incomparable from two items which are just equally good

---

[262] Later we shall use the strict revealed preference relation defined by $x < y$ iff $x \in \sigma(\{x, y\})$ and $y \notin \sigma(\{x, y\})$. In a subset S of $\overline{R}$, all elements are minimal under this relation $<$, so liberal minimization is not applicable here, since it would give us $\sigma(S) = S$ rather than $\sigma(S) = \emptyset$.

[263] For several variants of modularity or rankedness, compare Lewis (1981, Section 5), Ginsberg (1986, Section 4.4), Lehmann and Magidor (1992, Section 3.6), Rott (1992b, Section 4), and Makinson (1994, Section IV.1).

under $<$? The following definition allows to express ties, and it captures exactly what we want, at least if the relation in question is transitive. We will say that x and y are *tied*[264] under $<$, in symbols $x \sim y$, if and only if for all z in X, $x < z$ just in case $y < z$, and $z < x$ just in case $z < y$; that is, if x and y have exactly the same sets of dominating and dominated elements.[265] This condition seems to be sufficient since it is the strongest relevant condition expressible in terms of $<$. How could two items x and y possibly satisfy this condition and yet be called incomparable *in terms of* $<$? On the other hand, the condition seems also necessary, for if there is an item z which is preferable to x, say, but not to y, then there must at least be one respect which prevents x and y from being 'equally good'. Intuitively, we would say in this case that either y is preferable to x, or else they are incomparable. We must not assume that differences in various respects offset one another. For example, it seems plausible to suppose that Harry would prefer his present job with a salary increased by 10 per cent (z) to his present job under the actual conditions (x), but he may find it impossible to compare either of these situations with a completely different kind of job at the other end of the world (y). Having introduced $\sim$, we can define $x \leq y$ iff $x < y$ or $x \sim y$.[266] Finally, x and y are said to be *incomparable* under $<$ if neither $x < y$ nor $y < x$ nor $x \sim y$ holds.

If $<$ is modular, the relation $\asymp$, defined by $x \asymp y$ iff neither $x < y$ nor $y < x$, coincides with the relation $\sim$ and is thus an equivalence relation. In this case—and only in this case—$x \asymp y$ expresses the fact that x and y are tied-or-identical rather than incomparable. Intuitively, the relation of being tied-or-identical must be an equivalence relation, while incomparability is neither reflexive nor transitive.

6.4　Two kinds of revealed preferences

If the observed choice behaviour of an agent is coherent, then it is possible to recover his preferences. 'Underlying' preferences are said to be *revealed* by 'manifest' behaviour. One of the most commonly discussed types of non-strict revealed preference relations are the *Samuelson preferences*.[267]

We formulate a slightly adapted version that is suitable for minimization rather than maximization and takes into account that an agent may refuse to

---

[264]More exactly, *tied-or-identical*.

[265]Technically speaking, indifference then is a *coincidence relation* for preference. Herzberger (1973, Section 5) refers to Leibniz's doctrine of the identity of indiscernibles and his conception of indifference in this connection.

[266]If $<$ has been obtained as the asymmetric part of a non-strict relation $\leq$, this suggested (re-)construction of a non-strict from the strict relation does not in general give back the original $\leq$. For us, however, this no reason to worry since we want to take strict relations $<$ as primitive.

[267]Herzberger (1973, p. 211) quotes a passage of Samuelson (1950) as 'one of [the] more recent formulations' of this preference concept; it has been in Samuelson's writings much before that paper.

choose. For any choice function σ over $\mathcal{X}$ with taboo set $\overline{R} = X - \bigcup \{\sigma(S) : S \in \mathcal{X}\}$, we define

$$x \leq y \quad \text{iff} \quad \text{there is an } S \in \mathcal{X} \text{ such that } y \in S \text{ and } x \in \sigma(S), \text{ or } y \in \overline{R}$$

$$\leq_\sigma \;=\; \bigcup \{\sigma(S) \times S : S \in \mathcal{X}\} \;\cup\; (X \times \overline{R})$$

Samuelson preferences (for maximization and without the clause concerning taboos) are used in Rott (1993). In the present book, we always work with strict preferences which are to be used for liberal minimization. We will not construct the strict version of Samuelson preferences by taking their asymmetric parts. We rather consider the following relation $<$ as 'the right' idea from which Samuelson preferences can be obtained by taking the converse complement, so that $x \leq y$ has means 'y is not less than x' rather than 'x is less than or equal to y'. For our official definition, we proceed by taking the converse complement of the relation just introduced.

Definition 4 Let σ be a choice function over some field $\mathcal{X}$ in X. Then the relation $< \,= \mathcal{P}(\sigma)$ expressing the *Samuelson preferences* in X is given by

$$x < y \quad \text{iff} \quad \text{for every } S \in \mathcal{X} \text{ such that } x \in S \text{ it holds that } y \notin \sigma(S),$$
$$\text{and } x \notin \overline{R}$$

Another one of the most commonly discussed types of non-strict 'revealed preference' relations are the so-called *base preferences* (Uzawa 1956, Arrow 1959; the terminology is again taken from Herzberger 1973, where many other possibilities of defining notions of non-strict revealed preference are discussed). As before, we formulate a version that suits minimization and takes into account refusal to choose.

$$x \leq_2 y \quad \text{iff} \quad x \in \sigma(\{x, y\}) \;\text{ or }\; y \in \overline{R} \;\text{ or }\; \{x, y\} \text{ is not in } \mathcal{X}$$

$$\leq_2 \;=\; \bigcup \{\sigma(S) \times S : S \in \mathcal{X} \text{ and } S \text{ has at most two elements}\}$$
$$\bigcup (X \times \overline{R}) \;\cup\; \{\langle x, y \rangle : \{x, y\} \notin \mathcal{X}\}$$

As in the case of Samuelson preferences, we transfer the basic idea into the framework where strict preferences are taken as primitive.

Definition 5 Let σ be a choice function over some field $\mathcal{X}$ in X. Then the relation $<_2 \,= \mathcal{P}_2(\sigma)$ expressing the *base preferences* in X is given by

$$x <_2 y \quad \text{iff} \quad \{x, y\} \text{ is in } \mathcal{X}, \; x \in \sigma(\{x, y\}) \text{ and } y \notin \sigma(\{x, y\})$$

Notice that $\mathcal{P}(\sigma)$ is guaranteed to be irreflexive, and $\mathcal{P}_2(\sigma)$ is guaranteed to be asymmetric. $\mathcal{P}_2(\sigma)$ is defined for arbitrary σs, but the definition makes good sense only for 2-covering ones for which every pair $\{x, y\}$ is in $\mathcal{X}$. In this case, $\mathcal{P}(\sigma)$ is also guaranteed to be asymmetric (if we assume ($\emptyset 1$) and ($\emptyset 2$)).

Lemma 13 *Let σ be a choice function and $< = \mathcal{P}(\sigma)$ and $<_2 = \mathcal{P}_2(\sigma)$. Then*
    (i) If σ is 12-covering and satisfies ($\emptyset 1$) and ($\emptyset 2$), then $x < y$ implies $x <_2 y$.
    (ii) If σ satisfies (I), then $x <_2 y$ implies $x < y$.
    (iii) If σ is 12-covering and satisfies ($\emptyset 1$) and (I), then $\mathcal{P}(\sigma) = \mathcal{P}_2(\sigma)$.

The following lemmas list a number of important facts the essence of which is well-known in the general theory of rational choice (cf. Aizerman 1985, Aizerman and Malishevski 1981, Herzberger 1973, Moulin 1985, Sen 1982, Suzumura 1983). They explain why it is justified to regard the conditions (I) – (IV) as *the* central conditions of the classical theory of rational choice. Our formulations are slightly more complex than the conditions usually given in the literature since we devote special care and attention to domain conditions and to refusals of choice. Although the gist of the results is not new, then, we give detailed proofs for the sake of completeness and precision.

The first lemma is concerned with the consequences and preconditions of a choice function's being relational with respect to either Samuelson or base preferences.

Lemma 14 *(i) If σ is relational then it satisfies (I) and (II^∞), and thus ($\emptyset 2$), but not necessarily ($\emptyset 1$) or ($\emptyset 2^+$). If σ is in addition smooth, then it also satisfies ($\emptyset 1$) and ($\emptyset 2^+$).*
    (ii) If σ is 12-covering and satisfies ($\emptyset 1$), (I) and (II^∞), then it is relational with respect to $\mathcal{P}_2(\sigma)$, i.e., $\sigma = \mathcal{S}(\mathcal{P}_2(\sigma))$.
    (iii) If σ is 12-covering and satisfies ($\emptyset 1$), (I) and (II^∞), then it is relational with respect to $\mathcal{P}(\sigma)$, i.e., $\sigma = \mathcal{S}(\mathcal{P}(\sigma))$.
    (iv) If σ is 12-covering and relational and satisfies ($\emptyset 1$), then it is relational with respect to both $\mathcal{P}(\sigma)$ and $\mathcal{P}_2(\sigma)$, i.e., $\sigma = \mathcal{S}(\mathcal{P}(\sigma)) = \mathcal{S}(\mathcal{P}_2(\sigma))$.
    (v) If σ is additive and satisfies ($\emptyset 2^+$), (I) and (II^∞), then it is relational with respect to $\mathcal{P}(\sigma)$, i.e., $\sigma = \mathcal{S}(\mathcal{P}(\sigma))$.

The proof of Lemma 14 shows that if we restrict our demand for relationality to finite menus, then we can replace additivity by finite additivity and the infinitary condition (II^∞) by its finitary version (II). If for all *finite* menus S it holds that $\sigma(S)$ is the same as $\sigma'(S)$, where $\sigma' = \mathcal{S}(<)$ for a strict relation $<$, we will say that σ is *finitely relational* with respect to $<$.

Corollary 15 *(i) If σ is 12-covering and satisfies ($\emptyset 1$), (I) and (II), then it is finitely relational with respect to both $\mathcal{P}_2(\sigma)$ and $\mathcal{P}(\sigma)$.*
    (ii) If σ is finitely additive and satisfies ($\emptyset 2^+$), (I) and (II), then it is finitely relational with respect to $\mathcal{P}(\sigma)$.

The next lemma deals with various properties of revealed preference relations.

Lemma 16 *(i) If σ is 12n-covering and satisfies ($\emptyset 1$) and (I), then $\mathcal{P}(\sigma) = \mathcal{P}_2(\sigma)$ is n-acyclic. If σ is ω-covering and satisfies ($\emptyset 1$) and (I), then $\mathcal{P}(\sigma) = \mathcal{P}_2(\sigma)$ is acyclic.*
    (ii) If σ is 123-covering and satisfies ($\emptyset 1$), (I) and (III), then $\mathcal{P}(\sigma) = \mathcal{P}_2(\sigma)$ is transitive.

(iii) If σ is 123-covering and satisfies ($\emptyset$1), (I) and (IV), then $\mathcal{P}(\sigma) = \mathcal{P}_2(\sigma)$
is modular.
(iv) If σ is finitely additive and satisfies ($\emptyset$1) and (IV), then $\mathcal{P}(\sigma)$ is modular.

It is worth noting that there is no analogue of part (iv) for the base preference relation $\mathcal{P}_2(\sigma)$. To see this, consider the choice function over $\mathcal{X} = Pow(\{x, y, z\})$ which is defined by $\sigma\{x, y\} = \{x\}$ and $\sigma(S) = S$ for every other set S in $\mathcal{X}$. It is clear that this function satisfies (IV). However, for the base preferences $<$ we have $x < y$, but neither $x < z$ nor $z < y$, so Modularity is violated. Part (iii) of Lemma 16 tells us that this cannot happen if σ satisfies (I).

In partial converse to the Lemma 16, the next lemma tells us about the properties of choice functions which are determined by transitive or modular preference relations.

Lemma 17 *(i) If $<$ is transitive and well-founded and $\mathcal{S}(<)$ satisfies ($\emptyset$1), then $\mathcal{S}(<)$ satisfies (III).*
(ii) If $<$ is transitive and smooth, $\mathcal{X}$ is subtractive and $\mathcal{S}(<)$ satisfies ($\emptyset$1), then $\mathcal{S}(<)$ satisfies (III).
(iii) If $<$ is modular, then $\mathcal{S}(<)$ satisfies (IV).

The next observation summarizes the results of this section as they apply to the important special case of finitary choice functions in a more convenient way.

Observation 18 *Let σ be a choice function which satisfies ($\emptyset$1) and can take all and only the finite subsets of a given domain X as arguments.*
(i) σ is rationalizable iff it is rationalizable by the preference relation $\mathcal{P}_2(\sigma)$ defined by

$$x < y \quad \text{iff} \quad x \in \sigma(\{x, y\}) \text{ and } y \notin \sigma(\{x, y\})$$

(ii) σ is rationalizable iff it satisfies (I) and (II).
(iii) σ is transitively rationalizable iff it satisfies (I), (II), and (III).
(iv) σ is modularly rationalizable iff it satisfies (I) and (IV).

An interesting concept in the theory of choice is that of *collected extremal choice* (Aizerman 1985, Aizerman and Malishevski 1981) or *pseudo-rationalizability* (Moulin 1985). Adapting the relevant definitions to our context, we call a choice function σ *pseudo-rationalizable* by a finite sequence of modular preference relations $<_1, \ldots, <_n$, if for all menus S in $\mathcal{X}$ it holds that $\sigma(S)$ is the set of all options that are minimal under at least one of the preference relations $<_i$, in symbols, if $\sigma(S) = \bigcup \{\min_{<_i}(S) : i = 1, \ldots, n\}$.[268] Intuitively, the different preference relations represent different *criteria* under which the options are evaluated or different *members* of a group in a collective choice problem. If σ_i is a

---

[268] Actually Moulin works with anti-symmetric orderings here for which $x \sim y$ implies $x = y$. (There is a misprint in the definition of an ordering on p. 149 of Moulin (1985), where 'asymmetric' should be replaced by 'anti-symmetric'.) As in doxastic contexts we typically encounter ties, we stick to the account of Aizerman and Malishevski (1985, 1981) who do not impose anti-symmetry.

criterion's or a member's choice function, the one rationalized by $<_i$, then clearly $\sigma = \bigcup \sigma_i$.[269] So according to collected extremal choice, an option is considered eligible for choice if it is 'best' with regard to at least one aspect or for at least one member of the group.

Obviously, each rationalizable choice function is pseudo-rationalizable (put $n = 1$). But the converse is not valid. We have seen that every rationalizable choice function satisfies (II), and we now give an example that shows that pseudo-rationalizable choice functions do not in general satisfy (II). Let X be $\{x, y, z\}$ and $\mathcal{X} = Pow(X)$. Let $<_1$ and $<_2$ be transitive preference relations with $x <_1 y <_1 z$ and $z <_2 y <_2 x$. We see that $<_2$ just reverses the preferences of $<_1$. Now a moment's reflection shows that $\sigma(\{x, y, z\}) = \{x, z\}$ and that $\sigma(\{x, y\}) = \{x, y\}$ and $\sigma(\{y, z\}) = \{y, z\}$. We find that y is in $\sigma(\{x, y\}) \cap \sigma(\{y, z\})$ but not in $\sigma(\{x, y\} \cup \{y, z\})$. Thus condition (II) is violated.[270]

Now the question is which conditions characterize pseudo-rationalizability. The following important theorem is due to Aizerman and Malishevski (1981):

Observation 19 *Let the set X of conceivable options be finite and let $\mathcal{X}$ be the set of all non-empty subsets of X. Then a choice function over $\mathcal{X}$ is pseudo-rationalizable if and only if it satisfies conditions (I) and (III).*[271]

There is yet another concept we ought to mention here, the concept of *path independence* made prominent by Plott (1973) and later studied by Blair, Bordes, Kelly and Suzumura (1976) among others. The condition comes in two equivalent versions:

(PI) $\sigma(S \cup S') = \sigma(\sigma(S) \cup S')$
(PI') $\sigma(S \cup S') = \sigma(\sigma(S) \cup \sigma(S'))$

It is shown by Plott that (PI) and (PI') are equivalent for finitary and ω-covering choice functions. Notice that the very formulation of Path Independence requires that σ is iterative and finitely additive. In fact these conditions are sufficient for the equivalence of (PI) and (PI').

Path Independence was appealed to as a basic intuition in the seminal monograph of Arrow (1951/63, second edition, p. 120). The condition is very appealing intuitively: It suggests that a choice in a large menu can always be decomposed into choices in smaller, more manageable submenus (which need not be disjoint), followed by a final choice in the union of all the choice sets thus obtained. Or, in Plott's (1973, pp. 1079–1080) own words:

---

[269] This is of course short for $\sigma(S) = \bigcup \{\sigma_i(S) : i = 1, \ldots, n\}$, for all S in $\mathcal{X}$.

[270] The example is taken from Moulin (1985, p. 156). Also compare the scenario given in Nehring (1997, p. 404).

[271] A proof of this observation can be found in Moulin (1985, p. 157). I do not know whether it generalizes to infinite alternative sets X, even if one was ready to allow infinitely many relations $<_i$ (representing infinitely many criteria or agents) to be used for the pseudo-rationalizing. From the point of view of the 'syntactic' application discussed in the next two chapters, the most interesting case would be the one in which X is countably infinite and $\mathcal{X}$ is the set of all finite subsets of X (i.e., where the choice function is finitary and ω-covering).

... the process of choosing, from a dynamic point of view, frequently proceeds in a type of 'divide and conquer' manner. The alternatives are 'split up' into smaller sets, a choice is made over each of these sets, the chosen elements are collected, and then a choice is made from them. Path independence, in this case, would mean that the final result would be independent of the way the alternatives were initially divided up for consideration.

The following observation was first stated for finitary and ω-covering choice functions by Aizerman and Malishevski, but a similar result, with the Superset axiom instead of (III), was obtained earlier for finitely additive and iterative choice functions by Blair, Bordes, Kelly and Suzumura (1976, Theorem 1). We take down Aizerman and Malishevski's version, but remove the restriction to finite menus.

Observation 20 *A finitely additive and iterative choice function satisfies path independence if and only if it satisfies conditions (I) and (III).*

The results about pseudo-rationalizable choice (*alias* collective extremal choice) and path-independent choice are particularly interesting to us for the following reason. We shall find in Chapter 7 that if the theory of rational choice is applied to the theoretical realm of belief revision and nonmonotonic reasoning, conditions (I) and (III) yield rather more fundamental and plausible principles than (II). In the light of Observations 19 and 20, it thus appears that 'from a logical point of view', pseudo-rationalizability and path independence have a privileged status and are more plausible than plain rationalizability by a single (possibly intransitive) preference relation. We recall, however, that path independence presupposes that choice functions are *iterative*, i.e., that $\sigma(S)$ is in $\mathcal{X}$ for all S in $\mathcal{X}$. Later we will see that this condition is satisfied for syntactic choice functions (which are finitary and ω-covering), but it is violated for semantic choice functions (which are in general neither finitary nor ω-covering).

6.5   More conditions on choice functions

We have good reason to look at a few more conditions that do not have a strong independent standing in the theory of rational choice.

(I⁻)   If $S \subseteq S'$ and $\sigma(S') \subseteq S$, then $\sigma(S') \subseteq \sigma(S)$

(II⁺)   If $x \in \sigma(S)$ and $y \in \sigma(S')$, then $x \in \sigma(S \cup S')$ or $y \in \sigma(S \cup S')$

(IV⁺)   If $S \subseteq S'$, then $\sigma(S) \subseteq \sigma(S')$

Although (I) is a very basic constraint in the theory of choice, it will turn out that its weakening (I⁻) which is called *Cut* by Lindström (1991) is interesting in itself.[272] It is still stronger than the idempotence of σ (saying that $\sigma(\sigma(S)) = \sigma(S)$ for all S).

---

[272]The condition appears in Delgrande's (1987, pp. 113–114) definition of *NP*-models but it is redundant there.

Condition (II$^+$) implies (II), and is implied by (IV) together with ($\emptyset$1). It does not follow from conditions (I) – (III), as is shown by the following counterexample. Let $X = \{x, y, z, z'\}$ and $\mathcal{X} = Pow(X)$. Suppose there is a rationalizing preference relation $<$ consisting only of $z < x$ and $z' < y$. Notice that $<$ is (vacuously) transitive. Since $\sigma = \mathcal{S}(<)$ is rationalized by a transitive preference relation, it follows that σ satisfies (I) – (III). But we have $\sigma(\{x, z'\}) = \{x, z'\}$ and $\sigma(\{y, z\}) = \{y, z\}$ and yet $\sigma(\{x, y, z, z'\}) = \{z, z'\}$. Thus σ violates (II$^+$). I have not detected condition (II$^+$) or some equivalent variation of it in the literature.

Condition (IV$^+$) is a strengthening of (IV), and it is a very strong condition indeed. Read contrapositively, it says that if an option is not considered worth choosing in a large menu, then it should not be considered worth choosing in any smaller one. So if something is not considered best among, say, a hundred alternative options, it should not be considered best even when the number of alternative options has decreased to two. That is, an option is either eligible for choice or it is not, irrespective of the context of available alternatives. As a general choice condition, (IV$^+$) is certainly undesirable. We shall keep it, however, as a limiting case.

(IV$^+$) is implied by the Indiscrimination condition (Indis). Taken together with (I), it is strong enough to imply (Indis), provided that $\mathcal{X}$ is closed under finite unions or intersections. Remembering our above interpretation of the indiscrimination condition as one of 'satisficing', we can say that (IV$^+$) is essentially a condition of bounded rationality (see again Section 6.1.1).

6.6   Remainder functions

A concept in a sense dual to that of a choice function is that of a remainder function. A *remainder function* ρ over $\mathcal{X}$ can be defined from a choice function by putting, for all S in $\mathcal{X}$,

$$\rho(S) = S - \sigma(S)$$

The set $\rho(S)$ consists of all elements which are ineligible for choice within S, which remain disregarded if S is the issue. One can of course take the concept of a remainder function as primitive and consider it independently of any prior choice function. Remainder functions can be approached in a way analogous to the way choice functions have first been approached in this chapter, namely by means of abstract constraints. In fact it is fairly straightforward to translate the central rationality constraints for choice functions into rationality constraints for remainder functions. The following list is the result of such a translation, with the obvious correspondences: ρ satisfies (...$^\rho$) iff σ satisfies (...). The coherence constraints for remainder functions take a markedly different form from, and are in a sense dual to, those for choice functions.

(Faith$^\rho$) $\exists R \, \forall S :$ if $S \cap R \neq \emptyset$ then $\rho(S) = S - R$

(Indis$^\rho$) $\exists R \, \forall S : \rho(S) = S - R$

(Success$^\rho$)   If $\rho(S) = S$, then $S = \emptyset$

$(\emptyset^{\rho})$	If $\rho(S) = S$, then $S \cap S' \subseteq \rho(S')$
$(\mathbf{I}^{\rho})$	If $S \subseteq S'$, then $\rho(S) \subseteq \rho(S')$
$(\mathbf{II}^{\rho})$	$\rho(S \cup S') \cap S \cap S' \subseteq \rho(S) \cup \rho(S')$
$(\mathbf{III}^{\rho})$	If $S \subseteq S'$ and $S' \subseteq S \cup \rho(S')$, then $S \cap \rho(S') \subseteq \rho(S)$
$(\mathbf{IV}^{\rho})$	If $S \subseteq S'$ and $S \not\subseteq \rho(S')$, then $S \cap \rho(S') \subseteq \rho(S)$
$(\mathbf{I}^{-\rho})$	If $S \subseteq S'$ and $S' \subseteq S \cup \rho(S')$, then $\rho(S) \subseteq \rho(S')$
$(\mathbf{II}^{+\rho})$	If $x \in S \cap \rho(S \cup S')$ and $y \in S' \cap \rho(S \cup S')$, then $x \in \rho(S)$ or $y \in \rho(S')$
$(\mathbf{IV}^{+\rho})$	If $S \subseteq S'$, then $S \cap \rho(S') \subseteq \rho(S)$

Except from $(\mathbf{I}^{\rho})$, the formulations of coherence constraints in terms of remainder functions do not appear to be more appealing than those in terms of choice functions. $\overline{R}$ is a taboo set of ρ if for all menus S, $\overline{R} \subseteq \rho(S)$ and $\rho(S) = S$ only if $S \subseteq \overline{R}$.

It is instructive to put the following question: Given some function over $\mathcal{X}$ yielding a subset of S when applied to any S in $\mathcal{X}$, how do we know whether we are dealing with a choice function or a remainder function? First, one can mention the monotonicity condition $(\mathbf{I}^{\rho})$ which makes no sense for choice functions but does make good sense for remainder functions. Another important idea is that choice functions aim at returning non-empty subsets $\sigma(S)$ of S while remainder functions aim at returning proper subsets $\rho(S)$ of S. In the case of a choice function, it is of primary importance to reach a decision, i.e., to actually yield at least one selected element. The case of a remainder function just reverses the intuition: It should not be the case that all elements offered are returned as ineligible for choice. These are the respective notions of *success* for the two types of functions. Although we said that we want to allow for choice or remainder functions that are not successful, such functions should be regarded as limiting cases. In principle, one should choose *something* and not dismiss *every* available alternative. In contrast, there is nothing wrong with the idea that all alternatives in an issue S are eligible for choice, or equivalently, that no alternative must be disregarded.

Remainder functions will become interesting when we turn to syntactic choices involved in belief revision and nonmonotonic reasoning (Sections 7.3 and 7.4). There a choice function will actually pick the 'worst' beliefs or expectations, the ones that are most conveniently given up. So, in a way, the remainder functions leave us with the 'good' beliefs or expectations, the ones that remain untouched. In fact an obvious idea for modelling a belief contraction would be to put $K \dot{-} \phi = \rho_{\phi}(K)$, with the remainder function ρ_{ϕ} being dependent on the sentence ϕ to be withdrawn. The reason why we want choice functions in belief change to give us the 'worst' rather than the 'best' elements is just the basic intuition of what it means for such a function to be successful. It must be said, though, that there is still a discrepancy between the respective notions of success. Success in belief contraction requires that ϕ is not in $K \dot{-} \phi$; success for choice functions only demands that *some* sentence from K should not be in $K \dot{-} \phi$.[273] We have to take measures to guarantee that it is actually ϕ which is included in K but missing in $(K \dot{-} \phi)$. Our solution to this problem will be that choices are made not directly in K, but primarily in $Cn(\phi)$, with a special logical constraint guaranteeing that ϕ will be in the choice set of $Cn(\phi)$ (unless the choice set is empty).

---

[273] Conservativity pulls in the converse direction and sees to it that belief contractions are not 'too successful'. For instance, one might think that the case $\sigma(S) = S$ or $\rho(S) = \emptyset$ is usually ruled out by conservativity. In the common conception of belief contraction, however, violations of conservativity are deemed much more pardonable than violations of success (cf. Rott 2000b).

7

COHERENTIST BELIEF CHANGE AS A PROBLEM OF RATIONAL CHOICE

I have always thought of Parker, Quincy and Rice as honourable men. Parker, Quincy and Rice are neighbours of mine. Every Thursday evening they meet with Smith, a man living next street, for a few hours of poker. Yesterday—it was a Thursday—Smith was killed. The only persons who could have committed the crime are his poker partners. Now consider the following scenarios.

(1) As usual, Parker, Quincy and Rice were all participating in the poker game.
(2) Parker and Quincy were present, but Rice had to attend an important business meeting.
(3) Parker and Rice were present, but Quincy visited his mother in the country for the celebration of her 70th birthday.

Suppose I believe that situation (1) is the actual one. Considering all I know about my neighbours, as well as my opinions about them, I come to the conclusion that Parker may have been the assassin. This cautious conclusion neither excludes that I have doubts about Quincy or Rice as well, nor does it imply that I think that Parker actually *is* the crook. But it has implications for scenarios (2) and (3). If the number of suspects was limited to two while still including Parker, then I should still be ready to maintain that he may have been the murderer. Otherwise my judgments would justly be called incoherent.

Now change the scenario. Let us suppose that at first I believe that (2), and that I later believe that (3) correctly describes the actual situation. This may happen as a result of accumulating evidence or only hypothetically, as a thought experiment. Assume further that in both situations I find myself prepared to admit that Parker may have been the perpetrator. Then I should acknowledge this possibility in situation (1) as well. Otherwise my judgment would again be open to the charge of incoherence.

In order to make things shorter, let my original beliefs that Parker, Quincy and Rice are honourable men be symbolized by p, q and r respectively. Situation (1) demands that one of p, q and r be withdrawn, situations (2) and (3) similarly demand that either p or q, or respectively, that either p or r, be withdrawn. I have argued that if p is withdrawn in situation (1), then it should be withdrawn in situations (2) and (3) as well. Conversely, I said that if p is withdrawn both in situation (2) and in situation (3), it should be withdrawn in situation (1) as well. These constraints are, we claimed, *constraints of coherence*. The problem is to determine exactly the meaning and impact of coherence.

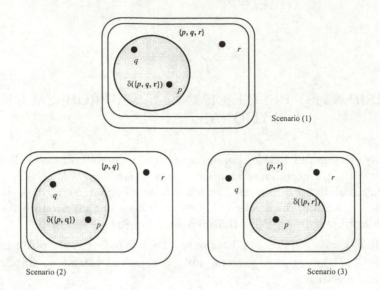

FIG. 7.1. Choosing least plausible beliefs

One can look at examples like this from two different angles. On the one hand, we clearly face a *problem of belief change*, i.e., a problem of what beliefs to give up in the light of conflicting evidence. So we should be able to apply the theories of belief change of the kind expounded in Chapters 3 – 5. A second way of viewing the problem presented by our example is to consider it as a problem of *choice* or *decision*. Viewed from this angle, we may expect help from the theory of rational choice as developed in Chapter 6 rather than from the theory of belief revision.

In scenarios (1) – (3), I have to make up my mind about which of the three beliefs p, q and r to drop. At least one of them must be wrong, and if there are 'ties' regarding my trust in Parker, Quincy and Rice, that is, if I do not possess a decisive criterion for choosing the culprit, I will have to give up even more than just *one* of p, q and r. For each of the surviving poker players, it is conceivable that he is dishonourable. Figure 7.1 illustrates an example in which the selection of the culprit in various scenarios is rational in the sense that the choices made in different situations cohere with each other. In the particular case depicted in Fig. 7.1 I suspect that both Parker and Quincy might have been the murderer, but I am convinced that Rice is innocent. Lightly shaded areas in the figure mark the menus, darkly shaded areas mark the choice sets of the respective situations.

Instead of choosing the sentences that are considered most likely to be false, the agent may reflect on which world could most plausibly be the real one. Assuming I have the same doxastic attitudes as depicted in Fig. 7.1, it is most

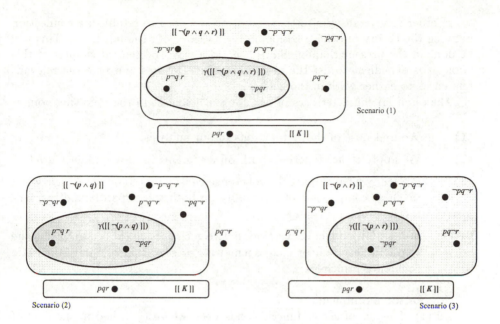

FIG. 7.2. Choosing most plausible worlds

likely for me that one of the worlds $\neg pqr$ and $p\neg qr$ is the real world.[274] The suitable choices in sets of possible worlds are then represented by Fig. 7.2.

The fact that p and q are considered by the agent to be less plausible (more questionable) than r, is now reflected by the fact that worlds satisfying $\neg p$ or $\neg q$ are considered more plausible (less far-fetched) than worlds satisfying $\neg r$.[275]

It is an aim of this chapter to unify the two ways of looking at the example. The questions we are going to ask are the following. Do the constraints of rational choice, or even the mere interpretation of belief change in terms of choice, impose interesting restrictions for revision functions? Are the usual postulates for belief revision justified by such constraints? Conversely, which constraints for rational choice can be retrieved from the postulates for rational belief change, and what is their status in the theory of choice?

---

[274]Notation: Here and in Fig. 7.2, the expression $\neg pqr$, for instance, stands for 'the possible world' in which p is false and q and r are true. For the sake of simplicity we use 'small' worlds which are uniquely identifiable by the truth values of the three propositional variables we are interested in.

[275]Actually Fig. 7.2 gives an answer to a problem which is not backed by anything we have said so far. In none of the three cases is the world $\neg p\neg qr$ among the most plausible worlds falsifying the presumption set $\{p, q, r\}$. I neglect the possibility that the crime was committed jointly. The intuitive reason is that $\neg p\neg qr$ falsifies two presumptions, while the worlds $\neg pqr$ and $p\neg qr$ falsify only one. A way of justifying the rejection of $\neg p\neg qr$ along these lines is discussed in Rott (1993, Section 6).

In order to give answers to these question, we need to establish a connection between the theory of belief revision and the theory of rational choice. This will be done in the present chapter. Since it is, in a way, the central chapter of this book, it is expedient to list the most important features and give the relevant references to earlier work in the field.

The essential characteristics of our approach consist in the following points.

(1)   We make use of choice functions in our analysis.

(2)   We apply choice functions both on a semantic and a syntactic level.

(3)   We study what impact the coherence constraints for choices have when taken as components of rationality on both the semantic and the syntactic level.

(4)   We establish direct connections between choices on the semantic and the syntactic level, under component-wise preservation of choice-theoretic coherence principles.

Now for some comments.

Ad (1). The use of choice functions has been well-established in the field of belief revision and nonmonotonic reasoning. In fact choice functions have figured as semantic structures for a comparatively long time in the related area of conditional logics. Here selection functions took the role as a paradigm instrument for semantic analyses in the seminal work of Robert Stalnaker (1968), David Lewis (1973, Section 2.7) and Brian Chellas (1975). Stalnaker's functions select single worlds, while Lewis's and Chellas's functions are more general in that they allow that sets of worlds are chosen. About ten years later, set-valued selection functions were the semantic tools utilized in the classical paper in belief revision by Alchourrón, Gärdenfors and Makinson (1985).

As we have seen in Chapter 6, the application of choice functions gives rise to a much more general approach than the application of preference relations. Not every reasonable choice function can be rationalized by an underlying preference relation.

Ad (2). The objects of choice on the semantic level are worlds, models, interpretations, or maximal consistent sets of sentences. The objects of choice on the syntactic level are single sentences; here (but not on the semantic level) choices have to respect the content or the logical structure of the sentences.

In all the work mentioned under (1) choice functions were applied on the semantic level. As far as I know, there are only two quite recent places in the literature where choices on the syntactic level are briefly discussed. Both are in the connection with generalizations or variants of the so-called safe contractions as introduced and studied by Alchourrón and Makinson (1985). The suggestions in point have been made by Sven Ove Hansson (1994a, 1999a) and Abhaya Nayak (1994a). What Hansson calls a *kernel contraction function* and Nayak calls a *rejector* bears considerable similarity to the choice functions on the syntactic level that we will consider below.[276] Like Alchourrón and Makinson, Hansson studies the impact of rationalizability by an acyclic preference relation (that may in addition be logically well-behaved[277] and modular). Nayak studies the impact of coherence criteria for choices, but he focuses on singleton-valued (successful and 'credulous') choice functions satisfying (I). As Nayak observes, such functions satisfy *all* interesting constraints for coherent choices,[278] thus leading to a case which would trivialize the whole undertaking of the present chapter. Singleton-valued choice functions are applicable in the context of safe contractions, but not for the approach to be advocated below. The syntactic approach taken in this chapter is different from the Alchourrón-Makinson approach; a preliminary reports about it is given in Rott (1994, 1998).

Ad (3). The first appeal to the theory of choice in the context of doxastic change was made by Doyle (1991). He referred to specific *problems* (Arrow's theorem), so his appeal was made with a sceptical and 'deconstructive' rather than constructive intention. The first fully systematic and constructive attempts to exploit the knowledge accumulated in the abstract theory of choice for the concerns of belief revision and nonmonotonic reasoning were made by Lindström (1991) and Rott (1993). These papers are similar to one another in that they both present one-to-one correspondences of coherence criteria or rationality postulates in the different fields of practical and theoretical reason by applying choice functions to the semantic entities used in belief revision and nonmonotonic reasoning. The criteria are *components* or *modules of rationality* in the sense that one or the other may be taken away or added without diminishing or disturbing the general idea. The residual differences between the approaches of Lindström and Rott are briefly commented upon in Rott (1993, pp. 1427–1428).

Ad (4). As far as I know, Section 7.8 below offers the first study that directly connects choices in a semantic domain with choices in a syntactic domain. With a slightly different twist, however, suggestions to relate the semantic and the syntactic level have been put forward for preference relations. Preferences between worlds, models or maximal sets of sentences have been linked with preferences between single sentences several times. The idea can be traced back to Shackle (1961) and has been followed, with differences in details but great resemblances in the overall picture, by Lewis (1973), who connects 'comparative similarity" and 'comparative possibility', by Zadeh (1978) and Dubois and Prade (1980, 1988, 1991) who connect 'possibility distributions' with 'possibility measures' and/or 'necessity measures', by Grove (1988), Spohn (1988), Rott (1991a),

---

[276]In Hansson's words: 'Intuitively, the kernel selection function s can be thought of as selecting for deletion the least valuable elements of each kernel. ... $s(X)$ consists of all those elements of F that are "worst" in the sense of not being epistemically "better" than any other element.' (Hansson 1999a, p. 93) Similarly, a rejector in Nayak's (1994a, p. 501) sense 'picks out [the] most rejectable elements' of any set of basic beliefs.

[277]More precisely, that may in addition satisfy conditions (EE2↑) and (EE2↓) of Definition 18 of the next chapter.

[278]Except of course the conditions (IV⁺) and (Indiscrimination). As pointed out in Chapter 6, these conditions are interesting only as limiting case conditions.

Freund (1993), Gärdenfors and Makinson (1994), and many others. The principal idea is that the position of a sentence ϕ in a syntactic preference relation $<$ is determined by the positions in the semantic preference relation $\prec$ of the worlds that satisfy ϕ. Conversely the position of a world w in a semantic preference relation $\prec$ is determined by the positions in the syntactic preference relation $<$ of the sentences that are satisfied by w.

Implicitly, any rule how to make use of preference relations in theoretical reasoning defines a choice function. For instance, if all the models minimal under some relation $<$ are said to be relevant for an operation of revision or reasoning, then all these models are selected for further semantic evaluation.

Besides being considerably less general than the choice functions approach, however, the approach working with preferences forgoes the advantage of a neat analysis of components of rationality. Logical properties get merged with properties that can be derived from coherent choice, and all the latter properties are lumped together in a few simple properties of the underlying preferences. Basing our considerations on choice functions, we shall see that almost all coherence criteria for choices *taken individually* are preserved under the transition from the syntactic to the semantic level, and vice versa.

As has been mentioned several times, I will investigate two ideas of constructing a contraction with respect to an opinion (belief or expectation), one operating on a semantic level (this section) and one operating on a syntactic level (the next section). Between these two main themes we insert a few comments on the original account of Alchourrón, Gärdenfors and Makinson which utilizes maximal non-implying subsets of a theory and has turned out to be intimately connected with the semantic approach (Section 7.2). In Sections 7.5 and 7.6, we use the correspondences found in Chapter 4 to transform the analysis of belief contraction into an analysis of belief revision and nonmonotonic reasoning. Section 7.7 presents the central representation results, and Section 7.8 explains a remarkable confluence in these results by showing that almost all constraints for coherent choice are preserved in the transitions between the semantic and the syntactic level.

7.1   How to give it up: Choices between models

The first idea we study is the following. When retracting a sentence ϕ from the set of one's opinions (beliefs or expectations), one takes into account the *most plausible* or *best* models that *falsify* ϕ.

In order to formally develop this idea, we need a number of preparatory definitions. The basic ingredients of this approach are *models m* and a classical *satisfaction relation* $\models$. Models should not be confused with worlds. Every individual world is just like a model as regards its 'satisfaction profile'. A crucial difference, however, emerges if we look at classes of models. If two models satisfy the same set of sentences, they are identical, while the same does not hold true for worlds. Two possible worlds may differ in respects that are not expressible with the linguistic means available.[279] We need not specify the set of models we are dealing with; this is always supposed to be the set of *all* models for a given language L, denoted by $\mathcal{M}_L$. The set $\widehat{m} = \{\phi : m \models \phi\}$ is the set of sentences satisfied by or, as we also say, *verified* by m. For any set M of models, $\widehat{M} = \{\phi : m \models \phi \text{ for every } m \in M\}$ is the set of sentences verified by all elements of M, also called the *theory of M*. For any set of sentences F, $[\![F]\!]$ denotes the set of models m verifying the set F, i.e., models for which $m \models \phi$ for every ϕ in F, and $]\!]F[\![$ denote the set of models m *falsifying* the set F, i.e. models for which $m \not\models \phi$ for some ϕ in F. We usually drop set brackets within double square brackets and write, for instance, $]\!]\phi_1, \ldots, \phi_n[\![$ instead of $]\!]\{\phi_1, \ldots, \phi_n\}[\![$. Clearly, $F \subseteq \widehat{m}$ if and only if $m \in [\![F]\!]$ if and only if $m \notin]\!]F[\![$.

As before, we let Cn and $\vdash$ denote a fixed standard consequence operation and its associated consequence relation. The properties of Cn and $\vdash$ are listed in Section 1.2.4. The relation between $\vdash$ and $\models$ is such that $F \vdash \phi$ holds if and only if $m \models F$ implies $m \models \phi$, for all models m. So Cn and $\vdash$ are not meant to refer to specific proof calculi, but rather to a semantic concept. We now take down a preparatory lemma.

Lemma 21 *For all sentences ϕ and all sets F and G of sentences, we have*

(i) $]\!]\phi[\![=]\!]Cn(\phi)[\![$;

(ii) $]\!]F[\![\cup]\!]G[\![=]\!]F \cup G[\![$;

(iii) $F \vdash \phi$ *if and only if* $[\![F]\!] \cap]\!]\phi[\![= \emptyset$;

(iv) $]\!]F[\![\subseteq]\!]G[\![$ *if and only if* $Cn(F) \subseteq Cn(G)$.

Now we introduce the concept of a semantic choice function. To mark the distinction from general choice functions σ, we denote semantic choice functions by the letter γ.[280] If $\mathcal{X}$ is any class of subsets of $\mathcal{M}_L$, a choice function with domain $\mathcal{X}$ will be called a *semantic choice function*. In most of the following, however, we focus on a particular $\mathcal{X}$: the class of all elementary subsets of $\mathcal{M}_L$, i.e., the class of all model sets S such that $S = [\![\phi]\!]$ for some sentence ϕ. Since the language of L is supposed to be closed under Boolean connectives, this domain is finitely additive, subtractive, closed under finite intersections and compact.[281] Sometimes we consider semantic choice functions which take all Σ-elementary subsets of $\mathcal{M}_L$, i.e., the class of all model sets S such that $S = \bigcup \{[\![\phi]\!] : \phi \in H\}$ for some set of sentences H; their domains are additive, but neither subtractive

---

[279] For a thorough technical discussion of the consequences of injectiveness in the context of non-monotonic reasoning, see Freund (1993) and Bochman (2000a). It is also worth mentioning that we do not work with partial models, as Kraus, Lehmann and Magidor (1990) do.

[280] This continues the traditional way of denoting selection functions for belief contraction, see Alchourrón, Gärdenfors and Makinson (1985), Hansson (1993a, 1993b), Rott (1993).

[281] Note that $[\![\phi]\!] \cup [\![\psi]\!] = [\![\phi \vee \psi]\!]$, $[\![\phi]\!] \cap [\![\psi]\!] = [\![\phi \wedge \psi]\!]$ and $[\![\phi]\!] - [\![\psi]\!] = [\![\phi \wedge \neg \psi]\!]$. Compactness in the sense of Section 6.1 follows from the compactness of the standard consequence operation Cn.

nor closed under intersections.[282] The semantic choice functions we consider are not in general iterative, since there is no guarantee that $\gamma(\,[\![\phi]\!]\,)$ is elementary or that $\gamma(\,]\!]F[\![\,)$ is Σ-elementary.

The *completion* γ^+ of a semantic choice function γ is defined by

$$\gamma^+(S) = S \cap \widehat{[\![\gamma(S)]\!]}$$
$$= \{m \in S : \text{ for all } \phi, \text{ if } m' \models \phi \text{ for all } m' \in \gamma(S) \text{ then } m \models \phi\}$$

This is the analogue of a definition of Alchourrón, Gärdenfors and Makinson (1985, p. 519). Clearly $\gamma(S) \subseteq \gamma^+(S)$, but the converse is not valid. A model m of S is in $\gamma^+(S)$ if $\bigcap\{\widehat{m'} : m' \in \gamma(S)\} \subseteq \widehat{m}$. A semantic choice function γ is called *complete* if it is identical with its own completion, i.e., if $\gamma = \gamma^+$. It is easy to show that a choice function is complete if and only if for every S the set $\gamma(S)$ is characterizable as the intersection of S with a Δ-elementary set, i.e., the set of models of some set of sentences.[283]

Here is our official definition of *semantic choice-based contraction functions*:

Definition 6 The contraction function $\dot{-}$ over K is generated by a choice function γ over models, in symbols $\dot{-} = \mathcal{C}(\gamma)$, if and only if for every ψ,

$$\psi \in K \dot{-} \phi \ \text{ iff } \ \psi \in K \text{ and } \gamma(\,]\!]\phi[\![\,) \subseteq [\![\psi]\!]$$

So a belief ψ of K remains accepted in $K \dot{-} \phi$ if and only if it is satisfied by all the most plausible worlds that falsify ϕ. A alternative and shorter way to express this is

$$K \dot{-} \phi = K \cap \widehat{\gamma(\,]\!]\phi[\![\,)}$$

We invoke as a background assumption that γ is satisfies (Faith1) with respect to $[\![K]\!]$.[284]

The limiting cases $\phi \in Cn(\emptyset)$ and $\phi \notin K$ do not call for a special treatment. The AGM postulates entail that in both cases $K \dot{-} \phi$ should equal K. If $\phi \in Cn(\emptyset)$, then $\gamma(\,]\!]\phi[\![\,) \subseteq]\!]\phi[\![\, = \emptyset$, so $K \dot{-} \phi = K$, as desired. And if $\phi \notin K$, i.e. if $[\![K]\!] \cap]\!]\phi[\![\, \neq \emptyset$, then since γ is supposed to satisfy (Faith1) with respect to $[\![K]\!]$, we get $\gamma(\,]\!]\phi[\![\,) \subseteq [\![K]\!]$. But $\psi \in K = Cn(K)$ already entails $[\![K]\!] \subseteq [\![\psi]\!]$, so clearly $\gamma(\,]\!]\phi[\![\,) \subseteq [\![\psi]\!]$, and again $K \dot{-} \phi = K$, as desired.

Notice that clearly $\mathcal{C}(\gamma) = \mathcal{C}(\gamma^+)$, so any choice function γ over models may be replaced, for the purposes of belief contraction, by its more well-behaved completion γ^+.

---

[282] Compare Bell and Slomson (1969, p. 141). It is easy to see that S is Σ-elementary if and only if $S =]\!]F[\![$ for some set of sentences F: $\bigcup\{\,[\![\phi]\!] : \phi \in F\} =]\!]\{\neg\phi : \phi \in F\}[\![$. Note that $]\!]F \cup]\!]G[\![\, =]\!]F \cup G[\![$, $]\!]F[\![\, -]\!]G[\![\, \subseteq]\!]F - G[\![$, and $]\!]F[\![\, \cap]\!]G[\![\, \supseteq]\!]F \cap G[\![$. The converses of the latter inclusions, however, are not valid.

[283] Compare Bell and Slomson (1969, pp. 140–142) and Rott (1993, pp. 1434–1436).

[284] In principle, Definition 6 is suitable for constructing two-place contraction functions (with K as the first argument). Compare the discussion in Sections 3.6 and 7.10.

The idea of a choice function for models easily generalizes to the problem of pick contractions, that is, the problem of giving up at least one of several beliefs.[285]

Definition 7 The pick contraction function $\dot{-}_{()}$ over K is generated by a choice between models, in symbols $\dot{-} = \mathcal{C}_{()}(\gamma)$, if and only if for every ψ in K

$$\psi \in K \dot{-} \langle F \rangle \quad \text{iff} \quad \psi \in K \text{ and } \gamma(\,]F[\,) \subseteq [\![\psi]\!]$$

A belief ψ in K survives the contraction of K with respect to F if all most plausible models falsifying F verify ψ. If the set F is finite, then a pick contraction with respect to F can be reduced to a contraction by the single sentence $\bigwedge F$, since a model falsifies F if and only if it falsifies $\bigwedge F$.

7.2   Partial meet contraction

The most influential paper in the theory of belief change has been Alchourrón, Gärdenfors and Makinson (1985). It is the joint work of the three authors that initiated the by now classic *AGM paradigm* or *AGM approach* to belief change. The theory of *partial meet contraction* developed in this paper makes extensive use of choice functions. The objects of choice are not models but maximal non-implying sets. More precisely, the idea is that for a contraction of a theory K with respect to a sentence ϕ, the first step is to look for all the maximal subsets of K that fail to imply ϕ. The set of all such maximal non-implying subsets is called $K \perp \phi$. It has turned out that it is no good to pick a single element of $K \perp \phi$ (which intuitively preserves 'too many' beliefs of K) or to take the intersection of all elements of $K \perp \phi$ (which preserves 'too few' beliefs of K). So AGM employ a choice function γ to select the best elements $\gamma(K \perp \phi)$ of $K \perp \phi$, and the contracted belief set $K \dot{-} \phi$ is then defined to be the intersection of all elements of $\gamma(K \perp \phi)$.

Adam Grove (1988) first noticed that for every ϕ which is in K but not in $Cn(\emptyset)$, there is a bijective correspondence between the elements of $K \perp \phi$ and the models satisfying $\neg \phi$.

For an arbitrary set of sentences H, let $H \perp \phi = \{M \subseteq H : \phi \notin Cn(M)$ and $\phi \in Cn(N)$ for all N with $M \subset N \subseteq H\}$. Note that the elements of $H \perp \phi$ are closed under Cn if H is closed under Cn. If $\phi \notin Cn(\emptyset)$, then $H \perp \phi$ is non-empty, by the compactness of Cn. We define $H \perp = \{H \perp \phi : \phi \in Cn(H) - Cn(\emptyset)\}$ and $U_H = \bigcup (H \perp)$.[286] If $Cn(H) \neq Cn(\emptyset)$, then $H \perp$ is a non-empty subset of $Pow(Pow(H)) - \{\emptyset\}$. The case $Cn(H) = Cn(\emptyset)$ will be handled separately in our reconstruction of partial meet contraction.

---

[285] See footnote 107 above. When speaking of 'pick contractions', I do not intend to distinguish picking from choosing in the way Ullmann-Margalit and Morgenbesser (1977) do. What I mean is really choosing rather than picking the sentence(s) to give up, but I want to avoid employing the term 'choice' for too many purposes at the same time and stick to the terminology of Rott (1992c).

[286] Here we deviate a little from the definitions in Alchourrón, Gärdenfors and Makinson (1985) that also include ϕs from $Cn(\emptyset)$.

Lemma 22 *Let K be a theory and $\phi \in K$. Then $M \in K \perp \phi$ if and only if there is a model m in $]\phi[$ such that $M = K \cap \hat{m}$.*

For the proof, compare Section 4 of Grove (1988). Now let K again be a theory. For every maximal non-implying set $M \in U_K$, put $mod(M) = m$ for the unique model m in $[\![M \cup \{\neg\phi\}]\!]$, where ϕ is any proposition in $K - M$. For every model m in $]\!K[\![$, put $mns(m) = \hat{m} \cap K$. The reader is invited to verify: mod is well-defined, mod is a bijection from U_K to $]\!K[\![$, and mns is the converse of mod, i.e., $mns(mod(M)) = M$ and $mod(mns(m)) = m$.

It has become clear that the concept of partial meet contraction of Alchourrón, Gärdenfors and Makinson (1985) is essentially equivalent to our semantic modelling of the previous subsection. There are two points of philosophical interest, however, which make a difference.

The equivalence holds only as long as the belief state of an agent is represented by a logically closed theory. If the set of beliefs or opinions is not closed, then the two accounts differ. The concept of a partial meet contraction of such a non-theory, i.e., a belief *base*, makes perfectly good sense and has indeed been studied extensively, most notably in a long series of papers by Sven Ove Hansson.[287] It is no longer true, however, that each maximal non-implying subset corresponds to a falsifying world, and so it is not at all obvious how to connect changes of a non-theory to a possible worlds modelling. The syntactic structure of the belief base encodes information which is relevant for potential changes of the belief state, and this kind of information has to be captured by some kind of selection mechanism for possible worlds or maximal non-implying subsets.[288]

Second, partial meet contractions are committed, by their reference to *maximal* non-implying subsets of a theory, to the idea of minimal change or informational economy. This is the only motivation to focus on the set $K \perp \phi$ in the first place. There is no such commitment in our choice-theoretic approach applied to possible worlds, in fact the concept of minimality is not even mentioned in any of our construction recipes.[289] For this reason, it is doubtful that the bijective correspondence between $K \perp \phi$ and $]\phi[$ is sufficient to warrant the same construction recipes for belief contractions. In partial meet contractions, it is very natural to take the intersection of all the elements of $\gamma(K \perp \phi)$—these simply are the very best maximal non-implying subsets. However, it is not clear that a choice function over worlds should be applied only to worlds falsifying ϕ.

---

[287] See the references given in footnote 142 of Chapter 3.

[288] The situation is now quite well-understood, thanks to papers by Lewis (1981), Nebel (1989, 1992), Rott (1993) and del Val (1994).

[289] If the choice functions employed satisfy coherence constraints (I) and (II), they allow a reconstruction through liberal minimization with respect to revealed preferences (see Chapter 6). In a sense, the idea of coherence entails the idea of minimality, although not minimality with respect to subset inclusion. I argue in Rott (2000b) that it is doubtful whether this idea of minimality has anything to do with Gärdenfors's (1988) motivation of belief revision in terms of 'informational economy' or with epistemological conservatism as surveyed by Christensen (1994).

In general there will be $[\![\phi]\!]$-worlds which are at least as plausible as the best $]\!\phi[\!$-worlds, and it is hard to justify that the former worlds are disregarded if the latter worlds are admitted for consideration. These doubts as to whether the AGM concept of contraction is really the most convincing one and whether the principle of minimum mutilation should not be subordinated to certain principles of preference and indifference are turned into an alternative theory of ('severe') belief withdrawals in Pagnucco and Rott (1999).[290] This paper uses yet another semantic modelling, the systems of spheres approach in the tradition of Lewis (1973, Section 1.3) and Grove (1988). It is not difficult to prove, and it has actually been proved by Lewis (1973, Section 2.7), that the systems of spheres approach is equivalent to a choice functional approach. Systems of spheres, however, capture only a very limited variety of choice functions, namely those that satisfy all interesting constraints (I) – (IV) for coherent choice.[291] There is no possibility to analyse the individual impact of weaker conditions in a system of spheres modelling.

7.3 How to give it up: Choosing worst sentences

The second idea of contracting a set of opinions (beliefs or expectations) in the face of counterevidence against ϕ involves choices between (previously accepted) sentences rather than choices between (previously rejected) models. This idea was the one that sprang to mind immediately when we discussed the murder example at the beginning of this chapter. For the coherentist, however, it is essential that the set of opinions is represented as a theory, that is, closed under the monotonic logic Cn. Thus we have to make our model somewhat more sophisticated than was evident in the example.

Prima facie the choice problems generated by different sentences with respect to which a set of beliefs or opinions is to be contracted seem to be so different as to require different choice functions. The opinions one chooses to give up in a contraction with respect to ϕ are in general very different from the ones one chooses to give up in a contraction with respect to ψ. It is true that the choices have to be made *within the same menu*, namely the theory K. But since we cannot say anything about the connection between different choice functions, we seem to face serious difficulties in arriving at any interesting results about theory change. However, in the study of belief change different injunctions—to contract by ϕ or to contract by ψ—are supposed to meet with the same cognitive state of the agent. This is why there are interesting connections between contractions

---

[290] We follow the suggestion of Makinson (1987) and reserve the term 'contraction' only for belief removal operations that satisfy the Recovery postulate (−5). By taking intersections in the last step, the account of AGM itself compromises the principle of minimum mutilation. The coherentist account in Pagnucco and Rott (1999) violates Recovery, in contrast to AGM and the coherentist suggestions made in the present chapter.

[291] Condition (4) of Lewis (1973, p. 58) is precisely Arrow's axiom (see page 154 above) applied to a possible worlds setting. This very strong axiom is equivalent to the conjunction of (I) and (IV), from which (II) and (III) are also derivable, provided (∅1) is also satisfied.

with respect to different sentences, connections which find expression in the variants of the AGM postulates ($\dot{-}7$) and ($\dot{-}8$) (see Chapter 4).

Fortunately, it is possible to model the problem of belief contraction vis-à-vis a single cognitive state as a problem that involves a single all-purpose choice function, associated with the given belief set K. Rather than applying *different choice functions* for the contraction with respect to ϕ and the contraction with respect to ψ, we can reformulate the problem as one involving the application of this single choice function to *different menus* which are, of course, dependent on the sentence to be contracted. We have done this on the semantic level above. Let us look how the idea can be set to work on a syntactic level. This is done in two steps.

First, if a belief set is contracted with respect to a sentence ϕ, the agent must figure out which *consequences* of ϕ to discard. Some consequences of ϕ should probably be given up when retracting ϕ because their being accepted may crucially depend on the acceptance of ϕ.[292] But we can make an even stronger claim. It is necessary that some consequences of ϕ be given up—unless the agent refuses to give up ϕ itself. Both $\phi \vee \psi$ and $\phi \vee \neg\psi$ follow from ϕ, but they cannot both be retained in a successful contraction of K with respect to ϕ, since the contraction of a theory should result in a theory again, and $\phi \vee \psi$ and $\phi \vee \neg\psi$ taken together would allow one to recover ϕ. The question of course is: Which consequences of ϕ should be given up? I will not attempt to give material recommendations. All I wish to emphasize here is that the agent has to make a choice *among the elements of* $Cn(\phi)$.

A principal idea is that a consequence ψ of ϕ, i.e., an element of $Cn(\phi)$, is in $K\dot{-}\phi$ if and only if ψ is not selected for removal in K given that ϕ has to be retracted, and this in turn reduces to the condition that ψ is not selected for removal within $Cn(\phi)$. Now we introduce the concept of a syntactic choice function. To mark the distinction from both general choice functions σ and semantic choice functions γ, we denote syntactic choice functions by the letter δ.[293] If $\mathcal{X}$ is any class of subsets of L, a choice function δ with domain $\mathcal{X}$ is called a *syntactic choice function*. In the following, however, we usually focus on a particular domain $\mathcal{X}$: the class of all subsets of L, i.e., the class of all sets of sentences. This domain is additive, subtractive and closed under intersections. Sometimes we will consider finitary and ω-covering syntactic choice functions which take all and only finite sets of sentences; their domains are finitely additive, subtractive and closed under intersections. Both kinds of syntactic choice functions we consider are iterative. Intuitively, syntactic choice functions have a somewhat 'negative' character since the chosen elements are, from the point of view of the believer, the *worst* elements in a belief set, the *most dispensable* ones. Now suppose that the belief set K contains the sentence ϕ, and thus all

---

[292] Cf. Fuhrmann's (1991) 'Filtering' and Rott's (2000a) 'Simple Filtering' conditions.

[293] The letter 'δ' is also meant to be reminiscent of the word '*difference*', since the members of $\delta(F)$ are *taken away* from a set F of opinions or beliefs.

consequences of ϕ. Then a basic idea of how to construct a contraction of K with respect to ϕ is this:

(UCRM) For all ψ in $Cn(\phi)$, $\psi \in K \dot{-} \phi$ iff $\psi \notin \delta_\phi(K)$ iff $\psi \notin \delta(Cn(\phi))$

Let us call this thesis *Unified Choice in Restricted Menus*. As in the case of semantic choices, we are going to use one and the same choice (or remainder) function δ for different ϕ's. The it differently, the choice function as a representation of the agent's mental state is regarded as independent of the particular sentence ϕ in question. By the same token, we want the choices made across different issues of the form $Cn(\phi)$ to be coherent.[294]

This connection between contractions with respect to ϕ and choices in $Cn(\phi)$ is also motivated by the central idea of *success* in both cases. In belief change, $K \dot{-} \phi$ is successful if ϕ is eliminated; this is the AGM Success postulate ($\dot{-}4$). In the theory of choice, δ is successfully applied to $Cn(\phi)$ if something in this set is actually chosen; this is the criterion of Success for rational choice. In the next section we introduce a logical constraint on syntactic choice functions that entails that if $\delta(Cn(\phi))$ is non-empty, then ϕ is in that set. So according to this interpretation, too, being successful means eliminating ϕ.

However, the idea elaborated so far works fine only for sentences ψ which are among the logical consequences of ϕ. We said that a major concern of an agent when withdrawing ϕ is with $Cn(\phi)$, but now we should move on and discuss the more delicate question of what happens to sentences which are not consequences of ϕ, and may in fact be entirely unrelated in content to ϕ. This is the normal case in belief contraction since usually $Cn(\phi)$ is a *proper* subset of K.

We now take the *second step* to motivate our syntactic-choice account of belief contraction. When eliminating ϕ, we have to pay attention to the connections that any element of K has with ϕ. More precisely, we have to focus on the *reasons for ϕ* and *consequences of ϕ*. *Reasons* for ϕ may be thought of as sentences that figure as premises in non-trivial derivations of ϕ, and that *consequences* of ϕ may be thought of as sentences that figure as conclusions in non-trivial derivations starting from a premise set containing ϕ (thus 'consequence' is now understood in a looser sense than the one given by Cn). Now observe that if ψ is a reason for ϕ, then $\phi \vee \psi$ is just as well a reason for ϕ. Observe further that if ψ is a consequence of ϕ, then $\phi \vee \psi$ is just as well a consequence of ϕ. Substituting $\phi \vee \psi$ for ψ in a derivation of the above-mentioned sorts will—perhaps with a little modification—again result in a derivation of the same sort. So we can conclude that in all respects that are relevant for a contraction with respect to ϕ, the sentence ψ plays exactly the same role as the disjunction $\phi \vee \psi$. In

---

[294] I remind the reader that varying choices are made only potentially. In reality only one sentence ϕ will actually have to be withdrawn. After the contraction has been performed the agent's choice function may have changed. The constraints for rational choice just mean that at any given moment the pattern underlying potential choices, or the *disposition* to make choices in various menus, should be coherent or 'consistent'.

an important sense, the sentences ψ and $\phi \vee \psi$ are functionally equivalent. We therefore advance the following *Disjunction thesis*:

(DT)   For all sentences ϕ, $\psi \in K \dot{-} \phi$ iff $(\phi \vee \psi) \in K \dot{-} \phi$

Since we make reference to the presence or absence of $\phi \vee \psi$ in a (contracted) belief state, condition (DT) makes sense only if K is a theory.[295] This indicates that the approach using 'syntactic' choice functions belongs to the coherentist paradigm. Condition (DT) follows formally from the AGM conditions of Closure and Recovery, ($\dot{-}1$) and ($\dot{-}5$), but I do not want to give this formal argument too much weight in the present context.[296]

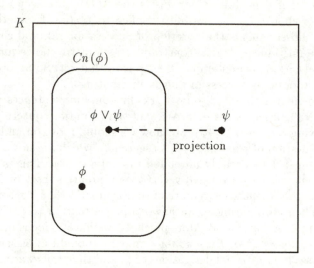

FIG. 7.3. 'Contracting K with respect to ϕ' means 'Choosing to give up at least one element of $Cn(\phi)$'

Now we realize that $\phi \vee \psi$ is included in $Cn(\phi)$! So the Disjunction thesis can be combined with the Unified-choice-in-restricted-menus thesis, and this allows us to focus exclusively on the elements of $Cn(\phi)$ if we contract our beliefs with respect to ϕ (see Fig. 7.3). We arrive at the following official definition of *syntactic choice-based contraction functions*:

[295]Or that it is at least disjunctively closed, cf. Hansson (1999a, Chapter 4, Sections 5 and 12[+]).
[296]Remember that the Recovery postulate of Alchourrón, Gärdenfors and Makinson is very controversial.

Definition 8 The contraction function $\dot{-}$ over K is generated by a choice function δ over sentences, in symbols $\dot{-} = \mathcal{C}(\delta)$, if and only if for every ψ,

$$\psi \in K \dot{-} \phi \text{ iff } \psi \in K \text{ and } \phi \vee \psi \notin \delta(Cn(\phi))$$

If $\dot{-} = \mathcal{C}(\delta)$, we also say that the contraction function $\dot{-}$ is *based on the syntactic choice function* δ. Let me repeat once more that δ is a choice function that gives us the most *implausible* or the *worst* (least entrenched) elements $\delta(F)$ of some issue F. The symbol δ will always be used for denoting choice functions which pick *worst* elements. We invoke as a background assumption that δ is satisfies (Faith1) with respect to $L - K$. It should also be noted that Definition 8 requires a choice function that is capable of taking infinite sets.[297] The essential step involved in withdrawing a belief ϕ from a belief set K consists in choosing what to retract from $Cn(\phi)$.

As in the semantic case, the limiting cases $\phi \in Cn(\emptyset)$ and $\phi \notin K$ do not call for a special treatment. The AGM postulates entail that in both cases $K \dot{-} \phi$ should equal K. If $\phi \in Cn(\emptyset)$, then $\phi \vee \psi \in Cn(\emptyset)$ and by the logical constraint (LP1) to be introduced in the next section, $Cn(\emptyset) \cap \delta(F) = \emptyset$ for any F, so $K \dot{-} \phi = K$, as desired. And if $\phi \notin K$, we need to get that $\psi \in K$ implies $\phi \vee \psi \notin \delta(Cn(\phi))$. From $\phi \notin K$, we get that $Cn(\phi) \cap (L - K) \neq \emptyset$. Then since δ is supposed to satisfy (Faith1) with respect to $L - K$, which gives us $\delta(Cn(\phi)) \subseteq L - K$. But since by supposition ψ is in K and K is a theory, so $\phi \vee \psi$ is in K, and hence $\phi \vee \psi$ is not in $\delta(Cn(\phi))$, as desired.

It often happens that one is confronted with the task of giving up at least one of several beliefs. The example at the beginning of this chapter is a case in point. We noted that a solution to the problem is straightforward if we use semantic choice functions. But it is far less clear how to extend the idea of syntactic choice functions to the case of pick contractions. The right way to do this is the following.

Definition 9 The pick contraction function $\dot{-}_{\langle\rangle}$ over K is generated by a choice between sentences, in symbols $\dot{-} = \mathcal{C}_{\langle\rangle}(\delta)$, if and only if,

$$\psi \in K \dot{-} \langle F \rangle \text{ iff } \psi \in K \text{ and } \psi \vee \chi \notin \delta(Cn(F)) \text{ for every } \chi \in F$$

Clearly, if $F = \{\phi\}$, then Definition 9 reduces to Definition 8. In the next section we show that Definition 9 reduces to some very plausible condition in two more special cases (finite F, and ϕ in F). However, we shall not deal systematically with the theory of pick contractions in this essay. The term 'contraction function' will only refer to singleton contractions in what follows.

---

[297]The shortest way of expressing Definition 8 is this: $K \dot{-} \phi = K \cap Cn(\rho(Cn(\phi)) \cup \{\neg\phi\})$, where ρ is the remainder function corresponding to δ (cf. Section 6.6).

7.4   Logical constraints for choice functions over sentences

In this section we take account of the fact that the domain of δ consists of sentences which possess an internal logical structure. There is no equivalent structure to be considered of the objects of choice in the semantic case since all models are treated here as points and are as such unrelated to each other.[298] In the syntactic case, however, there is a dependency relation between doxastic items (opinions, beliefs, expectations) which must be respected by ideally rational agents: the relation of logical consequence.

Considering the fact that we are using choice functions in order to determine which expectations are to be given up by a rational, logically competent reasoner, I propose following conditions which relate choice functions to the monotonic consequence operation Cn.

Logical intra-menu condition

> If ϕ is selected to be withdrawn in the menu F, then for every subset G of F that logically entails ϕ at least one element of G must also be withdrawn in the menu F.

Logical inter-menu condition

> If two menus F and G are logically equivalent, then every ϕ which is contained in both F and G is selected to be withdrawn in the menu F just in case it is withdrawn in the menu G.

The first principle constrains the choices made in a single issue, while the second principle constrains choices across varying menus. The first condition says that if ϕ is selected to be withdrawn, one *really* wants to get rid of ϕ, without the possibility of later rederiving ϕ from the set of remaining beliefs.[299] The second condition says that if one selects ϕ for removal, this should not be due to the contingent syntactic way of how the questionable beliefs are presented. The vulnerability of a belief is determined in the context of the beliefs one is *committed to*, not in the context of the beliefs one happens to have explicitly represented. This condition has a strong coherentist flavour; its spirit is hardly compatible with the foundationalist (Maxim B) (cf. Section 1.2.1). We can formulate the two new conditions more succinctly:

(LP1)   If $G \subseteq F$ and $\delta(F) \cap G = \emptyset$, then $\delta(F) \cap Cn(G) = \emptyset$

(LP2)   If $Cn(F) = Cn(G)$, then $\delta(F) \cap G \subseteq \delta(G)$

Both of these conditions reflect the fact that the choices are not made on a purely syntactic, surface-grammatical level. The *content* of the sentences, in

---

[298] This is not necessarily so. One can, for instance, argue that models possess internal structure and can be compared to each other in terms of the sets or numbers of atoms on which they disagree. I do not want to commit myself to such a form of logical atomism which I think is much more open to controversy than the requirement that beliefs should respect logical interconnections.

[299] Similar considerations motivate the discussion of smooth incision functions in Hansson (1994a, p. 850) and (1999a, pp. 90–92).

so far as it can be captured by the monotonic 'background' logic Cn, is what really matters. The things to be chosen by δ possess an internal structure—a fact that is not exploited by any of the purely choice-theoretic postulates that we discussed in the Chapter 6. So we alert the reader that when we speak of syntactic choice operations, this is meant to indicate only that choices are made *among sentences*, but the operations in question are thought to be semantically grounded and thus constrained by the intra-menu and inter-menu conditions (LP1) and (LP2). Syntactically different ways of expressing the same proposition will always be treated the same.

If in (LP1) we put $G = \rho(F) = F - \delta(F)$, we see that (LP1) is equivalent to

(LP1$^\rho$) $Cn(\rho(F)) \cap F \subseteq \rho(F)$

That is, (LP1) effectively requires that the remainder $\rho(F) = F - \delta(F)$ is closed under Cn relative to F. If F is a theory, so is $\rho(F)$.

It follows from (LP1) that $\delta(F) \cap Cn(\emptyset) = \emptyset$ for all F. This principle expresses the fact that we should never consider withdrawing logically valid sentences from the set of our expectations. All elements ϕ of $Cn(\emptyset)$ are necessarily taboo (under δ), since by (LP1), $\delta(\{\phi\}) = \emptyset$ for all such ϕ (take $F = \{\phi\}$ and $G = \emptyset$). (LP1) implies that the taboo set $\overline{R}$ of δ is closed under Cn and hence in particular includes $Cn(\emptyset)$. (The taboo set $\overline{R}$ is uniquely determined, since $\mathcal{X}$ is 1-covering and we presume that δ to satisfies condition $(\emptyset)$ of Chapter 6; remember Lemma 10.) So choice functions satisfying (LP1) are not entirely 'successful': They may yield $\delta(F) = \emptyset$ even if $F \neq \emptyset$.

If only elements of $Cn(\emptyset)$ are taboo under δ, i.e., if $\overline{R} = Cn(\emptyset)$, then we call the choice function δ *resolute* or *virtually successful*.

(Virtual Success)   If $\delta(F) = \emptyset$, then $F \subseteq Cn(\emptyset)$

Another consequence of (LP1) is that if the set $\delta(Cn(\phi))$ used in Definition 8 is non-empty, then it contains ϕ. Thus, if anything from K is given up in $K \dot{-} \phi$ at all, then ϕ is among the discarded sentences.

(LP2) is a more disputable condition than (LP1). It says that if ϕ is selected for withdrawal in F and it is also an element of a set G with the same content as F, then ϕ must be selected for withdrawal in G as well. If a belief or expectation ϕ is untenable in the context of information contradicting F, then it is untenable in the context of (the same piece of) information contradicting G.

If δ is an infinitary function, i.e., if it can take infinite sets as arguments, a simpler and equivalent form of (LP2) is

(LP2^{inf}) $\delta(F) = F \cap \delta(Cn(F))$

This condition expresses the idea that the choice is 'really' made in the theory $Cn(F)$ and we select for removal exactly those 'axioms' in F that are selected for removal in the theory generated by F. The drawback of the condition is that in contrast to (LP2), it is not applicable to finitary choice functions.[300] A

---

[300] More conditions equivalent with (LP2) are.

very important feature of (LP2) is that it allows us to give equivalent *finitary* reformulations of Definition 8:

Definition 10 The contraction function $\dot{-}$ over K is generated by a (possibly finitary) choice function δ over sentences, in symbols $\dot{-} = \mathcal{C}_{fin}(\delta)$, if and only if for every ψ,

$$\psi \in K \dot{-} \phi \ \text{ iff } \ \psi \in K \text{ and } \phi \vee \psi \notin \delta(\{\phi, \phi \vee \psi\})$$

or equivalently,

$$\psi \in K \dot{-} \phi \ \text{ iff } \ \psi \in K \text{ and } \phi \vee \psi \notin \delta(\{\phi \vee \psi, \phi \vee \neg\psi\})$$

When $\dot{-} = \mathcal{C}_{fin}(\delta)$, we say that the contraction function $\dot{-}$ is *finitarily based on* δ. (LP2) comes as a relief, since it reduces the 'big decisions' involved in Definition 8 to choices in very small menus. It guarantees that every contraction operation based on a syntactic choice function δ is also finitarily based on δ, but of course the converse is not true: If the choice function δ is itself finitary, then Definition 8 is simply not applicable. In the proofs of the representation theorems below, we shall in fact construct only finitary syntactic choice functions δ, so we indeed need to have recourse to Definition 10. In contexts where it is clear that δ is finitary and that it satisfies (LP2), we shall apply the latter definition, often simply say (with a slight abuse of terminology) that $\dot{-}$ is based on δ, and write $\dot{-} = \mathcal{C}(\delta)$.

Given (LP1) and (LP2), it is easy to see that $\delta(F) = \emptyset$ if and only if $\delta(G) = \emptyset$, provided that $Cn(F) = Cn(G)$.

Another consequence of (LP1) that if δ satisfies (Faith) or (Indis), then the complement of the set R of absolutely satisfactory options mentioned in these conditions must be closed under Cn.

Lemma 23 *(LP1) and (Faith) for δ imply that $L - R$ is a theory.*
(LP1) and (Indis) for δ imply that $L - R$ is a theory.

Suppose that we have a choice function δ over some set F of sentences and that δ satisfies some of the choice-theoretic constraints we have mentioned in Chapter 6. Then it is easy to verify that one can extend δ to a choice function δ^+ over the whole language L which satisfies the same constraints as δ. This can be done by means of the following definition:

$$\delta^+(G) = \begin{cases} \delta(G) & , \text{ if } G \subseteq F \\ G - F & , \text{ if } G \not\subseteq F \end{cases}$$

Clearly, δ^+ is faithful to $L - F$, and it has the same taboo set as δ.

If $Cn(F) = Cn(G)$, then $\delta(F) = \delta(F \cup G) \cap F$
If $F \subseteq G \subseteq Cn(F)$, then $\delta(F) = F \cap \delta(G)$

We now turn to pick contraction functions that are generated by syntactic choice functions.

If the set of sentences with respect to which we have to contract is finite, then pick contractions reduce to the contraction with respect to conjunctions. This is not only a plausible point of view, but is provable in our context.

Lemma 24 *Let δ be a choice function over sentences satisfying (LP1) and (LP2), and let F be a finite set of sentences. Furthermore, let $\dot{-} = \mathcal{C}(\delta)$ be the contraction function and $\dot{-}_{\langle\rangle} = \mathcal{C}_{\langle\rangle}(\delta)$ be the pick contraction function over K that are generated by δ. Then*

(i) For ϕ in $F \cap K$, we have $\phi \in K \dot{-} \langle F \rangle$ if and only if $\phi \notin \delta(F)$.

(ii) $K \dot{-} \langle F \rangle = K \dot{-} \bigwedge F$.

Part (i) of the Lemma shows that pick contractions allow a straightforward representation of the syntactic choices on which they are based. That ϕ is chosen from F just means that ϕ is withdrawn from K in the construction of $K \dot{-} \langle F \rangle$. (If δ satisfies (Faith2) with respect to $L - K$, we can drop the restriction to ϕ's in K.) Part (ii) shows that if we are asked to give up (at least) one belief of a finite menu, this is exactly the same task as giving up a single sentence, viz., the conjunction of the sentences in the menu—provided that we rely on our constructions based on syntactic choice functions. This result parallels a similar result for semantic choice functions. (See the remark after Definition 7.)

In the literature on belief revision and nonmonotonic reasoning we find the idea of a reconstruction of 'epistemic entrenchments' *alias* 'expectation orderings' of propositions from the observed contraction behaviour of a doxastic agent. The characteristic definition due to Gärdenfors and Makinson (1988, 1994; compare Rott 1992b) is essentially this:

$$\phi < \psi \;\; \text{iff} \;\; \phi \notin K \dot{-} (\phi \wedge \psi) \text{ and } \psi \in K \dot{-} (\phi \wedge \psi)$$

In Chapter 8 I shall work out the idea that an entrenchment relation $<$ can also be interpreted as a base preference in the sense of Definition 5:[301]

$$\phi < \psi \;\; \text{iff} \;\; \delta(\{\phi, \psi\}) = \{\phi\}$$

I argue that the instruction 'remove $\phi \wedge \psi$' that underlies the Gärdenfors-Makinson condition should be regarded as an instruction to remove ϕ *or* remove ψ, where the agent holding theory K has free choice which proposition(s) out of $\{\phi, \psi\}$ to remove. It thus corresponds exactly to the pick contraction with respect to $\{\phi, \psi\}$ and ultimately to a choice in this two-element menu. Lemma 24 confirms this interpretation.

7.5   Revision

We now introduce direct ways of constructing revisions which parallel the semantic and syntactic choice-based constructions of belief contraction. Although good

[301] This was first hinted at in Rott (1992b, p. 61) and (1993, p. 1431, footnote 2).

motivation can be drawn from the connection with contractions, the definitions provide an independent way of looking at revisions which can thus be studied as objects in their own right. Most of the choice-theoretic analyses will give parallel results. There is one interesting exception which concerns the coherence criterion (Faith2). This condition is without any force when applied to contractions.[302] However, in the case of revisions we shall see that (Faith2) installs condition $(*3)$ which says that $K * \phi \subseteq Cn(K \cup \{\phi\})$. If revisions are defined from contractions with the help of the Levi identity, $(*3)$ follows trivially from $(\dot{-}2)$ and the impact of (Faith2) on revisions (as opposed to its impact on contractions) gets obscured. Forgetting about the Levi identity when looking at revisions has the additional advantage that two properties imported by the very idea of the identity, $(*1)$ and $(*2)$, may be treated as separate features.

Here are the crucial definitions of *choice-based revision functions*.

Definition 11 The revision function $*$ over K is generated by a choice function γ over models, in symbols $* = \mathcal{R}(\gamma)$, if and only if for every ψ in L,

$$\psi \in K * \phi \ \text{ iff } \ \gamma(\llbracket \phi \rrbracket) \subseteq \llbracket \psi \rrbracket$$

So a sentence ψ is accepted in $K * \phi$ if and only if it is satisfied by all the most plausible worlds that verify ϕ. A short way of expressing this is

$$K * \phi = \bigcap \hat{\gamma}(\rrbracket \phi \llbracket \,))$$

Comparing Definition 11 with Definition 6, we immediately recognize that semantic choice-based contractions and revisions are related by the Harper identity.

For Definition 11 it is again sufficient that the domain of γ is the set of all elementary sets of models. Δ-elementary sets enter the picture with bunch revisions (accept all elements of an infinite set F), which we will not study here.

Definition 12 The revision function $*$ over K is generated by a choice function δ over sentences, in symbols $* = \mathcal{R}(\delta)$, if and only if for every ψ in L,

$$\psi \in K * \phi \ \text{ iff } \ \phi \supset \psi \notin \delta(Cn(\neg\phi))$$

or equivalently (to be used if δ is finitary)

$$\psi \in K * \phi \ \text{ iff } \ \phi \supset \psi \notin \delta(\{\phi \supset \psi, \phi \supset \neg\psi\})$$
$$\psi \in K * \phi \ \text{ iff } \ \phi \supset \psi \notin \delta(\{\neg\phi, \phi \supset \psi\})$$

---

[302]It *would* have the effect of yielding Inclusion, $(\dot{-}2)$, $K \dot{-} \phi \subseteq K$, were this not already guaranteed by other parts of Definitions 6 and 8.

If the material conditional $\phi \supset \psi$ is not discarded when certain consequences of $\neg\phi$ have to be discarded, then this can be taken to mean that the conditional is accepted not just because the negation of its antecedent is accepted. So if the agent with choice function δ comes to learn that ϕ is in fact true, this will not destroy his or her belief in $\phi \supset \psi$, and consequently ψ is believed in the revised belief set.

Comparing Definition 12 with Definition 8, we can recognize that syntactic choice-based contractions and revisions are related by the Levi identity.

Revision functions are usually defined with respect to some arbitrary but fixed belief set K. This fact finds expression in the AGM postulates (∗3) and (∗4) which revision functions are intended to satisfy; both (∗3) and (∗4)—but no other postulate—make essential reference to the belief set K. But in the above definitions there is no reference to K. We shall see that these postulates are fulfilled if and only if the semantic and syntactic choice functions γ and δ are faithful with respect to $[\![K]\!]$ and $L - K$, respectively. We shall assume that our choice functions meet the constraints (Faith1) and (Faith2) at the appropriate places, but it is perhaps better not to make them an integral part of the definitions of $\mathcal{R}(\gamma)$ or $\mathcal{R}(\delta)$.

7.6 Nonmonotonic inference

Finally we introduce direct ways of constructing nonmonotonic inferences which parallel the semantic and syntactic choice-based constructions of belief change. Motivation can be gained from the Makinson-Gärdenfors identity (see Section 4.4), but the definitions may be understood as an independent way of looking at inference relations which are studied as objects in their own right. Here are the central definitions of *choice-based inference relations*:

Definition 13 The inference relation $\vdash\!\!\!\sim$ is generated by a choice function γ between models, in symbols $\vdash\!\!\!\sim\, = \mathcal{I}(\gamma)$, if and only if for every ψ in L,

$$\phi \vdash\!\!\!\sim \psi \quad \text{iff} \quad \gamma([\![\phi]\!]) \subseteq [\![\psi]\!]$$

Definition 14 The inference relation $\vdash\!\!\!\sim$ is generated by a choice function δ between sentences, in symbols $\vdash\!\!\!\sim\, = \mathcal{I}(\delta)$, if and only if for every ψ in L,

$$\phi \vdash\!\!\!\sim \psi \quad \text{iff} \quad \phi \supset \psi \notin \delta(Cn(\neg\phi))$$

or equivalently (to be used if δ is finitary)

$$\phi \vdash\!\!\!\sim \psi \quad \text{iff} \quad \phi \supset \psi \notin \delta(\{\phi \supset \psi, \phi \supset \neg\psi\})$$
$$\phi \vdash\!\!\!\sim \psi \quad \text{iff} \quad \phi \supset \psi \notin \delta(\{\neg\phi, \phi \supset \psi\})$$

So semantically speaking, a sentence ψ can be inferred from the sentence ϕ if and only if ψ is satisfied by all the most plausible worlds that verify ϕ. On the syntactical side, if the material conditional $\phi \supset \psi$ is not discarded when certain consequences of the expectation $\neg\phi$ have to be discarded, then this means that the conditional is still usable if we take as a premise that ϕ is true, and consequently ψ may be inferred from ϕ.

Evidently, there is a very close similarity between these definitions and the ones for revisions. The main difference is that while conditions of (Faith) for choice functions γ and δ are important for belief revision, there is no obvious place for them in the definition of nonmonotonic inference relations. It is evident that changes of belief start from a given belief set K, but how do our expectations or opinions enter the picture?

We can get help through arguments from the general theory of choice. In Section 6.1.1 we saw that if a choice function is faithful to some set, then this set is the choice set that results if the menu is the universal set of alternatives (if that set is in the domain of the choice function; compare condition (Faith')). In the present case, this means that γ is faithful to $\gamma(\mathcal{M}_L) = \gamma(\llbracket\top\rrbracket)$ and that δ is faithful to $\delta(L) = \delta(Cn(\neg\top))$, where it is important to note that $\llbracket\top\rrbracket$ is in the domain of γ and $Cn(\neg\top)$ is taken to be in the domain of δ.

If we start from a given inference operation Inf that is rationalizable by a choice function that is faithful to some set, can we characterize this set in term of Inf? We can indeed verify from the Definitions 13 and 14 that $\gamma(\llbracket\top\rrbracket) \subseteq \widehat{\llbracket\gamma(\llbracket\top\rrbracket)\rrbracket} = \llbracket Inf(\top))\rrbracket$ [303] and $\delta(Cn(\neg\top)) = L - Inf(\top)$.[304] Hence, the set $Inf(\top)$, the set of 'expectations', plays exactly the same role in non-monotonic reasoning as the belief set K plays in operations of belief contraction or revision.[305]

7.6.1   An example: Composers and compatriots

Let us discuss a counterexample to rational monotony due to Ginsberg (1986). In his story (which is modelled on a story of Quine's) the failure of rational monotony can be attributed to the fact that our reasoning proceeds from belief bases rather than from theories.[306] We can show that the example can be neatly analysed and reinterpreted in the coherentist framework as a failure of the choice

---

[303]Remember here that if Inf is generated by γ, it is also generated by the completion γ^+ of γ; compare Section 7.1.

[304]If Inf is generated by a finitary syntactic choice function δ, then of course the set L of all sentences is not in δ's domain any more, but thanks to (LP2) $Inf(\top)$ can be characterized as $L - \bigcup\{\delta(\phi, \perp) : \phi \in L\}$ or as $L - \bigcup\{\delta(\phi, \neg\phi) : \phi \in L\}$.

[305]As pointed out in Rott (1991a, footnote 6), one could dispense with an explicit reference to K in the case of belief change as well and 'reconstruct' the belief set from unary functions representing the dynamics of belief, by defining $K = \dot{-}(\perp)$ or $K = *(\top)$. Notice that this invariably yields a consistent theory K.

[306]It has long been known that conditional reasoning based on minimal changes of premise sets violates rational monotony—see Pollock (1976), Veltman (1976), Kratzer (1981) and Lewis (1981)—and that belief base revision does not satisfy the eighth axiom of AGM—see Nebel (1989, 1992), Rott (1993) and del Val (1994, 1997).

condition (IV).

Let the agent's body of beliefs or expectations contain information about the nationality of the three famous composers Bizet (b, 1838–1875), Satie (s, 1866–1925) and Verdi (v, 1813–1901). The agent believes or expects (with equal strength) that

$$Fb \quad \text{Bizet was French}$$
$$Fs \quad \text{Satie was French}$$
$$\neg Fv \quad \text{Verdi was not French}$$

The agent's set of beliefs or expectations is $Cn(\{Fb, Fs, \neg Fv\})$. We consider the following sentences that can be inferred from the above premises:

$$Cbs \quad \text{Bizet and Satie were compatriots}$$
$$\neg Cbv \quad \text{Bizet and Verdi were not compatriots}$$

Here, Cxy is an abbreviation for $(Fx \wedge Fy) \vee (\neg Fx \wedge \neg Fy)$. What would have been the case if Bizet and Verdi had been compatriots? Well either Bizet and Verdi would have been French, or they would both have been Italian. Satie's nationality is not affected. In order to represent this, we use the formalism of nonmonotonic reasoning, but a representation in terms of belief revision would work just as well.

$$Cbv \mathrel{|\!\sim} Fs \quad \text{and}$$
$$Cbv \mathrel{|\!\not\sim} \neg Cbs$$

But now, what would have been the case if all three composers had been compatriots? Ginsberg claims (and I agree on this point) that intuitively we would reason as follows.

$$Cbv \wedge Cbs \mathrel{|\!\not\sim} Fs$$

This is justified because if all three composers were compatriots, they might as well all have been Italian. Thus Rational Monotony is violated. How can this be explained in our framework?

Let us use an abbreviation for $Cbv \wedge Cbs$:

$$Cbsv \quad \text{Bizet and Satie and Verdi were compatriots}$$

The following analysis in terms of semantic and syntactic choice functions reflects what I think a normal agent would choose as most plausible worlds or as least plausible sentences on the hypothesis that Bizet and Verdi were compatriots, or respectively, on the hypothesis that all three composers were compatriots.

As illustrated in Fig. 7.4, the analysis in terms of semantic choice functions goes like this: We have $[\![Cbsv]\!] \subseteq [\![Cbv]\!]$ and

$$\gamma([\![Cbv]\!]) \cap [\![Cbsv]\!] \neq \emptyset$$

but

$$\gamma([\![Cbsv]\!]) \not\subseteq \gamma([\![Cbv]\!])$$

Thus we have a clear violation of (IV).

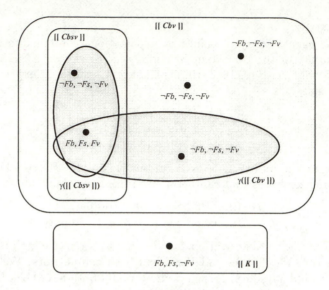

FIG. 7.4. Compatriots example: Semantic choice analysis

As illustrated in Fig. 7.5, the analysis in terms of syntactic choice functions goes like this: We have $Cn(\{\neg Cbsv\}) \subseteq Cn(\{\neg Cbv\})$ and

$$\delta(Cn(\{\neg Cbv\})) \cap Cn(\{\neg Cbsv\}) \neq \emptyset$$

but

$$\delta(Cn(\{\neg Cbsv\})) \not\subseteq \delta(Cn(\{\neg Cbv\}))$$

Thus again, we have a clear violation of (IV).

The example exhibits a remarkable structural similarity between semantic and syntactic choices. Although the ideas motivating the semantic approach are fundamentally different from those motivating the syntactic approach, there is only very little difference between the γ-formulae and the δ-formulae displayed a few lines above: Exactly at the places where we have the menu '$[\![\phi]\!]$' in the former case, we have the menu '$Cn(\{\neg\phi\})$' in the latter case. The ϕ's themselves are the same in both cases. We will now investigate more systematically the link between semantic and syntactic choices on the one hand and belief change and nonmonotonic reasoning on the other hand.

7.7 Representation theorems for semantically and syntactically choice-based contraction functions: A remarkable confluence of results

In the last sections we have subsumed three different but closely related fields of theoretical reason under a framework which has been developed in the more comprehensive area of practical reason. Figure 7.6 shows the six ways in which the theory of rational choice may be applied to processes of belief contraction, revision and nonmonotonic reasoning.

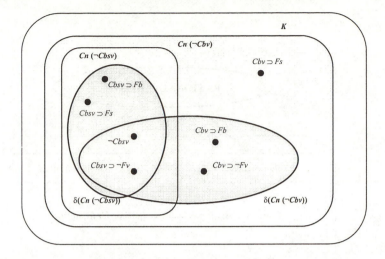

FIG. 7.5. Compatriots example: Syntactic choice analysis

The vertical arrows in this figure indicate the connections that have been discussed in Chapter 4 where the relevant postulates have been dealt with. The Levi and Harper identities serve as a link between contractions and revisions, the Makinson-Gärdenfors identity as a link between revisions and nonmonotonic inference operations. Our next task is to study the effects of coherence constraints for choice functions on the 'rationality' of contraction and revision functions and nonmonotonic inference operations—where this rationality is gauged in terms of the postulates presented in Chapter 4. Since both semantic and syntactic choices can be applied directly to construct not only contractions, but also revisions and nonmonotonic inferences, we could in principle perform independent analyses of six cases, where each case in turn consists of an array of soundness and completeness theorems. We shall state all the relevant theorems, but we shall actually prove only *two* exemplary cases, and for the rest rely on the connections between the postulates that were established in Chapter 4. The two cases we consider in detail are: Contractions based on semantic choice functions, and nonmonotonic inferences based on syntactic choice functions. These are indicated by the the darkened areas in Fig. 7.6. All the other cases are be strictly analogous.

As an additional reassurance that the results achieved here are in fact correct, and as an explanation of the striking structural parallel that we shall observe, we furnish and prove connections through direct transitions between semantic and syntactic choice functions in Section 7.8. Once these horizontal connections are established, it turns out that it would actually have sufficed to prove only *one* out of the six representation theorems stated. But due to the independent interest of the representation theorems and the completely different methods of proof involved, we are going to work through the proofs of the two representation theorems mentioned. The horizontal arrows represent direct transitions from se-

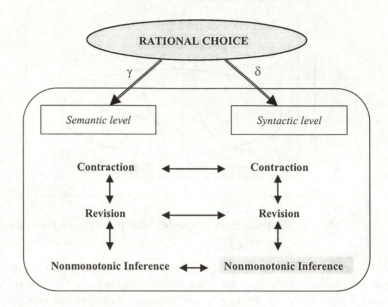

FIG. 7.6. Applying choice functions in belief change and
nonmonotonic reasoning

mantic choice functions to syntactic choice functions and vice versa. The idea
is that horizontal links relate equivalent choice operations, in the sense that related choice functions lead to identical contractions, revisions, and nonmonotonic
inferences.

7.7.1 *Representation theorems for contractions*

The following two observations have early ancestors in Rott (1993, Lemmas 7
and 8). The results given there are now translated from the context of an AGM-
style partial meet contraction to that of semantically choice-based contraction,
and extended to nine or ten (rather than only five) cases which are considered
in parallel.

Observation 25 *For every semantic choice function γ which satisfies*

$$\left\{\begin{array}{l} \textit{(Faith1) wrt } [\![K]\!] \\ \textit{(Success)} \\ \textit{(I)} \\ \textit{(I}^-) \\ \textit{(II) and is complete} \\ \textit{(II}^+) \\ \textit{(III)} \\ \textit{(IV)} \\ \textit{(IV}^+) \\ \textit{(Ø1)} \\ \textit{(Ø2)} \end{array}\right\} \text{, the contraction function } \dot{-} = \mathcal{C}(\gamma) \textit{ over } K \textit{ based on}$$

γ *satisfies* $(\dot-1) - (\dot-2)$, $(\dot-5) - (\dot-6)$ *and*
$$\left\{\begin{array}{l} (\dot-3) \\ (\dot-4) \\ (\dot-7) \\ (\dot-7c) \\ (\dot-8vwd) \\ (\dot-8d) \\ (\dot-8c) \\ (\dot-8) \\ (\dot-8m) \\ (\dot-8rc) \\ (\dot-7r) \end{array}\right\} \text{, respectively.}$$

Notice that (Faith2) for γ with respect to $]\!]K[\![$ takes no effect in the context of this theorem, since $K \dot- \phi \subseteq K$ is guaranteed by the construction of $K \dot- \phi$ anyway (Definition 6).

Given a contraction function $\dot-$ over K, we can derive from it a semantic choice function $\gamma = \mathcal{G}(\dot-)$ over the *elementary* sets of models. The idea is that a model falsifying ϕ is a best choice in $]\!]\phi[\![$ if and only if it satisfies everything that is contained in the contraction of K with respect to ϕ. So we define, for every sentence ϕ

$$\boxed{\gamma(]\!]\phi[\![) =]\!]\phi[\![\cap [\![K \dot- \phi]\!] = \{m \in]\!]\phi[\![: K \dot- \phi \subseteq \widehat{m}\}}$$

We have to check for well-definedness, that is, we have to show that $\gamma(]\!]\phi[\![) = \gamma(]\!]\psi[\![)$ whenever $]\!]\phi[\![=]\!]\psi[\![$. Let $]\!]\phi[\![=]\!]\psi[\![$. By Lemma 21, this entails that $Cn(\phi) = Cn(\psi)$. But from this and $(\dot-6)$ it follows that $K \dot- \phi = K \dot- \psi$, so by the definition of $\gamma = \mathcal{G}(\dot-)$, $\gamma(]\!]\phi[\![) = \gamma(]\!]\psi[\![)$, as required. Notice that by construction, the choice function $\gamma = \mathcal{G}(\dot-)$ is complete.

We say that the contraction function $\dot{-}$ *reveals*, or is a *manifestation of*, the underlying semantic choice function $\mathcal{G}(\dot{-})$. The next observation shows that interesting properties of $\dot{-}$ lead to corresponding interesting properties of γ.

Observation 26 *Every contraction function $\dot{-}$ over a theory K which satisfies*

$$(\dot{-}1) - (\dot{-}2), (\dot{-}5) - (\dot{-}6) \text{ and } \left\{\begin{array}{c} \text{—} \\ (\dot{-}3) \\ (\dot{-}4) \\ (\dot{-}7) \\ (\dot{-}7c) \\ (\dot{-}8vwd) \\ (\dot{-}8d) \\ (\dot{-}8c) \\ (\dot{-}8) \\ (\dot{-}8m) \\ (\dot{-}8rc) \\ (\dot{-}7r) \end{array}\right\} \text{ can be represented as the contrac-}$$

tion function $\mathcal{C}(\gamma)$ based on a semantic choice function γ which is complete and

$$\text{satisfies (Faith2) w.r.t. } [\![K]\!] \text{ and } \left\{\begin{array}{c} \text{—} \\ \text{(Faith1) w.r.t. } [\![K]\!] \\ \text{(Success)} \\ (I) \\ (I^-) \\ (II) \\ (II^+) \\ (III) \\ (IV) \\ (IV^+) \\ (\emptyset 1) \\ (\emptyset 2) \end{array}\right\} \text{, respectively.}$$

We now turn to choices between sentences. The following two theorems for contractions can be stated and proved in a way entirely analogous to Observations 33 and 34 for nonmonotonic inferences (for which proofs are given in the Appendix). The Levi and Makinson-Gärdenfors identities can serve as the bridge.[307]

---

[307] Alternatively one could exploit the results from Section 7.8.

Observation 27 *For every syntactic choice function δ which satisfies (LP1),*

$$(LP2) \text{ and} \begin{cases} \overline{} \\ (Faith1)\ wrt\ L-K \\ (Virtual\ Success) \\ (I) \\ (I^-) \\ (II) \\ (II^+) \\ (III) \\ (IV) \\ (IV^+) \\ (\emptyset 1) \\ (\emptyset 2) \end{cases}, \text{ the contraction function } \dot{-} = \mathcal{C}(\delta) \text{ over } K$$

based on δ satisfies $(\dot-1) - (\dot-2)$, $(\dot-5) - (\dot-6)$ and $\begin{cases} \overline{} \\ (\dot-3) \\ (\dot-4) \\ (\dot-7) \\ (\dot-7c) \\ (\dot-8wd) \\ (\dot-8d) \\ (\dot-8c) \\ (\dot-8) \\ (\dot-8m) \\ (\dot-8rc) \\ (\dot-7r) \end{cases}$ *, respectively.*

Similarly to the semantic case, (Faith2) for δ with respect to $L - K$ has no effect in the context of this theorem.

Given a contraction function $\dot{-}$ over K, we can derive from it a *finitary* and *ω-covering* syntactic choice function $\delta = \mathcal{D}(\dot-)$ for K.[308] The idea is that a sentence ϕ_i is chosen as a *best* (or more properly, a *worst*) element of $\{\phi_1, \ldots, \phi_n\}$ if it is in fact withdrawn in the contraction of K with respect to $\phi_1 \wedge \cdots \wedge \phi_n$. We build upon the idea that the task of giving up at least one element of a finite set $\{\phi_1, \ldots, \phi_n\}$ may be identified with the task of giving up the conjunction $\phi_1 \wedge \cdots \wedge \phi_n$. So we define, for all sentences $\phi_1, \ldots, \phi_n$

$$\delta(\{\phi_1, \ldots, \phi_n\}) = \{\phi_i : \phi_i \notin K \dot- (\phi_1 \wedge \cdots \wedge \phi_n)\}$$

---

[308] This restriction to finitary arguments in the syntactic case corresponds to the restriction to elementary arguments in the semantic case.

The choice set $\delta(\{\phi_1, \ldots, \phi_n\})$ contains just those ϕ_is which are *not* retained in $K \dot{-} (\phi_1 \wedge \cdots \wedge \phi_n)$.[309]

We say that the contraction function $\dot{-}$ *reveals*, or is a *manifestation of*, the underlying syntactic choice function $\mathcal{D}(\dot{-})$. The next observation shows that interesting properties of $\dot{-}$ lead to corresponding interesting properties of δ. As with Observation 27, a straightforward adaptation of the proofs we carry through for nonmonotonic inferences proves the following observation:

Observation 28 *Every contraction function $\dot{-}$ over a theory K which satisfies*

$$
(\dot{-}1) - (\dot{-}2),\ (\dot{-}5) - (\dot{-}6) \ and \
\left\{
\begin{array}{l}
\underline{\quad} \\
(\dot{-}3) \\
(\dot{-}4) \\
(\dot{-}7) \\
(\dot{-}7c) \\
(\dot{-}8wd) \\
(\dot{-}8d) \\
(\dot{-}8c) \\
(\dot{-}8) \\
(\dot{-}8m) \\
(\dot{-}8rc) \\
(\dot{-}7r)
\end{array}
\right\}
\ can \ be \ represented \ as \ the \ contraction
$$

function $\mathcal{C}(\delta)$ finitarily based on a syntactic choice function δ which satisfies

$$
(LP1),\ (LP2),\ (Faith2) \ w.r.t. \ L{-}K \ and \
\left\{
\begin{array}{l}
\underline{\quad} \\
(Faith1) \ w.r.t. \ L - K \\
(Virtual\ Success) \\
(I) \\
(I^-) \\
(II) \\
(II^+) \\
(III) \\
(IV) \\
(IV^+) \\
(\emptyset 1) \\
(\emptyset 2)
\end{array}
\right\}
, \ respectively.
$$

7.7.2 *Representation theorems for revisions*

Given the parallels drawn in Chapter 4, it is clear that the following observations on revisions are entirely analogous to the representation theorems for contractions we listed in the last section. For this reason, and because of a similar

---

[309] Or that are not in K in the first place.

connection with nonmonotonic inferences (which will be dealt with in the next section) we state them without proofs.

Observation 29 *For every semantic choice function* γ *which satisfies*

$$
\left\{
\begin{array}{c}
\overline{\;\;\;}\\
(Faith1)\ wrt\ [\![K]\!]\\
(Faith2)\ wrt\ [\![K]\!]\\
(Success)\\
(I)\\
(I^-)\\
(II)\ and\ is\ complete\\
(II^+)\\
(III)\\
(IV)\\
(IV^+)\\
(\emptyset 1)\\
(\emptyset 2)
\end{array}
\right\}
,\ \text{the revision function}\ *\ =\ \mathcal{R}(\gamma)\ \text{over}\ K\ \text{based on}\ \gamma
$$

satisfies (*1) – (*2), (*6) and
$$
\left\{
\begin{array}{c}
\overline{\;\;\;}\\
(*4)\\
(*3)\\
(*5)\\
(*7)\\
(*7c)\\
(*8vwd)\\
(*8d)\\
(*8c)\\
(*8)\\
(*8m)\\
(*8rc)\\
(*7r)
\end{array}
\right\}
,\ \text{respectively.}
$$

Given a revision function $*$ over K, we can derive from it a semantic choice function $\gamma = \mathcal{G}(*)$ over the elementary classes of models. The appropriate one which can be used for the proof of the following theorem is defined by

$$
\boxed{\gamma(\,[\![\phi]\!]\,) = [\![K * \phi]\!] = \{m : K * \phi \subseteq \widehat{m}\}}
$$

The obvious idea here is that a model verifying ϕ is a best element of $[\![\phi]\!]$ just in case it satisfies everything that is contained in the revision of K by ϕ.

Observation 30 *Every revision function* $*$ *over a theory* K *which satisfies*

$$(*1) - (*2), (*6) \text{ and } \left\{ \begin{array}{c} \overline{} \\ (*3) \\ (*4) \\ (*5) \\ (*7) \\ (*7c) \\ (*8vwd) \\ (*8d) \\ (*8c) \\ (*8) \\ (*8m) \\ (*8rc) \\ (*7r) \end{array} \right\} \text{ can be represented as the revision func-}$$

tion $\mathcal{R}(\gamma)$ *based on a semantic choice function* γ *which is complete and satisfies*

$$\left\{ \begin{array}{c} \overline{} \\ (Faith2) \ w.r.t. \ [\![K]\!] \\ (Faith1) \ w.r.t. \ [\![K]\!] \\ (Success) \\ (I) \\ (I^-) \\ (II) \\ (II^+) \\ (III) \\ (IV) \\ (IV^+) \\ (\emptyset 1) \\ (\emptyset 2) \end{array} \right\} , \ respectively.$$

Now we turn to revisions generated from syntactic choice functions.

Observation 31 *For every syntactic choice function δ which satisfies (LP1),*

$$(LP2)\ and\ \begin{Bmatrix} \overline{(Faith1)\ wrt\ L-K} \\ (Faith2)\ wrt\ L-K \\ (Virtual\ Success) \\ (I) \\ (I^-) \\ (II) \\ (II^+) \\ (III) \\ (IV) \\ (IV^+) \\ (\emptyset 1) \\ (\emptyset 2) \end{Bmatrix}\ ,\ the\ revision\ function\ *\ over\ K\ based\ on\ \delta$$

satisfies $(*1) - (*2)$, $(*6)$ and $\begin{Bmatrix} \overline{(*4)} \\ (*3) \\ (*5) \\ (*7) \\ (*7c) \\ (*8wd) \\ (*8d) \\ (*8c) \\ (*8) \\ (*8m) \\ (*8rc) \\ (*7r) \end{Bmatrix}$ *, respectively.*

Given a revision function $*$ over K, we can derive from it a *finitary* and ω-*covering* syntactic choice function $\delta = \mathcal{D}(*)$ for K. We define, for all sentences $\phi_1, \ldots, \phi_n$

$$\delta(\{\phi_1, \ldots, \phi_n\}) = \{\phi_i : \phi_i \notin K * \neg(\phi_1 \wedge \cdots \wedge \phi_n)\}$$

The idea is that a sentence ϕ_i is chosen as a *best*—or more properly, a *worst*—element of $\{\phi_1, \ldots, \phi_n\}$ if it is not maintained in the revision of K by $\neg(\phi_1 \wedge \cdots \wedge \phi_n)$. We build upon the idea that the task of revising by the negated conjunction $\neg(\phi_1 \wedge \cdots \wedge \phi_n)$ involves giving up at least one of the finite set $\{\phi_1, \ldots, \phi_n\}$. The choice set $\delta(\{\phi_1, \ldots, \phi_n\})$ contains just those just those ϕ_is which are *not* retained in $K * \neg(\phi_1 \wedge \cdots \wedge \phi_n)$.

Observation 32 *Every revision function ∗ over a theory K which satisfies (∗1)*

$$- (*2),\ (*6)\ and\ \left\{ \begin{array}{c} \overline{} \\ (*3) \\ (*4) \\ (*5) \\ (*7) \\ (*7c) \\ (*8wd) \\ (*8d) \\ (*8c) \\ (*8) \\ (*8m) \\ (*8rc) \\ (*7r) \end{array} \right\}\ can\ be\ represented\ as\ the\ revision\ function\ \mathcal{R}(\delta)$$

finitarily based on a syntactic choice function δ which satisfies (LP1), (LP2) and

$$\left\{ \begin{array}{c} \overline{} \\ (Faith2)\ w.r.t.\ L - K \\ (Faith1)\ w.r.t.\ L - K \\ (Virtual\ Success) \\ (I) \\ (I^-) \\ (II) \\ (II^+) \\ (III) \\ (IV) \\ (IV^+) \\ (\emptyset 1) \\ (\emptyset 2) \end{array} \right\}\ ,\ respectively.$$

In contrast to the case of contractions, (Faith2) plays a significant role in revisions. This is due to the fact that the belief set K is not mentioned at all in the defining clauses in Definitions 11 and 12. So the reference to the belief set K has to be built into the choice function.

7.7.3 *Representation theorems for nonmonotonic inferences*

If a nonmonotonic inference operation *Inf* is generated by a semantic choice function γ, then coherence constraints for γ will affect the conditions met by *Inf*. We are now going to state the relevant representation theorems. As explained above, we refrain from proving these theorems here, since they are entirely parallel to Theorems 25 and 26 for belief contractions.

Observation 33 *For every semantic choice function γ which satisfies*

$$\left\{
\begin{array}{c}
\text{---} \\
\text{(Faith1) wrt } \gamma(\mathcal{M}_L) \\
\text{(Faith2) wrt } \gamma(\mathcal{M}_L) \\
\text{(Success)} \\
\text{(I)} \\
\text{(I}^-\text{)} \\
\text{(II) and is complete} \\
\text{(II}^+\text{)} \\
\text{(III)} \\
\text{(IV)} \\
\text{(IV}^+\text{)} \\
(\emptyset 1) \\
(\emptyset 2)
\end{array}
\right\}$$

, the finitary inference operation $\mathit{Inf} = \mathcal{I}(\gamma)$ based on

γ *satisfies (Ref), (LLE), (RW), (And) and*

$$\left\{
\begin{array}{c}
\text{---} \\
\text{(WRMon)} \\
\text{(WCond)} \\
\text{(CP)} \\
\text{(Or)} \\
\text{(VWDRat)} \\
\text{(CMon)} \\
\text{(RMon)} \\
\text{(Cut)} \\
\text{(DRat)} \\
\text{(Mon)} \\
(\emptyset \text{CMon}) \\
(\emptyset \text{Cond})
\end{array}
\right\}$$

, respectively.

Observation 34 *Every finitary inference operation Inf which satisfies (Ref),*

(LLE), (RW), (And) and
$$
\left\{
\begin{array}{c}
\overline{} \\
(WCond) \\
(WRMon) \\
(CP) \\
(Or) \\
(Cut) \\
(VWDRat) \\
(DRat) \\
(CMon) \\
(RMon) \\
(Mon) \\
(\emptyset CMon) \\
(\emptyset Cond)
\end{array}
\right\}
\quad \text{can be represented as the inference}
$$

operation $\mathcal{I}(\gamma)$ based on a semantic choice function γ which is complete and

satisfies
$$
\left\{
\begin{array}{c}
\overline{} \\
(Faith2)\ w.r.t.\ [\![Inf(\top)]\!] \\
(Faith1)\ w.r.t.\ [\![Inf(\top)]\!] \\
(Success) \\
(I) \\
(I^-) \\
(II) \\
(II^+) \\
(III) \\
(IV) \\
(IV^+) \\
(\emptyset 1) \\
(\emptyset 2)
\end{array}
\right\}
\quad,\ respectively.
$$

The appropriate semantic choice function $\gamma = \mathcal{G}(Inf)$ over elementary classes of models that can be used for the proof of this theorem is defined by

$$
\boxed{\ \gamma(\,[\![\phi]\!]\,) = [\![Inf(\phi)]\!] = \{m : Inf(\phi) \subseteq \widehat{m}\}\ }
$$

The obvious idea here is that an element of $[\![\phi]\!]$ is a best—*most plausible* or *most normal*—model in $[\![\phi]\!]$ just in case it satisfies everything which may be inferred from ϕ.

Now we turn to nonmonotonic inference relations generated from syntactic choice functions. This now is the second case of the six ones depicted in Fig. 7.6 for which we give proofs of the representation theorems (see Appendix D).

Observation 35 *For every syntactic choice function δ which satisfies (LP1),*

$$(LP2) \text{ and } \left\{ \begin{array}{c} \overline{} \\ (Faith1) \ wrt \ \delta(L) \\ (Faith2) \ wrt \ \delta(L) \\ (Virtual \ Success) \\ (I) \\ (I^-) \\ (II) \\ (II^+) \\ (III) \\ (IV) \\ (IV^+) \\ (\emptyset 1) \\ (\emptyset 2) \end{array} \right\} , \text{ the inference operation Inf based on } \delta \text{ satis-}$$

$$\text{fies (Ref), (LLE), (RW), (And) and } \left\{ \begin{array}{c} \overline{} \\ (WRMon) \\ (WCond) \\ (CP) \\ (Or) \\ (Cut) \\ (WDRat) \\ (DRat) \\ (CMon) \\ (RMon) \\ (Mon) \\ (\emptyset CMon) \\ (\emptyset Cond) \end{array} \right\} , \text{ respectively.}$$

From the point of view of proof techniques, the parts concerning (Cut) and (CMon) are the most interesting ones because we are forced there to 'localize' choices in very small sets: We step back from choices made in whole theories to choices made in finite 'axiomatizations', apply our choice conditions there, and ascend to the theory level again.

Given a finitary nonmonotonic inference relation Inf, we can derive from it a *finitary and ω-covering* syntactic choice function $\delta = \mathcal{D}(Inf)$. The idea is that a sentence ϕ_i is chosen as a *best*—or more properly, a *worst*—element of $\{\phi_1, \ldots, \phi_n\}$ if it cannot be inferred from the premise $\neg(\phi_1 \wedge \cdots \wedge \phi_n)$. We build upon the idea that the task of drawing inferences from the negated conjunction $\neg(\phi_1 \wedge \cdots \wedge \phi_n)$ involves giving up at least one of the finite set of expectations $\{\phi_1, \ldots, \phi_n\}$. In other words, the choice set $\delta(\{\phi_1, \ldots, \phi_n\})$ contains just those expectations ϕ_i which are *not* in $Inf(\neg(\phi_1 \wedge \cdots \wedge \phi_n))$:

$$\delta(\{\phi_1, \ldots, \phi_n\}) = \{\phi_i : \phi_i \notin \mathit{Inf}(\neg(\phi_1 \wedge \cdots \wedge \phi_n))\}$$

We say that the nonmonotonic inference relation *Inf reveals*, or is a *manifestation* of, the underlying syntactic choice function $\mathcal{D}(\mathit{Inf})$. The next observation shows that interesting properties of *Inf* lead to corresponding interesting properties of δ.

Observation 36 *Every finitary inference operation Inf which satisfies (Ref),*

$$
\mathit{(LLE),\ (RW),\ (And)\ and}
\left\{
\begin{array}{c}
\text{---} \\
\mathit{(WCond)} \\
\mathit{(WRMon)} \\
\mathit{(CP)} \\
\mathit{(Or)} \\
\mathit{(Cut)} \\
\mathit{(WDRat)} \\
\mathit{(DRat)} \\
\mathit{(CMon)} \\
\mathit{(RMon)} \\
\mathit{(Mon)} \\
\mathit{(\emptyset CMon)} \\
\mathit{(\emptyset Cond)}
\end{array}
\right\}
\mathit{can\ be\ represented\ as\ the\ inference}
$$

operation finitarily based on a syntactic choice function δ which satisfies (LP1),

$$
\mathit{(LP2)\ and}
\left\{
\begin{array}{c}
\text{---} \\
\mathit{(Faith2)\ w.r.t.\ L - Inf(\top)} \\
\mathit{(Faith1)\ w.r.t.\ L - Inf(\top)} \\
\mathit{(Virtual\ Success)} \\
\mathit{(I)} \\
\mathit{(I^-)} \\
\mathit{(II)} \\
\mathit{(II^+)} \\
\mathit{(III)} \\
\mathit{(IV)} \\
\mathit{(IV^+)} \\
\mathit{(\emptyset 1)} \\
\mathit{(\emptyset 2)}
\end{array}
\right\}
\mathit{,\ respectively.}
$$

The role of the Faith conditions in the last theorems is again somewhat different from the cases of contractions and revisions. For since the latter are always understood as contractions and revisions *with respect to some given belief set K*, no designated set is given in nonmonotonic reasoning. Therefore we have to construct afresh the 'point of reference' for Faith, starting from *Inf*, γ or δ.

Nonmonotonic inference operations satisfying Rational Monotony, (RMon), have been said to represent 'lazy' reasoning (see Satoh 1990). Monotonic logics, then, reflect the reasoning habits of agents that are much lazier; we might call them 'bone-lazy'. We have seen that (Mon) corresponds to the choice-theoretic condition (IV$^+$) and thus essentially to totally indiscriminating choice behaviour, the choice behaviour determined by Simon's 'bounded rationality'. (Compare our brief discussion on page 147.[310]) It is interesting that arguments for monotonic logics can now be viewed as arguments for the rationality of *satisficing*, while arguments for nonmonotonic logics appear to be arguments for the rationality of *maximizing*. With a little exaggeration, one might say that standard logics correspond to a non-standard recipe for choices, and that standard recipes for choices correspond to non-standard logics. From the present point of view, monotonic reasoning which has until recently dominated all of logic, appears as a rather crude and awkward kind of reasoning: that which is shown by bone-lazy or by most severely resource-bounded agents.

7.7.4 *The quest for explanation*

It is striking that the logical postulates for belief revision and nonmonotonic reasoning that correspond to the central choice-theoretic constraints in the syntactic approach are almost exactly the same as those in the semantic approach. On page 206, there is a table using the postulates for nonmonotonic reasoning as representative of the side of theoretical reason (contraction and revision would serve just as well). Nonmonotonic reasoning systems satisfying (Ref), (LLE), (RW), (And) and a given combination of the logical conditions mentioned in the table are exactly the systems representable as inference relations based on (semantic or syntactic) choice functions which satisfies (LP1), (LP2) and the corresponding combination of choice constraints. Further connections on the side of theoretical reason with belief contractions and revisions may be obtained by conjoining this table with the similarly structured one on page 118.

In six out of eleven or twelve cases[311] the logical postulates corresponding to the choice constraints are identical on the syntactic and the semantic approach. There are two minor differences resulting from the different kinds of options between which to choose. First, the two constraints of Faith refer to $[\![K]\!]$ in the semantic and to $L - K$ in the syntactic approach.[312] Second, while we can directly employ the Success constraint without restrictions in the semantic approach, we have to weaken it to Virtual Success in the syntactic approach, due to the

---

[310] Herbert Simon is the 1978 Nobel laureate in economics and one of the outstanding pioneers in cognitive psychology and AI. But apparently there is little connection in his work between the two aspects of rationality—choice and reasoning—, except for the fact 'bounded rationality' is a common theme.

[311] Eleven for contractions, twelve for revisions and nonmonotonic reasoning—as mentioned above, Faith2 makes the difference.

[312] We can neglect the point whether we deal with beliefs or expectations. In the latter case, substitute '$Inf(\top)$' for 'K'.

Nonmonotonic Inferences	Choices
(Cn-closure) $Inf(\phi) = Cn(Inf(\phi))$	(LP1) synt. If $\phi \in Cn(G), G \subseteq F$ and $\phi \in \delta(F)$, then $\delta(F) \cap G \neq \emptyset$
(RW) If $\psi \in Inf(\phi)$ and $\chi \in Cn(\psi)$, then $\chi \in Inf(\phi)$	
(And) If $\psi \in Inf(\phi)$ and $\chi \in Inf(\phi)$, then $\psi \wedge \chi \in Inf(\phi)$	
(Ref) $\phi \in Inf(\phi)$	
(WCond) $Inf(\phi) \subseteq Cn(Inf(\top) \cup \{\phi\})$	(Faith2) $\forall S \in \mathcal{X} : S \cap R \subseteq \sigma(S)$
(WRMon) If $\neg\phi \notin Inf(\top)$, then $Inf(\top) \subseteq Inf(\phi)$	(Faith1) $\forall S \in \mathcal{X} :$ if $S \cap R \neq \emptyset$ then $\sigma(S) \subseteq R$
(LLE) If $Cn(\phi) = Cn(\psi)$, then $Inf(\phi) = Inf(\psi)$	(LP2) synt. If $Cn(F) = Cn(G)$ and $\phi \in \delta(F) \cap G$, then $\phi \in \delta(G)$
(CP) If $Cn(\phi) \neq L$, then $Inf(\phi) \neq L$	(Success) sem./ (Virtual Success) synt. If $S \neq \emptyset$, then $\gamma(S) \neq \emptyset$ If $S - Cn(\emptyset) \neq \emptyset$, then $\delta(S) \neq \emptyset$
(Cut) If $\psi \in Inf(\phi)$, then $Inf(\phi \wedge \psi) \subseteq Inf(\phi)$	(I^-) If $S \subseteq S'$ and $\sigma(S') \subseteq S$, then $\sigma(S') \subseteq \sigma(S)$
($\emptyset$Cond) If $\perp \in Inf(\phi \wedge \psi)$, then $\neg\psi \in Inf(\phi)$	($\emptyset 2$) If $S \subseteq S'$ and $\sigma(S) = \emptyset$, then $S \cap \sigma(S') = \emptyset$
(Or) / (Cond) $Inf(\phi) \cap Inf(\psi) \subseteq Inf(\phi \vee \psi)$ $Inf(\phi \wedge \psi) \subseteq Cn(Inf(\phi) \cup \{\psi\})$	(I) / (I') If $S \subseteq S'$, then $S \cap \sigma(S') \subseteq \sigma(S)$ $\sigma(S \cup S') \subseteq \sigma(S) \cup \sigma(S')$
($\emptyset$CMon) If $\perp \in Inf(\phi)$, then $\perp \in Inf(\phi \wedge \psi)$	($\emptyset 1$) If $S \subseteq S'$ and $\sigma(S') = \emptyset$, then $\sigma(S) = \emptyset$
(CMon) If $\psi \in Inf(\phi)$, then $Inf(\phi) \subseteq Inf(\phi \wedge \psi)$	(III) If $S \subseteq S'$ and $\sigma(S') \subseteq S$, then $\sigma(S) \subseteq \sigma(S')$
(DRat) $Inf(\phi \vee \psi) \subseteq Inf(\phi) \cup Inf(\psi)$ ($DRat^+$) $Cn(\phi \vee \psi) \subseteq Inf(\phi)$ or $Cn(\phi \vee \psi) \subseteq Inf(\psi)$	(II^+) If $x \in \sigma(S)$ and $y \in \sigma(S')$, then $x \in \sigma(S \cup S')$ or $y \in \sigma(S \cup S')$
(WDRat) $Inf(\phi \vee \psi) \subseteq Cn(Inf(\phi) \cup \{\psi\})$ $\cup Cn(Inf(\psi) \cup \{\phi\})$ ($WDRat^+$) $Inf(\phi \vee \psi) \subseteq Cn(Inf(\phi) \cup \{\psi\})$ or $Inf(\phi \vee \psi) \subseteq Cn(Inf(\psi) \cup \{\phi\})$	(II) synt. $\delta(S) \cap \delta(S') \subseteq \delta(S \cup S')$
(VWDRat) $Inf(\phi \vee \psi) \subseteq Cn(Inf(\phi) \cup Inf(\psi))$	(II) sem. $\gamma(S) \cap \gamma(S') \subseteq \gamma(S \cup S')$
(NRat) $Inf(\phi) \subseteq Inf(\phi \wedge \psi) \cup Inf(\phi \wedge \neg\psi)$	
(RMon) If $\neg\psi \notin Inf(\phi)$, then $Inf(\phi) \subseteq Inf(\phi \wedge \psi)$	(IV) If $S \subseteq S'$ and $\sigma(S') \cap S \neq \emptyset$, then $\sigma(S) \subseteq \sigma(S')$
(Mon) $Inf(\phi) \subseteq Inf(\phi \wedge \psi)$	(IV^+) If $S \subseteq S'$, then $\sigma(S) \subseteq \sigma(S')$

fact that we are not allowed to withdraw logical truths, by virtue of the logical constraint (LP1).

The most important and the only substantial discrepancy in the list of postulates for nonmonotonic reasoning is that between those corresponding to the basic choice constraint (II). Here (VWDRat) is the corresponding postulate in the semantic approach. In the syntactic approach, however, we have (WDRat) as a matching condition which is substantially stronger than (VWDRat).

The strength of (WDRat) in fact presents us with a puzzle. We observed in Lemma 4(vi) that (WDRat) and (CMon) taken together imply the more well-known condition (DRat), which is obviously stronger than (WDRat). This makes us expect that the analogous implication holds for the choice-theoretic counterparts of the postulates (as postulates for the syntactic approach), i.e., that (II) and (III) imply (II$^+$). However, in Section 6.5 we gave a counterexample against that very implication. How is this possible?

The solution to this question is that for syntactic choices, we have imposed two substantive *logical constraints*, over and above the choice-theoretic constraints under consideration. And it turns out that for the choice functions satisfying these constraints there can be no counterexample to the implication mentioned:

Observation 37 *Let δ be a finitary and ω-covering syntactic choice function which satisfies (LP1) and (LP2). If δ satisfies (II) and (III), then it satisfies (II$^+$).*[313]

Now we have seen that, first, there is a far-reaching parallel between the two lists of corresponding conditions for the methods that apply choices on the semantic and on the syntactic level, but that, second, a single striking difference shows up that concerns the counterparts of the Expansion Axiom (II). Both these findings are surprising. At first, it seems strange that there should be such a parallel, but once one has become used to the idea that there is such a parallel, it is again peculiar that the line containing the choice-theoretic coherence constraint (II) turns out to be an exception to the rule. One would like to have an explanation of the parallel as well as its singularity. However, it is hard to imagine what could count as an explanation of the things we want to explain here. Might the correspondence and its major exception be just brute matters of fact?

I think that one can account for the parallel and its singularity. In the next section I try to meet the quest for explanation.

---

[313] As the proof shows, this observation can be slightly generalized. The choice function δ satisfying the logical constraints need not be finitary and ω-covering. It is enough that δ is finitely additive and *singleton-expansible*, in the sense that for every S in its domain $\mathcal{X}$ and every x in X it holds that $S \cup \{x\}$ is in $\mathcal{X}$ as well.

7.8   Linking choices between models and choices between sentences

We first rephrase the problem, in order to have a more compact formulation of our question. Every set of coherence constraints for choices of the sort discussed in Chapter 6 will be called a *choice profile*, and similarly, every set of rationality postulates for belief contractions or revisions or nonmonotonic inferences of the sort discussed in Chapter 4 is a *logic profile*. We neglect for a moment the exception regarding (II) emphasized above. What we wish to explain, then, are the following observations:

If a certain choice profile of a semantic choice function γ implies that the contraction/revision/inference operations generated by γ have a certain logic profile, *then* the same choice profile enforces the same logic profile for syntactic choice functions, too. And vice versa, *if* a certain choice profile of a syntactic choice function δ implies that the contraction/revision/inference operations generated by δ have a certain logic profile, *then* the same choice profile enforces the same logic profile for semantic choice functions, too.

Why do the same choice profiles guarantee the same logic profiles if applied, in very different ways, for choices of 'best' models and choices of 'worst' sentences? The explanation for the surprising parallel in the representation theorems that I will offer is this. Any choice between models *engenders* or *is engendered by* an equivalent choice between sentences, under preservation of the choice profile of the semantic choice function (with the one notable exception that I mentioned). And conversely, any choice between sentences *engenders* or *is engendered by* an equivalent choice between models, under preservation of the choice profile of the syntactic choice function (without any exception). Here two choice functions γ and δ are called *equivalent* if they generate the same contraction function, that is, if $\mathcal{C}(\gamma) = \mathcal{C}(\delta)$.

It is clear already from our representation theorems that semantically constructed contractions (or revisions, or nonmonotonic inference operations) are at the same time syntactically constructed contractions (or revisions, or nonmonotonic inference operations, respectively). If we start with a semantic choice function γ, we can take the generated contraction function $\dot{-} = \mathcal{C}(\gamma)$ and then construct the syntactic contraction function $\delta = \mathcal{D}(\dot{-})$ which, as we know from Observation 28, generates $\dot{-}$. Hence γ and $\delta = \mathcal{D}(\mathcal{C}(\gamma))$ are equivalent. Similarly, if we start with a syntactic choice function δ, we can take the generated contraction function $\dot{-} = \mathcal{C}(\delta)$ and then construct the semantic contraction function $\gamma = \mathcal{G}(\dot{-})$ which, as we know from Observation 26, generates $\dot{-}$. Hence δ and $\gamma = \mathcal{G}(\mathcal{C}(\delta))$ are equivalent. What is more, our representations theorems in Sections 7.7.1 through 7.7.3 show that in all cases but one the coherence constraints obeyed by the choice functions in question will be preserved.

However, for a serious conceptual comparison of semantic choice functions with syntactic choice functions it will not do just to graft one construction studied in the previous chapter onto another. Since we take choices to be more fundamental than contractions (or revisions, or nonmonotonic inference operations), such a roundabout way via contractions, say, would mean putting the cart in front of the horse. What we need is a natural, transparent and direct connection between the two sorts of choices—one of which we can gain an easy intuitive grasp. This is what I shall try to supply in this section. The interest of the following does not lie in the bare demonstration that semantic and syntactic choices can somehow be related but in the fact that the relevant transitions are perspicuous, that the choice functions involved can be linked directly to each other (even in the infinite case), and that the transitions (in all cases but one) preserve choice-theoretic rationality profiles.

The following account is somewhat more general than required in that it allows the semantic choice functions to take Σ-elementary (rather than just elementary) model sets as arguments. Syntactic choice functions are conceived of as taking arbitrary (rather than just finite) sets of sentences as arguments. No finiteness assumptions are needed for the results of this section.[314] Moreover, we shall avoid any reference to a fixed belief set K.

We now state the mappings which establish the desired bridge between the two approaches.

Definition 15 Let a choice function γ over Σ-elementary sets of models be given. Then we can define a corresponding choice function $\delta = \mathcal{D}(\gamma)$ over sentences by putting

$$\phi \in \delta(F) \text{ iff } \phi \in F \text{ and } \gamma(]F[) \cap]\phi[\neq \emptyset$$

If at least one of the sentences in F has to be given up, then ϕ from F is among the discarded sentences if and only if at least one of the most plausible models falsifying F also falsify ϕ. A more concise way of putting this is

$$\delta(F) = F - \widehat{\gamma(]F[)}$$

Now for the converse transition.

Definition 16 Let a choice function δ over arbitrary sets of sentences be given. Then we can define a corresponding choice function $\gamma = \mathcal{G}(\delta)$ over models by putting

$$m \in \gamma(]F[) \text{ iff } m \in]F[\text{ and } Cn(F) - \widehat{m} \subseteq \delta(Cn(F))$$

That is, m is among the most plausible models falsifying F if and only if all the sentences in $Cn(F)$ which are falsified by m are given up when it comes to discarding the worst sentences in $Cn(F)$.[315] In other words, m is among the

---

[314] It can be seen from the formulation of the next definitions, however, that the restriction of the domain of semantic choice functions to elementary sets corresponds precisely to the restriction of the domain of syntactic choice functions to finite sets.

[315] Let us add a few useful remarks on the structure of sets of the form $Cn(F) - \widehat{m}$. Clearly, $m \in]F[$ means that $Cn(F) - \widehat{m} \neq \emptyset$. If $F \subseteq m$, then $Cn(F) - \widehat{m} = \emptyset$; if $F \not\subseteq \widehat{m}$, then

most plausible models falsifying F if and only if m satisfies all the sentences in $Cn(F)$ which are not given up when it comes to discarding the worst sentences in $Cn(F)$. Using the concept of a remainder function defined by $\rho(F) = F - \delta(F)$ (see Section 6.6), we can put Definition 16 more concisely as

$$\gamma(\,]F[\,) =]F[- [\rho(Cn(F))]$$

It is clear that $\gamma = \mathcal{G}(\delta)$ is well-defined, since $]F[=]G[$ implies $Cn(F) = Cn(G)$.

Definition 16 needs a syntactic choice function that can take infinite arguments. One may wonder how one can go about defining the transition from δ to γ, if the former is finitary. In this case the sets F need to be finite. But in order to do the work of Definition 16, it is not sufficient to check whether $F - \widehat{m} \subseteq \delta(F)$. The choices in F do not determine the choices on $Cn(F)$, and we need to know the latter. Nor will the conjunctive and/or disjunctive closure of F do any better (just think of a singleton F). The correct definition for finitary syntactic choice functions is defined by putting, for every finite F

$$m \in \gamma(\,]F[\,) \quad \text{iff} \quad m \in]F[\text{ and } \{\phi \in Cn(F) : \phi \notin \delta(F \cup \{\phi\})\} \subseteq \widehat{m}$$

or equivalently

$$m \in \gamma(\,]F[\,) \quad \text{iff} \quad m \in]F[\text{ and } \{\phi : \bigwedge F \vee \phi \notin \delta(\bigwedge F, \bigwedge F \vee \phi)\} \subseteq \widehat{m}$$

We shall not pursue this topic further in this section, but rather base our considerations on syntactic choice functions δ that can take full theories.

As a first result, we show that these direct transitions can in fact be viewed as generalized and condensed formulations of the indirect ones mentioned in the fourth paragraph of this section.

Observation 38 *(i) Let δ be a syntactic choice function, let $\gamma = \mathcal{G}(\delta)$ and let $\gamma' = \mathcal{G}(\mathcal{C}(\delta))$. Then for all sentences ϕ*

$$\gamma(\,]\phi[\,) \subseteq \gamma'(\,]\phi[\,)$$

and if δ satisfies (Faith1) and (Faith2) with respect to $L - K$, then

$$\gamma'(\,]\phi[\,) \subseteq \gamma(\,]\phi[\,)$$

(ii) Let γ be a semantic choice function, let $\delta = \mathcal{D}(\gamma)$ and let $\delta' = \mathcal{D}(\mathcal{C}(\gamma))$. Then for all finite sets $\{\phi_1, \ldots, \phi_n\}$ of sentences

$$\delta(\{\phi_1, \ldots, \phi_n\}) \subseteq \delta'(\{\phi_1, \ldots, \phi_n\})$$

and if γ satisfies (Faith2) with respect to $[\![K]\!]$, then

$$\delta'(\{\phi_1, \ldots, \phi_n\}) \subseteq \delta(\{\phi_1, \ldots, \phi_n\})$$

$Cn(Cn(F) - \widehat{m}) = Cn(F)$. Finally, notice that the set $Cn(F) - \widehat{m}$ is closed under conjunction and disjunction, but it does not continue up or down with respect to Cn (i.e., $\phi \in Cn(F) - \widehat{m}$ together with $\phi \vdash \psi$ or $\psi \vdash \phi$ does not imply $\psi \in Cn(F) - \widehat{m}$).

Observation 38 tells us that the choice functions that can be obtained indirectly via the contraction behaviour of the agent are restrictions (to the case of elementary sets of models or respectively, to finite sets of sentences) of the functions that directly connected to each other with the help of the new definitions 15 and 16.

We now begin to state the first central result of this section that paves the way to the explanation I promised. The following observation shows that it makes sense to talk of semantic and syntactic choice functions 'corresponding' to each other.

Observation 39 *(i) For every semantic choice function γ, the corresponding syntactic choice function $\delta = \mathcal{D}(\gamma)$ satisfies (LP1) and (LP2).*

(ii) For every syntactic choice function δ that satisfies (LP1) and (LP2), the corresponding semantic choice function $\gamma = \mathcal{D}(\delta)$ is complete.

(iii) For every syntactic choice function δ that satisfies (LP1) and (LP2), the syntactic choice function corresponding to the semantic choice function corresponding to δ is identical with δ, that is

$$\mathcal{D}(\mathcal{G}(\delta)) = \delta$$

(iv) For every semantic choice function γ, the semantic choice function corresponding to the syntactic choice function corresponding to γ is the completion γ^+ of γ, that is

$$\mathcal{G}(\mathcal{D}(\gamma)) = \gamma^+$$

(v) Corresponding choice functions lead to identical contraction functions, that is, it holds both that

$$\mathcal{C}(\mathcal{D}(\gamma)) = \mathcal{C}(\gamma)$$

and

$$\mathcal{C}(\mathcal{G}(\delta)) = \mathcal{C}(\delta)$$

Remember from Section 7.1 that a semantic choice function γ is equivalent with its own completion, i.e., $\mathcal{C}(\gamma^+) = \mathcal{C}(\gamma)$.

Next we show that in almost all cases, the choice profiles are preserved in the transitions from the semantic to the semantic level and vice versa.

Observation 40 *For every syntactic choice function δ which satisfies (LP1), (LP2) and*

$$\left\{ \begin{array}{l} \textit{(Faith1) w.r.t. } \delta(L) \\ \textit{(Faith2) w.r.t. } \delta(L) \\ \textit{(Virtual Success)} \\ \textit{($\emptyset$1)} \\ \textit{($\emptyset$2)} \\ \textit{(I)} \\ \textit{(II)} \\ \textit{(III)} \\ \textit{(IV)} \\ \textit{(I^-)} \\ \textit{(II^+)} \\ \textit{(IV^+)} \end{array} \right\} , \textit{ the corresponding semantic choice function } \gamma = \mathcal{G}(\delta)$$

$$\textit{satisfies} \left\{ \begin{array}{l} \textit{(Faith1) w.r.t. } \gamma(\mathcal{M}_L) \\ \textit{(Faith2) w.r.t. } \gamma(\mathcal{M}_L) \\ \textit{(Virtual Success)} \\ \textit{($\emptyset$1)} \\ \textit{($\emptyset$2)} \\ \textit{(I)} \\ \textit{(II)} \\ \textit{(III)} \\ \textit{(IV)} \\ \textit{(I^-)} \\ \textit{(II^+)} \\ \textit{(IV^+)} \end{array} \right\} , \textit{ respectively.}$$

We have now verified that the transfer from the syntactic to the semantic level works perfectly. The transfer from the semantic to the syntactic level is almost equally smooth. There is, however, an important exception concerning condition (II) which explains the asymmetry in the representation theorems for contractions, revisions and nonmonotonic inference operations with respect to this condition.

Observation 41 *For every semantic choice function γ which satisfies*

$$
\left\{
\begin{array}{c}
(Faith1)\ wrt\ \gamma(\mathcal{M}_L) \\
(Faith2)\ wrt\ \gamma(\mathcal{M}_L) \\
(Success) \\
(\emptyset 1) \\
(\emptyset 2) \\
(I) \\
(I^-) \\
(II^+) \\
(III) \\
(IV) \\
(IV^+)
\end{array}
\right\}
\quad the\ corresponding\ syntactic\ choice\ function\ \delta = \mathcal{D}(\gamma)
$$

satisfies (LP1) and (LP2) and
$$
\left\{
\begin{array}{c}
(Faith)\ wrt\ \delta(L) \\
(Faith)\ wrt\ \delta(L) \\
(Virtual\ Success) \\
(\emptyset 1) \\
(\emptyset 2) \\
(I) \\
(I^-) \\
(II^+) \\
(III) \\
(IV) \\
(IV^+)
\end{array}
\right\}
,\ respectively.
$$

However, if γ satisfies (II), it does not follow that the corresponding syntactic choice function $\delta = \mathcal{D}(\gamma)$ satisfies (II), even in the finite case and when γ in addition satisfies (I) and (III).

7.9 What have we achieved?

Let us summarize this chapter's picture of how to give up beliefs. We have said that coherentist belief changes involves choices, either choices between models or choices between sentences. The choices should be coherent in the sense that there is a single underlying choice function γ or δ that takes care for all potential 'inputs' ϕ, and the choice function should satisfy some selection of coherence constraints of the sort investigated in the classical theory of choice.

The crucial definition for contractions made by choosing 'best models' is this. The contraction function $\dot{-}$ over K is generated by a choice function γ over models if and only if for every ψ in K

$\mathcal{C}(\gamma) \qquad \psi \in K \dot{-} \phi \ \ \text{iff} \ \ \gamma(\,]\!]\phi[\![\,) \subseteq [\![\psi]\!]$

where $]\!]\phi[\![$ is the set of models falsifying ϕ and $[\![\psi]\!]$ is the set of models verifying ψ. The alternative definition for contractions made by choosing 'worst sentences' runs as follows. The contraction function $\dot{-}$ over K is generated by a choice function δ between sentences if and only if for every ψ in K,

$\mathcal{C}(\delta)$ $\qquad \psi \in K \dot{-} \phi$ iff $\phi \vee \psi \notin \delta(Cn(\phi))$

$\qquad\qquad\qquad$ iff $\phi \vee \psi \notin \delta(\{\phi, \phi \vee \psi\})$

$\qquad\qquad\qquad$ iff $\phi \vee \psi \notin \delta(\{\phi \vee \psi, \phi \vee \neg\psi\})$

The syntactic choice functions δ must satisfy two logical constraints in order to make sure that it is the content of the sentences (and not their syntactic form) that matters:

(LP1) If $G \subseteq F$ and $\delta(F) \cap G \neq \emptyset$, then $\delta(F) \cap Cn(G) = \emptyset$

(LP2) If $Cn(F) = Cn(G)$, then $\delta(F) \cap G \subseteq \delta(G)$

Condition (LP2) was our basis for stating the three equivalent forms of the definition of syntactically choice-based contractions $\mathcal{C}(\delta)$.

We demonstrated how constraints for rational or coherent choice transfer from the model level to the sentential level, and vice versa, almost without any changes. The constraints are shown to give rise to corresponding lists of conditions for contraction, revision and inference operations. I take this to be strong evidence for the unity of theoretical and practical reason, with the principles for the former being special cases of principles for the latter. Choices are viewed as having extra-logical determinants, some kinds of preferences that have a 'logic' of their own. They are constrained by this logic, so to speak, but not determined by logic. I think it is justified to speak of the *primacy of practical over theoretical reason* here—not in the sense that the former competes with or overcomes the latter, but that practical reason comprises much of what would prima facie be conceived of as 'purely theoretical'.

In terms of the correspondence between choice constraints and systems of nonmonotonic reasoning, it has turned out that (I) corresponds to the condition Or, (II) to Weak or Very Weak Disjunctive Rationality, (III) to Cumulative Monotony, and (IV) to Rational Monotony. It is interesting that viewed from the perspective of coherent choices, (Very) Weak Disjunctive Rationality is more plausible or fundamental than Cumulative Monotony. This is so, at least, if we insist—as many people do—that the most important property of choice functions is their 'rationalizability', that is, their being based on an underlying preference relation. We have seen that conditions (I) and (II)[316] are decisive for rationalizability. Thus Or and (Very) Weak Disjunctive Rationality enter at the most basic stage. Cumulative and Rational Monotony play their role only afterwards, if preference relations are to be transitive or modular.

Among logicians, it is usually thought that Cumulative Monotony is more plausible or basic than Disjunctive Rationality. If we need choice functions tailored to comply to this intuitive judgment on the side of theoretical reason, then we may argue that Rationalizability is not the most important requirement. We

---

[316] We here neglect the fact that in some contexts we actually need the infinitary version (II^∞) of (II).

may hold that the leading part should be played by Pseudo-rationalizability and Path Independence, which, as we saw in Chapter 6 (Observations 19 and 20), are both equivalent to (I) and (III), given certain domain conditions. In the finite case at least (when Cn partitions L into finitely many cells), Pseudo-rationalizability and Path Independence thus correspond precisely to the preferential inference operations (in the sense of Section 4.4). We conclude that from the viewpoint of theoretical reason, Pseudo-rationalizability and Path Independence appear to be more fundamental ideas than Rationalizability.

There is nothing in preferential belief change or inference operations that corresponds to the second relationality condition (II). But Rationalizability surely is a natural and important requirement. Hence (Very) Weak Disjunctive Rationality should be regarded as a natural and important principle as well. One result of our choice-theoretic analysis may be seen in the recommendation to put more emphasis on the study of preferential systems that satisfy (Very) Weak Disjunctive Rationality without satisfying Rational Monotony. On the semantic level we get Very Weak Disjunctive Rationality which is not very well-known in the literature.[317] On the syntactic level we get Weak Disjunctive Rationality which is, in the context of basic reasoning with Cumulative Monotony, sufficient to derive Disjunctive Rationality.[318] The latter condition has been studied rather carefully in the literature, especially by Freund (1993).[319] Of course the case of belief change is analogous to the one for nonmonotonic inference operations: Belief change as characterized by a transitively relational choice functions γ or δ lies strictly between preferential belief change[320] and (fully) rational belief change.[321]

The results furnished in Section 7.8 parallel and extend the work done in Rott (1991a) where the connection between preference relations over maximal non-implying sets and preference relations over sentences were studied. The former ones correspond to preference relations over models (cf. Section 7.2), the latter ones are usually called relations of epistemic entrenchment and are relations over sentences (cf. Chapter 8). I have studied the connection between (preference-guided) choices and (revealed) preferences on both the semantic and the syntactic level in earlier work: For the former see Rott (1993), for the latter Rott (1994).

One can establish a direct bridge between semantic and syntactic preferences $\mathcal{P}_{sem}$ and $\mathcal{P}_{syn}$, by linking the arrows in Fig. 7.7. The resulting definitions are

---

[317]But compare condition (R8) of Katsuno and Mendelzon (1991), which is taken over by del Val (1994), conditions (Gamma) and (BC7) of Lindström (1991), and condition ($\div$8r) in Rott (1993).

[318]Compare Lemma 4 and discussion in Section 4.4.

[319]Freund gives a semantic characterization of preferential reasoning with disjunctive rationality in terms of a certain class of 'preferential models' in the style of Kraus, Lehmann and Magidor (1990).

[320]In the sense analogous to preferential inference operations, cf. Chapter 4, and also Kraus, Lehmann and Magidor (1990) and Rott (1992b).

[321]In the sense analogous to rational inference operations, cf. Chapter 4, and also Alchourrón, Gärdenfors and Makinson (1985) and Lehmann and Magidor (1992).

$$\mathcal{P}_{sem}(\mathcal{P}_{synt}) := \mathcal{P}(\mathcal{G}(\mathcal{S}(\mathcal{P}_{synt})))$$
$$\mathcal{P}_{synt}(\mathcal{P}_{sem}) := \mathcal{P}_2(\mathcal{D}(\mathcal{S}(\mathcal{P}_{synt})))$$

where $\mathcal{P}$ and $\mathcal{P}_2$ are the operations of taking the revealed preferences (strict Samuelson and base preferences respectively) and $\mathcal{S}$ is the operation of liberal minimization, as explained in Chapter 6.[322] The present results are deeper than

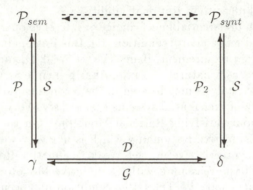

FIG. 7.7. Linking preferences over models and preferences over sentences

those of Rott (1991a) in several respects. First, the present approach makes provision for contractions, revisions and nonmonotonic inference operations that do not satisfy the full set of rationality postulates, including in particular (÷8) alias (*8) alias Rational Monotony. Thus the approach taken here is much more general and allows one to go for one's favourite set of postulates. In addition, it is modular in the sense that it traces a one-to-one correspondence between these logical postulates and constraints for rational choice, both on a semantic and a syntactic level. Finally, it shows that in almost all cases fulfilment of a coherence constraint for rational choice on the semantic level entails fulfilment of *the same* coherence constraint on the syntactic level, and vice versa (the exception is constraint (II)). Of the various relationships proved in Section 7.8, not a single one depends on the assumption that the set of beliefs or expectations be logically finite.

The advantages of the present approach can be summarized as follows:

- Features of theoretical coherence and rationality can be derived from features of practical coherence and rationality.

- Precise representation theorems can be given which establish a one-to-one correspondence between prominent conditions of theoretical and practical reason.

---

[322] In order to guarantee that $\mathcal{P}$ and $\mathcal{S}$ (or $\mathcal{P}_2$ and $\mathcal{S}$) are inverse operations, we have to assume that some of the antecedent conditions mentioned in Lemma 14 are satisfied. I return to this topic below, in Section 8.6.

- One can apply choices on both the semantic and the syntactic level, and gets an almost perfect parallelism between the two kinds of modelling.
- Nonmonotonic logic can thus be decomposed into monotonic logic and choice.
- The theory of epistemic entrenchment can be reconstructed as the theory of revealed syntactic preferences.[323]
- Via the choice-theoretic account, various decision-theoretic concepts and results promise to be applicable to the investigation of belief revision and nonmonotonic reasoning.
- The approach provides a precise and instructive point of view from which to investigate the philosophical tenability of the doctrine of doxastic voluntarism: Do the choice functions featuring in the formal models represent free and willful choices on the side of the agent, or are they compatible with materialist or deterministic philosophy? (Cf. Section 1.1)

A pictorial representation of the connections mentioned is given in Fig. 7.8. The chests of drawers stand for assortments of constraints or postulates.

7.10   Iterated belief change in the coherence-constrained mode

In this chapter, belief change has been called rational in so far as it makes use of choice functions that are independent of the particular input sentence ϕ (more precisely, independent of the menus $]\phi[$ or $Cn(\phi)$). Inspecting the definitions for the contraction and revision of a given belief set K, however, it is clear that the choice functions do not depend on K either. It is a natural question to ask whether we might not argue in a similar way that belief change is rational in so far as the choice functions used are independent of the current set of beliefs.

In fact, the seminal results of Alchourrón, Gärdenfors and Makinson's refer to unary contraction functions over a fixed belief set K.[324] Our theory relieves us of this dependency. There is nothing in the above definitions that prevents us from using one and the same choice function γ or δ to define contraction functions $\mathcal{C}(\gamma)$ or $\mathcal{C}(\delta)$ for *varying belief sets* K, rather than just one for a fixed belief set K. In other words, our choice-functions formally generate two-place rather than just unary contraction functions. Thus it is possible to model iterated changes of belief in the coherentist setting. The information encoded in a choice-function exceeds the information encoded in a unary contraction function over some fixed consistent K.[325] It is rather equivalent to a family of unary contraction functions, one contraction function for every belief set K, or, to put it differently, to a belief

---

[323] This will be the topic of Chapter 8.

[324] Actually, this is not quite true. See footnote 153 of Chapter 3.

[325] This claim gives a description of the situation that is just the converse of what Nayak *et al.* (1996) claim. The reason that they end up with a different diagnosis is that (a) they work with two-place revision functions and (b) they apply an entrenchment relation $<$ in such a way that it is suitable only for its 'epistemic content' $\{\phi : \perp < \phi\}$.

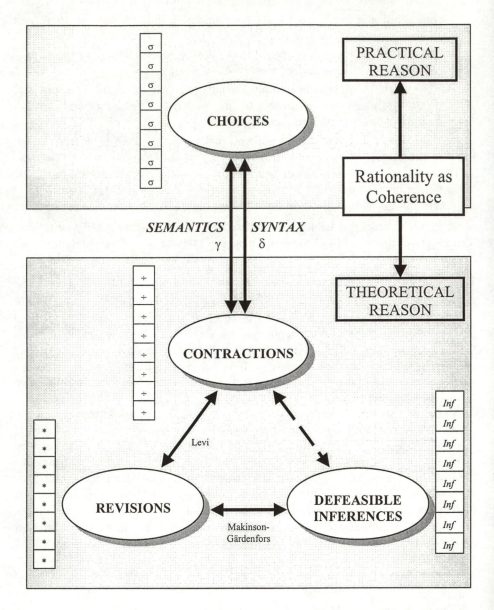

FIG. 7.8. The unity of practical and theoretical reason in the coherentist modelling

contraction model. For a general discussion of the dialectical relation between the statics and dynamics of belief compare Section 3.6 above.

If it is consistent, the current belief set K of a doxastic subject may receive a new interpretation from this perspective. It can be regarded as the result of removing inconsistency from the multiplicity of information received—and in practice this multiplicity is very likely to be inconsistent. For any semantic or syntactic choice function γ or δ, K may therefore be characterized as $L \dot{-} \bot$, yielding $\gamma(\widehat{\mathcal{M}_L})$ or $L - \delta(L)$, according to Definitions 6 and 8. On this interpretation, belief contraction models implicitly characterize the set of currently held beliefs.[326]

It is worthwhile to take down some of the most basic properties of the belief contraction models generated by a single semantic or syntactic choice function. Let $\dot{-}$ now stand for a *two-place* contraction function $\mathcal{C}(\gamma)$ or $\mathcal{C}(\delta)$. A look at Definitions 6 and 8, taken as definitions that are applicable to varying belief sets K, immediately shows that

$$K \dot{-} \phi = K \cap (L \dot{-} \phi)$$

It turns out that all we need to know for the contraction of an arbitrary belief set K is how to contract the 'absurd theory' L.[327] Belief contraction models providing contraction functions for infinitely many belief sets reduce to unary contraction functions over the single, degenerate belief set L.[328] Notice, however, a strange property of such contraction functions. For the vast majority of belief sets K, the contraction functions over K thus defined will not satisfy the Vacuity postulate ($\dot{-}3$) with respect to K. Before commenting on this, let us first have a look at further properties that appear to be somewhat strange.

Secondly, if two-place choice functions are held fixed and are never revised in the light of incoming evidence, then our definitions support a universally valid Principle of Monotony:[329]

$$\text{If } K \subseteq K', \text{ then } K \dot{-} \phi \subseteq K' \dot{-} \phi$$

Gärdenfors (1988, pp. 59–60 and 65) argues forcefully against this principle. The counterexample he puts forward is a case in which $\phi \notin K$ but $\phi \in K'$. The point

---

[326] For a contrasting interpretation, see Section 3.6, where arguments are given that belief revision strategies may be independent of the current belief set.

[327] Alchourrón and Makinson (1985, Obs. 7.5) prove an equivalent result for safe contractions based in belief-independent hierarchies that satisfy the conditions (EE↑) and (EE↓) discussed in Section 8.3 below. For a good survey of a variety of related ideas, see Areces and Becher (2001).

[328] Note that the Recovery postulate ($\dot{-}5$) for L entails that $\neg\phi$ has to be in $L \dot{-} \phi$. The other postulates for preferential contractions do not seem to have a particularly interesting impact on the contraction of L. ($\dot{-}2$) and ($\dot{-}3$) are trivial for $K = L$. For the absurd theory L, contraction with respect to ϕ coincides with revision by $\neg\phi$, if the revision is defined as $\mathcal{R}(\gamma)$ or $\mathcal{R}(\delta)$, respectively.

[329] For a discussion of the significance of this unqualified principle in the theory of *updates*, that is, changes of belief states in response to changes in the world, see Katsuno and Mendelzon (1992) and Morreau and Rott (1991).

then is that K does not have to be changed for the removal of ϕ whereas K', which includes ϕ, should of course be contracted. And there is no reason to think that the result is in any way related to K—which may be chosen as just *any* closed subset of K' that does not contain ϕ.

Thirdly, similar considerations apply to another principle validated by our definition (again, if the underlying choice function is two-place and remains constant). The principle concerns iterated contractions. It is easy to verify that

$$(K \dot{-} \phi) \dot{-} \psi = K \dot{-} \phi \cap K \dot{-} \psi$$

Again this Intersection Principle is intuitively implausible. The problem arises when ψ is already given up in $K \dot{-} \phi$. In that case it is natural not to change $K \dot{-} \phi$ any more, that is, to put $(K \dot{-} \phi) \dot{-} \psi = K \dot{-} \phi$. However, the above equation makes us intersect $K \dot{-} \phi$ with $K \dot{-} \psi$ even when ψ is not in $K \dot{-} \phi$.[330]

The reason why Definitions 6 and 8 yield awkward results lies in their excessive generality: They do not require choice functions to be *faithful* to given belief sets. In the case of the iterated belief contraction $(K \dot{-} \phi) \dot{-} \psi$, we would expect that the choice function γ (or δ) is faithful to $[\![K]\!]$ (or respectively, $L - K$) in the first contraction and faithful to $[\![K \dot{-} \phi]\!]$ (or respectively, $L - (K \dot{-} \phi)$) in the second contraction. If ψ is not contained in $K \dot{-} \phi$, then a contraction of $K \dot{-} \phi$ with respect to ψ in the second step does not require considering $\neg\psi$-worlds outside $[\![K \dot{-} \phi]\!]$, nor does it require rejecting consequences of ψ which are contained in $K \dot{-} \phi$.[331]

The situation looks even more drastic if we consider two-place revisions $\mathcal{R}(\gamma)$ and $\mathcal{R}(\delta)$ constructed according to Definitions 11 and 12 (see Section 7.5). Without the requirement that choice functions must be faithful (to $[\![K]\!]$ and $L - K$ for semantic and syntactic choices respectively) there is no dependence at all of the revised set upon the original belief set K. $K * \phi$ is identical with $K' * \phi$ for arbitrary K and K'—provided that the same γ or the same δ are applied to K and K'.

A general moral to be drawn from all this is that we should take the constraints of Faith seriously. Faith relates choices across varying menus to a set of absolutely satisfactory options. These options are tied to the current belief set K, more exactly they are given by $[\![K]\!]$ for γ and by $L - K$ for δ. If we wish to satisfy Faith, it becomes evident that different belief sets *must* be accompanied by different choice functions. The idea of the unity of choice which is characteristic of the coherentist account of belief change (as construed in this chapter) does not extend to the diachronic dimension.[332] Doxastic input, we conclude, does not only affect the propositional content of the agents's beliefs, but also

---

[330]The difference would vanish if we had a principle guaranteeing that $K \dot{-} \phi$ is a subset of $K \dot{-} \psi$ whenever $\psi \notin K \dot{-} \phi$. This condition is invalid in AGM-style belief contraction, but we will discuss a method of belief withdrawal that satisfies it in Section 8.7.

[331]This argument also intuitively blocks the satisfaction of the order independence condition $(K \dot{-} \phi) \dot{-} \psi = (K \dot{-} \psi) \dot{-} \phi$.

[332]The talk of dimensions here is explained in Section 3.1.

his dispositions what to choose in cases of evidential conflict. Choice functions do not remain stable, because they are not independent of the beliefs held by the agent—as is witnessed by $(\dot{-}2)$ and $(\dot{-}3)$, the conditions corresponding to the choice constraints (Faith2) and (Faith1) respectively. But more than that, choice functions should not be regarded as fully determined by the agent's plain beliefs. As was pointed out in Section 3.6, two agents may fully agree as regards the propositions they accept and yet have entirely different strategies for revising their beliefs in the case of conflict.

Those who regard postulates $(\dot{-}2)$ and $(\dot{-}3)$ as indispensable conditions, but hold that the choice function of an agent should remain essentially stable, will suggest to change Definitions 6 and 8 in such a way that their main clauses apply only when ϕ is in K and put $K \dot{-} \phi = K$ otherwise. More precisely, the modified conditions guaranteeing $(\dot{-}3)$ for constant γ's and δ's and varying K's are these:

$$\psi \in K \dot{-} \phi \quad \text{iff} \quad \psi \in K \text{ and } (\phi \notin K \text{ or } \gamma(\,]\phi[\,) \subseteq [\![\psi]\!]\,)$$

$$\psi \in K \dot{-} \phi \quad \text{iff} \quad \psi \in K \text{ and } (\phi \notin K \text{ or } \phi \lor \psi \notin \delta(Cn(\phi)))$$

On both the semantic and the syntactic account, this gives a two-place contraction function that puts for every belief set K

$$K \dot{-} \phi = \begin{cases} K & , \text{ if } \phi \notin K \\ K \cap (L \dot{-} \phi) & , \text{ if } \phi \in K \end{cases}$$

These *faithful variants* of our central Definitions 6 and 8 validate a qualified monotony principle for contractions of varying theories

$$\text{If } K \subseteq K' \text{ and } \phi \in K, \text{ then } K \dot{-} \phi \subseteq K' \dot{-} \phi$$

as well as a qualified intersection principle for iterated contractions

$$(K \dot{-} \phi) \dot{-} \psi = \begin{cases} K \dot{-} \phi & , \text{ if } \psi \notin K \dot{-} \phi \\ K \dot{-} \phi \cap K \dot{-} \psi & , \text{ otherwise} \end{cases}$$

Turning now from contractions to revisions, we can easily see that the corresponding faithful variants of Definitions 11 and 12 determine two-place revision functions by putting, for every belief set K, $K * \phi = Cn(K \cup \{\phi\})$ if $\neg\phi$ is not in K, and $K * \phi = L * \phi$ otherwise. Depending on the properties of the choice functions involved, these faithful revision functions satisfy at least two and at most three of the four well-known postulates for iterated revision suggested by Darwiche and Pearl (1994, 1997):[333]

---

[333] A similar result was obtained independently by Areces and Becher (2001). Compare it with the discussion in a foundationalist context in Section 5.2.3 where all four Darwiche-Pearl postulates were found to be satisfied.

Observation 42 *The faithful variants of Definitions 11 and 12 satisfy*

(DP3)   *If* $\phi \in K * \psi$, *then* $\phi \in (K * \phi) * \psi$

(DP4)   *If* $\neg\phi \notin K * \psi$, *then* $\neg\phi \notin (K * \phi) * \psi$

If the choice functions γ and δ satisfy (I) and (IV), then the faithful variants also satisfy

(DP1)   *If* $\phi \in Cn(\psi)$, *then* $(K * \phi) * \psi = K * \psi$

However, they do not satisfy

(DP2)   *If* $\neg\phi \in Cn(\psi)$, *then* $(K * \phi) * \psi = K * \psi$

Thus if we take the Darwiche-Pearl postulates as a measure of rationality for iterated belief revision, we may be quite satisfied with the faithful versions of our main definitions. I do not think that the violation of (DP2) must be considered a grave shortcoming. So by and large, there are no serious objections from a formal point of view against such a proposal (that surely improves things a lot as compared with models satisfying the unqualified monotony and intersecton principles). Philosophically, however, this seems to be an advance in the wrong direction. I consider it to be intuitively misguided to insist on stable choice functions for belief states evolving in time under the influence of incoming information.[334] There is no reason to suppose that belief change strategies are in any respect less sensitive to new experiences or new information than beliefs. Intelligent believers are adaptable to new experiences in various sorts of ways. In this respect, the moral to be drawn from iterated belief change in the coherentist setting is not at all different from the moral of iterated belief revision in the direct mode (as presented in Section 5.2.3).

---

[334]For a possible counterexample against the qualified monotony principle see Lindström and Rabinowicz (1992).

8

REVEALED PREFERENCES: UNDERSTANDING THE THEORY OF EPISTEMIC ENTRENCHMENT

The notion of the *epistemic entrenchment* has attracted quite some attention of philosophers, logicians and researchers in knowledge representation and reasoning.[335] It was introduced, first under the name 'epistemic importance', by Peter Gärdenfors (1984, 1988). He specified two possible 'origins' of epistemic entrenchment: an *information-theoretic approach* and a *paradigm approach*. However, neither of these approaches squares well with the logical constraints which are placed on entrenchment relations in the later and more mature 1988 work of Gärdenfors and Makinson (1988). Although it has since then become well-known how epistemic entrenchment can be put to work technically, the problem of providing a *systematic justification of the postulates for entrenchment relations* is still in need of a systematic clarification. The present chapter offers a reconstructive solution to this problem which is faithful to the literal meaning of the term 'entrenchment'. It is worth emphasizing that for most of the present chapter we shall *reconstruct epistemic entrenchment* from belief change or choice behaviour rather than use epistemic entrenchment for *constructing* belief changes. A 'complementary', more constructive approach has been the background of several proposals to extract entrenchment relations from other information structures[336] which are surveyed in Rott (2000a).

We again follow the philosophical strategy of deriving logical principles from abstract principles of rationality for choices and decisions which we consider to be more fundamental than the former. We are going to interpret the standard theory of epistemic entrenchment (Gärdenfors and Makinson 1988, Rott 1992b) in terms of the general theory of rational choice. The following questions are being addressed in the course of this chapter. Are the usual postulates for entrenchment justified by the classical constraints of rational choice? Do the latter constraints, or the mere interpretation of entrenchments in terms of choice, impose novel constraints on entrenchment relations? Conversely, which constraints of rational

---

[335] According to what we said in Chapter 1, we had better speak of 'doxastic entrenchment'. For the sake of terminological continuity, however, we keep the widely used term 'epistemic entrenchment'. Among the papers on 'epistemic' entrenchment are Cantwell (2001), Dubois and Prade (1991), Gärdenfors and Makinson (1988, 1994), Georgatos (1997), Lindström and Rabinowicz (1991), Meyer *et al.* (2000), Nayak (1994b), Nayak, Nelson and Polansky (1996), Nebel (1992), Rott (1991a, 1991b, 1992a, 1992b, 1992c, 1994), and Williams (1994, 1995).

[336] Rott (1991a, 1992a, 1992c).

choice can be retrieved from the entrenchment postulates, and what is their status in the theory of choice?

Not every logical constant is equally important for the study of belief revision. Conjunction will be regarded as the primary operation on opinions (beliefs and expectations, things accepted), since it allows us to tie together pieces of belief. Conjunction plays a dual role in belief change. In belief revisions (as well as in nonmonotonic reasoning), conjunctions help to capture the notion of a *bunch revision* (also called 'package revision'): Accepting all sentences of a finite set simultaneously essentially means accepting their conjunction. In belief contractions they help to capture the notion of *choice contraction* (also called 'pick contraction'): Giving up at least one of the sentences in a finite set essentially means giving up their conjunction. It is the usage of conjunctions in the latter context that leads us into the theory of 'epistemic entrenchment'.

Conjunctions help us to lay bare the finer structure of the meaning of 'entrenchment'. It is one of the principal aims of the present chapter to analyse how the properties of epistemic entrenchment that have been specified in the literature can be motivated. We will see that as long as we are using exclusively conjunctions in our object language, we remain completely within the realm of the pure theory of rational choice. It is only when other connectives are used that genuinely logical relationships come into play. It will be particularly interesting to analyse the contribution made by the logical constraints that we imposed on the syntactic choice functions discussed in the previous chapter. Some of the properties of epistemic entrenchment derive from the mere fact that these relations are preferences revealed by syntactic choices; some of them are consequences of the fact that the relevant choices are assumed to be rational in one way or other; and the fact that entrenchments respect logical equivalences is due to our assumption that syntactic choices satisfy the logical constraints (LP1) and (LP2). These constraints induce additional structure but also additional complications for choices on the syntactic level.

The notion of epistemic entrenchment thus turns out to be constructible as a preference relation revealed by an agent's actual or potential choices between propositional beliefs. In my opinion, this way of viewing entrenchment provides us not only with an extremely versatile and adaptable way of analysing an agent's syntactic choices, but also with a most natural interpretation of the basic intuition behind 'entrenchment'.[337] Entrenchment, as applied in belief change and nonmonotonic reasoning, provides us with a means of extending 'local decisions' (concerning 'small' sets with only two elements) in a coherent way into 'global revisions' or 'global inferences' (concerning 'large' sets of beliefs or expectations, represented in general by infinitely many sentences).

As discussed in Section 6.3, there are two ways of talking about preferences: One can phrase one's statements in terms of strict or in terms of non-strict pref-

---

[337] 'Entrenchment' is applied here to sentences or propositions, not—like Goodman's (1955) famous notion of entrenchment—to the predicates of a certain language.

erence relations. I will regard the non-strict epistemic entrenchment relation $\leq$ of Gärdenfors and Makinson (1988) as the converse complement of a more intuitive and useful strict relation $<$. Although both ways are equivalent mathematically, it has turned out that the systematic place of some conditions within the theory is much more natural when they are formulated in strict rather than non-strict terms (see the discussion in Rott 1992b, Section 4). For instance, transitivity imposed on non-strict relations has a impact very different from transitivity imposed on strict relations. There is a certain danger to get confused about these issues in the relational setting which is absent when all work is done using choice functions in the first place.

Some general aspects of revealed preferences were described in Section 6.4. One can of course define revealed preferences from choices on the semantic as well as the syntactic level. The upshot of the present chapter is that the preferences revealed by a *syntactic* choice function δ of the sort discussed in Chapter 7 give us precisely what has been studied in the theory of epistemic entrenchment. I start my discussion from a choice-theoretic point of view, then I present one of the standard accounts of epistemic entrenchment and finally relate the two presentations to each other with the help of a carefully designed translation procedure.

8.1   Basic principles

The present chapter presupposes a propositional language that is at least equipped with the binary connective '$\wedge$' for conjunction and a constant (nullary connective) '$\top$' for 'the truth'. Large parts of our discussion need no other connective. A doxastic state will again be partially represented by the set K of sentences that are believed or accepted by an agent in that doxastic state. Revision and contraction functions are used for the representation of changes of belief. The general format used is this: For a fixed initial belief state, we have a function

$$\text{set of beliefs to retract} \longmapsto \text{new belief state}$$

The retract set F consists of the set of sentences *at least one of which* has to be given up. We are going to work with the concept of multiple contraction that is more general than contraction with respect to only a single sentence. The pick contraction of K with respect to F, denoted $K \dot{-} \langle F \rangle$, is the change of the belief set K that is required in order to discard at least one element of F. It is left to the agent's decision which elements of F are best to give up.

For a large part of this chapter, I do not assume that the set of beliefs of an agent is logically closed. For reasons that will soon become clear, I only require that the beliefs are closed under conjunction:[338]

(A) $\{\phi_1, \ldots, \phi_n\} \subseteq K$ iff $\phi_1 \wedge \cdots \wedge \phi_n \in K$

Clearly conjunctive closure is a weakening of the usual AGM postulate $(\dot{-}1)$ according to which belief sets must be closed under Cn. Since the empty set is

---

[338] Our agents thus suffer from conjunctivitis (Kyburg 1970).

a subset of K, $\top = \bigwedge \emptyset$ is an element of each belief set K. The constant $\top$ has no informational content, nor is it important or interesting in any special way, but it is convenient to have it included in our notation.

Call two sentences *variants* of each other if they are accepted in exactly the same doxastic states. For the start, we are interested only in $\wedge$-*variants*. From (A), we infer that $\wedge$ satisfies associativity, commutativity, $\wedge$-contraction and $\wedge$-expansion, in the sense that

$$(\phi \wedge \psi) \wedge \chi \quad \text{and} \quad \phi \wedge (\psi \wedge \chi)$$
$$\phi \wedge \psi \quad \text{and} \quad \psi \wedge \phi$$
$$\phi \quad \text{and} \quad \phi \wedge \phi$$

are $\wedge$-variants. This allows us to delete parentheses and write, for instance, $(\phi \wedge \psi) \wedge (\phi \wedge \chi)$ just as $\phi \wedge \psi \wedge \chi$. Conjunction inherits its structural properties from properties of sets; let us call this the *variability of conjunctions*. Conjoining the truth constant with a sentence ϕ leads to a variant of ϕ; that is, $\phi \wedge \top$ is a $\wedge$-variant of ϕ.

In this chapter, belief contractions are taken as the basic or paradigmatic changes of belief. Our second assumption says that $\wedge$-variants as inputs lead to identical belief states. The condition of conjunctive equivalence is a weakening of the AGM contraction postulate ($\dot{-}6$):

(B) If ϕ and ψ are $\wedge$-variants, then $K \dot{-} \phi = K \dot{-} \psi$

$\wedge$-variants are interchangeable in all contexts that will be considered.

The following two principles on which this chapter will be based concern multiple contractions of which there are two different kinds. The task of a pick contraction is to discard *at least one* element of a set F, whereas the task of a bunch contraction is to discard *each* element of a set F.[339] In this chapter, we need only consider pick contractions, where the goal is to see to it that $F \not\subseteq Cn(K \dot{-} \langle F \rangle)$. We know from (A) that in order to retract one element of a finite set $\{\phi_1, \ldots, \phi_n\}$, it is necessary and sufficient to give up the conjunction $\phi_1 \wedge \cdots \wedge \phi_n$. What is much more, the latter task seems intuitively even *identical* with the former. This is the fundamental idea underlying our third principle:

(C) $K \dot{-} \langle \{\phi_1, \ldots, \phi_n\} \rangle = K \dot{-} (\phi_1 \wedge \cdots \wedge \phi_n)$

This condition that identifies pick contractions and contractions with respect to conjunctions is validated by the constructive suggestions to define pick contractions that I made in Chapter 7 (Definitions 7 and 9). However, the considerations in the present chapter do not depend on any concrete idea of how to construct pick contractions. Notation: We delete set brackets within pointed brackets and write '$K \dot{-} \langle \phi, \psi \rangle$' instead of '$K \dot{-} \langle \{\phi, \psi\} \rangle$' etc.

Pick contractions present by their very nature problems of rational choice. For any non-empty set of sentences F, the agent must make a decision which

---

[339] Still the best survey on pick and bunch contractions, there called 'choice contractions' and 'package contractions' is given by Fuhrmann and Hansson (1994).

sentences of F to give up in $K \dot- \langle F \rangle$. He must retract at least one element of F, but may—and should—in the case of ties retract several of them at the same time. We now connect pick contractions with the device of syntactic choice functions δ. We require that for every $\phi \in F$,

(D) $\phi \in \delta(F)$ iff $\phi \notin K \dot- \langle F \rangle$

This condition embodies a new idea of using syntactic choice functions in pick contractions. But it is easy to verify, that like (C), (D) is provable for our constructive Definition 9 of pick contractions by means of syntactic choice functions (see Lemma 24).

As before, we call F the *issue* or *menu*, and $\delta(F)$ the *choice set* of F (with respect to δ). It is required here that $\delta(F) \subseteq F$. We do not require $\delta(F)$ to be empty, since the agent may refuse to withdraw any sentence in the issue. Intuitively, δ picks the elements of F which are *best to withdraw* from K, or *least secure* in F. If neither ϕ nor ψ is in $\delta(\{\phi, \psi\})$, say, this means that the agent refuses to suspend judgment with respect to *both* ϕ *and* ψ, so $K \dot- \langle \{\phi, \psi\} \rangle = K \dot- \phi \wedge \psi = K$ continues to include ϕ, ψ and $\phi \wedge \psi$.[340]

If, on the other hand, both ϕ and ψ are in $\delta(\{\phi, \psi\})$, this means that the agent suspends judgment with respect to both ϕ and ψ because he can reach no decision as to which of them is more secure. We restrict ourselves in this chapter to the case of finite menus F in order to ensure that (C) is applicable. More notation: We delete set brackets within round brackets and write $\delta(\phi, \psi)$ instead of $\delta(\{\phi, \psi\})$ etc.

The last basic principle we are going to invoke is precisely the idea of our official Definition 5 of base preferences $< = \mathcal{P}_2(\delta)$ in Chapter 6, now applied to the choice of retractible sentences.

(E) $\phi < \psi$ iff $\phi \in \delta(\{\phi, \psi\})$ and $\psi \notin \delta(\{\phi, \psi\})$

This is how agents are supposed to reveal their doxastic preferences: We say that ψ is *more firmly entrenched* (or simply, *more entrenched*) in K than ϕ iff the agent would jettison ϕ but keep ψ when facing the need to give up (at least) one of ϕ and ψ. The notation '$\phi < \psi$' will, for the rest of this chapter, always be used as a shorthand for 'ψ is more entrenched than ϕ'.

The connections established by our five basic principles are summarized in Fig. 8.1.

The *differentia specifica* that marks off the following from the *genus proximum* of the general theory of choice and preference is that syntactic choices are not made between unstructured objects but between sentences that possess *inner structure*: they can be split into parts at the occurrences of $\wedge$. Since there are constraints on the acceptance and the retractability of conjunctions, pick contractions—and hence the choice functions involved—become constrained accordingly. As we saw in Chapter 6, the general theory of rational choice imposes multifarious constraints on choices across varying issues. However, additional

---

[340] In particular it follows from what has been said that $K \dot- \top = K \dot- \langle \emptyset \rangle = K$.

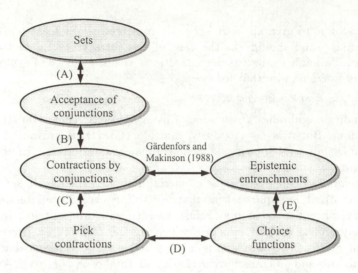

FIG. 8.1. Connecting choice and entrenchment

constraints are necessary for syntactic choice functions. We have made extensive use of the logical constraints (LP1) and (LP2) in Chapter 7. In this chapter I am going to divert attention from the specifics that may be imported by the background logic Cn, and do as much as possible with reference to conjunctions only. The effect, however, will be pretty much the same.

We should actually be more precise here. Let $L_\wedge$ be the propositional language with conjunction as the only logical operator, and with parentheses as auxiliary symbols. Let, for two sentences ϕ and ψ from $L_\wedge$, $\phi \sqsubseteq \psi$ stand for the fact that ϕ is part of ψ in the sense that the propositional variables occurring in ϕ are a subset of the propositional variables occurring in ψ. Let $CCn(G) = \{\phi : \phi \sqsubseteq \bigwedge G\}$. Instead of the logical constraints (LP1) and (LP2) for syntactic choice functions, we shall have the following ones now:

(CP1)   If $G \subseteq F$ and $\delta(F) \cap G = \emptyset$, then $\delta(F) \cap CCn(G) = \emptyset$

(CP2)   If $CCn(F) = CCn(G)$, then $\delta(F) \cap G \subseteq \delta(G)$

In the setting of $L_\wedge$ and CCn, logic reduces to checking the occurrence of propositional variables.

Lemma 43 *The constraints (CP1) and (CP2) can be derived from the basic principles (A) – (E).*

The theory of rational choice alone does not tell us anything about how the choice made in $\{\phi \wedge \psi, \chi\}$, say, is related to the choice made in $\{\phi, \psi \wedge \chi\}$. For this theory, the two menus are just different and incomparable. The principles (A) – (E) above, however, heavily constrain the choices in these menus. They entail, for instance, that if you choose χ and disregard $\phi \wedge \psi$ in the first menu, you *must* choose $\psi \wedge \chi$ and disregard ϕ in the second. We shall see how this works presently.

In their seminal paper on epistemic entrenchment, Gärdenfors and Makinson (1988) study the following fundamental condition:

$$\phi \leq \psi \quad \text{iff} \quad \phi \notin K \dot{-} (\phi \wedge \psi) \text{ or } \phi \wedge \psi \in Cn(\emptyset)$$

If we connect conditions (C), (D) and (E) with one another, we immediately get a condition which is in all principal cases (though not in all limiting cases) equivalent with Gärdenfors and Makinson's condition:

Definition 17 The entrenchment relation $<$ over L is revealed by a contraction function $\dot{-}$, in symbols $< = \mathcal{P}_2(\dot{-})$, iff for all ϕ and ψ,

$$\phi < \psi \quad \text{iff} \quad \phi \notin K \dot{-} (\phi \wedge \psi) \text{ and } \psi \in K \dot{-} (\phi \wedge \psi)$$

Gärdenfors and Makinson work with non-strict relations of epistemic entrenchment which may be related to our strict relations by taking the converse complement.[341] On this transformation, their account has it that $\phi < \psi$ iff $\phi \wedge \psi \notin Cn(\emptyset)$ and $\psi \in K \dot{-} (\phi \wedge \psi)$. Given the AGM postulates ($\dot{-}1$) and ($\dot{-}4$), this is equivalent to our definition. However, as should be clear from Chapters 6 and 7, I want to treat ($\dot{-}4$) as an optional extra and to model reasoners who may refuse to eliminate beliefs which are not logical truths. Definition 17 has the advantage of being a purely choice-theoretic condition, without any reference to a background logic Cn. For these two reasons we prefer Definition 17 to the definition used by Gärdenfors and Makinson.

We *explain* the content of Gärdenfors and Makinson's suggestion (a) by connecting contractions with respect to conjunctions with multiple contractions, (b) by subsuming belief change under the rational choice and (c) by interpreting entrenchments as revealed preferences.

8.2 Entrenchments as revealed preferences: Translations

8.2.1 *The essential connection*

Let us put together our basic principles. The general argument to be used in our translations is this:

---

[341] See Rott (1992b). Philosophically speaking, Gärdenfors and Makinson's relation '$\phi \leq \psi$' does not mean 'ψ is at least as entrenched as ϕ' but really means 'ψ is at least as entrenched as ϕ, *or incomparable with it*'. Since everything is comparable in the context of Gärdenfors and Makinson, this distinction does not matter in their paper.

$\phi_1 \wedge \cdots \wedge \phi_n < \psi_1 \wedge \cdots \wedge \psi_m$ by (E)

    iff $\phi_1 \wedge \cdots \wedge \phi_n \in \delta(\{\phi_1 \wedge \cdots \wedge \phi_n, \psi_1 \wedge \cdots \wedge \psi_m\})$

        and $\psi_1 \wedge \cdots \wedge \psi_m \notin \delta(\{\phi_1 \wedge \cdots \wedge \phi_n, \psi_1 \wedge \cdots \wedge \psi_m\})$ by (CP2)

    iff $\phi_1 \wedge \cdots \wedge \phi_n \in \delta(CCn(\{\phi_1, \ldots, \phi_n, \psi_1, \ldots, \psi_m\}))$

        and $\psi_1 \wedge \cdots \wedge \psi_m \notin \delta(CCn(\{\phi_1, \ldots, \phi_n, \psi_1, \ldots, \psi_m\}))$ by (CP1)

    iff $\exists i \in \{1, \ldots, n\} : \phi_i \in \delta(CCn(\{\phi_1, \ldots, \phi_n, \psi_1, \ldots, \psi_m\}))$

        and $\forall j \in \{1, \ldots, m\} : \psi_j \notin \delta(CCn(\{\phi_1, \ldots, \phi_n, \psi_1, \ldots, \psi_m\}))$ by (CP2)

    iff $\exists i \in \{1, \ldots, n\} : \phi_i \in \delta(\{\phi_1, \ldots, \phi_n, \psi_1, \ldots, \psi_m\})$

        and $\forall j \in \{1, \ldots, m\} : \psi_j \notin \delta(\{\phi_1, \ldots, \phi_n, \psi_1, \ldots, \psi_m\})$ by $\delta(X) \subseteq X$

    iff $\forall j \in \{1, \ldots, m\} : \psi_j \notin \delta(\{\phi_1, \ldots, \phi_n, \psi_1, \ldots, \psi_m\}) \neq \emptyset$

We see that although we could in principle translate a preference of $\psi_1 \wedge \cdots \wedge \psi_m$ over $\phi_1 \wedge \cdots \wedge \phi_n$ into a choice between them in one step (using principle (E)), this would not reflect the full information encoded in the syntactic structure of the sentences of our language. Conjunctions allow us to make packages of sentences and thus express preferences between sets of sentences, so to speak. Our basic principle allow us to exploit the (conjunctive) structuring of the objects of syntactic choice functions.

In short, we have for any two finite sets of sentences F and G that

$$\boxed{\;\; \bigwedge F < \bigwedge G \;\; \text{iff} \;\; \delta(F \cup G) \neq \emptyset \text{ but } G \cap \delta(F \cup G) = \emptyset \;\;}$$

This is the key to all the translations between postulates we shall make in this chapter.[342]

8.2.2 *From entrenchments to choices*

In order to keep the promises made at the beginning of this chapter we have to translate talk about entrenchments into talk about (logically constrained) syntactic choices. The translation process is described by the above argument read from the top to the bottom. The procedure is actually ambiguous, due to the possibility of leaving some conjunctions unbroken. We insist, however, that the *standard translation* must break up every conjunction into its atomic parts. But we are not dealing with real atoms of the object language here. We must keep

---

[342]Can we derive a preference relation between finite *sets* of beliefs from a preference (entrenchment) relation between single beliefs? Yes, we can: $F < G$ iff $\bigwedge F < \bigwedge G$ i.e., iff $\delta(F \cup G) \neq \emptyset$ but $G \cap \delta(F \cup G) = \emptyset$, i.e., iff there is a $\phi \in F$ such that $\bigwedge(F \cup G) \not< \phi$, but for all $\psi \in G$ it holds that $\bigwedge(F \cup G) < \psi$. Can we express the last condition without reference to $\bigwedge(F \cup G)$? Yes, if $<$ satisfies the conditions (E&C), (EI), (EII) and (EIII) below (modularity is not necessary), then $\bigwedge F < \bigwedge G$ if and only if there is a $\phi \in F$ such that for all $\psi \in G$ it holds that $\phi < \psi$.

in mind that all sentential metavariables may be instantiated by conjunctions. Hence

$$\phi \wedge \psi < \chi$$

for example, must not be converted simply into

$$\chi \notin \delta(\{\phi, \psi, \chi\}), \text{ and either } \phi \in \delta(\{\phi, \psi, \chi\}) \text{ or } \psi \in \delta(\{\phi, \psi, \chi\})$$

but instead into the more general

$$H \cap \delta(F \cup G \cup H) = \emptyset \text{ and } \delta(F \cup G \cup H) \neq \emptyset$$

where ϕ, ψ and χ are thought of as (conjunctive variants of) the conjunctions of the elements of F, G and H, respectively.

8.2.3   *From choices to entrenchments*

In the converse direction, we translate talk about syntactic choices in finite menus into the language of entrenchments. The most general choice conditions we shall want to translate are of the general form $G \cap \delta(F \cup G) = \emptyset$ where F and G are finite sets of sentences. The key to the translation process is again described by the longish argument of Section 8.2.1, now read from the bottom to the top. Due to the variability of conjunctions, however, the translation process is *essentially* ambiguous. The fully general formulation of the translation procedure is a little complicated. Let two sets F and G be given. Then take two arbitrary series of sets of sentences $F_1, \ldots, F_k$ and $G_1, \ldots, G_k$ such that $G_1 \cup \cdots \cup G_k = G$ and $F_i \cup G_i = F \cup G$, for every $i \leq k$. Then we can translate the condition

$$G \cap \delta(F \cup G) = \emptyset$$

into

$$\text{either } \bigwedge(F \cup G) \not< \top \text{ or } \bigwedge F_i < \bigwedge G_i \text{ for every } i$$

This is the *most general translation* of $G \cap \delta(F \cup G) = \emptyset$. One can verify that this actually translates $G \cap \delta(F \cup G) = \emptyset$ by using the converse translation described in the last section. Since $\bigwedge \emptyset = \top$, the clause $\bigwedge(F \cup G) \not< \top$ encodes the case that $\delta(F \cup G) = \emptyset$.

As a special case, we translate $\phi \in \delta(F)$ into $\bigwedge F < \top$ and $\bigwedge F \not< \phi$. As an important example for the phenomenon of essential ambiguity, let us suppose that we want to translate the fact that

$$\delta(\{\phi, \psi, \chi\}) = \{\phi\} \tag{†}$$

Obviously $\delta(\{\phi, \psi, \chi\}) \neq \emptyset$. The identity can now be expressed in many different ways. A straightforward way would be to say that $\{\psi, \chi\}$ does not intersect with $\delta(\{\phi, \psi, \chi\})$ and hence, by the above argument, that

$$\phi \wedge \psi \wedge \chi < \psi \wedge \chi \tag{a}$$

or equivalently,

$$\phi < \psi \wedge \chi \tag{b}$$

Notice that both conditions must be equivalent, if the basic principles (A) – (E) are to hold. But this is not the only way of expressing (†). One could alternatively say that neither $\{\psi\}$ nor $\{\chi\}$ intersects with $\delta(\{\phi, \psi, \chi\})$. This translates into

$$\phi \wedge \psi \wedge \chi < \psi \quad \text{and} \quad \phi \wedge \psi \wedge \chi < \chi \tag{c}$$

or equivalently, into

$$\phi \wedge \chi < \psi \quad \text{and} \quad \phi \wedge \psi < \chi \tag{d}$$

We see that we have found four equally legitimate, but different translations of the single condition (†). Our translation is indeed a one-many translation, there is a genuine indeterminacy involved here. And again, by the above argument, the latter two conditions must be equivalent with the first two, if our basic principles are to hold.

But this is not the end of the story. The translation is also many-one. Each of the conditions (A)–(E) is not only a translation of $\delta(\{\phi, \psi, \chi\}) = \{\phi\}$, but at the same time a translation of the more general condition

$$\delta(F \cup G \cup H) \neq \emptyset \quad \text{and} \quad (G \cup H) \cap \delta(F \cup G \cup H) = \emptyset \tag{‡}$$

where ϕ, ψ and χ can again be understood as representing (conjunctive variants of) the conjunctions of the elements of finite sets F, G and H, respectively. We have decided above to take (‡) as the standard formulation, when we translate from the entrenchment conditions (a)–(d) into statements about choices.[343]

Since the most general translation of $G \cap \delta(F \cup G) = \emptyset$ is fairly complex, we would rather want to replace it by a simpler one. In the next section it will turn out that it is in fact sufficient to translate this condition into

$$\text{either} \quad \bigwedge(F \cup G) \nless \top \quad \text{or} \quad \bigwedge F < \bigwedge G$$

We call this simple translation the *standard translation* of $G \cap \delta(F \cup G) = \emptyset$.

---

[343] What is an adequate translation? A translation from one language into another ought to *preserve meanings*. It is not easy to say what exactly is meant by this, but if the source or target languages are equally expressive, the following is surely a minimal constraint:

$$(\Phi \text{ implies } \Psi) \quad \text{iff} \quad (Trans(\Phi) \text{ implies } Trans(\Psi))$$

Back-and-forth translations $Trans_{\rightarrow}$ and $Trans_{\leftarrow}$ also ought to be *compatible with each other* in the sense that

$$\text{If } Trans_{\leftarrow}(Trans_{\rightarrow}(\Phi)) = \Psi, \text{ then } \Psi \text{ is equivalent with } \Phi$$

and vice versa. Implication and equivalence are to be understood relative to the logics that govern the source and the target language.

8.2.4 *Admissible relations*

Not every preference relation over sentences is suitable for representing a syntactic choice function—where 'representing' means 'being revealed by'.[344] It makes even good sense to define the notion of entrenchment in this way: We say that $<$ is an *entrenchment relation* if and only if it is the base preference revealed by some syntactic choice function.[345]

We are now going to specify constraints that every preference relation that can possibly represent a syntactic choice function has to meet. Which relations $<$ over the set of all sentences in our language can possibly qualify as entrenchment relations? Our basic principles limit the field of candidate relations considerably. First, we get the condition of *substitutivity of variants*:

(SV)   If ϕ and ψ are $\wedge$-variants, then: $\phi < \chi$ iff $\psi < \chi$, and $\chi < \phi$ iff $\chi < \psi$

Then, by principle (E), entrenchment relations must clearly be irreflexive.

As already pointed out, the ambiguity inherent in the translation of conditions of epistemic entrenchment into choice conditions is harmless for our concerns. The ambiguity involved in the converse direction, on the other hand, is important. Many conditions on the side of epistemic entrenchment correspond to essentially a single condition on the side of rational choice. For this reason some nontrivial necessary conditions on entrenchment relations are to be imposed by the very idea that entrenchment conditions reflect choices in a structured domain. Entrenchment relations take account of the internal structure of the sentences related. Our aim is to make entrenchment relations $<$ consistently translatable into (and a consistent translation of) syntactic choice functions δ. The key to this project is the following condition. Prima facie, the condition seems hard to understand, but it just embodies, in a nutshell, the connection between *entrenchment and choice*:

(E&C) $\phi < \psi \wedge \chi$ iff $\phi \wedge \psi < \chi$ and $\phi \wedge \chi < \psi$

As seen in our discussion of example (†) above, both sides of (E&C) are ways of expressing that $\delta(\{\phi, \psi, \chi\}) = \{\phi\}$. Hence (E&C) is a necessary condition for all relations $<$ representing syntactic choice functions.

We call an arbitrary binary relation $<$ between sentences *admissible* (for the representation of entrenchments) if and only if it is irreflexive and satisfies (SV) and (E&C). Only admissible relations can possibly qualify as relations of epistemic entrenchment. We shall use these properties freely, without necessarily indicating the use of Irreflexivity and (SV).

If we substitute $\top = \wedge \emptyset$ for χ in (E&C) and apply (SV), we get that $\phi < \psi$ implies $\phi \wedge \psi < \top$. This is a relational way of expressing the fact that $\delta(\{\phi, \psi\}) = \{\phi\}$ implies $\delta(\{\phi, \psi\}) \neq \emptyset$.

---

[344] Formally, a relation $<$ represents a choice function σ if $< = \mathcal{P}_2(\sigma)$ (compare Definition 5 in Chapter 6).

[345] Because it is based on choices in two-element sets, this is basically a 'competitive interpretation' of entrenchment. In some contexts it is equivalent with a 'minimal change interpretation'. See Rott (1992c).

As further consequences of (SV) and (E&C), we get the following conditions which will prove to be useful for the derivation of more interesting results.

(1) $\phi \wedge \psi < \psi$ iff $\phi < \psi$ *(Conjunctiveness)*

(2) $\phi \wedge \psi < \psi \wedge \chi$ iff $\phi < \psi \wedge \chi$

(3) $\phi \wedge \psi < \chi \wedge \xi$ and $\phi \wedge \chi < \psi \wedge \xi$ iff $\phi < \psi \wedge \chi \wedge \xi$

(4) $\phi \not< \phi \wedge \psi$

Our choice-theoretic interpretation had made it clear already that (1) – (4) must hold. Let ϕ be interpreted as the conjunction of the elements of F, and let $\psi = \bigwedge G$, $\chi = \bigwedge H$ and $\xi = \bigwedge H'$. In (1), both sides encode that $\delta(F \cup G) \neq \emptyset$ and $G \cap \delta(F \cup G) = \emptyset$; in (2), both sides encode that $\delta(F \cup G \cup H) \neq \emptyset$ and $(G \cup H) \cap \delta(F \cup G \cup H) = \emptyset$; in (3), both sides encode that $\delta(F \cup G \cup H \cup H') \neq \emptyset$ and $(G \cup H \cup H') \cap \delta(F \cup G \cup H \cup H') = \emptyset$; condition (4) means that it is impossible that $\delta(F \cup G) \neq \emptyset$ and both $(F \cup G) \cap \delta(F \cup G) = \emptyset$. But now we are looking for an axiomatization of choice-based entrenchments, and for this reason we have to verify whether (1) – (4) are actually derivable (SV) and (E&C). (A proof of this is given in Appendix E.)

Let us further take to the records that all admissible relations are asymmetric. For suppose for reductio that both $\phi < \psi$ and $\psi < \phi$. Then, by the Conjunctiveness condition (1) and by (SV), we have both $\phi \wedge \psi \wedge \phi < \psi$ and $\phi \wedge \psi \wedge \psi < \phi$. Thus, by (E&C), $\phi \wedge \psi < \psi \wedge \phi$, contradicting Irreflexivity.

The following conditions of non-dominance are also valid on our choice-theoretic interpretation of $<$:

(ND)   There is an i such that $\phi_1 \wedge \cdots \wedge \phi_n \not< \phi_i$

or equivalently,

(ND')   There is an i such that $\phi_1 \wedge \cdots \wedge \phi_{i-1} \wedge \phi_{i+1} \wedge \cdots \wedge \phi_n \not< \phi_i$

That (ND) is valid can be seen as follows. For each i, $\phi_1 \wedge \cdots \wedge \phi_n < \phi_i$ means that $\delta(\phi_1, \ldots, \phi_n) \neq \emptyset$ but $\phi_i \notin \delta(\phi_1, \ldots, \phi_n)$. And obviously this cannot be the case for every i. The case for (ND') is analogous. By (1) and (SV), (ND') is equivalent with (ND) and thus one of them is superfluous. Furthermore, we can say this:

Lemma 44 *Given (E&C), (ND') is equivalent with the asymmetry of $<$.*

We are now going to verify that the conditions of admissibility are sufficient to prove that the complicated most general translation of $G \cap \delta(F \cup G) = \emptyset$ is equivalent with the much simpler standard translation.

Lemma 45 *Let $<$ be an admissible relation, and let two sets of sentences F and G be given. The most general translation of $G \cap \delta(F \cup G) = \emptyset$ says that for any two sequences $F_1, \ldots, F_k$ and $G_1, \ldots, G_k$ such that $G_1 \cup \cdots \cup G_k = G$ and $F_i \cup G_i = F \cup G$ for every $i \leq k$, the following condition holds:*

$$\bigwedge(F \cup G) \not< \top \quad \text{or} \quad \bigwedge F_i < \bigwedge G_i \text{ for every } i \tag{†}$$

This translation is equivalent with the standard translation stating that

$$\bigwedge(F \cup G) \not< \top \quad \text{or} \quad \bigwedge F < \bigwedge G \qquad\qquad (\ddag)$$

8.2.5   *Refusing to choose*

In Section 6.1.2 we have fixed rationality criteria for agents that may refuse to choose even when presented with a non-empty menu. This can happen when all elements in the menu are absolutely unacceptable for an agent. Conditions concerning such 'taboo sets' were the first criteria involving varying menus, and the only criteria of that sort we accepted as universally valid. Only as the limiting case can agents be expected to be 'virtually successful', what means that the only sentences that are never discarded are the logical truths (i.e., $\delta(F)$ is a non-empty subset of F unless $F \subseteq Cn(\emptyset)$).

The relevant constraints translate into the following postulates for entrenchment relations.

Observation 46 *The conditions ($\emptyset$1), ($\emptyset$2) and (Virtual Success) for the syntactic choice function δ translate into*

(E$\emptyset$1)   *If $\phi < \top$, then $\phi \wedge \psi < \top$*

(E$\emptyset$2)   *If $\phi \wedge \psi < \top$, then $\phi < \psi$ or $\psi < \top$*

(EVSuccess)   *If $\phi \notin Cn(\emptyset)$, then $\phi < \top$*

A way of expressing the combined constraint ($\emptyset$) for δ in the language of entrenchments is

(E$\emptyset$)   If $\phi \wedge \psi < \top$, then $\phi < \psi$ or $\psi \wedge \chi < \top$

Here is another set of principles useful for further derivations. Any admissible relation that satisfies (E$\emptyset$1) and (E$\emptyset$2)—we call such relations *fully admissible relations*—validates the following conditions.

(5)   If $\phi < \psi$, then $\phi < \top$

(6) $\phi \wedge \psi < \top$ if and only if $\phi < \top$ or $\psi < \top$

(7)   If $\phi < \top$ and $\psi \not< \top$, then $\phi < \psi$

Condition (7) implies that all sentences ψ which are not less than $\top$ under $<$ are comparable with one another in the sense that they have identical sets of dominated and dominating elements. They are both dominated by no sentences at all.[346] Condition (7) is called *Top equivalence* in Rott (1992b) and treated as an optional extra in that paper. But it seems more natural to take it as a standard assumption, given the present interpretation of epistemic entrenchment.[347]

8.2.6   *Not having to choose*

In Section 6.1.1 we have specified rationality criteria for an agent that does not really have to choose, because some of the options offered in the menu are absolutely satisfactory.

---

[346] $\top$ is less to nothing. Clearly $\top = \bigwedge \emptyset$. Let $\phi = \bigwedge F$. Given our standard translation, $\top < \phi$ means that there is some element of $\delta(F \cup \emptyset)$ which is not in F. Since $\delta(F) \subseteq F$, this is not possible. Can a sentence ψ such that $\psi \not< \top$ be less than anything else? No, by (5).

[347] Compare the remark on the top of p. 55 of Rott (1992b).

Let us adopt the standard assumption of Chapter 7 that choice function δ is faithful to $L - K$. (Faith1) says that if F is not a subset of K, then $\delta(F)$ does not intersect with K. By the above translation, this means

$$\text{If } F \not\subseteq K, \text{ then } \bigwedge F \not< \top \text{ or } \bigwedge F < \bigwedge(F \cap K)$$

Taking $\phi = \bigwedge(F - K)$, $\psi = \bigwedge(F \cap K)$, and using the fact that $\phi \wedge \psi < \phi$ iff $\psi < \phi$, we can reformulate this as

(EFaith1)   If $\phi \notin K$ and $\psi \in K$, then $\phi \not< \top$ or $\phi < \psi$

(Faith2) with respect to $L - K$ says that every element of F which is not in K gets selected from F by δ. The translation is

$$\text{If } \phi \in F - K, \text{ then } \bigwedge F < \top \text{ and } \bigwedge F \not< \phi$$

which may be simplified, using the same abbreviation as before, to

(EFaith2)   If $\phi \notin K$, then $\phi < \top$ and $\psi \not< \phi$

Observation 47 *The conditions (Faith1) and (Faith2) with respect to $L - K$ for the syntactic choice function δ translate into (EFaith1) and (EFaith2)*

The combined constraint (Faith) for δ simply corresponds to

(EFaith)   If $\phi \notin K$ and $\psi \in K$, then $\phi < \psi$

Condition (EFaith) is called K-*representation* in Rott (1992b) and treated as an optional extra in that paper. We shall point out where we use conditions of Faith in what follows.

8.3   Postulates for epistemic entrenchment

As I have announced before, my aim in this chapter is to show that relations of epistemic entrenchment are really just the preferences revealed by syntactic choice functions with logical constraints. Originally, however, entrenchment was not introduced from a choice-theoretic point of view. In this section I present the 'rationality postulates' for entrenchment relations that have been given, on entirely independent grounds, by Gärdenfors (1988), Gärdenfors and Makinson (1988), Rott (1992b) and Gärdenfors and Rott (1995).[348] Later I address the question of how this independently issued set of postulates relates to postulates that can be motivated directly through choice-theoretic rationality.

The following presentation is essentially based on Rott (1992b). As mentioned before, Gärdenfors and Makinson's account in terms of a non-strict relation $\leq$ can be recovered by taking the converse of complement of our strict relation $<$, i.e. by defining $\phi \leq \psi$ as $\psi \not< \phi$.

---

[348] Gärdenfors and Makinson (1994) give an account of nonmonotonic reasoning on the basis of *expectation orderings* the axioms of which are almost identical with the axioms for epistemic entrenchment relations for full beliefs. See footnote 351 below.

Definition 18 A *GEE-relation* (or *relation of generalized epistemic entrench-ment*) is a relation $<$ over the set L of all sentences satisfying the following basic set of postulates.

(EE1)	$\phi \not< \phi$	*(Irreflexivity)*
(EE2$^\uparrow$)	If $\phi < \psi$ and $\psi \vdash \chi$, then $\phi < \chi$	*(Continuing up)*
(EE2$^\downarrow$)	If $\phi < \psi$ and $\chi \vdash \phi$, then $\chi < \psi$	*(Continuing down)*
(EE3$^\uparrow$)	If $\phi < \psi$ and $\phi < \chi$, then $\phi < \psi \wedge \chi$	*(Conjunction up)*
(EE3$^\downarrow$)	If $\phi \wedge \psi < \psi$, then $\phi < \psi$	*(Conjunction down)*

The relation $<$ is called a GEE-relation *with comparability* (or *relation of stan-dard epistemic entrenchment*, in short *SEE-relation*) if it in addition satisfies

(EE4)	If $\phi < \psi$, then $\phi < \chi$ or $\chi < \psi$	*(Modularity)*

A justification of these postulates will be given later. Just notice now that we have three different kinds of postulates here. (EE1) and (EE4) are *purely struc-tural* conditions. (EE3$^\uparrow$) and (EE3$^\downarrow$) may also be called *structural conditions*. They refer to the internal structure of the items (= sentences) related by the relation $<$, but as we have seen, the operation of forming conjunctions can be interpreted as just packing items together in a menu for choices. The Continuing up and down conditions (EE2$^\uparrow$) and (EE2$^\downarrow$) in contrast are not suitable for an interpretation in terms of rational choice. They are genuinely *logical conditions*. (EE2$^\uparrow$) and (EE2$^\downarrow$) were used by Alchourrón and Makinson (1985, 1986) in the context of so-called safe contractions.

Note that the conditions (EE1) – (EE3$^\downarrow$) are in Horn form, and thus the class of GEE-relations is closed under intersections and under unions of $\subseteq$-chains. If we wished to avoid any reference to particular connectives in the object language, we could alternatively employ the following versions of the postulates:

(EE$^\uparrow$) If $\phi < \psi$ for every ψ in a non-empty set H and $H \vdash \chi$, then $\phi < \chi$

(EE$^\downarrow$) If $\phi < \psi$ and $\{\psi, \chi\} \vdash \phi$, then $\chi < \psi$

It is easy to check that (EE$^\uparrow$) is equivalent to the combination of (EE2$^\uparrow$) and (EE3$^\uparrow$) and that (EE$^\downarrow$) is equivalent to the combination of (EE2$^\downarrow$) and (EE3$^\downarrow$) (Rott 1992b, Lemma 3). Here I continue using the language based on conjunc-tions for a while since it allows us to tie together beliefs to packages that can serve as the menus for syntactic choices.

We note the following simple consequences of the postulates for GEE-relations:

Lemma 48 *Suppose the relation $<$ over L satisfies (EE1) – (EE3$^\downarrow$). Then it also has the following properties:*

(i)   *If $Cn(\phi) = Cn(\psi)$, then: $\phi < \chi$ iff $\psi < \chi$, and $\chi < \phi$ iff $\chi < \psi$*
                                                                      (Extensionality)

(ii)   *If $\phi \wedge \chi < \psi \wedge \chi$, then $\phi < \psi$*

(iii)   *If $\phi < \psi$ and $\chi < \xi$, then $\phi \wedge \chi < \psi \wedge \xi$*

(iv)   *If $\phi < \psi$ then not $\psi < \phi$*   *(Asymmetry)*

(v)   *If $\phi < \psi$ and $\psi < \chi$, then $\phi < \chi$*   *(Transitivity)*

(vi) $\phi < \phi \lor \psi$ *iff* $\phi \lor \neg\psi < \phi \lor \psi$

The GEE-relations of Gärdenfors and Makinson (1988) are required to satisfy a number of postulates that go beyond (EE1) – (EE3$^\downarrow$). Most importantly, we must add to the basic set of postulates the postulate of Modularity (also known as Virtual Connectedness), that is, Gärdenfors and Makinson use entrenchment relations with comparability. It is not difficult to specify concrete instances of relations that satisfy (EE1) – (EE4),[349] so clearly this set of postulates is consistent.

Modularity guarantees that any two sentences in our language are *comparable* with respect to $<$.[350] We have defined in Section 6.3 that two items ϕ and ψ are *tied-or-identical* (or *equally good*) with respect to a strict preference relation $<$ just in case they stand in the same $<$-relationship to all other items in the field, that is, just in case $\phi < \chi$ iff $\psi < \chi$, and $\chi < \phi$ iff $\chi < \psi$, for all χ (compare p. 157). We call two items ϕ and ψ *comparable* with respect to a strict preference relation $<$ if and only if either ϕ is preferable to ψ, or the reverse, or they are tied-or-identical, all of this understood of course with respect to $<$. If $<$ is modular, this makes sure that any two items ϕ and ψ are comparable in exactly that sense.

Modularity is a very powerful condition. We will see that it corresponds to the AGM conditions ($\dot{-}8$) and ($*8$) for belief change, to Rational Monotony in nonmonotonic reasoning, and to condition (IV) for syntactic choices. For the theory of GEE-relations itself, it has the following consequences:

Observation 49 *Suppose the relation $<$ satisfies (EE1) and (EE4). Then*
  (i) $<$ satisfies (EE2$^\uparrow$) iff it satisfies (EE2$^\downarrow$);
  (ii) if $<$ satisfies (EE3$^\uparrow$), then it satisfies (EE3$^\downarrow$); if $<$ is asymmetric and satisfies (EE3$^\downarrow$), then it also satisfies (EE3$^\uparrow$).

GEE-relations with comparability can thus be characterized succinctly, e.g., by postulates (EE1), (EE2$^\uparrow$), (EE3$^\uparrow$) and (EE4). There are in addition minimality and maximality conditions in Gärdenfors and Makinson (1988):

(EE5)   If $K \neq L$, then $\phi \in K$ iff $\psi < \phi$ for some ψ *(Minimality)*

(EE6)   If $\phi \notin Cn(\emptyset)$, then $\phi < \psi$ for some ψ *(Maximality)*

---

[349]Manageable finite representations by means of 'bases' for SEE-relations are introduced in Rott (1991b, Section 3), a more general point of view is taken in Rott (2000a).

[350]The advantages of working with strict relations seem to me overwhelming; see Rott (1992b, Section 4; 1993, footnote 1). I base my generalization of the theory of epistemic entrenchment developed by Gärdenfors and Makinson on a strict relation $<$ that is the converse complement of their non-strict relation $\leq$. This is possible because according to their theory all items are comparable with respect to $\leq$ (and their equivalence relation $\leq \cap \leq^{-1}$ really is the relation of being-tied-or-incomparable). If $\leq$ were not connected, the converse complement $<$ would end up with cycles, i.e., cases where both $\phi < \psi$ and $\psi < \phi$ hold. Obviously, the theory presented in this chapter does not at all presuppose full comparability.

These conditions relate $<$ to the set of currently held beliefs K and the background logic Cn and are thus not purely structural conditions. (EE5) and (EE6) are considerably stronger than (EFaith) and (Top Equivalence), respectively. For the present analysis only one half of (EE5), namely

(EE5′)   If $\psi < \phi$ for some ψ, then $\phi \in K$

is interesting, because the converse direction follows from (EFaith).

The liberalization achieved by the generalized concept of epistemic entrenchment meets some serious objections against the strength of the original Gärdenfors-Makinson postulates.[351] The following features of their account appear to be unnecessarily restrictive.

(i)   Since Virtual Connectivity follows from (EE1) – (EE4), it is forbidden to classify two sentences ϕ and ψ as incomparable.

(ii) Due to (EE5), it is impossible to have a plausibility gradation of the sentences which are not included in the current belief set K.

(iii) Due to (EE6), it is impossible to assign to sentences that are not logically true a degree of epistemic entrenchment which is as high as the degree of logical truth.

A rejoinder to the first and perhaps most important of these objections was first offered by Lindström and Rabinowicz (1991) who replace Gärdenfors and Makinson's conjunctiveness condition, according to which $\phi \leq \phi \wedge \psi$ or $\psi \leq \phi \wedge \psi$ for every ϕ and ψ, by a weaker condition stating that if $\phi \leq \psi$ and $\phi \leq \chi$ then $\phi \leq \psi \wedge \chi$. However, Lindström and Rabinowicz cannot apply a contraction recipe based on entrenchments that is as simple as that of Gärdenfors and Makinson. In contrast, we shall see that the relations defined above—which are *not* required to satisfy (EE4) – (EE6)—are appropriate to meet the three objections against the Gärdenfors-Makinson approach, use simple construction recipes and at the same time allow us to model (a 'sceptical' interpretation of) relational belief changes in the style of Lindström and Rabinowicz, as well as a form of iterated belief change. A thoroughgoing comparison between the Lindström-Rabinowicz approach and the approach of Rott (1992b) which is similar to the one presented here has recently been provided by Cantwell (2001).

One important advantage of the Gärdenfors-Makinson concept of epistemic entrenchment is that the *amount of information* that needs to be specified in order to determine a SEE-ordering is comparatively small if the belief set in question is logically finite. A logically finite belief set can be described via its set of *atoms* (maximal conjunctions of possibly negated sentential variables) or via its set of *dual atoms* (maximal disjunctions of possibly negated sentential variables).

---

[351] The expectation orderings utilized in Gärdenfors and Makinson (1994) are located 'between' generalized relations of epistemic entrenchment and the more demanding ones of Gärdenfors and Makinson (1988) in that they are connected, i.e., satisfy (EE4), but are not required to satisfy the minimality and maximality conditions (EE5) and (EE6).

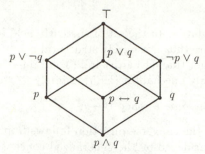

FIG. 8.2. The eight element Boolean algebra

For example consider the propositional language L with two atoms p and q, let Cn be classical propositional logic and the belief set $K = Cn(\{p,q\})$. The logic Cn divides this set into equivalence classes which can be described as an 8-element Boolean algebra. This algebra can be depicted by a Hasse diagram as in Fig. 8.2 (lines upwards mean logical implication). The dual atoms are the topmost elements except for $\top$, in this case $p \vee \neg q$, $p \vee q$ and $\neg p \vee q$, or more exactly, the corresponding equivalence classes in K under Cn. Gärdenfors and Makinson (1988) show that the postulates (EE1) – (EE6) introduce constraints on the SEE-ordering $<$ in such a way that the number of orderings satisfying these postulates will be fairly small. The following result shows that the number of total orderings of epistemic entrenchment over a Boolean algebra with 2^n elements is only $n!$:

Observation 50 *Let K be a logically finite belief set, and let $DA(K)$ be the set of all dual atoms of K. Then any two relations $<$ and $<'$ over L that satisfy (EE1) – (EE6) and agree on all pairs of elements in $DA(K)$ are identical.*

The theorem implies that if the SEE-ordering of the dual atoms $p \vee \neg q$, $p \vee q$ and $\neg p \vee q$ in the Hasse diagram in Fig. 8.2 is specified, then the relationships between the remaining elements in $K = Cn(\{p,q\})$ is fully determined. For example, if we assume that $(p \vee \neg q) < (p \vee q) < (\neg p \vee q)$, then it follows from (EE1) – (EE6) that the ordering of the eight elements is $p \wedge q \sim p \sim (p \vee \neg q) \sim (p \leftrightarrow q) < q \sim (p \vee q) < (\neg p \vee q) < \top$. Each element ϕ in K is as entrenched as the minimal element of $DA(\phi)$ is, where $DA(\phi)$ is the set of dual atoms of K that are implied by ϕ.

The computational interpretation of this result is that in order to specify a SEE-relation over a belief set K containing 2^n elements, one needs only feed in the (transitive and connected) ordering of n elements from K.[352]

Observation 50, however, does not hold any more if the generalized notion of epistemic entrenchment as characterized by the reduced postulate set (EE1) – (EE3$^\downarrow$) is used. As an example, consider again $K = Cn(\{p,q\})$ and a generalized entrenchment relation $<$ of which we just know that no element of $\{p \vee q, p \vee \neg q, \neg p \vee q\}$ is better entrenched than any other. This information is compatible

---

[352]But cf. Doyle (1992, p. 43).

not only with $p \not< \neg p \vee q$, but also with $p < \neg p \vee q$.[353] In Gärdenfors and Makinson's theory the latter would be impossible.

The SEE-relations of Gärdenfors and Makinson (1988), as well as the GEE-relations of Rott (1992b), have nice axiomatizations. The motivation of the axioms, however, is not immediately transparent and derives mainly from the representation theorems relating these axioms to certain desired collections of postulates for belief contraction. The pertinent representation theorems available for SEE-relations (Gärdenfors and Makinson 1988, Theorems 4 and 5) and GEE-relations (Rott 1992b, Theorem 2, parts (i) and (ii)) relate *packages* of axioms: All axioms for SEE-relations taken together are shown to correspond to all axioms for AGM contractions taken together, and similarly, all axioms for GEE-relations taken together are shown to correspond to all axioms for 'preferential contractions' taken together. This is certainly nice, but without an independent justification of the axioms for entrenchments, it does not yield a philosophical underpinning for AGM-style or preferential belief change. In contrast, the view taken in this chapter is that the nature of entrenchment relations consists their being revealed preferences of syntactic choice functions, and that their properties should be derived from the properties of the latter. I aim at giving a direct and perspicuous interpretation of axioms for entrenchment, and at establishing correspondences not of packages of axioms, but of single axioms *taken one by one*. From now on, I use the labels *GEE-relation* and *SEE-relation* to refer to the axiomatizations of entrenchment given in this section. When I simply talk of *entrenchment* (without any more technical labelling), I have in mind relations that 'come from' or 'encode' syntactic choices of the sort investigated in Chapter 7.

8.4 The postulates translated

8.4.1 *From coherent choice to epistemic entrenchment*

Let us first find out how the widely accepted coherence constraints for choice functions that we discussed in Chapter 6 translate into the language of entrenchments. After this, we can see whether these conditions are already taken account of by the usual requirements for GEE- or SEE-relations. Applying the translation specified in Section 8.2.3, we get the following result.

Observation 51 *The coherence criteria (I) – (IV), (I⁻), (II⁺) and (IV⁺) for rational choice translate into the following constraints on fully admissible preference relations.*

---

[353] How does the latter possibility square with (LP1) and (LP2)? On the one hand, $p < \neg p \vee q$ means that $\delta(\{p, \neg p \vee q\}) = \{p\}$. From this and (LP1) and (LP2), we get that $\delta(\{p \vee q, p \vee \neg q, \neg p \vee q\})$ must contain at least one of $p \vee q$ and $p \vee \neg q$, but must not contain $\neg p \vee q$. On the other hand, that the dual atoms are not related by $<$ means that $\delta(\{p \vee q, p \vee \neg q\}) = \{p \vee q, p \vee \neg q\}$, $\delta(\{p \vee q, \neg p \vee q\}) = \{p \vee q, \neg p \vee q\}$ and $\delta(\{p \vee \neg q, \neg p \vee q\}) = \{p \vee \neg q, \neg p \vee q\}$ (assuming that nothing is taboo). Thus it seems that $\delta(\{p \vee q, p \vee \neg q, \neg p \vee q\})$ must be $\{p \vee q, p \vee \neg q, \neg p \vee q\}$. But we cannot infer this, since we have neither (II) nor (IV) in the context of 'preferential reasoning' (cf. Section 4.4 above and Rott 1992b).

(EI)	If $\phi < \psi$, then $\phi \wedge \chi < \psi$
(EII)	If $\phi \wedge \psi < \chi$, then $\phi < \chi$ or $\psi < \chi$
(EIII)	If $\phi < \psi \wedge \chi$, then $\phi < \psi$
(EIV)	If $\phi \wedge \psi < \chi$ and $\psi \not< \phi \wedge \chi$, then $\phi < \chi$
(EI$^-$)	If $\phi \wedge \psi < \chi$ and $\phi < \psi$, then $\phi \wedge \chi < \psi$
(EII$^+$)	If $\phi \wedge \psi < \chi \wedge \xi$ and $\chi < \top$ and $\xi < \top$, then $\phi < \chi$ or $\psi < \xi$
(EIV$^+$)	If $\phi < \top$, then $\psi \not< \phi$

Having found out about the correct translations,[354] we have a short look of the relative strengths of these conditions.

Lemma 52 (i) *Asymmetry, (E&C) and (EIV) taken together imply (EII);*

(ii) *Asymmetry, (E&C) and (EIV) taken together imply (EIII);*

(iii) *(EI) implies (EI$^-$);*

(iv) *Asymmetry, (SV), (E&C), (E∅2) and (EII$^+$) taken together imply (EII);*

(v) *(EII) and (EIII) taken together imply (EII$^+$);*

(vi) *(E&C), (E∅2) and (EIV$^+$) taken together imply (EIV);*

(vii) *(SV), (E&C), (EI) and (EIII) taken together imply Transitivity;*

(viii) *Asymmetry, (SV), (E&C), (EI) and (EIV) taken together imply Modularity;*

(ix) *Asymmetry, (SV), (E&C) and Modularity taken together imply (EIV).*

It should not worry us that although (I) implies (∅2) and (III) implies (∅1), their translations into the language of entrenchment do not mirror these implications. We have used the conditions for refusals of choice (which we have been regarding as standard, obligatory conditions) in order to simplify (EI) and (EIII). It is not to be expected, that our streamlined versions of (EI) and (EIII) still imply the conditions which have been presupposed as valid when transforming the original, more awkward translations of (I) and (III).

As far as I am aware, (EII) is a simple condition that has not been discussed in the literature on entrenchment before. It looks plausible, but it is doubtful whether it should be satisfied in all situations; see the problematic example in Rott (1992b, pp. 56–57).

It is perhaps worth pointing out that just the two postulates that are part of both the standard and the choice-theoretic account of entrenchment, viz., (EI) (*alias* (EE2$^{\downarrow*}$)) and (EIII) (*alias* (EE2$^{\uparrow*}$)), imply transitivity when taken together with (E&C).

Parts (viii) and (ix) of the lemma show that the role of (EIV) comes close to that of Modularity. However, since (EI) is indispensable in part (viii), we can

---

[354] A consequence of the observation is this: If δ satisfies (∅1), (∅2), (CP1), (CP2) and (x), then $< = \mathcal{P}_2(\delta)$ satisfies (Ex). Here 'x' stands for 'I', 'II', etc., and (CP1) and (CP2) are the conditions for syntactic choices stated in note 8.1. As pointed out earlier, a prerequisite of our translation is that the special choice functions δ employed here satisfy the two conjunctive principles (CP1) and (CP2).

see that in the context of admissible relations, Modularity is even stronger than (EIV).

(EIV$^+$) is an extremely strong condition. It in effect tells us that there are only two levels under the relation $<$. In the upper level we find $\top$ and all other sentences which are 'taboo', in the lower level we find everything else. If (5), (7), and (EIV$^+$) hold, then $\phi < \psi$ just in case $\phi < \top$ and $\psi \not< \top$.

Relations satisfying—at least some interesting selection from—these postulates can be taken to exhibit features of rational choice. Notice that nothing in what we have used for the derivation of the entrenchment properties so far makes any reference to logic.

As we discussed at length in the last chapter, the coherentist way of changing beliefs involves choices, either choices between models or choices between sentences. It is essential that the agent's mental state is represented by a single choice function that determines the agent's changes of mind in response to all potential inputs. For these *changes* to be coherent, the *choices* involved should be coherent. And for this it is particularly important that we can say in which cases choices between sentences are in fact being made in a relational way. Given that a syntactic choice function δ satisfies ($\emptyset$1) (compare Lemma 14, part (iv)), this means: In which cases can we use the base preferences $< = \mathcal{P}_2(\delta)$, that is, *entrenchments*, to reconstruct the choices as being determined by just these preferences? We need to state properties of the base preferences $<$ which guarantee that $\delta = \mathcal{D}(\mathcal{P}_2(\delta))$.

We have always presupposed that δ is ω-covering and that it satisfies the limiting case conditions ($\emptyset$1) and ($\emptyset$2) for refusals of choice; we will neglect an explicit mentioning of these requirements for a while. The results of Chapter 6 that are most relevant now are Lemma 14 and Corollary 15. They tell us that if a choice function δ satisfies (I) and (II$^\infty$), then it is relational with respect to $< = \mathcal{P}_2(\delta)$, that is, it holds that for all $F \in \mathcal{X} - Pow(\overline{R})$

$$\delta(F) = \min_<(F) \tag{\dagger}$$

This equation characterizes the method of liberal minimization. Remember that $\overline{R}$ is the set of all beliefs that are 'taboo' for δ, that are never given up. Moreover, if δ satisfies (I) and (II), then it is finitely relational with respect to $<$, that is, ($\dagger$) holds for all finite $F \in \mathcal{X} - Pow(\overline{R})$.

It makes a difference how exactly we employ syntactic choice functions. Remember that we have used two ways to check whether a belief ψ in K remains in the contraction of K with respect to ϕ. The principal Definition 8 looks whether $\phi \lor \psi$ is contained in $\delta(Cn(\phi))$, and the finitary version, Definition 10—equivalent by virtue of (LP2)—looks whether $\phi \lor \psi$ is contained in $\delta(\{\phi, \phi \lor \psi\})$ or in $\delta(\{\phi \lor \psi, \phi \lor \neg\psi\})$.

Now if we want to use the principal Definition 8 for syntactically choice-based contractions, the domain of δ must include infinite sets of sentences, and hence we need (II$^\infty$) in order to guarantee the relationality of the choice function. However, (II$^\infty$) is not translatable into a condition in terms of entrenchments, since we do

not have infinite conjunctions in our language. So we take recourse to the finitary
Definition 10. Although this definition itself refers only to choices in two-element
sets, our coherentist account takes the view that the very same choice function
δ needed for this definition should be appropriate for choices in sets of arbitrary
finite cardinality. It does not make sense to think of mental states of agents that
can choose rationally only in situations representable by two-element menus. So
we insist that the domain of the syntactic choice function must not be restricted
to sets with exactly two elements. The results of Chapter 6 then tell us that
the syntactical choice function δ has to satisfy (I) and (II) in order to be finitely
relational. But this means, as we have seen in this chapter, that the revealed base
preferences $<$ have to satisfy (EI) and (EII), or equivalently and more compactly,
that $<$ has to satisfy

(EFinRelationality) $\phi \wedge \psi < \chi$ iff $\phi < \chi$ or $\psi < \chi$

Only if $<$ satisfies this condition, can we be sure that it is suitable for determining, via (†), the same selections as the choice function δ from which it has
originally been derived.

8.4.2 *Logical constraints for entrenchments*

Up to now we have been neglecting in this chapter the connection between choice
and logic—in so far as logic goes beyond the trivial logic of conjunctions.[355] But
remember that syntactic choices are subject to logical constraints over and above
the constraints of rational choice. Let us now introduce corresponding logical
conditions for revealed preferences and hope that they match up with postulates
that are known from the existing literature on epistemic entrenchment. I want
to keep the constraints induced by logic wholly separate from the constraints
induced by choice rationality, and will therefore only rely on background conditions which characterize fully admissible relations between sentences, namely
Irreflexivity, (SV), (E&C), (E∅1) and (E∅2). I am now going to translate the
central logical constraints (LP1) and (LP2) of Section 7.4.

Observation 53 *The logical constraints (LP1) and (LP2) for syntactic choice
functions translate into the following constraints for entrenchment relations.*

(ELP1)   *If $\phi \in Cn(\psi)$ and $\phi \wedge \chi < \psi$, then $\psi \wedge \chi < \phi$*

(ELP2)   *If $Cn(\phi \wedge \psi) = Cn(\phi \wedge \chi)$ and $\chi < \phi$, then $\psi < \phi$*

Also note the following fact, which can be recognized as a consequence of
(ELP1) if we put $\chi = \top$ and recall that the relation $<$ is asymmetric.

(ELP1⁻) If $\phi \in Cn(\psi)$, then $\phi \not< \psi$

This entails that if $\phi \in Cn(\emptyset)$, then $\phi \not< \top$.

---

[355]This is not quite true. Virtual Success is a constraint that mentions Cn.

The logical constraints for syntactic choices in Chapter 7.4 are quite nice and handy. Unfortunately, their translations into the language of entrenchments do not look particularly appealing. I now want to show how they connect with the nicer logical postulates of the existing theories of epistemic entrenchment. The most interesting question thus is how (ELP1) and (ELP2) relate to (Extensionality). The surprising answer is that provided the admissibility conditions are satisfied, the two conditions taken together are precisely equivalent with Extensionality.

Observation 54 *For all admissible relations $<$, the conjunction of (ELP1) and (ELP2) is equivalent with (Extensionality).*

Since Extensionality implies (SV), we can call irreflexive relations $<$ that are extensional and satisfy (E&C) *logically admissible* preference relations. Logically admissible relations are required to meet far less demanding conditions than the standard entrenchment relations of Gärdenfors and Makinson (1988) or the generalized ones of Rott (1992b). Still they have a number of non-trivial properties:

Lemma 55 *Let $<$ satisfy Irreflexivity, Extensionality and (E&C). Then it also satisfies the following properties:*

(i) $\phi < \psi$ iff $\phi \wedge \psi < \psi$ *(Conjunctiveness)*

(ii) $\phi < \psi$ iff $\psi \supset \phi < \psi$ ·   *(Conditionalization)*

(iii) If $\phi \vdash \psi$ then $\psi \not< \phi$ *(GM-Dominance)*

(iv) Not both $\phi \wedge \psi < \phi$ and $\phi \wedge \psi < \psi$ *(GM-Conjunctiveness)*

(v)   If $\phi < \psi$ and $\phi < \chi$ and $\phi \vdash \psi \wedge \chi$ then $\phi < \psi \wedge \chi$

                                                  (Weak Conjunction Up)

(vi) If $\phi < \psi$ and $\psi \vdash \chi$ and $\phi \wedge \chi \vdash \psi$ then $\phi < \chi$ *(Weak Continuing Up)*

(vii) If $\phi < \psi$ and $\chi \vdash \phi$ and $\phi \vdash \psi \supset \chi$ then $\chi < \psi$ *(Weak Continuing Down)*

The logical constraints for syntactic choice functions that played a central role in Chapter 7 have now turned out to be captured precisely by the Extensionality condition for their revealed base preferences. This single condition thus encodes the full 'logic component' of the constraints for syntactic choices. All other conditions derive from coherence criteria for rational choices that have nothing whatsoever to do with the dictates of a background logic (be it classical or non-classical), but only with the fact that items of belief can be packed together into menus by means of conjunctions. We have accomplished a neat separation of the contribution of *logic* and the contribution of *choice* to the concept of epistemic entrenchment—and to the problem of belief change in general.

8.4.3 *From epistemic entrenchment to coherent choice*

By straightforward application of the translation specified in Section 8.2.2, we obtain the following constraints on choice functions.

Observation 56 *The constraints (EE1) – (EE4) on entrenchment relations translate into the following coherence criteria for rational choice.*

(CC1) $\delta(F) = \emptyset$ *or* $F \cap \delta(F) \neq \emptyset$

(CC2$^\uparrow$) *If* $G \cap \delta(F \cup G) = \emptyset$ *and* $H \subseteq Cn(G)$, *then* $H \cap \delta(F \cup H) = \emptyset$

(CC2$^\downarrow$) *If* $G \cap \delta(F \cup G) = \emptyset$ *and* $F \subseteq Cn(H)$, *then* $G \cap \delta(G \cup H) = \emptyset$

(CC3$^\uparrow$) *If* $G \cap \delta(F \cup G) = \emptyset$ *and* $H \cap \delta(F \cup H) = \emptyset$, *then*
$\quad\quad (G \cup H) \cap \delta(F \cup G \cup H) = \emptyset$

(CC3$^\downarrow$) *If* $F \cap \delta(F \cup G \cup G) = \emptyset$, *then* $F \cap \delta(F \cup G) = \emptyset$

(CC4) *If* $G \cap \delta(F \cup G) = \emptyset$, *then* $H \cap \delta(F \cup H) = \emptyset$ *or* $G \cap \delta(G \cup H) = \emptyset$.

8.5 Implications

We now study the relative strengths of the postulates for choice and entrenchment. We do this both in the context of the theory of choice and in the context of the theory of entrenchment.

8.5.1 *Implications between choice postulates*

We begin by considering the status of the translations of entrenchment conditions into the theory of (syntactic) choice functions.

Observation 57 *(CC1) and (CC3$^\downarrow$) are vacuously satisfied. Furthermore, given (LP1) and (LP2), the following implications hold true for every choice function* δ:

(i)   (CC2$^\uparrow$) is equivalent with (III);

(ii) (CC2$^\downarrow$) is equivalent with (I);

(iii) (CC3$^\uparrow$) follows from (I);

(iv) (CC4) implies (IV), and follows from the conjunction of (I) and (IV).

Now some of the things that have emerged are striking. Although (EE3$^\downarrow$) is a non-trivial and indeed an important postulate in the theory of epistemic entrenchment, its translation is trivial. Part (iii) of Observation 57 shows that although (EE3$^\uparrow$) is logically independent from (EE2$^\downarrow$) in the theory of epistemic entrenchment, the translations are not. The explanation for these surprising facts is that *given the admissibility of* $<$, (EE3$^\downarrow$) does not add anything new, and (EE3$^\uparrow$) is just a weakening of (EE2$^\downarrow$). In the usual axiomatizations of epistemic entrenchment, however, there are no conditions corresponding directly to the admissibility of $<$.

Obviously, since (EE2$^\uparrow$) and (EE2$^\downarrow$) involve the logical consequence operation Cn, we need to make use of (LP1) and (LP2). Alternatively, we could have taken apart the logical and the choice-theoretic content of (EE2$^\uparrow$) and (EE2$^\downarrow$). The logical part can be assigned to (Extensionality) (see Observation 54). The choice-theoretic part then consists in the following versions of (EE2$^\uparrow$) and (EE2$^\downarrow$):

(EE2$^{\uparrow *}$) If $\phi < \psi \wedge \chi$, then $\phi < \psi$

(EE2$^{\downarrow *}$) If $\phi < \psi$, then $\phi \wedge \chi < \psi$

If we take it for granted that ϕ is in $Cn(\phi \wedge \psi)$, as is customary, then it is clear that $(EE2^{\uparrow *})$ and $(EE2^{\downarrow *})$ follow from their unstarred cousins $(EE2^{\uparrow})$ and $(EE2^{\downarrow})$. For the converse direction one has to employ (Extensionality).

Lemma 58 *Translated into choice-theoretic conditions, $(EE2^{\uparrow *})$ and $(EE2^{\downarrow *})$ become:*

$(CC2^{\uparrow *})$ If $(G \cup H) \cap \delta(F \cup G \cup H) = \emptyset$, then $G \cap \delta(F \cup G) = \emptyset$

$(CC2^{\downarrow *})$ If $G \cap \delta(F \cup G) = \emptyset$, then $G \cap \delta(F \cup G \cup H) = \emptyset$

The following equivalences do not depend on the logical conditions (LP1) and (LP2) for their proof.

Lemma 59 *$(CC2^{\uparrow})$ is equivalent to (III) and $(CC2^{\downarrow})$ is equivalent to (I).*

We can thus say that the entrenchment conditions $(EE2^{\uparrow *})$ and $(EE2^{\downarrow *})$ translate into the choice constraints (III) and (I), respectively.

Together with Observation 18, the second part of Lemma 59 and part (iv) of Observation 57 reveal *a new axiomatization of modularly rationalizable choice functions*: A choice function is rationalizable by a modular preference relation iff it satisfies $(CC2^{\downarrow *})$ and (CC4).[356]

Since (II^{∞}) follows from (IV), it is also implied by (CC4). (II^{∞}) and its finitary version (II) are very basic properties in the theory of rational choice, but they have as yet no systematic standing in the theory of epistemic entrenchment and belief revision.

8.5.2 *Implications between entrenchment postulates*

We know from the previous section and the translatability of choice conditions into conditions for entrenchment relations that certain equivalences must hold between the entrenchment conditions originally set up in Gärdenfors and Makinson (1988) and Rott (1992b), and the ones we derived from the concept of coherent choice. Still it is instructive to study in detail the interrelations on the side of entrenchment relations as well. In this section we relate the standard entrenchment postulates found in the literature to the choice-theoretically inspired ones we have derived above.

Observation 60 *(i) (SV) follows from $(EE2^{\uparrow})$ and $(EE2^{\downarrow})$*

(ii) (E&C) follows from $(EE2^{\uparrow})$, $(EE2^{\downarrow})$, $(EE3^{\uparrow})$ and $(EE3^{\downarrow})$;

(iii) (Asymmetry) follows from (EE1), $(EE2^{\downarrow})$ and $(EE3^{\uparrow})$;

(iv) (E∅1) follows from $(EE2^{\downarrow})$;

(v) (E∅2) follows from $(EE3^{\downarrow})$ and (EE4);

---

[356]This again presupposes that δ is 12-covering and that it satisfies (∅1). The preference relation is again the one expressing base preferences, see principle (E) of Section 8.1.—What we call modular rationalizability is mostly referred to as *transitive* rationalizability in the literature, since most authors make use of a non-strict relation $\leq$ which may be obtained from our relation $<$ by taking the converse complement.—Our crucial condition (CC4) is similar in spirit to conditions studied by Ranade (1985, 1987).

(vi) (ELP1) follows from (EE2$^\top$) and (EE2$^\downarrow$);

(vii) (ELP2) follows from (EE2$^\downarrow$) and (EE3$^\downarrow$);

(h) (EVSuccess) follows from (EE2$^\top$) and (EE6);

(i) (EI) is identical with (EE2$^{\downarrow*}$) and follows from (EE2$^\downarrow$);

(x) (EII) follows from (EE1), (EE3$^\top$) and (EE4);

(xi) (EIII) is identical with (EE2$^{\top*}$) and follows from (EE2$^\top$);

(xii) (EIV) follows from (EE2$^\downarrow$), (EE3$^\downarrow$) and (EE4).

(xiii) (EI$^-$) follows from (EE2$^\downarrow$);

(xiv) (EII$^+$) follows from (EE1), (EE2$^\top$), (EE3$^\top$) and (EE4);

(xv) (EIV$^+$) does not follow from any combination of standard postulates for epistemic entrenchment.

Observation 60 shows that the usual axiomatizations of entrenchment relations are sufficiently strong to yield all interesting constraints for preferences related to syntactic choices, but that there is no natural grouping of the usual axioms from the choice-theoretic point of view.

Observation 61 *(i) Extensionality follows from (ELP1) and (ELP2);*

(ii) (EE1) is Irreflexivity;

(iii) (EE2$^{\top*}$) is identical with (EIII);

(iv) (EE2$^{\downarrow*}$) is identical with (EI);

(v) (EE3$^\top$) follows from (E&C) and (EI);

(vi) (EE3$^\downarrow$) follows from (E&C) and (SV);

(vii) (EE4) follows from Irreflexivity, (SV), (E&C), (EI) and (EIV).

In view of the many different conditions we have come across in this chapter, one may ask what is 'the best' axiomatization for entrenchment. For entrenchment with comparabilities, I recommend to take the following group of five: Irreflexivity, Modularity, Extensionality, (E&C) and (EI). The first two postulates are *purely structural*, the third postulate is *purely logical*, the fourth postulate is the key to our *choice-theoretic* interpretation of entrenchment, and finally, the fifth one is also choice-theoretic in character and expresses the requirement that syntactic choices should satisfy Sen's condition α (which is our condition (I)). For various weakenings of comparability, one may replace Modularity either by (EII) (thus paying tribute to relationality), by (EIII) (thus paying tribute to pseudo-rationalizability and path independence) or by both (EII) and (EIII) (thus paying tribute to transitive relationality).

Let us summarize in non-technical terms the results obtained so far in this chapter.

First, the entrenchment postulates (EE1), (EE2$^\downarrow$), (EE3$^\top$) and (EE3$^\downarrow$), translated into constraints on the choice of retractable sentences, are satisfied by all rationalizable choice functions. In addition, the translation of (EE2$^\top$) is satisfied by all transitively rationalizable choice functions, and the translation of (EE4) is satisfied by all modularly rationalizable choice functions.

Second, the general admissibility constraints imposed by the interpretation of entrenchment in terms of choices, as well as the choice postulates (I) and (III), translated into constraints on the entrenchment of sentences, are satisfied by all generalized relations of epistemic entrenchment. In addition, the choice postulates (II) and (IV) are satisfied by all standard relations of epistemic entrenchment.

Third, all admissible relations < satisfying the choice postulates (I) and (III), translated into constraints on the entrenchment of sentences, are generalized relations of epistemic entrenchment. If < in addition satisfies (IV), translated into a constraints on the entrenchment of sentences, it is a standard relation of epistemic entrenchment.

A detailed synopsis of conditions is given in the table on p. 250.

8.6 The semantics of epistemic entrenchment

In this chapter we have been working solely in syntactic terms. It is an interesting problem, however, to spell out the semantics of epistemic entrenchment. As I have interpreted entrenchments as the revealed preferences of syntactic choices, it is desirable to combine, reformulate and simplify the results of Chapters 6 and 7. We can build on the mutual transferability of semantic and syntactic choices that was investigated in Section 7.8. More specifically, the central Definitions 15 and 16 of that section can be transformed into a relational formulation. To do this I not only presume that choices between sentences are relational (with respect to a relation of epistemic entrenchment) but also that semantic choices are relational. This is an additional assumption. We know from Observation 40 that if δ satisfies (I) and (II), then the corresponding semantic choice function γ satisfies (I) and (II).[357] However, we do not yet know whether γ satisfies (II$^\infty$), and this is what is necessary to guarantee that γ is relational (γ will typically take infinite menus).

Let us first define the entrenchments of sentences in terms of preferences over models (worlds). We can do this by working our way through the idea that $< = \mathcal{P}_2(\mathcal{D}(\mathcal{S}(\prec)))$, where $\prec$ is an ordering of models.

---

[357]Recall from Observation 41 that in general the reverse is not true. If δ is generated from a semantic choice function that satisfies (II), it does not necessarily itself satisfy (II). In this section, however, we are interested in the semantics of entrenchment, so we start from a syntactic choice function δ which is rationalizable by <. Let the semantic choice function $\gamma = \mathcal{G}(\delta)$ be generated from some syntactic choice function δ which satisfies (II). Then, as I said, γ will itself satisfy (II). Moreover, the syntactic choice function $\mathcal{D}(\gamma)$ that can be obtained from this particular function γ will also satisfy (II). This is so simply because it is generally true that $\mathcal{D}(\mathcal{G}(\delta)) = \delta$ (Observation 39(iii)).

Choices	Entrenchments
(LP1) synt. If $\phi \in Cn(G), G \subseteq F$ and $\phi \in \delta(F)$, then $\delta(F) \cap G \neq \emptyset$	(ELP1) If $\phi \in Cn(\psi)$ and $\phi \wedge \chi < \psi$, then $\psi \wedge \chi < \phi$
(Faith2) $\forall S \in \mathcal{X} : S \cap R \subseteq \sigma(S)$	(EFaith2) If $\phi \notin K$, then $\phi < \top$ and $\psi \not< \phi$
(Faith1) $\forall S \in \mathcal{X} :$ if $S \cap R \neq \emptyset$ then $\sigma(S) \subseteq R$	(EFaith1) If $\phi \notin K$ and $\psi \in K$, then $\phi \not< \top$ or $\phi < \psi$
(LP2) synt. If $Cn(F) = Cn(G)$ and $\phi \in \delta(F) \cap G$, then $\phi \in \delta(G)$	(ELP2) If $Cn(\phi \wedge \psi) = Cn(\phi \wedge \chi)$ and $\psi < \phi$, then $\chi < \phi$
(Success) sem./ (Virtual Success) synt. If $S \neq \emptyset$, then $\gamma(S) \neq \emptyset$ If $F - Cn(\emptyset) \neq \emptyset$, then $\delta(F) \neq \emptyset$	(EVSuccess) If $\phi \notin Cn(\emptyset)$ then $\phi < \top$
(I^-) If $S \subseteq S'$ and $\sigma(S') \subseteq S$, then $\sigma(S') \subseteq \sigma(S)$	(EI^-) If $\phi \wedge \psi < \chi$ and $\phi < \psi$, then $\phi \wedge \chi < \psi$
$(\emptyset 2)$ If $S \subseteq S'$ and $\sigma(S) = \emptyset$, then $S \cap \sigma(S') = \emptyset$	$(\mathrm{E}\emptyset 2)$ If $\phi \wedge \psi < \top$, then $\phi < \psi$ or $\psi < \top$
(I) / (I') If $S \subseteq S'$, then $S \cap \sigma(S') \subseteq \sigma(S)$ $\sigma(S \cup S') \subseteq \sigma(S) \cup \sigma(S')$	(EI) If $\phi < \psi$, then $\phi \wedge \chi < \psi$
$(\emptyset 1)$ If $S \subseteq S'$ and $\sigma(S') = \emptyset$, then $\sigma(S) = \emptyset$	$(\mathrm{E}\emptyset 1)$ If $\phi < \top$, then $\phi \wedge \psi < \top$
(III) If $S \subseteq S'$ and $\sigma(S') \subseteq S$, then $\sigma(S) \subseteq \sigma(S')$	(EIII) If $\phi < \psi \wedge \chi$, then $\phi < \psi$
(II^+) If $x \in \sigma(S)$ and $y \in \sigma(S')$, then $x \in \sigma(S \cup S')$ or $y \in \sigma(S \cup S')$	(EII^+) If $\phi \wedge \psi < \chi \wedge \xi$ and $\chi < \top$ and $\xi < \top$, then $\phi < \chi$ or $\psi < \xi$
(II) synt. $\delta(F) \cap \delta(G) \subseteq \delta(F \cup G)$	(EII) If $\phi \wedge \psi < \chi$, then $\phi < \chi$ or $\psi < \chi$
(II) sem. $\gamma(S) \cap \gamma(S') \subseteq \gamma(S \cup S')$	
(IV) If $S \subseteq S'$ and $\sigma(S') \cap S \neq \emptyset$, then $\sigma(S) \subseteq \sigma(S')$	(EIV) If $\phi \wedge \psi < \chi$ and $\psi \not< \phi \wedge \chi$, then $\phi < \chi$
(IV^+) If $S \subseteq S'$, then $\sigma(S) \subseteq \sigma(S')$	(EIV^+) If $\phi < \top$, then $\psi \not< \phi$

$\phi < \psi$

> iff $\phi \in \delta(\phi, \psi)$ and $\psi \notin \delta(\phi, \psi)$
>
> iff $]\phi[\cap \gamma(]\{\phi, \psi\}[) \neq \emptyset$ and $]\psi[\cap \gamma(]\{\phi, \psi\}[) = \emptyset$
>
> iff $\exists m \in \min_{\prec}]\{\phi, \psi\}[$ such that $m \not\models \phi$, and $\forall m \in \min_{\prec}]\{\phi, \psi\}[: m \models \psi$
>
> iff $\forall m \in \min_{\prec}]\{\phi, \psi\}[: m \models \psi$, and $\min_{\prec}]\{\phi, \psi\}[\neq \emptyset$

This can serve as the semantic foundation of epistemic entrenchment. We can turn it into the following definition:

Definition 19 Let an ordering $\prec$ over the set $\mathcal{M}_L$ of models be given. Then the corresponding entrenchment relation $<$ over the set L of sentences is defined by

$$\phi < \psi \quad \text{iff} \quad \emptyset \neq \min_{\prec}]\{\phi, \psi\}[\subseteq [\![\psi]\!]$$

The definition says that ϕ is *less entrenched* than ψ (more easily given up than ψ) if the models which are minimal in the set of all models falsifying either ϕ or ψ without exception satisfy ψ (and thus falsify ϕ), and that in fact there are such minimal models. Notice that in contrast to the sphere semantics that can be given for epistemic entrenchment,[358] this definition is well-suited for the case of incomparabilities between worlds and incomparabilities between sentences.

We now give a reformulation of the above definition for the case where $\prec$ is stoppered. A preference relation $\prec$ is called *stoppered* with respect to $\mathcal{X}$ if for every $S \in \mathcal{X}$ and every x in S, either x itself is minimal in S under $\prec$ or there is a y in S such that $y \prec x$ and y is minimal in S under $\prec$.[359]

Observation 62 *For all sentences ϕ and ψ, the following conditions are equivalent, provided that ϕ is not in $Cn(\emptyset)$ and $\prec$ is stoppered:*

(i) For all m in $\min_{\prec}]\{\phi, \psi\}[$ it holds that $m \models \psi$, and $\min_{\prec}]\{\phi, \psi\}[\neq \emptyset$;

(ii) For all m in $]\psi[$ there is an n in $]\phi[$ such that $n \prec m$.

Line (ii) of this Observation states a condition in the spirit of the similar Definition 4 of Rott (1991a) where preferences between maximal non-implying sets are connected with entrenchments (cf. Section 7.2).[360] Related work on the

---

[358] See Grove (1988), Gärdenfors (1988, Section 4.8), Peppas and Williams (1995) and Pagnucco and Rott (1999). Lindström and Rabinowicz (1991) generalize this semantic account decisively by allowing incomparabilities. Compare page 239 above.

[359] Stopperedness is different from what was called smoothness in Chapter 6. Recall that $\prec$ has been called smooth with respect to $\mathcal{X}$ if there are no infinite descending $\prec$-chains in S, for every $S \in \mathcal{X}$. Stopperedness follows from smoothness if $\prec$ is transitive; it does not imply smoothness, even when $\prec$ is modular.

[360] This is Definition 4 of Rott (1991a): If $\preceq$ is a preference relation on $Pow(L)$ (with respect to a knowledge set K) then the associated entrenchment relation $\leq$ on L (with respect to K) is given by

$\phi \leq \psi$ if and only if for all M in U_K such that $\psi \notin M$ there is an M' in U_K
    such that $\phi \notin M'$ and $M \preceq M'$.

connection between preferences over worlds and preferences over propositions—the former unconstrained, the latter meeting some logical constraints—was done by Lewis (1973, Section 2.3 and 2.5, 'comparative similarity' vs. 'comparative possibility'), Freund (1993) and Gärdenfors and Makinson (1994, Section 3.4, 'nice preferential models' vs. 'expectation orderings') among others. All of these accounts, however, presuppose full comparability.

The relation $<$ generated by Definition 19 exhibits a number of well-known structural and logical properties. Our previous results show that $\gamma = \mathcal{S}(\prec)$ satisfies (I) and (II$^\infty$), hence $\delta = \mathcal{D}(\mathcal{S}(\prec))$ satisfies (LP1), (LP2), (I) and (II), hence $< = \mathcal{P}_2(\mathcal{D}(\mathcal{S}(\prec)))$ satisfies (E&C), (ELP1), (ELP2), (EI) and (EII).

Now let us address the converse problem and define the relation $\prec$ in terms of entrenchment. Here we unroll the idea that $\prec = \mathcal{P}(\mathcal{G}(\mathcal{S}(<)))$, with $<$ being an entrenchment relation over sentences.

$$m \prec n$$

$$\text{iff} \quad \forall S: \text{ if } m \in S \text{ then } n \notin \gamma(S) \quad \text{and} \quad \exists S' : m \in \gamma(S')$$

$$\text{iff} \quad \forall F: \text{ if } m \in]F[, \text{ then } n \notin]F[\text{ or } Cn(F) - \widehat{n} \not\subseteq \delta(Cn(F))$$
$$\text{and} \quad \exists G : m \in]G[\text{ and } Cn(G) - \widehat{m} \subseteq \delta(Cn(G))$$

$$\text{iff} \quad \forall F: \text{ if } m, n \in]F[, \text{ then } Cn(F) - \widehat{n} \not\subseteq \min_<(Cn(F))$$
$$\text{and} \quad \exists G : m \in]G[\text{ and } Cn(G) - \widehat{m} \subseteq \min_<(Cn(G))$$

Using the abbreviation $\rho_<(F)$ for $F - \min_<(F)$, we can turn the above argument into the following definition:

Definition 20 Let an entrenchment relation $<$ over over the set L of sentences be given. Then the corresponding ordering $\prec$ over the set of models is defined by

$$m \prec n \text{ iff } \text{ for all } F \text{ such that } m, n \in]F[, \ \rho_<(Cn(F)) \not\subseteq \widehat{n}$$
$$\text{and there is a } G \text{ such that } m \in]G[\text{ and } \rho_<(Cn(G)) \subseteq \widehat{m}$$

The last condition is hard to understand. It is therefore nice to have another reformulation in a style akin to Definition 5 of Rott (1991a).[361] We call a relation $<$ *conversely well-founded* if and only if there are no infinite ascending $<$-chains in the domain of $<$.

---

[361]This is Definition 5 of Rott (1991a): If $\leq$ is an entrenchment relation on L (with respect to a knowledge set K) then the associated preference relation $\preceq$ on $Pow(L)$ (with respect to K) is given by

$M \preceq N$ if and only if for all $\phi \notin N$ there is a $\psi \notin M$ such that $\phi \leq \psi$.

Observation 63 *For all models m and n, the following conditions are equivalent, provided that $<$ is an admissible relation which is conversely well-founded and satisfies Extensionality, (EI), (EII), and (EIII).*

(i) For all F, if m and n are in $]\!]F[\![$, then $Cn(F) - \hat{n} \not\subseteq \min_{<}(Cn(F))$, and there is an G such that m is in $]\!]G[\![$ and $Cn(G) - \hat{m} \subseteq \min_{<}(Cn(G))$;

(ii) For every ϕ falsified by m there is a ψ falsified by n such that $\phi < \psi$.

We have found two convenient ways of linking preferences $\prec$ between models and preferences $<$ between sentences. The simpler conditions mentioned in Observations 62 and 63, however, are not universally applicable. They depend on the assumptions that $\prec$ is stoppered and that $<$ is conversely well-founded.

8.7 Applying epistemic entrenchment in coherentist belief change: Two construction recipes

The fundamental idea of our interpretation of entrenchment relations is that they are revealed preferences (more precisely, revealed base preferences) of syntactic choice functions, $< = \mathcal{P}_2(\delta)$. We have also seen (in Section 7.7.1) how syntactic choice functions can be retrieved from belief contractions, $\delta = \mathcal{D}(\dot{-})$. Connecting this, we have got Definition 17

$$\phi < \psi \quad \text{iff} \quad \phi \notin K \dot{-} (\phi \wedge \psi) \text{ and } \psi \in K \dot{-} (\phi \wedge \psi)$$

which we introduced in Section 8.1. Definition 17 tells us that ϕ is less entrenched than ψ if and only if an agent who has to give up $\phi \wedge \psi$, that is, to give up at least one of ϕ and ψ, gives up ϕ and keeps ψ. This condition, which is called the competitive interpretation of entrenchment in Rott (1992c), seems to be perfectly well-motivated. It is not in need of justification, but, on the contrary, may serve to justify other ideas. I will use it for the justification of construction recipes for contraction functions using entrenchment relations.

But now we take the reverse point of view which seems actually more in line with what one expects from a theory of epistemic entrenchment. Let us assume that an entrenchment relation is somehow represented in the mental state of the agent, and this relation determines his disposition to change his mind. The question we have to answer is: How can entrenchment relations be applied in belief change? Intuitively, belief change guided by an entrenchment relation may be regarded as a judicious expansion of very many 'small decisions'—choices in menus consisting of only two beliefs[362]—into a single 'big revision' of an infinite belief set.

The idea in lline with the architecture of this book is first to let the entrenchment relation determine a syntactic choice function, by liberal minimization (see Definition 3 in Chapter 6). Second, one can then use this choice function in the way followed in the last chapter, namely by using either Definition 8 or its finitary version, Definition 10 of Chapter 7. In short, one can apply entrenchment

---

[362]See Definition 5 of Chapter 6, alias principle (E) in this chapter.

relations by defining $\mathcal{C}(<) = \mathcal{C}(\mathcal{S}(<))$ or $\mathcal{C}(<) = \mathcal{C}_{fin}(\mathcal{S}(<))$.[363] The following construction corresponds precisely to the finitary Definition 10, since we can translate $\phi \vee \psi \notin \delta(\{\phi, \phi \vee \psi\})$ into $\phi \not< \top$ or $\phi < \phi \vee \psi$.[364]

Definition 21 The contraction function $\dot{-}$ over K is generated by an entrenchment relation $<$ between sentences, in symbols $\dot{-} = \mathcal{C}(<)$, iff for every ψ,

$$\psi \in K \dot{-} \phi \quad \text{iff} \quad \psi \in K \text{ and } (\phi \not< \top \text{ or } \phi < \phi \vee \psi)$$

In essence this is identical with the construction recipe proposed and studied by Gärdenfors and Makinson (1988). For an independent motivation, we repeat the argument given by them, adding in parentheses the new limiting case in which the agent refuses to withdraw his beliefs. If ϕ is not in K, then we do not need to change K at all. So let ϕ be in K. According to Definition 17, $\phi < \psi$ means that $\psi \in K \dot{-} \phi \wedge \psi$ (and $\phi \notin K \dot{-} \phi \wedge \psi$). If we replace ψ by $\phi \vee \psi$, we get that $\phi < \phi \vee \psi$ means that $\phi \vee \psi \in K \dot{-} \phi \wedge (\phi \vee \psi)$ (and $\phi \notin K \dot{-} \phi \wedge (\phi \vee \psi)$). But by $(\dot{-}6)$, $K \dot{-} \phi \wedge (\phi \vee \psi)$ is the same as $K \dot{-} \phi$. And, given the understanding that the contraction operation should satisfy the Recovery postulate $(\dot{-}5)$, we have also that $\neg \phi \vee \psi \in K \dot{-} \phi$, for any ψ that is in K. Hence, for any $\psi \in K$, we have $\phi \vee \psi \in K \dot{-} \phi$ iff $\psi \in K \dot{-} \phi$, by $(\dot{-}1)$. Putting this together gives: If $\phi \in K$, then for any ψ in K, it holds that $\psi \in K \dot{-} \phi$ (and $\phi \notin K \dot{-} \phi$) iff $\phi < \phi \vee \psi$.[365]

By Extensionality, (E&C) and (EI), $\phi < \phi \vee \psi$ is equivalent to $\neg \phi \supset \neg \psi < \neg \phi \supset \psi$. Thus we can understand the principal clause of Definition 21 in the following way. Sentence ψ in K is retained in the contraction of K with respect to ϕ just in case ψ *under the condition* $\neg \phi$ is better entrenched than $\neg \psi$ *under the condition* $\neg \phi$. Roughly, retaining ψ in $K \dot{-} \phi$ means believing that ψ is true even if it were the case that $\neg \phi$ is true.

So there is a lot of support for Definition 21. Prima facie, however, it seems much more natural and straightforward to compare sentences without recourse to disjunctions, if we are given some relation of entrenchment. The simple idea is this:

---

[363] An alternative idea is that the contraction determined by an entrenchment operation $<$ should be the contraction determined by that choice function the revealed preferences of which are represented by the entrenchment relation $<$ (more exactly, by one of those choice functions). That means, one should define the contraction operation $\mathcal{C}(<)$ as $\mathcal{C}(\delta)$ for some δ such that $< = \mathcal{P}_2(\delta)$. In this case one has to make sure that the contraction is well-defined, that is, that $\mathcal{P}_2(\delta) = \mathcal{P}_2(\delta')$ implies $\mathcal{C}(\delta) = \mathcal{C}(\delta')$. This is easy to verify, provided that δ satisfies (LP1) and (LP2).

[364] Extensionality is used here to ensure harmony with our interpretation of entrenchments, I set the taboo set $\overline{R}$ of the choice function $\mathcal{S}(<)$ to $\{\phi \in L : \phi \not< \top\}$.

[365] Gärdenfors and Makinson (1988, pp. 89–90). Note that this argument does not stand completely on its own feet, since it presumes Definition 17 and the validity of several of the basic postulates for contraction including most conspicuously $(\dot{-}5)$. Conversely, the disjunction in Definition 21 guarantees satisfaction of the Recovery postulate $(\dot{-}5)$.

Definition 22 The withdrawal function $\dot{-}$ over K is generated by an entrenchment relation $<$ between sentences, in symbols $\dot{-} = \ddot{\mathcal{C}}(<)$, iff for every ψ,

$$\psi \in K \ddot{-} \phi \text{ iff } \psi \in K \text{ and } (\phi \not< \top \text{ or } \phi < \psi)$$

The idea of withdrawing a belief from theories in this way was first ventilated in Rott (1991a). The first axiomatizations were given by Pagnucco (1996) and Fermé and Rodriguez (1998). Independent arguments in favour of this definition are adduced by Levi (1998) who calls $K \dot{-} \phi$ a 'mild contraction'. Recently, the idea has been closely investigated and defended, using general philosophical principles and a semantic modelling in terms of systems of spheres, by Pagnucco and Rott (1999). If we presuppose Extensionality and (EIII), then $\phi < \psi$ implies $\phi < \phi \vee \psi$, so $K \ddot{-} \phi$ is a subset of $K \dot{-} \phi$. Another feature of $K \ddot{-} \phi$ is that it violates the Disjunction Thesis we used in the motivation of Definition 8 in Chapter 7, namely that ψ should be in $K \dot{-} \phi$ if and only if $\phi \vee \psi$ is. Satisfaction of the thesis is guaranteed by Extensionality for $\dot{-} = \mathcal{C}(<)$ but it does not hold true for $\dot{-} = \ddot{\mathcal{C}}(<)$, because clearly $\phi < \phi \vee \psi$ does not imply $\phi < \psi$.

The choice-theoretic justification of Definition 8 is attractive in so far as it relates the problem of withdrawing a belief ϕ to the problem of choice in a single menu, viz., in $Cn(\phi)$. Choice according to this recipe is, so to speak, holistic. The finitary version, though convenient and even necessary if only finitary choice functions are available, loses this attractive feature, since it advises us to proceed in a piecewise manner. The agent is construed as choosing in the set $\{\phi, \phi \vee \psi\}$, for every ψ in K individually. Choice is atomistic here. Notice that if only choices in two-element menus are considered, then all the central constraints of the classical theory of rational choice run idle, since they are effective only if applied to menus of varying cardinality.[366] We still maintain the coherentist point of view, however, by stipulating that the choice function underlying the choices in two-element menus is in principle applicable to menus of all, or at least all finite, cardinalities.

Although the holistic character of $\mathcal{C}(\delta)$ is lost in the Definition $\mathcal{C}_{fin}(\delta)$, and thus in the above definition of $\mathcal{C}(<)$, we have at least an account of this latter definition in terms of rational choices in binary menus: $\phi \not< \top$ or $\phi < \phi \vee \psi$ taken together just mean that $\phi \vee \psi \notin \delta(\phi, \phi \vee \psi)$. Is there a similarly compact choice-theoretic account of the definition of $\ddot{\mathcal{C}}(<)$? The answer is: Yes, if we allow an exception for the limiting case when the agent refuses to withdraw the belief he is supposed to withdraw, but strictly speaking, no. Let me explain this. The characteristic clause of Definition 22 is of course $\phi < \psi$. This means that in the choice between ϕ and ψ, ϕ gets selected and ψ does not get selected. But then

---

[366] Of course, this is not to say that these constraints do not have consequences for sets with two elements, *if* sets with more elements are available in the domain $\mathcal{X}$ of σ. For instance, (I) and (III) together determine that $\sigma(\{x,y\}) = \{y\}$ and $\sigma(\{y,z\}) = \{z\}$ together imply $\sigma(\{x,z\}) = \{z\}$. This corresponds to the transitivity of $<$, compare Lemmas 16(ii) and 52(vii).

there is the clause for the case when ϕ is taboo: $\phi \not< \top$. In contrast to Definition 21, this limiting clause cannot neatly be combined with the main clause. In that definition, we have $\phi \vee \psi \notin \delta(\phi, \phi \vee \psi)$ as the translation of both clauses combined. The obvious idea for Definition 22 is $\psi \notin \delta(\phi, \psi)$. But a careful translation in terms of preferences shows that this means

$$(\phi \not< \top \text{ and } \psi \not< \top) \text{ or } \phi < \psi$$

As the clause for the definition of $K \dot{-} \phi$, this would determine the contracted set $K \dot{-} \phi$ for a belief ϕ with $\phi \not< \top$ to be the set of all taboo sentences in K, $\{\psi \in K : \psi \not< \top\}$. Our official definition, however, has $K \dot{-} \phi = K$ in such a case. There is only a rather complicated choice-theoretic characterization of the definition of $\ddot{C}(<)$, one for which there is little, if any intuitive support: ψ is in $K \dot{-} \phi$ if and only if ψ is in K and either $\phi \notin \delta(\phi)$ or $\delta(\phi, \psi) = \{\phi\}$. More importantly and in contrast to the Gärdenfors-Makinson definition of $C(<)$, the definition of $\ddot{C}(<)$ does not derive from a holistic choice problem in a single menu.

From the philosophical point of view taken in this chapter, Definition 22 is somewhat less well-founded than Definition 21. Still the former is more handy than the latter. One may ask what else could assist us in reaching a decision as to which of the two construction recipes is the better one. The following abstract argument seems promising. The most solid intuition we have come across is the competitive interpretation of the entrenchment relation, that is, the interpretation of entrenchments as revealed base preferences, and that is, ultimately, Definition 17. Now we can call $C(<)$ or $\ddot{C}(<)$ more adequate than the other, if it harmonizes better with this most solid intuition. Two ways of making this argument more precise suggest themselves. First, one can start out from an entrenchment relation (leaving its exact properties as open as possible) and use it in either of the two definitions. Then take $C(<)$ and $\ddot{C}(<)$ and look at their respective revealed preferences as obtained through Definition 17. A construction recipe for contractions is in harmony with the solid intuition, if the two steps lead us back to the original entrenchment relation $<$. Alternatively, one can start out from a belief removal operation $\dot{-}$ and look at its revealed base preferences, i.e., the corresponding entrenchments. Then use this entrenchment relation to construct another belief contraction operation, either $C(<)$ or $\ddot{C}(<)$. A construction recipe for contractions is in harmony with the solid intuition, if the two steps lead us back to the original belief removal operation $\dot{-}$.

We first check how the Gärdenfors-Makinson account can be justified in this way. The following observation is a refinement of Gärdenfors and Makinson (1988, Corollary 6) and of Rott (1992b, Theorem 2, parts (iii) and (iv)).

Observation 64 *(i) Let $<$ satisfy Irreflexivity, (E&C), (Ext) and (E02), and let $<^* = \mathcal{P}_2(C(<))$. Then we have $\phi <^* \psi$ if and only if $\psi \in K$ and $((\phi \notin K$ and $\phi \wedge \psi \not< \top)$ or $\phi < \psi)$.*

If $<$ satisfies in addition (EFaith), then we have $\phi <^* \psi$ if and only if $\psi \in K$ and $\phi < \psi$.

If moreover $<$ satisfies (EE5'), then we have $\phi <^* \psi$ if and only if $\phi < \psi$.

(ii) Let $\dot{-}$ satisfy $(\dot{-}1) - (\dot{-}2)$, $(\dot{-}5)$ and $(\dot{-}6)$, and $\dot{-}^* = \mathcal{C}(\mathcal{P}_2(\dot{-}))$. Then we have $\psi \in K \dot{-}^* \phi$ if and only if $\psi \in K \dot{-} \phi$.

I comment on this result only after seeing how the rival definition of $\ddot{\mathcal{C}}(<)$ passes the same test. In order to assess this properly, we need to introduce a new kind of belief removal operation.

A *severe withdrawal function* is an operation $\ddot{-}$ that satisfies $(\ddot{-}\emptyset 1)$, $(\ddot{-}\emptyset 2)$, $(\ddot{-}1) - (\ddot{-}3)$, $(\ddot{-}5^0)$, $(\ddot{-}6)$, $(\ddot{-}7a)$ and $(\ddot{-}8)$ (for these conditions, see Section 4.2). Severe withdrawal functions are introduced and studied in Pagnucco and Rott (1999). They satisfy $(\ddot{-}7c)$ and $(\ddot{-}8c)$, but they do not satisfy the Recovery postulate $(\ddot{-}5)$, on pain of triviality. The Antitony postulate $(\ddot{-}7a)$ is very strong and seems objectionable in many contexts and for many purposes. It in effect leads to the loss of very many beliefs in belief removal (that is why we call these withdrawals 'severe'). For instance, Antitony entails that when performing severe withdrawals, one cannot remove two beliefs 'independently': Either ϕ is lost in $K \ddot{-} \psi$, or else ψ is lost in $K \ddot{-} \phi$ (provided that neither ϕ nor ψ is taboo, i.e., that $\phi \notin K \ddot{-} \phi$ and $\psi \notin K \ddot{-} \psi$).[367] Pagnucco and Rott (1999), however, argue that the concept of severe withdrawal is still interesting and well-motivated. Notational convention: Whenever it is clear that we are dealing with severe withdrawals, we use the symbol $\ddot{-}$ rather than $\dot{-}$.

The following conditions have been formulated by Pagnucco (1996):

($\ddot{-}$9)   If $\phi \notin K \ddot{-} \psi$, then $K \ddot{-} \psi \subseteq K \ddot{-} \phi$

($\ddot{-}$10)   If $\phi \notin K \ddot{-} \phi$ and $\phi \in K \ddot{-} \psi$, then $K \ddot{-} \phi \subseteq K \ddot{-} \psi$

These conditions entail that the results all potential contractions are ordered by subset inclusion. Neither of $(\ddot{-}9)$ and $(\ddot{-}10)$ is valid for AGM contractions. They are, however, among the postulates satisfied by severe withdrawals.

Lemma 65 *(i) If $\ddot{-}$ satisfies $(\ddot{-}\emptyset 2)$, $(\ddot{-}1)$, $(\ddot{-}2)$, $(\ddot{-}5^0)$ and $(\ddot{-}7a)$, then it satisfies $(\ddot{-}7c)$;*

(ii) If $\ddot{-}$ satisfies $(\ddot{-}\emptyset 1)$, $(\ddot{-}1)$, $(\ddot{-}2)$, $(\ddot{-}5^0)$ and $(\ddot{-}8)$, then it satisfies $(\ddot{-}8c)$;

(iii) Severe withdrawals satisfy $(\ddot{-}9)$;

(iv) Severe withdrawals satisfy $(\ddot{-}10)$;

(v) The only severe withdrawal function $\ddot{-}$ that satisfies Recovery $(\ddot{-}5)$ is the trivial one satisfying

$$K \ddot{-} \phi = \begin{cases} \overline{R} & \text{if } \phi \notin \overline{R} \\ K & \text{otherwise} \end{cases}$$

where $\overline{R} = \{\chi : \chi \in K \ddot{-} \chi\}$ is the taboo set of $\ddot{-}$. As a consequence of this, for all sentences ϕ and ψ in K which are not taboo for $\ddot{-}$, their biconditional $\phi \equiv \psi$ is taboo.

---

[367]This observation, or rather its analogue for $\ddot{\mathcal{C}}(<)$, was made by Sven Ove Hansson (1999a, p. 102), who calls this property *Expulsiveness* and takes it as definitely showing the inappropriateness of severe withdrawals.

Many more properties of severe withdrawal functions are investigated in Pagnucco and Rott (1999). What is most important in our present context, however, is the following. If one can draw on the postulates for severe withdrawal, the process of retrieving entrenchments from withdrawal behaviour can be simplified considerably. This is the content of the following

Lemma 66 *Consider the following two conditions:*

(†) $\phi \notin K \dot{-}(\phi \wedge \psi)$ and $\psi \in K \dot{-}(\phi \wedge \psi)$
(‡) $\phi \notin K \dot{-}\phi$ and $\psi \in K \dot{-}\phi$

If $\dot{-}$ satisfies ($\dot{-}7c$) and ($\dot{-}8c$), then (†) *implies* (‡). *If $\dot{-}$ satisfies ($\dot{-}7a$) and ($\dot{-}8c$), then* (‡) *implies* (†).

The basic condition employed in Definition 17 can thus be replaced by a much simpler condition—provided the contraction function in question satisfies ($\dot{-}7c$), ($\dot{-}8c$) and the strong Antitony condition ($\dot{-}7a$). This means, essentially, that the main clause of Definition 22, $\phi < \psi$, can be used in both directions; not only in the construction of contractions from entrenchment, but also in the reconstruction of entrenchments from patterns of contraction behaviour.

We are now ready to make the test how well Definition 22 can be justified by its being in harmony with Definition 17:

Observation 67 *(i) Let $<$ satisfy Irreflexivity, (E&C) and (Ext) and let $<^* = \mathcal{P}_2(\ddot{\mathcal{C}}(<))$. Then we have $\phi <^* \psi$ if and only if $\psi \in K$ and (($\phi \notin K$ and $\phi \wedge \psi \not< \top$) or $\phi < \psi$).*

If $<$ satisfies in addition (EFaith), then we have $\phi <^* \psi$ if and only if $\psi \in K$ and $\phi < \psi$.

If moreover $<$ satisfies (EE5′), then we have $\phi <^* \psi$ if and only if $\phi < \psi$.

(ii) Let $\ddot{-}$ satisfy ($\dot{-}1$) – ($\dot{-}2$), ($\dot{-}5^0$), ($\dot{-}6$), ($\dot{-}7a$) and ($\dot{-}8c$), and let $\ddot{-}^* = \ddot{\mathcal{C}}(\mathcal{P}_2(\ddot{-}))$. Then we have $\psi \in K \ddot{-}^* \phi$ if and only if $\psi \in K \ddot{-}\phi$.

Comparing the results for $\ddot{\mathcal{C}}(<)$ with those for $\mathcal{C}(<)$, we can sum up the situation as follows. In the justification method under part (i) of Observations 64 and 67, it does not matter at all whether we use the the recipe for Gärdenfors-Makinson contraction or for severe withdrawal. The same conditions for $<$ do the justification to the same extent! Weak conditions—surprisingly weak conditions!—are already sufficient for an agreement of $\mathcal{P}_2(\mathcal{C}(<))$ and $\mathcal{P}_2(\ddot{\mathcal{C}}(<))$ with $<$, as long as only sentences in K are compared; and a full agreement can be reached if conditions of Faith and Minimality with respect to the belief set K are supposed to be satisfied.

The justification method under part (ii) of Observations 64 and 67 gives a somewhat different picture. On the one hand, full agreement is no longer dependent on limiting case conditions relative to K. On the other hand, the preconditions for the justification are more demanding, and they differ considerably for the two construction recipes for entrenchment-based belief removal. The justification of the Gärdenfors-Makinson recipe $\mathcal{C}(<)$ is dependent on the controversial Recovery postulate ($\dot{-}5$). That condition is not needed for the alternative recipe

$\ddot{C}(<)$, but in return one has to put up with ($\dot{-}$8c) and especially the strong and questionable Antitony postulate ($\dot{-}$7a). In the coherentist view expounded in Chapter 7 Recovery is always accepted while Antitony is not, but it is perhaps not unfair to say that we have found no Archimedean point to depend upon.

We conclude that neither of our methods of determining which contraction recipe is more natural by means of a harmony constraint with the revealed preference idea $< = \mathcal{P}_2(\dot{-}) = \mathcal{P}_2(\mathcal{D}(\dot{-}))$ has been victorious. In both cases the justification itself is successful in so far as harmony can be achieved quite satisfactorily. But the first method of evaluation, the one starting from entrenchments, does not have any discriminating power between $\mathcal{C}(<)$ and $\ddot{C}(<)$. The second method, the one starting from belief removals, does have a discriminating effect, but unfortunately only at the expense of properties of belief removal operations that have to be antecedently accepted—and these properties are different, but very demanding for either of the two construction recipes.

The program of the last three chapters of this book may roughly be described by saying that belief change and nonmonotonic inference are attempted to be proven 'rational' by making them 'relational'. We have followed this route in two steps by first construing belief change as generated by choice functions and then construing choice functions as deriving from some underlying preference relations. Although the program has turned out to work well, there are some limits. Choice functions can only be rationalized by a preference relation if they satisfy (I) and (II), and choice functions satisfy these constraints only if the generated belief change functions satisfy some non-trivial postulates. In the case of contractions generated by syntactic choice functions, postulates ($\dot{-}$7) and ($\dot{-}$8wd) are the relevant conditions (see Observations 27 and 28).

But now one of the results of the present section opens up a puzzle. Look at the second part of Observation 64. Here we learn that a contraction function can already be generated from a preference relation if it satisfies conditions ($\dot{-}$1), ($\dot{-}$2), ($\dot{-}$5) and ($\dot{-}$6). No additional assumptions are necessary!

Let us call a contraction function $\dot{-}$ *directly relationalizable* if $\dot{-} = \mathcal{C}(\mathcal{P}_2(\dot{-}))$. In contrast, we call it *indirectly relationalizable* (via a syntactic choice function) if $\dot{-} = \mathcal{C}(\mathcal{D}(\mathcal{P}_2(\mathcal{D}(\dot{-}))))$. Using these terms, what we have just noticed is that there are more contraction functions that are directly relationalizable than functions that can be generated from a relationalizable syntactic choice function δ. This is the *puzzle of direct relationalizability*. It is very remarkable indeed that the idea of direct relationalizability which was first expounded by Gärdenfors and Makinson (1988) in the context of the full force of all the AGM postulates ($\dot{-}$1) $-$ ($\dot{-}$8) generalizes to contexts where only half of the postulates are satisfied.

In order to explain how the puzzle comes about, let us use the operators $\mathcal{C}$, $\mathcal{D}$ and $\mathcal{P}_2$ to denote the direct transitions between contractions, syntactic choice functions and base preferences (i.e., entrenchments) in the way indicated in Fig. 8.3 (this continues the notation as used so far). What we have from earlier observations is this:

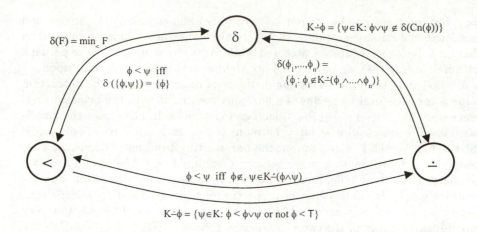

$\delta(F) = \min_< F$

$\phi < \psi$ iff
$\delta(\{\phi,\psi\}) = \{\phi\}$

$K \dot{-} \phi = \{\psi \in K: \phi \vee \psi \notin \delta(Cn(\phi))\}$

$\delta(\phi_1,...,\phi_n) =$
$\{\phi_i: \phi_i \notin K \dot{-}(\phi_1 \wedge ... \wedge \phi_n)\}$

$\phi < \psi$ iff $\phi \notin, \psi \in K \dot{-}(\phi \wedge \psi)$

$K \dot{-} \phi = \{\psi \in K: \phi < \phi \vee \psi \text{ or not } \phi < \top\}$

FIG. 8.3. Direct and indirect relationalizability

For any contraction function $\dot{-}$
  satisfying $(\dot{-}1)$, $(\dot{-}2)$, $(\dot{-}5)$ and $(\dot{-}6)$, $\mathcal{C}(\mathcal{P}_2(\dot{-})) = \dot{-}$.   (Obs. 64)
  satisfying $(\dot{-}1)$, $(\dot{-}2)$, $(\dot{-}5)$ and $(\dot{-}6)$, $\mathcal{C}(\mathcal{D}(\dot{-})) = \dot{-}$.   (Obs. 28)
  $\mathcal{P}_2(\dot{-}) = \mathcal{P}_2(\mathcal{D}(\dot{-}))$.   (obvious)
For any choice function δ
  satisfying (I) and (II$^\omega$), $\mathcal{D}(\mathcal{P}_2(\delta)) = \delta$.   (Lemma 14, Cor. 15)
  satisfying (LP1), (LP2), (I) and (II$^\omega$),
      $\mathcal{C}(\delta)$ satisfies $(\dot{-}7)$ and $(\dot{-}8wd)$.   (Obs. 27)
For any preference relation $<$
  $\dot{-} = \mathcal{C}(\mathcal{D}(<))$ is defined by
      $K \dot{-} \phi = \{\psi \in K: \text{there is a } \chi \in Cn(\phi) \text{ s.th. } \chi < \phi \vee \psi\}$ (obvious)
  $\mathcal{C}(<) \subseteq \mathcal{C}(\mathcal{D}(<))$.   (obvious)
  satisfying Continuing Down (EE2$^\downarrow$), $\mathcal{C}(\mathcal{D}(<)) \subseteq \mathcal{C}(<)$ (obvious)
  $\mathcal{D}(<)$ satisfies (I) and (II$^\omega$).   (Lemma 14)
  $\mathcal{C}(\mathcal{D}(<))$ satisfies $(\dot{-}7)$ and $(\dot{-}8wd)$.   (from the above)
For any contraction function $\dot{-}$
  $\mathcal{C}(\mathcal{D}(\mathcal{P}_2(\dot{-})))$ satisfies $(\dot{-}7)$ and $(\dot{-}8wd)$.   (from the above)

From all this we can see that if a contraction function $\dot{-}$ does not satisfy $(\dot{-}7)$ and $(\dot{-}8wd)$, then either $\mathcal{C}(\mathcal{D}(\mathcal{P}_2(\dot{-})))$ must be different from $\dot{-}$, or else $\mathcal{D}(\mathcal{P}_2(\dot{-}))$ does not satisfy (LP1) and (LP2).

Let us look at an example. It is enough to consider the theory $K = Cn(\{p, q\})$ in the language with only two propositional variables p and q. In Fig. 8.4, $\binom{\phi}{\psi}$ is an abbreviation for $K \dot{-} \phi = Cn(\psi)$.[368] It is easily verified that the contraction function $\dot{-}$ over K defined by the Hasse diagram satisfies $(\dot{-}1) - (\dot{-}7)$ and $(\dot{-}8c)$.

---

[368] The same example appears for different purposes in Rott (1993, p. 1438).

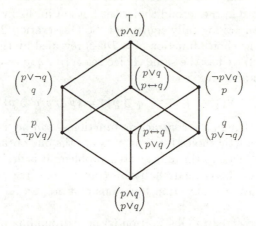

FIG. 8.4. A puzzling example

But it does not satisfy ($\dot{-}$8vwd), because $K\dot{-}(p \wedge q) = Cn(p \vee q) \not\subseteq Cn(p \leftrightarrow q) = Cn(\{\neg p \vee q, p \vee \neg q\}) = Cn((K\dot{-}p)\cup(K\dot{-}q))$. Since ($\dot{-}$8wd) is stronger than ($\dot{-}$8vwd), the contraction function does not satisfy ($\dot{-}$8wd) either.

It is a simple exercise to find out that the entrenchment relation $< = \mathcal{P}_2(\dot{-})$ retrievable from this contraction function relates only few sentences in K.[369] Every sentence that is not a logical truth is less $<$-entrenched than $\top$. Besides, we have $p \wedge q < p \vee q$ and $p \leftrightarrow q < p \vee q$, but no other pairs in the belief set K are related by $<$. It is also easy to verify that this relation $<$ satisfies all conditions of GEE-relations, it in particular satisfies Continuing Down, (EE2$^\downarrow$). Therefore we know that $\mathcal{C}(\mathcal{D}(<)) = \mathcal{C}(<) = \dot{-}$. However, by construction, $\dot{-}$ does not satisfy ($\dot{-}$8vwd), while the results listed above seem to imply that all indirectly relationalizable contractions must satisfy ($\dot{-}$8vwd). What has gone wrong?

Fortunately, nothing has gone wrong. As regards the entrenchment relation $< = \mathcal{P}_2(\dot{-})$, we can see that it does not satisfy (EII), because we have

$$(p \supset q) \wedge (q \supset p) < p \vee q$$

but

$$\text{neither } p \supset q < p \vee q \text{ nor } q \supset p < p \vee q$$

The crucial point, however, is that the choice function $\delta = \mathcal{D}(<)$ obtained from $<$ by liberal minimization does not satisfy (LP2) (it does satisfy (LP1)). We have

$$\delta(\{p \vee q, p \leftrightarrow q\}) = \{p \leftrightarrow q\}$$

but

---

[369]This is made explicit in the appendix.

$$\delta(\{p \vee q, p \supset q, q \supset p\}) = \{p \vee q, p \supset q, q \supset p\}$$

so $p \vee q$ is contained in the second choice set but not in the first, although the two menus in question are logically equivalent. So Observation 27 is not applicable.

In contrast, the choice function $\delta' = \mathcal{D}(\dot{-})$ revealed by the choice function $\dot{-}$ does satisfy (LP2). It has the same choice set $\delta(\{p \vee q, p \leftrightarrow q\})$ but it differs for the second menu and gives us

$$\delta'(\{p \vee q, p \supset q, q \supset p\}) = \{p \supset q, q \supset p\}$$

instead. So even though the contraction function $\dot{-}$ can be 'rationalized' by both $<$ and δ', the syntactic choice function δ' is not the one obtained by liberal minimization from $<$. Due to this unexpected problem, it is desirable to add a separate observation on direct relationalizable contractions, regarding the connection between properties of contraction functions and properties of entrenchment relations.

The result presented in Observation 64 needs rounding off. Although a large part has been implicit in previous results, I add a more conventional-style theorem stating the direct transfer of properties between entrenchments and contraction functions, based on Definitions 17 and 21. I do so in order to give a componentwise characterization of contraction properties (which is not yet found in the literature), and to underscore the central role of the axiom *Entrenchment and Choice*, (E&C). (I refrain now from explicitly repeating all the limiting conditions that we have decided to accept as we went along, such as (EFaith), (E∅1) and (E∅2), and the corresponding postulates for belief contraction, ($\dot{-}$7r) and ($\dot{-}$8rc).)

Observation 68 *(i) For every logically admissible entrenchment relation* $<$, *i.e., for every preference relation that satisfies Irreflexivity, Extensionality and (E&C), the contraction function* $\dot{-} = \mathcal{C}(<)$ *satisfies* ($\dot{-}$1), ($\dot{-}$2), ($\dot{-}$5) *and*

$$($\dot{-}$6).\ \textit{If} < \textit{in addition satisfies} \left\{ \begin{array}{c} Minimality\ (EE5') \\ EFaith\ wrt\ K \\ Maximality \\ Continuing\ Down\ (EE2^{\downarrow}) \\ Continuing\ Up\ (EE2^{\uparrow}) \\ (EII) \\ (EII^{+}) \\ (EIV) \end{array} \right\},\ \textit{then}\ \dot{-}\ \textit{sat-}$$

$$\textit{isfies} \left\{ \begin{array}{c} ($\dot{-}$2) \\ ($\dot{-}$3) \\ ($\dot{-}$4) \\ ($\dot{-}$7) \\ ($\dot{-}$8c) \\ ($\dot{-}$8wd) \\ ($\dot{-}$8d) \\ ($\dot{-}$8) \end{array} \right\}.$$

(ii) For every contraction function $\dot{-}$ that satisfies $(\dot{-}1)$ and $(\dot{-}6)$, the preference relation $< = \mathcal{P}_2(\dot{-})$ satisfies Irreflexivity, Extensionality and (E&C).

$$
\text{If } \dot{-} \text{ in addition satisfies}
\left\{
\begin{array}{c}
(\dot{-}2) \\
(\dot{-}3) \\
(\dot{-}4) \\
(\dot{-}7) \\
(\dot{-}8c) \\
(\dot{-}7) \text{ and } (\dot{-}8c) \\
(\dot{-}8wd) \\
(\dot{-}8d) \\
(\dot{-}8)
\end{array}
\right\}
, \text{ then } < \text{ satisfies}
$$

$$
\left\{
\begin{array}{c}
Minimality\ (EE5') \\
EFaith\ wrt\ K \\
Maximality \\
Continuing\ Down\ (EE2^{\downarrow}) \\
Continuing\ Up\ (EE2^{\uparrow}) \\
Transitivity \\
(EII) \\
(EII^{+}) \\
(EIV)
\end{array}
\right\} .
$$

Notice the central role played by the conditions (Continuing Up) and (Continuing Down) as counterparts of $(\dot{-}7)$ and $(\dot{-}8c)$.[370]

One could now formulate a corresponding representation theorem for the severe withdrawals generated by Definitions 22 instead of 21. That is, a corresponding theorem stating how characteristic properties of entrenchment relations lead to characteristic properties of withdrawal operations ('soundness'), and showing that all withdrawal operations with certain characteristic properties can be represented as based on entrenchments, via Definition 22, where the entrenchment relation has the corresponding characteristic properties ('completeness'). For $\ddot{C}(<)$, one can take advantage of the convenient bridge mentioned in Lemma 66. Some results are reported in Pagnucco and Rott (1999), but I refrain from further presenting this material here.

The last thing that remains to be furnished in order to make the undertaking of the present section successful is perhaps a constructive account of entrenchments. In this chapter we have mainly been following a reconstructive route, one retrieving entrenchments from belief changes, more exactly, one taking the idea of base preferences revealed by an underlying disposition to choose as the most important perspective on entrenchments. But if one actually wants to *apply* in belief change processes, the converse idea makes much more sense. Then entrenchments should be prior to contractions, not the other way round. And

---

[370]Compare the effects of the properties (Continuing Up) and (Continuing Down) for 'hierarchies' used in Alchourrón and Makinson's (1985, Observations 4.3 and 5.3) safe contractions: Either one guarantees satisfaction of $(\dot{-}7)$.

entrenchments, with their rather demanding formal structure, must come from somewhere. Again I refrain from laying down strategies for doing this here. A survey of some of the most promising approaches—approaches that link the coherentist device of an entrenchment ordering to the foundationalist structure of a belief base is given in Rott (2000a). The upshot of the reflections in this paper is that there are non-trivial ways of generating entrenchments from other data structures, but that there remains a gap in the reconstructive capabilities of entrenchments that cannot be closed with the means so far available. More research is needed to bring more light into these largely unknown areas.

I close this section with a word about revisions. One may be tempted to think that a decision can be made as to the superiority of $\mathcal{C}(<)$ or $\ddot{\mathcal{C}}(<)$ by finding out about which of these contractions may serve as the better basis for revisions. Recall that revisions $* = \mathcal{R}(\dot{-})$ are obtained from contractions and withdrawals $\dot{-}$ by means of the Levi identity: $K * \phi = Cn((K \dot{-} \neg \phi) \cup \{\phi\})$. It turns out, however, that the two ways of defining entrenchment-based belief removals are 'revision-equivalent':

Observation 69 *Let $<$ be an entrenchment relation satisfying Extensionality, (E&C), (EI) and (EIII). Then $\mathcal{R}(\mathcal{C}(<)) = \mathcal{R}(\ddot{\mathcal{C}}(<))$.*

The proof of this observation shows that a condensed and direct route of defining revisions $* = \mathcal{R}(<)$ from entrenchment relations can be motivated by either way of constructing entrenchment-based contractions:

$$K * \phi = \begin{cases} \{\psi : \neg\phi < \phi \supset \psi\} & \text{if } \neg\phi < \top \\ L & \text{otherwise} \end{cases}$$

By Extensionality, (E&C) and (EI), $\neg\phi < \phi \supset \psi$ is equivalent to $\phi \supset \neg\psi < \phi \supset \psi$. Thus we can understand the principal clause of this representation in the following way. Sentence ψ is believed in the revision of K by ϕ just in case ψ *under the condition $\neg\phi$ is better entrenched than $\neg\psi$ under the condition $\neg\phi$*. Very roughly, accepting ψ in $K * \phi$ means believing that ψ is true if ϕ is true. Compare this with the conditions given in Section 7.5.

8.8   Functional and relational belief change

Belief change as discussed so far is *functional* in that it produces, for every belief set K and every sentence ϕ, exactly one revised belief set $K \dot{-} \phi$. Lindström and Rabinowicz (1991) have advocated the idea of *relational* belief revision. If a theory is to be changed in response to new information, they argue, there may be various rational ways to do this, without any one of these possibilities being better than any other. This is, of course, a very interesting and well-taken point. But in one respect we are not satisfied if all we know is that, say, K_1, K_2, K_3, etc., are all rational ways of removing ϕ from K. One answer we would like to get from a model for belief revision is an answer to the question what we should *actually* do when forced to retract ϕ. What should our epistemic state *actually* look like after retracting ϕ? A parallel processing of multiple alternative revised belief sets is psychologically unrealistic and computationally intractable, in particular if iterated belief changes are taken into account that would lead to an exponential growth of the number of belief sets. The most reasonable advice one can get from the point of view of relational belief revision seems to be the *sceptical* one: Retain, in the actual contraction $K \dot{-} \phi$, only those sentences which are left unquestioned in each of the possible ways to perform a rational contraction, i.e., take the intersection $K_1 \cap K_2 \cap K_3 \cap \dots$. We thus let the Principle of Indifference take priority over the Principle of Minimal Change.

One way of conceiving the problem is to assume that each of the candidate belief sets is itself generated by a belief change operation of the kind considered so far. For the problem of contraction, this means that each of the K_i's is obtained by a contraction function $\dot{-}_i$ of its own. Then the sceptical solution to the problem of relational belief change is defined by putting $\psi \in K \dot{-} \phi$ if and only if $\psi \in K \dot{-}_i \phi$ for every i in the index set I. Assuming that each of the candidate contractions is in turn based on some entrenchment relation $<_i$, we have in the principal case where $\phi < \top$

$$K \dot{-} \phi =$$
$$\bigcap \{K \dot{-}_i \phi : i \in I\} =$$
$$K \cap \{\psi : \phi <_i \phi \vee \psi \text{ for every } i \in I\} =$$
$$K \cap \{\psi : \phi < \phi \vee \psi\} , \text{ with } < = \bigcap\{<_i : i \in I\}.$$

If we can guarantee that the class of all entrenchment relations is closed under intersections, then the sceptical interpretation of a relational belief contraction function based on a collection of entrenchment relations $<_i$ results in a simple (functional) contraction based on a new, more coarsely grained entrenchment relation, viz. on $< = \bigcap \{<_i : i \in I\}$. In particular, every sceptical contraction based on entrenchment relations *with full comparability* can be represented as a simple contraction based on an entrenchment relation without comparability. Normally the full comparability of all sentences gets lost in this process: Modularity is a property that is not preserved under intersection.

It is natural to ask a converse question. If we have a contraction based on some entrenchment relation, can it always be represented as the 'sceptical' meet of a multitude of entrenchment-based contractions, where the latter entrenchment relations have full comparability? This question about contractions reduces to a similar question concerning the underlying relations of epistemic entrenchment. Can a given entrenchment relation $<$ always be represented as the meet of a suitably chosen set of entrenchment relations *with comparability* which extend $<$? Or more ambitiously, is it true that every entrenchment relation is the meet of all its extensions into entrenchments with full comparability?

In order to find an answer to this question, we first address another question which is interesting in itself. Suppose you have an arbitrary 'base relation' $R \subseteq L \times L$ consisting of pairs of sentences $\langle \phi, \psi \rangle$ such that, intuitively speaking, ϕ is epistemically less firmly entrenched than ψ. We may wish to interpret R as a partial specification of an unknown entrenchment relation. How can we know whether this base relation is suitable for serving as the germ of a fully-fledged relation of epistemic entrenchment with or without comparability? What we have to do in order to answer this question is investigate whether R can be *extended* to an entrenchment relation (with or without comparability). Initially specified particular relationships in terms of entrenchment must be taken as given, but in order to make them coherent, one has to comply to (some sample) of the axioms for entrenchment relations, that is, one has to somehow 'close up' the partial specification by *adding* more such relationships between pairs of sentences. For instance, if $\langle \phi, \psi \rangle$ is given, then we add all instances of $\langle \phi \wedge \chi, \psi \rangle$ in order to satisfy (Continuing Down). The following observation tells us how to check whether this can possibly be done. In order to keep things simple, let us now identify entrenchment relations with relations of generalized epistemic entrenchment (GEE-relations) in the sense of Definition 18.

Observation 70 *Let R be a binary relation over L. Then R can be extended to an entrenchment relation iff it satisfies the following condition of entrenchment consistency:*[371]

(EC)   *There is no non-empty finite set of pairs $\langle \phi_1, \psi_1 \rangle, \ldots, \langle \phi_n, \psi_n \rangle$ in R such that $\{\phi_1, \ldots, \phi_n\} \subseteq Cn(\{\psi_1, \ldots, \psi_n\})$.*

Let us say that a binary relation R over L is *entrenchment consistent*, or simply *consistent*, if it can be extended to a GEE-relation over L. For any entrenchment relation $<$, we say that a pair $\langle \phi, \psi \rangle$ is *consistent with* $<$ if and only if $< \cup \{\langle \phi, \psi \rangle\}$ is consistent, otherwise we say that $\langle \phi, \psi \rangle$ is *inconsistent with* $<$. We collect a number of useful facts in the following

Lemma 71 *(i) A pair $\langle \phi, \psi \rangle$ is inconsistent with an entrenchment relation $<$ if and only if $\phi \in Cn(\psi)$ or $\psi < \psi \supset \phi$.*

(ii) If $<$ is a relation that satisfies (EC) and $\phi \wedge \psi$ is not in $Cn(\emptyset)$, then either $\langle \phi, \psi \rangle$ or $\langle \psi, \psi \supset \phi \rangle$ is consistent with $<$.

(iii) If $<$ is a relation that satisfies (EC) and

(Completeness)   *For every pair of sentences ϕ and ψ such that $\phi \wedge \psi$ is not in $Cn(\emptyset)$, either $\phi < \psi$ or $\psi < \psi \supset \phi$.*

then $<$ is an entrenchment relation with comparability in the sense of Definition 18.

(iv) For a logically admissible relation $<$, (Completeness) is equivalent with

(Maxichoice)   *For every pair of sentences ϕ and ψ such that ϕ is not in $Cn(\emptyset)$, either $\phi < \phi \vee \psi$ or $\phi < \phi \vee \neg \psi$.*

Lemma 71 shows how to linearize an entrenchment relation by extending it 'as much as possible'. The condition (Completeness) is due to Rott (1992b, p. 65), the condition (Maxichoice) is due to Fermé (1999) who shows that contractions

---

[371] Also compare similar conditions proposed in Rescher (1973, p. 115) and (1976, p. 15), Rott (1991b, Observation 3 and Definition 12) and Williams (1994, p. 92).

generated by such finely grained entrenchment relations are exactly the 'maxi-choice contractions' of Alchourrón, Gärdenfors and Makinson (1985) or the 'saturatable sets' of Levi (1991, Sections 4.2–4.4; 1998).

For the next observation, we use the fact that L is supposed to be a countable language.

Observation 72 *Every entrenchment relation can be extended to an entrenchment relation with full comparability.*

An immediate consequence of Observations 70 and 72 is

Corollary 73 *Let R be a binary relation over L. Then R can be extended to an entrenchment relation with comparability iff it satisfies the condition of entrenchment consistency.*

At last it is no longer difficult to show that our aim concerning relational belief change can by achieved. Belief revision based on an entrenchment relation without full comparability may always be interpreted as the sceptical solution to relational belief revision (with multiple outcomes) based on entrenchments with full comparability.

Observation 74 *Every entrenchment relation $<$ is the meet of all entrenchment relations with full comparability that extend $<$.*

This result is similar in spirit to Lindström and Rabinowicz's (1991) Theorem 3.14. Due to different formalizations of epistemic entrenchment, however, our proof is entirely different from the proof given there. For a thoroughgoing study of the different ways of understanding entrenchment without comparability, see Cantwell (2001).

APPENDIX A: PROOFS FOR CHAPTER 4

Given $(\dot{-}2)$ and $(\dot{-}5)$, $(\dot{-}8\text{vwd})$ follows from $(\dot{-}8\text{wd})$.

Proof: We show that for all χ in K, if χ is in $Cn(K\dot{-}\phi \cup \{\neg\psi\}) \cup Cn(K\dot{-}\psi \cup \{\neg\phi\})$, then it is in $Cn(K\dot{-}\phi \cup K\dot{-}\psi)$. The claim then follows from $(\dot{-}2)$. Let χ be in K and in $Cn(K\dot{-}\phi \cup \{\neg\psi\})$. Hence $K\dot{-}\phi \vdash \neg\psi \supset \chi$. But also, by $(\dot{-}5)$, $K\dot{-}\psi \vdash \psi \supset \chi$, so $\chi \in Cn(K\dot{-}\phi \cup K\dot{-}\psi)$.

Lemma 1

(i) A contraction function $\dot{-}$ over a consistent theory K which satisfies $(\dot{-}8\text{m})$ in combination with $(\dot{-}1)$, $(\dot{-}2)$, $(\dot{-}3)$ and $(\dot{-}5)$ is trivial in the sense that it puts $K\dot{-}\phi = K$ for all sentences ϕ;

(ii) a contraction function $\dot{-}$ over K satisfies postulates $(\dot{-}7)$ and $(\dot{-}8\text{m})$ if and only if there is a theory H such that for every ϕ, $K\dot{-}\phi = H$.

(iii) a contraction function $\dot{-}$ over K satisfies postulates $(\dot{-}2)$, $(\dot{-}3)$, $(\dot{-}7)$ and $(\dot{-}8\text{wm})$ if and only if there is a subtheory H of K such that for every ϕ,

$$K\dot{-}\phi = \begin{cases} H & \text{if } \phi \in K \\ K & \text{otherwise} \end{cases}$$

Proof of Lemma 1.

(i) Let K be a consistent theory and let ϕ be an arbitrary element and ψ be an arbitrary non-element of K. By $(\dot{-}1)$, $\phi \wedge \psi$ is not in K, hence, by $(\dot{-}3)$, $K \subseteq K\dot{-}(\phi \wedge \psi)$, so ϕ is in $K\dot{-}(\phi \wedge \psi)$. Then $(\dot{-}8\text{m})$ gives us that ϕ is in $K\dot{-}\phi$. On the other hand, we know from $(\dot{-}5)$ and $(\dot{-}1)$ that $K \in Cn((K\dot{-}\phi) \cup \{\phi\}) = Cn(K\dot{-}\phi) = K\dot{-}\phi$, so by $(\dot{-}2)$, $K\dot{-}\phi = K$ for arbitrary ϕ in K. If ϕ is not in K, then $K\dot{-}\phi = K$ by $(\dot{-}2)$ and $(\dot{-}3)$.

(ii) RHS implies LHS: It is clear that $(\dot{-}1)$, $(\dot{-}2)$, $(\dot{-}7)$ and $(\dot{-}8\text{m})$ are satisfied by the contraction function specified in the RHS. (Notice that $(\dot{-}3)$ will in general be violated.)

LHS implies RHS: Take H to be the set $\{\phi \in K : K\dot{-}\phi = K\}$ of all sentences in K which cannot be successfully withdrawn. We first show that H is a theory. Suppose that $\phi_1, \ldots, \phi_n$ are in H and $\{\phi_1, \ldots, \phi_n\} \vdash \psi$. We have to show that ψ is in H. Since each ϕ_i is in H, we know that $K = \bigcap\{K\dot{-}\phi_i : \phi_i \in H\}$, so by $(\dot{-}7)$ and $(\dot{-}6)$ $K = K\dot{-}(\phi_1 \wedge \cdots \wedge \phi_n) = K\dot{-}(\phi_1 \wedge \cdots \wedge \phi_n) \wedge \psi$, so by $(\dot{-}8\text{m})$ and $(\dot{-}2)$, $K = K\dot{-}\psi$, so ψ is indeed in H.

Let $\psi \in K$. It remains to show that for arbitrary ϕ, ψ is in $K\dot{-}\phi$ iff it is in H.

Let $\psi \in K\dot{-}\phi$. Then, by $(\dot{-}6)$ and $(\dot{-}8\text{m})$, $\psi \in K\dot{-}\phi \vee \psi$. But we also have $\psi \in K\dot{-}\phi \vee \neg\psi$, since $(\phi \vee \neg\psi) \supset \psi \in K\dot{-}\phi \vee \neg\psi$ by $(\dot{-}5)$ and $K\dot{-}\phi \vee \neg\psi$ is a theory by $(\dot{-}1)$. So by $(\dot{-}7)$, $\psi \in K\dot{-}\psi$. By $(\dot{-}5)$ again and $(\dot{-}1)$, we get

$K \subseteq Cn((K \dot{-} \psi) \cup \{\psi\}) = Cn(K \dot{-} \psi) = K \dot{-} \psi$, and also, by $(\dot{-}2)$, $K \dot{-} \psi \subseteq K$, so $K \dot{-} \psi = K$, so $\psi \in H$ by definition.

Let $\psi \subseteq H$, so $K \subseteq K \dot{-} \psi \subseteq K \dot{-} \phi \vee \psi$, by $(\dot{-}8m)$, so $\psi \in K \dot{-} \phi \vee \psi$. Using the same argument as in the other direction, we also have $\psi \in K \dot{-} \phi \vee \neg \psi$. Hence $\psi \subseteq (K \dot{-} \phi \vee \psi) \cap (K \dot{-} \phi \vee \neg \psi) \subseteq K \dot{-} \phi$, by $(\dot{-}7)$.

(iii) RHS implies LHS: It is clear that $(\dot{-}1) - (\dot{-}3)$, $(\dot{-}7)$ and $(\dot{-}8wm)$ are satisfied by the contraction function specified in the RHS. (Notice that for theories H, a conjunction $\phi \wedge \psi$ is in H just in case both ϕ and ψ are in H.)

LHS implies RHS: Take H to be the set $\{\phi \in K : K \dot{-} \phi = K\}$ of all sentences in K which cannot be successfully withdrawn. We first show that H is a theory. Suppose that $\phi_1, \ldots, \phi_n$ are in H and $\{\phi_1, \ldots, \phi_n\} \vdash \psi$. As H is a subset of K and K is a theory, ψ is in K. We have to show that ψ is in H. Since each ϕ_i is in H, we know that $K = \bigcap\{K \dot{-} \phi_i : \phi_i \in H\}$, so by $(\dot{-}7)$ and $(\dot{-}6)$ $K = K \dot{-} (\phi_1 \wedge \cdots \wedge \phi_n) = K \dot{-} (\phi_1 \wedge \cdots \wedge \phi_n) \wedge \psi$, so by $(\dot{-}8wm)$ and $(\dot{-}2)$, $K = K \dot{-} \psi$, so ψ is indeed in H.

Let $\psi \in K$. If ϕ is not in K, then $K \dot{-} \phi = K$ by $(\dot{-}2)$ and $(\dot{-}3)$, so the RHS is fine. It remains to show that for arbitrary ϕ in K, ψ is in $K \dot{-} \phi$ iff it is in H.

Let $\psi \in K \dot{-} \phi$. Then, by $(\dot{-}6)$ and $(\dot{-}8wm)$, $\psi \in K \dot{-} \phi \vee \psi$. But we also have $\psi \in K \dot{-} \phi \vee \neg \psi$, since $(\phi \vee \neg \psi) \supset \psi \in K \dot{-} \phi \vee \neg \psi$ by $(\dot{-}5)$ and $K \dot{-} \phi \vee \neg \psi$ is a theory by $(\dot{-}1)$. So by $(\dot{-}7)$, $\psi \in K \dot{-} \psi$. By $(\dot{-}5)$ again and $(\dot{-}1)$, we get $K \subseteq Cn((K \dot{-} \psi) \cup \{\psi\}) = Cn(K \dot{-} \psi) = K \dot{-} \psi$, and also, by $(\dot{-}2)$, $K \dot{-} \psi \subseteq K$, so $K \dot{-} \psi = K$, so $\psi \in H$ by definition.

Let $\psi \in H$, so $K \subseteq K \dot{-} \psi \subseteq K \dot{-} \phi \vee \psi$, by $(\dot{-}8wm)$, so $\psi \in K \dot{-} \phi \vee \psi$. Using the same argument as in the other direction, we also have $\psi \in K \dot{-} \phi \vee \neg \psi$. Hence $\psi \subseteq (K \dot{-} \phi \vee \psi) \cap (K \dot{-} \phi \vee \neg \psi) \subseteq K \dot{-} \phi$, by $(\dot{-}7)$. $\qquad\square$

Lemma 2

Let the contraction function $\dot{-}$ satisfy the basic postulates $(\dot{-}1)$, $(\dot{-}2)$, $(\dot{-}5)$ and $(\dot{-}6)$. Each of the following conditions implies its successor.

(i) For all ψ, $\phi \in K \dot{-} \psi$

(ii) $\phi \in K$ and $K \dot{-} \phi = K$

(iii) $\phi \in K \dot{-} \phi$

(iv) There is a ψ such that $\phi \wedge \psi \in K \dot{-} (\phi \wedge \psi)$

If $\dot{-}$ satisfies in addition $(\dot{-}7)$ and $(\dot{-}8c)$, then (iv) implies (i). Hence in that case all conditions are equivalent.

Proof of Lemma 2. Let the contraction function $\dot{-}$ satisfy the basic postulates $(\dot{-}1)$, $(\dot{-}2)$, $(\dot{-}5)$ and $(\dot{-}6)$.

(i) implies (ii). From (i), we get as a particular case that $\phi \in K \dot{-} \phi$. Hence by $(\dot{-}2)$ and $(\dot{-}5)$ $\phi \in K = K \dot{-} \phi$.

(ii) implies (iii). Trivial.

(iii) implies (iv). Follows from $(\dot{-}1)$ and $(\dot{-}6)$, if we take $\psi = \phi$ or $\psi = \top$.

Now let $\dot{-}$ in addition satisfy $(\dot{-}7)$ and $(\dot{-}8c)$. Then (iv) implies (i). Let ψ be such that $\phi \wedge \psi \in K \dot{-} (\phi \wedge \psi)$, and suppose for reductio that there is a χ such that $\phi \notin K \dot{-} \chi$. By $(\dot{-}6)$ and $(\dot{-}7)$ it follows from $\phi \notin K \dot{-} \chi$ that either $\phi \notin K \dot{-} (\phi \vee \chi)$

or $\phi \notin K \dot{-} (\neg \phi \vee \chi)$. But the latter is impossible since by $(\dot{-}1)$ and $(\dot{-}5)$ we get that $(\neg \phi \vee \chi) \supset \phi$ and hence ϕ is in $K \dot{-} (\neg \phi \vee \chi)$. So $\phi \notin K \dot{-} (\phi \vee \chi)$. On the other hand, it follows from $\phi \wedge \psi \in K \dot{-} (\phi \wedge \psi)$ that $\phi \wedge \psi \in K \dot{-} ((\phi \vee \chi) \wedge (\phi \wedge \psi))$, by $(\dot{-}6)$. Hence, by $(\dot{-}8c)$, $K \dot{-} ((\phi \vee \chi) \wedge (\phi \wedge \psi)) \subseteq K \dot{-} (\phi \vee \chi)$, and thus, by $(\dot{-}1)$, $\phi \in K \dot{-} (\phi \vee \chi)$. This gives us a contradiction, and we are done. $\square$

Every basic inference operation satisfying (Or) also satisfies (Cut).
Proof (compare Kraus, Lehmann and Magidor 1990, p. 191).

Let $\phi \wedge \psi \hspace{0.1em}\vdash\hspace{-0.5em}\sim \chi$ and $\phi \hspace{0.1em}\vdash\hspace{-0.5em}\sim \psi$. From the former and $\phi \wedge \neg \psi \hspace{0.1em}\vdash\hspace{-0.5em}\sim \phi \wedge \neg \psi$, we get by applying (RW) twice, (OR) and (LLE) that $\phi \hspace{0.1em}\vdash\hspace{-0.5em}\sim \neg \psi \vee \chi$. From this and $\phi \hspace{0.1em}\vdash\hspace{-0.5em}\sim \psi$, we get by (AND) and (RW) that $\phi \hspace{0.1em}\vdash\hspace{-0.5em}\sim \chi$. $\square$

Every basic inference operation satisfies (Or) if and only if it satisfies (Cond).
Proof.

That every basic inference operation satisfying (Or) also satisfies (Cond) is proved in Kraus, Lehmann and Magidor 1990, Lemma 5.2. We repeat the proof for the sake of self-containedness. Suppose that $\phi \wedge \psi \hspace{0.1em}\vdash\hspace{-0.5em}\sim \chi$. Then, by (RW), $\phi \wedge \psi \hspace{0.1em}\vdash\hspace{-0.5em}\sim \psi \supset \chi$. Moreover, by (Ref) and (RW), we have $\phi \wedge \neg \psi \hspace{0.1em}\vdash\hspace{-0.5em}\sim \psi \supset \chi$. So by (Or) $(\phi \wedge \psi) \vee (\phi \vee \neg \psi) \hspace{0.1em}\vdash\hspace{-0.5em}\sim \psi \supset \chi$, that is, by (LLE), $\phi \hspace{0.1em}\vdash\hspace{-0.5em}\sim \psi \supset \chi$.

For the converse, suppose that $\phi \hspace{0.1em}\vdash\hspace{-0.5em}\sim \chi$ and $\psi \hspace{0.1em}\vdash\hspace{-0.5em}\sim \chi$. By (LLE), then, $(\phi \vee \psi) \wedge \phi \hspace{0.1em}\vdash\hspace{-0.5em}\sim \chi$ and $(\phi \vee \psi) \wedge \psi \hspace{0.1em}\vdash\hspace{-0.5em}\sim \chi$, so by (Cond), $\phi \vee \psi \hspace{0.1em}\vdash\hspace{-0.5em}\sim \phi \supset \chi$ and $\phi \vee \psi \hspace{0.1em}\vdash\hspace{-0.5em}\sim \psi \supset \chi$. But by (Ref), also $\phi \vee \psi \hspace{0.1em}\vdash\hspace{-0.5em}\sim \phi \vee \psi$. By repeated application of (And) and (RW), we get $\phi \vee \psi \hspace{0.1em}\vdash\hspace{-0.5em}\sim \chi$, as desired. $\square$

Lemma 4
(i) (DRat) implies (WDRat).

(ii) (WDRat), taken together with (Ref), implies (VWDRat).

(iii) (DRat), taken together with (RW) and (And), implies (SDRat).

(iv) (WDRat), taken together with (RW) and (And), implies

(WDRat$^+$) $Inf(\phi \vee \psi) \subseteq Cn(Inf(\phi) \cup \{\psi\})$ or $Cn(\phi \vee \psi) \subseteq Cn(Inf(\psi) \cup \{\phi\})$

(v) For any basic inference relation $\hspace{0.1em}\vdash\hspace{-0.5em}\sim$, (WDRat) is equivalent with

(Triplet) If $\phi \vee \psi \vee \chi \hspace{0.1em}\vdash\hspace{-0.5em}\sim \neg \phi$, then $\phi \vee \psi \hspace{0.1em}\vdash\hspace{-0.5em}\sim \neg \phi$ or $\phi \vee \chi \hspace{0.1em}\vdash\hspace{-0.5em}\sim \neg \phi$

(vi) For any basic inference relation $\hspace{0.1em}\vdash\hspace{-0.5em}\sim$, (WDRat) and (CMon) taken together imply (DRat).

Proof of Lemma 4.

(i) Immediate from the reflexivity of Cn.

(ii) Immediate from the monotonicity of Cn.

(iii) Suppose for reductio that neither $Inf(\phi \vee \psi) \subseteq Inf(\phi)$ nor $Inf(\phi \vee \psi) \subseteq Inf(\psi)$. Then there are χ_1 and χ_2 such that $\chi_1 \in Inf(\phi \vee \psi)$, $\chi_1 \notin Inf(\phi)$, $\chi_2 \in Inf(\phi \vee \psi)$ and $\chi_2 \notin Inf(\psi)$. By (And), $\chi_1 \wedge \chi_2 \in Inf(\phi \vee \psi)$, so by (DRat) either $\chi_1 \wedge \chi_2 \in Inf(\phi)$ or $\chi_1 \wedge \chi_2 \in Inf(\psi)$, so by (RW), either $\chi_1 \in Inf(\phi)$ or $\chi_2 \in Inf(\psi)$, and both cases lead to a contradiction.

(iv) The argument is similar to that of part (iii). Suppose for reductio that neither $Inf(\phi \vee \psi) \subseteq Cn(Inf(\phi) \cup \{\psi\})$ nor $Inf(\phi \vee \psi) \subseteq Cn(Inf(\psi) \cup \{\phi\})$. Then there are χ_1 and χ_2 such that $\chi_1 \in Inf(\phi \vee \psi)$, $\chi_1 \notin Cn(Inf(\phi) \cup \{\psi\})$, $\chi_2 \in Inf(\phi \vee \psi)$ and $\chi_2 \notin Cn(Inf(\psi) \cup \{\phi\})$. By (And), we get that $\chi_1 \wedge \chi_2 \in Inf(\phi \vee \psi)$, so by (WDRat) either $\chi_1 \wedge \chi_2 \in Cn(Inf(\phi) \cup \{\psi\})$ or $\chi_1 \wedge \chi_2 \in Cn(Inf(\psi) \cup \{\phi\})$, so by the properties of Cn (!) either $\chi_1 \in Cn(Inf(\phi) \cup \{\psi\})$ or $\chi_2 \in Cn(Inf(\psi) \cup \{\phi\})$, and both cases lead to a contradiction.

(v) (WDRat) implies (Triplet): Let $\phi \vee \psi \vee \chi \mathrel{\vvdash} \neg\phi$, or equivalently, by (LLE), $(\phi \vee \psi) \vee (\phi \vee \chi) \mathrel{\vvdash} \neg\phi$. By (WDRat), either $\phi \vee \psi \mathrel{\vvdash} (\phi \vee \chi) \supset \neg\phi$ or $\phi \vee \chi \mathrel{\vvdash} (\phi \vee \psi) \supset \neg\phi$. By (RW), this yields either $\phi \vee \psi \mathrel{\vvdash} \neg\phi$ or $\phi \vee \psi \mathrel{\vvdash} \neg\phi$, as desired.

(Triplet) implies (WDRat): Let $\phi \vee \psi \mathrel{\vvdash} \chi$. By (LLE) and (RW), we get $\neg(\neg\phi \vee \neg\psi \vee \chi) \vee \phi \vee \psi \mathrel{\vvdash} \neg\phi \vee \neg\psi \vee \chi$. Then (Triplet) gives us that either $\neg(\neg\phi \vee \neg\psi \vee \chi) \vee \phi \mathrel{\vvdash} \neg\phi \vee \neg\psi \vee \chi$ or $\neg(\neg\phi \vee \neg\psi \vee \chi) \vee \psi \mathrel{\vvdash} \neg\phi \vee \neg\psi \vee \chi$. Assume the former. Then (LLE), (Ref), (And) and (RW) yield $\phi \mathrel{\vvdash} \neg\psi \vee \chi$, so $\chi \in Cn(Inf(\phi) \cup \{\psi\})$. The latter case leads similarly to $\chi \in Cn(Inf(\psi) \cup \{\phi\})$, so we have proved (WDRat).

(vi) Let $\phi \vee \psi \mathrel{\vvdash} \chi$. We need to show that either $\phi \mathrel{\vvdash} \chi$ or $\psi \mathrel{\vvdash} \chi$. From the supposition, we get with (RW) that $\phi \vee \psi \mathrel{\vvdash} (\phi \vee \psi) \supset \chi$. From this we get with (LLE) that $\phi \vee \psi \vee \neg((\phi \vee \psi) \supset \chi) \mathrel{\vvdash} (\phi \vee \psi) \supset \chi$. Now we use (Triplet), which has already been shown to be equivalent with (WDRat), and get that either $\phi \vee \neg((\phi \vee \psi) \supset \chi) \mathrel{\vvdash} (\phi \vee \psi) \supset \chi$ or $\psi \vee \neg((\phi \vee \psi) \supset \chi) \mathrel{\vvdash} (\phi \vee \psi) \supset \chi$. Assume the former. With (Ref), (And) and (RW) we then get $\phi \vee \neg((\phi \vee \psi) \supset \chi) \mathrel{\vvdash} \phi$. From this and $\phi \vee \neg((\phi \vee \psi) \supset \chi) \mathrel{\vvdash} (\phi \vee \psi) \supset \chi$ we get with (CMon) and (LLE) that $\phi \mathrel{\vvdash} (\phi \vee \psi) \supset \chi$. Applying (Ref), (And) and (RW) again, we get that $\phi \mathrel{\vvdash} \chi$. The latter case leads similarly to $\psi \mathrel{\vvdash} \chi$, so we have proved (DRat). $\qquad\square$

(CMon) implies (WCMon), and (Or) implies (WOr), provided the basic conditions for nonmonotonic logics are satisfied.

Proof. For (WCMon), suppose that $\bot \in Inf(\phi)$. So by (RW), $\psi \in Inf(\phi)$. So by (CMon), $\bot \in Inf(\phi \wedge \psi)$. For (WOr), suppose that $\neg\phi \in Inf(\phi)$. Also, by (Ref) and (RW), $\neg\phi \in Inf(\neg\phi \wedge \psi)$. So by (Or) and (LLE), $\neg\phi \in Inf(\phi \vee (\neg\phi \wedge \psi)) = Inf(\phi \vee \psi)$.

Observation 5

(i) A basic inference operation Inf satisfies (Or) and (Mon) if and only if there is a fixed set of sentences F such that for every ϕ, $Inf(\phi) = Cn(F \cup \{\phi\})$;

(ii) a basic inference operation Inf satisfies (WRMon), (Or) and (WMon) if and only if there are two fixed sets of sentences F and G such that $F \subseteq G$ and for every ϕ,

$$Inf(\phi) = \begin{cases} Cn(G \cup \{\phi\}) \text{ if } G \cup \{\phi\} \not\mathrel{\vdash} \bot \\ Cn(F \cup \{\phi\}) \text{ otherwise} \end{cases}$$

Proof of Observation 5.

(i) Soundness: Notice that $F \cup \{\phi\} \vdash \chi$ and $F \cup \{\psi\} \vdash \chi$ imply $F \cup \{\phi \vee \psi\} \vdash \chi$, and $F \cup \{\phi\} \vdash \chi$ implies $F \cup \{\phi \wedge \psi\} \vdash \chi$.

Completeness: Suppose that Inf satisfies (Or) and (Mon) and put $F = \{\phi : Inf(\neg\phi) = L\}$. We first show that F is a theory under Cn. Suppose that $F \vdash \chi$ for some χ. Then, by the compactness of Cn, there are $\phi_1, \ldots, \phi_n$ in F such that $\{\phi_1, \ldots, \phi_n\} \vdash \chi$. By the properties of Cn, then, $Cn(\neg\chi) = Cn((\neg\phi_1 \vee \cdots \vee \neg\phi_n) \wedge \neg\chi)$. Since $\phi_i \in F$, we have $Inf(\neg\phi_i) = L$ for every i, hence by (Or), $Inf(\neg\phi_1 \vee \cdots \vee \neg\phi_n) = L$, so by (Mon) and (LLE) $Inf((\neg\phi_1 \vee \cdots \vee \neg\phi_n) \wedge \neg\chi) = Inf(\neg\chi) = L$, i.e., $\chi \in F$.

Now we have to show that $\phi \mathbin{\vert\!\sim} \psi$ iff $F \cup \{\phi\} \vdash \psi$. If $\phi \mathbin{\vert\!\sim} \psi$, then by (Mon), (Ref) and (And) $\phi \wedge \neg\psi \mathbin{\vert\!\sim} \psi \wedge \neg\psi$, so by (LLE) and (RW) $Inf(\neg(\phi \supset \psi)) = L$, that is $\phi \supset \psi \in F$, hence $F \cup \{\phi\} \vdash \psi$. Conversely, if $F \cup \{\phi\} \vdash \psi$, then $F \vdash \phi \supset \psi$. Since F was shown to be a theory, we get $\phi \supset \psi \in F$, that is, $Inf(\neg(\phi \supset \psi)) = L$, and in particular $\psi \in Inf(\neg(\phi \supset \psi))$. Since by (Ref) and (RW) $\psi \in Inf(\phi \wedge \psi)$, (Or) and (LLE) give us $\phi \mathbin{\vert\!\sim} \psi$.

(ii) Soundness: Let Inf be defined as stated in the Lemma. It is clear that Inf is basic.

For (WRMon), suppose that $\neg\phi \notin Inf(\top)$. It follows that $Inf(\top) \not\vdash \neg\phi$, by (And) and (RW) and the properties of Cn. Now by definition, $Inf(\top) = Cn(G)$ if G is consistent with respect to Cn, and $Inf(\top) = Cn(F)$ if G is inconsistent with respect to Cn. In the latter case $Inf(\top) = Cn(F) \subseteq Cn(F \cup \{\phi\}) = Inf(\phi)$, since ϕ is inconsistent with an inconsistent G. In the former case $Inf(\top) = Cn(G) \subseteq Cn(G \cup \{\phi\}) = Inf(\phi)$, since ϕ is consistent with $Cn(G) = Inf(\top)$. In both cases we have $Inf(\top) \subseteq Inf(\phi)$, as desired.

For (Or), suppose that $\phi \mathbin{\vert\!\sim} \chi$ and $\psi \mathbin{\vert\!\sim} \chi$. We want to show that $\phi \vee \psi \mathbin{\vert\!\sim} \chi$. If $G \cup \{\phi \vee \psi\} \vdash \bot$ (case 1), then both $G \cup \{\phi\} \vdash \bot$ and $G \cup \{\psi\} \vdash \bot$, so by hypothesis both $F \cup \{\phi\} \vdash \chi$ and $F \cup \{\psi\} \vdash \chi$. By the properties of Cn, then $F \cup \{\phi \vee \psi\} \vdash \chi$, so in this case $\phi \vee \psi \mathbin{\vert\!\sim} \chi$. If $G \cup \{\phi \vee \psi\} \not\vdash \bot$ (case 2), then we only need to show that $G \cup \{\phi \vee \psi\} \vdash \chi$. But since $F \subseteq G$, we have by hypothesis in any case that $G \cup \{\phi\} \vdash \chi$ and $G \cup \{\psi\} \vdash \chi$, so $G \cup \{\phi \vee \psi\} \vdash \chi$ by the properties of Cn.

For (WMon), let $\neg\phi$ be in $Inf(\top)$. Now by definition, $Inf(\top) = Cn(G)$ if G is consistent with respect to Cn, and $Inf(\top) = Cn(F)$ if G is inconsistent with respect to Cn. In the latter case $Inf(\phi) = Cn(F \cup \{\phi\}) \subseteq Cn(F \cup \{\phi \wedge \psi\}) = Inf(\phi \wedge \psi)$, and we are done. In the former case, $\neg\phi \in Inf(\top)$ means that $G \vdash \neg\phi$, so $G \cup \{\phi\}$ and also $G \cup \{\phi \wedge \psi\}$ are inconsistent with respect to Cn, and we again get, by the definition of Inf, $Inf(\phi) = Cn(F \cup \{\phi\}) \subseteq Cn(F \cup \{\phi \wedge \psi\}) = Inf(\phi \wedge \psi)$.

Completeness: Let Inf be a basic inference operation satisfying (WRMon), (Or) and (WMon). Put $F = \{\phi : Inf(\neg\phi) = L\}$, as in (i), and put $G = Inf(\top)$. We have shown above that F is a theory, and we now show that $F \subseteq G$. If ϕ is in F, then, by the definition of F, in particular $\neg\phi \mathbin{\vert\!\sim} \phi$. By Reflexivity, also $\phi \mathbin{\vert\!\sim} \phi$. Hence, by (Or) and (LLE) $\top \mathbin{\vert\!\sim} \phi$, so ϕ is in G.

Now we have to show that $\phi \mathbin{\vert\!\sim} \psi$ iff *either* $\neg\phi \notin G$ and $G \cup \{\phi\} \vdash \psi$, *or* $\neg\phi \in G$ and $F \cup \{\phi\} \vdash \psi$.

From left to right. Let $\phi \mathbin{\vert\!\sim} \psi$. Then by (Cond), which is equivalent to (Or), we get $\top \mathbin{\vert\!\sim} (\phi \supset \psi)$, that is $G \cup \{\phi\} \vdash \psi$. So we are ready if $\neg\phi \notin G = Inf(\top)$.

If $\neg\phi \in G = Inf(\top)$, then by (WMon), $Inf(\phi) \subseteq Inf(\phi \wedge \neg\psi)$, so we can infer from $\phi \mathrel{|\!\sim} \psi$ that $\phi \wedge \neg\psi \mathrel{|\!\sim} \psi$. But by (Ref) and (RW), also $\phi \wedge \neg\psi \mathrel{|\!\sim} \neg\psi$, so by (LLE), (And) and (RW), $Inf(\neg(\phi \supset \psi)) = Inf(\phi \wedge \neg\psi) = L$. That means, $\phi \supset \psi$ is in F, so $F \cup \{\phi\} \vdash \psi$, which is what we needed.

From right to left. For the first case, suppose that $\neg\phi \notin G = Inf(\top)$ and $G \cup \{\phi\} \vdash \psi$. From the former, we conclude, with (WRMon), that $Inf(\top) \subseteq Inf(\phi)$. But the latter yields that $\phi \supset \psi$ is in $G = Inf(\top)$, so we get $\phi \mathrel{|\!\sim} \phi \supset \psi$. Thus, by (Ref), (And) and (RW), we get $\phi \mathrel{|\!\sim} \psi$. For the second case, suppose that $\neg\phi \in G$ and $F \cup \{\phi\} \vdash \psi$. The latter gives us that $F \vdash \phi \supset \psi$, and since F is a theory, $\phi \supset \psi$ is in F. By the definition of F and (LLE), this means that $\phi \wedge \neg\psi \mathrel{|\!\sim} \bot$. We conclude by (Cond) which is equivalent with (Or) that $\phi \mathrel{|\!\sim} \neg\psi \supset \bot$, and with (RW) that $\phi \mathrel{|\!\sim} \psi$. $\qquad\square$

APPENDIX B: PROOFS FOR CHAPTER 5

Observation 6
Let H be a finite set of sentences and Inf be an arbitrary inference operation. Let $K = Inf(\bigwedge H)$, and define $K * \phi$ as $Inf(\bigwedge H \wedge \phi)$, for every sentence ϕ. Then the revision operation $*$ over K satisfies $\left\{ \begin{matrix} (*1) \\ (*2) \\ (*3) \\ (*4) \\ (*6) \\ (*7) \\ (*7c) \\ (*8) \\ (*8c) \end{matrix} \right\}$ if Inf satisfies

$\left\{ \begin{matrix} Cn\text{-Closure} \\ \text{(Ref) and (RW)} \\ \text{(Cond)} \\ \text{(RMon)} \\ \text{(LLE)} \\ \text{(Cond)} \\ \text{(Cut)} \\ \text{(RMon)} \\ \text{(CMon)} \end{matrix} \right\}$. However, the revision operation $*$ does not satisfy Consistency Preservation, $(*5)$.

Proof of Observation 6.

*Re (*1).* $K * \phi$ is a theory since $Cn\,(Inf(\bigwedge H \wedge \phi)) \subseteq Inf(\bigwedge H \wedge \phi)$, by (And) and (RW).

*Re (*2).* ϕ is in $K * \phi = Inf(\bigwedge H \wedge \phi))$ since by (Ref), $\bigwedge H \wedge \phi$ is in $Inf(\bigwedge H \wedge \phi)$ and thus by (RW) ϕ is in $Inf(\bigwedge H \wedge \phi)$.

*Re (*3).* Let $K = Inf(\bigwedge H)$. By definition, $K * \phi = Inf(\bigwedge H \wedge \phi)$ and $K + \phi = Cn\,(K \cup \{\phi\}) = Cn\,(Inf(\bigwedge H) \cup \{\phi\})$. So we have to show that

$$Inf(\bigwedge H \wedge \phi) \subseteq Cn\,(Inf(\bigwedge H) \cup \{\phi\})$$

This inclusion follows immediately from (Cond).

*Re (*4).* Let ϕ be Cn-consistent with $K = Inf(\bigwedge H)$. By definition, $K * \phi = Inf(\bigwedge H \wedge \phi)$ and $K + \phi = Cn\,(K \cup \{\phi\}) = Cn\,(Inf(\bigwedge H) \cup \{\phi\})$. So we have to show that

$$Cn\left(Inf(\bigwedge H) \cup \{\phi\}\right) \subseteq Inf(\bigwedge H \wedge \phi)$$

Suppose that ψ is in the LHS. So $\phi \supset \psi$ is in $Cn\left(Inf(\bigwedge\phi)\right)$. By ($Cn$-closure), $\phi \supset \psi$ is in $Inf(\bigwedge H)$. Since ϕ is consistent with $Inf(\bigwedge H)$, we can apply (RMon) and get that $\phi \supset \psi$ is in $Inf(\bigwedge H \wedge \phi)$. On the other hand, by (Ref) and (RW), we get that ϕ is in $Inf(\bigwedge H \wedge \phi)$. So by (And) and (RW) again, we finally get that ψ is in $Inf(\bigwedge H \wedge \phi)$.

Re (∗6). When $Cn(\phi) = Cn(\psi)$, then clearly $Cn(\bigwedge H \wedge \phi) = Cn(\bigwedge H \wedge \psi)$ and hence $Inf(\bigwedge H \wedge \phi) = Inf(\bigwedge H \wedge \psi)$, by (LLE).

Re (∗7). Let $K * \phi = Inf(\bigwedge H \wedge \phi)$. By definition and (LLE), $K * (\phi \wedge \psi) = Inf(\bigwedge H \wedge (\phi \wedge \psi)) = Inf((\bigwedge H \wedge \phi) \wedge \psi)$ and $(K * \phi) + \psi = Cn((K * \phi) \cup \{\psi\}) = Cn(Inf(\bigwedge H \wedge \phi) \cup \{\psi\})$. So we have to show that

$$Inf((\bigwedge H \wedge \phi) \wedge \psi) \subseteq Cn(Inf(\bigwedge H \wedge \phi) \cup \{\psi\})$$

This inclusion follows immediately from (Cond).

Re (∗7c). Let ψ be in $K * \phi = Inf(\bigwedge H \wedge \phi)$. By definition and (LLE), $K * (\phi \wedge \psi) = Inf(\bigwedge H \wedge (\phi \wedge \psi)) = Inf((\bigwedge H \wedge \phi) \wedge \psi)$ and $K * \phi = Inf(\bigwedge H \wedge \phi)$. So we have to show that

$$Inf((\bigwedge H \wedge \phi) \wedge \psi) \subseteq Inf(\bigwedge H \wedge \phi)$$

Since ψ is in $Inf(\bigwedge H \wedge \phi)$, this inclusion follows immediately from (Cut).

Re (∗8). Let ψ be consistent with $K * \phi = Inf(\bigwedge H \wedge \phi)$. By definition and (LLE), $(K * \phi) + \psi = Cn((K * \phi) \cup \{\psi\}) = Cn(Inf(\bigwedge H \wedge \phi) \cup \{\psi\})$ and $K * (\phi \wedge \psi) = Inf(\bigwedge H \wedge (\phi \wedge \psi)) = Inf((\bigwedge H \wedge \phi) \wedge \psi)$. So we have to show that

$$Cn(Inf(\bigwedge H \wedge \phi) \cup \{\psi\}) \subseteq Inf((\bigwedge H \wedge \phi) \wedge \psi)$$

Suppose that χ is in the LHS. So $\psi \supset \chi$ is in $Cn(Inf(\bigwedge H \wedge \phi))$. By ($Cn$-closure), $\psi \supset \chi$ is in $Inf(\bigwedge H \wedge \phi)$. Since ψ is consistent with $Inf(\bigwedge H \wedge \phi)$, we can apply (RMon) and get that $\psi \supset \chi$ is in $Inf((\bigwedge H \wedge \phi) \wedge \psi)$. On the other hand, by (Ref) and (RW), we get that ψ is in $Inf((\bigwedge H \wedge \phi) \wedge \psi)$. So by (And) and (RW) again, we finally get that χ is in $Inf((\bigwedge H \wedge \phi) \wedge \psi)$.

Re (∗8c). Let ψ be in $K * \phi = Inf(\bigwedge H \wedge \phi)$. By definition and (LLE), $K * \phi = Inf(\bigwedge H \wedge \phi)$ and $K * (\phi \wedge \psi) = Inf(\bigwedge H \wedge (\phi \wedge \psi)) = Inf((\bigwedge H \wedge \phi) \wedge \psi)$. So we have to show that

$$Inf(\bigwedge H \wedge \phi) \subseteq Inf((\bigwedge H \wedge \phi) \wedge \psi)$$

Since ψ is in $Inf(\bigwedge H \wedge \phi)$, this inclusion follows immediately from (CMon).

Consistency Preservation (∗5) fails immediately. Consider the base $H = \{\neg\phi\}$. Then $K * \phi$ equals $Inf(\phi \wedge \neg\phi)$ which is the inconsistent theory L, by (Ref) and (RW). $\qquad\square$

Observation 7

Let H be a finite set of sentences and Inf be an arbitrary inference operation. Let $K = Inf(\bigwedge H)$, and define $K * \phi$ as $Inf(\bigwedge(H *_0 \phi))$, for every sentence ϕ, where $*_0$ is the operation of full meet base revision. Then the revision operation

$$* \text{ over } K \text{ satisfies } \left\{ \begin{array}{l} (*1) \\ (*2) \\ (*3) \\ (*4) \\ (*5) \\ (*6) \end{array} \right\} \text{ if } Inf \text{ satisfies } \left\{ \begin{array}{c} Cn\text{-Closure} \\ (\text{Ref}) \text{ and } (\text{RW}) \\ (\text{Cond}) \\ (\text{RMon}) \\ (\text{CP}) \\ (\text{LLE}) \end{array} \right\}. \text{ However, the}$$

revision operation $*$ satisfies neither $(*7)$ nor $(*8)$, nor $(*8r)$, even when $Inf = Cn$. It satisfies both $(*7c)$ and $(*8c)$ when $Inf = Cn$, but fails to satisfy either of the two when Inf is different from Cn.

Proof of Observation 7.

$(*1)$, $(*2)$, $(*5)$, and $(*6)$ all follow immediately from their respective Inf-correlates.

(*3) and (*4). Let ϕ be consistent with $K = Inf(\bigwedge H)$. By (Ref), (RW) and (And), we get that ϕ is consistent with H. Hence $H \bot \neg \phi$ is $\{H\}$, and thus $H *_0 \phi = H \cup \{\phi\}$. Having said this, we reason as follows:

$$K * \phi =$$
$$Inf(\bigwedge(H *_0 \phi)) =$$
$$Inf(\bigwedge H \wedge \phi) = \quad \text{(by (Cond), (RMon), (Ref), (RW) and (And))}$$
$$Cn(Inf(\bigwedge H) \cup \{\phi\}) =$$
$$Cn(K \cup \{\phi\}) =$$
$$K + \phi.$$

Counterexample to (*7). Consider the information base H consisting of the two items $p \supset q \wedge r$ and $\neg p \vee \neg q$, and $Inf = Cn$. Then we have

$$\begin{aligned}
K &= Inf(\textstyle\bigwedge H) &&= Cn\,(\neg p) \\
H\bot\neg p &= \{\{p \supset q \wedge r\}, \{\neg p \vee \neg B\}\} \\
H\bot(\neg p \vee \neg q) &= \{\{p \supset q \wedge r\}\} \\
H\dot{-}_0\neg p &= \textstyle\bigcap H\bot\neg p &&= \emptyset \\
H\dot{-}_0(\neg p \vee \neg q) &= \textstyle\bigcap H\bot(\neg p \vee \neg q) &&= \{p \supset q \wedge r\} \\
H *_0 p &= (H\dot{-}_0\neg p) \cup \{p\} &&= \{p\} \\
H *_0 (p \wedge q) &= (H\dot{-}_0(\neg p \vee \neg q)) \cup \{p \wedge B\} &&= \{p \supset q \wedge r, p \wedge q\} \\
K * p &= Inf(\textstyle\bigwedge(H *_0 p)) &&= Cn\,(p) \\
K * (p \wedge q) &= Inf(\textstyle\bigwedge(H *_0 p \wedge q)) &&= Cn\,(p \wedge q \wedge r)
\end{aligned}$$

We see that $K * p \wedge q$ is not a subset of $Cn\,((K*p)\cup\{q\}) = Cn\,(p \wedge q)$, in violation of $(*7)$.

Counterexample to $(*8)$. Consider the information base H consisting of the three items $p \wedge q \supset r$, $p \supset s$ and $p \supset (\neg q \vee \neg r) \wedge \neg s$, and $Inf = Cn$. Then we have

$$\begin{aligned}
K &= Inf(\textstyle\bigwedge H) &&= Cn\,(\neg p) \\
H\bot\neg p &= \{\{p \wedge q \supset r, p \supset s\}, \{p \wedge q \supset r, p \supset (\neg q \vee \neg r) \wedge \neg s\}\} \\
H\bot(\neg p \vee \neg q) &= \{\{p \wedge q \supset r, p \supset s\}, \{p \supset (\neg q \vee \neg r) \wedge \neg s\}\} \\
H\dot{-}_0\neg p &= \textstyle\bigcap H\bot\neg p &&= \{p \wedge q \supset r\} \\
H\dot{-}_0(\neg p \vee \neg q) &= \textstyle\bigcap H\bot(\neg p \vee \neg q) &&= \emptyset \\
H *_0 p &= (H\dot{-}_0\neg p) \cup \{p\} &&= \{p, p \wedge q \supset r\} \\
H *_0 (p \wedge q) &= (H\dot{-}_0(\neg p \vee \neg q)) \cup \{p \wedge q\} &&= \{p \wedge q\} \\
K * p &= Cn\,(\emptyset) \underline{*} (\textstyle\bigwedge(H *_0 p)) &&= Cn\,(p \wedge (q \supset r)) \\
K * (p \wedge q) &= Inf(\textstyle\bigwedge(H *_0 p \wedge q)) &&= Cn\,(p \wedge q)
\end{aligned}$$

We see that $\neg q$ is not in $K * p$. Nevertheless, $Cn\,((K*p)\cup\{q\}) = Cn\,(p \wedge q \wedge r)$ is not a subset of $K * p \wedge q$, in violation of $(*8)$.

Re $(*7c)$ and $(*8c)$ for the case $Inf = Cn$. The monotonic inference operation $Inf = Cn$ does not allow for a counterexample to either $(*7c)$ or $(*8c)$. For in that case, $\psi \in K * \phi$ means that ψ is in $Cn\,(\textstyle\bigwedge(H\dot{-}_0\neg\phi) \wedge \phi)$, thus $\phi \supset \psi$ follows from $\textstyle\bigcap(H\bot\neg\phi)$, thus a fortiori $\phi \supset \psi$ follows from G for every G in $H\bot\neg\phi$. Then one can easily prove that $H\bot\neg\phi \vee \neg\psi$ is a subset of $H\bot\neg\phi$, and in fact $H\bot\neg\phi \vee \neg\psi = H\bot\neg\phi$. Hence clearly $H\dot{-}_0\neg\phi = H\dot{-}_0\neg\phi \vee \neg\psi$ and $Cn\,(\textstyle\bigwedge(H\dot{-}_0\neg\phi)\wedge\phi) = Cn\,(\textstyle\bigwedge(H\dot{-}_0\neg\phi\vee\neg\psi)\wedge\phi\wedge\psi)$. Hence $K*\phi = Inf(\textstyle\bigwedge(H *_0$

$\phi)) = Inf(\bigwedge(H \dot{-}_0 \neg \phi) \wedge \phi) = Inf(\bigwedge(H \dot{-}_0 \neg \phi \vee \neg \psi) \wedge \phi \wedge \psi) = Inf(\bigwedge(H *_0 \phi \wedge \psi)) = K * \phi \wedge \psi.$

However, there are simple inference nonmonotonic relations which show that both (*7c) and (*8c) are violated in the general case.

*Counterexample to (*7c) and (*8c).* Consider the information base H consisting of the two items $p \supset q \wedge r$ and $\neg p \vee \neg q$, and *Inf* defined by

$$Inf(\phi) = \begin{cases} Cn(\phi \wedge (p \supset q) \wedge \neg r) & \text{if } \phi \text{ is consistent with } (p \supset q) \wedge \neg r \\ Cn(\phi) & \text{otherwise} \end{cases}$$

It is easy to verify that *Inf* is a rational inference operation (compare Observation 5(ii)). Then we have

$$
\begin{array}{llll}
K & = Inf(\bigwedge H) & = Cn(\neg p \wedge \neg r) \\
H \perp \neg p & = \{\{p \supset q \wedge r\}, \{\neg p \vee \neg q\}\} \\
H \perp (\neg p \vee \neg q) & = \{\{p \supset q \wedge r\}\} \\
H \dot{-}_0 \neg p & = \bigcap H \perp \neg p & = \emptyset \\
H \dot{-}_0 (\neg p \vee \neg q) & = \bigcap H \perp (\neg p \vee \neg q) & = \{p \supset q \wedge r\} \\
H *_0 p & = (H \dot{-}_0 \neg p) \cup \{p\} & = \{p\} \\
H *_0 (p \wedge q) & = (H \dot{-}_0 (\neg p \vee \neg q)) \cup \{p \wedge q\} & = \{p \supset q \wedge r, p \wedge q\} \\
K * p & = Inf(\bigwedge(H *_0 p)) & = Cn(p \wedge q \wedge \neg r) \\
K * (p \wedge q) & = Inf(\bigwedge(H *_0 p \wedge q)) & = Cn(p \wedge q \wedge r)
\end{array}
$$

We see that $K * p \wedge q$ is in $K * p$. Nevertheless, $K * p \wedge q$ is neither a subset of $K * p$, nor is $K * p$ is not a subset of $K * p \wedge q$; they are in fact contradictory. Thus both (*7c) and (*8c) are violated.

*Counterexample to (*8r).* Consider the information base H consisting of the three items $p \supset (q \wedge r)$, $\neg p \vee \neg q$ and $q \supset (p \wedge \neg r)$, and $Inf = Cn$. Then we have

$$K \qquad\qquad = Inf(\bigwedge H) \qquad\qquad = Cn\,(\neg p \wedge \neg q)$$

$$H \perp \neg p \qquad\quad = \{\{p \supset (q \wedge r)\}, \{\neg p \vee \neg q, q \supset (p \wedge \neg r)\}\}$$

$$H \perp \neg q \qquad\quad = \{\{q \supset (p \wedge \neg r)\}, \{p \supset (q \wedge r), \neg p \vee \neg q\}\}$$

$$H \perp (\neg p \wedge \neg q) = \{\{p \supset (q \wedge r), \neg p \vee \neg q\}, \{\neg p \vee \neg q, q \supset (p \wedge \neg r)\}\}$$

$$H \dotminus_0 \neg p \qquad\quad = \bigcap H \perp \neg p \qquad\qquad = \emptyset$$

$$H \dotminus_0 \neg q \qquad\quad = \bigcap H \perp \neg q \qquad\qquad = \emptyset$$

$$H \dotminus_0 (\neg p \wedge \neg q) = \bigcap H \perp (\neg p \wedge \neg q) \qquad = \{\neg p \vee \neg q\}$$

$$H *_0 p \qquad\quad = (H \dotminus_0 \neg p) \cup \{p\} \qquad = \{p\}$$

$$H *_0 q \qquad\quad = (H \dotminus_0 \neg q) \cup \{q\} \qquad = \{q\}$$

$$H *_0 (p \vee q) \quad = (H \dotminus_0 (\neg p \wedge \neg q)) \cup \{p \vee q\} = \{p \vee q, \neg p \vee \neg q\}$$

$$K * p \qquad\quad = Inf(\bigwedge(H *_0 p)) \qquad = Cn\,(p)$$

$$K * q \qquad\quad = Inf(\bigwedge(H *_0 q)) \qquad = Cn\,(q)$$

$$K * (p \vee q) \quad = Inf(\bigwedge(H *_0 p \vee q)) \qquad = Cn\,(p \leftrightarrow \neg q)$$

We see that $\neg p \vee \neg q$ is in $K * (p \vee q)$. However, $\neg p \vee \neg q$ is not in $Cn\,((K * p) \cup (K * q)) = Cn\,(p \wedge q)$ (it is in fact inconsistent with it). Hence, $K * (p \vee q)$ is not a subset of $Cn\,((K * p) \cup (K * q))$, in violation of $(*8r)$. $\qquad\square$

Observation 8

Let $\mathcal{E}$ be a prioritized expectation base and let Inf be the inference operation on prioritized belief bases defined by $Inf(\mathcal{H}) = Consol(\mathcal{E} \circ \mathcal{H})$, for any prioritized information base $\mathcal{H}$. Take some such $\mathcal{H}$, let $K = Inf(\mathcal{H})$, and define $K * \phi$ as $Inf(\mathcal{H} \circ \langle \phi \rangle)$, for every sentence ϕ. Then the revision operation $*$ over K satisfies

$$\left.\begin{array}{l} (*1) \\ (*2^-) \\ (*3) \\ (*5^+) \\ (*6) \\ (*7) \\ (*8c) \\ (*8r) \end{array}\right\}.$$ However, the revision operation $*$ does not satisfy the Preservation

Principle (Pres), and *a fortiori*, it satisfies neither $(*4)$ nor $(*8)$.[372]

Proof of Observation 8.

Using the notation introduced in Section 1.2.7.2, we have $Inf(\mathcal{H}) = Consol(\mathcal{G}) = \bigcap \{Cn\,(G') : G' \in \mathcal{G} \parallel \perp\}$.

---

[372]Note that $K * \top = Consol(\mathcal{G} \circ \langle \top \rangle) = Consol(\mathcal{G}) = K$ in the present model.

Notice, as a Lemma, that for consistent ϕ, F is in $(\mathcal{G} \circ \langle \phi \rangle) \parallel \perp$ iff either F (if ϕ is in G) or $F - \{\phi\}$ (if not) is in $\mathcal{G} \parallel \neg\phi$. For inconsistent ϕ, F is in $(\mathcal{G} \circ \langle \phi \rangle) \parallel \perp$ iff it is in $\mathcal{G} \parallel \perp$.

(*1) follows from the fact that $Inf(\mathcal{H}) = Consol(\mathcal{E} \circ \mathcal{H})$ is a Cn-theory for any $\mathcal{H}$.

(*2⁻) and *(*5⁺)* follow from the fact that $Inf(\mathcal{H} \circ \langle \phi \rangle) = Consol(\mathcal{G} \circ \langle \phi \rangle)$ and every element of $(\mathcal{G} \circ \langle \phi \rangle) \parallel \perp$ is consistent and contains ϕ, if ϕ is consistent.

(*3) follows from *(*7)* which is proved below, and the facts that $K * \phi = K * (\top \wedge \phi)$ and $Cn(K \cup \{\phi\} = Cn((K * \top) \cup \{\phi\})$. (Notice that $K * \top = Consol(\mathcal{G} \circ \langle \top \rangle) = Consol(\mathcal{G}) = K$.)

(*6) follows immediately, since $Consol(\mathcal{G} \circ \langle \phi \rangle) = Consol(\mathcal{G} \circ \langle \psi \rangle)$ if $Cn(\phi) = Cn(\psi)$.

Re (*7). We have to show that $K * \phi \wedge q$ is a subset of $Cn((K * \phi) \cup \{\psi\})$.

Suppose that $\chi \in K * \phi \wedge \psi = Consol(\mathcal{G} \circ \langle \phi \wedge \psi \rangle)$. That is, $\chi \in Cn(G')$ for all G' in $(\mathcal{G} \circ \langle \phi \wedge \psi \rangle) \parallel \perp$.

Now suppose for reductio that $\chi \notin Cn((K * \phi) \cup \{\psi\})$. Thus $\psi \supset \chi \notin K * \phi$. Hence there is some G'' in $(\mathcal{G} \circ \langle \phi \rangle) \parallel \perp$ such that $\psi \supset \chi \notin Cn(G'')$.

If ϕ is inconsistent, then so is $\phi \wedge \psi$, and we have $(\mathcal{G} \circ \langle \phi \rangle) \parallel \perp = (\mathcal{G} \circ \langle \phi \wedge \psi \rangle) \parallel \perp = \mathcal{G} \parallel \perp$, and there cannot be a G'' in $(\mathcal{G} \circ \langle \phi \rangle) \parallel \perp$ such that $\psi \supset \chi \notin Cn(G'')$ since $\chi \in Cn(G'')$ by supposition.

So let ϕ be consistent. Then ϕ is in G''. Hence $\neg\phi \vee \neg\psi \notin Cn(G'')$. But then, since G'' is in $(\mathcal{G} \circ \langle \phi \rangle) \parallel \perp$, it is in $(\mathcal{G} \circ \langle \phi \rangle) \parallel \neg\phi \vee \neg\psi$ as well. But then either $G'' \cup \{\phi \wedge \psi\}$ (if ϕ is in G) or $(G'' - \{\phi\}) \cup \{\phi \wedge \psi\}$ (if not) is in $(\mathcal{G} \circ \langle \phi \wedge \psi \rangle) \parallel \perp$. Hence, by our supposition, $\chi \in Cn(G'' \cup \{\phi \wedge \psi\})$, hence $\psi \supset \chi \in Cn(G'' \cup \{\phi\})$, and since ϕ is in G'', we have $\psi \supset \chi \in Cn(G'')$, contradicting what we have been supposing.

Re (*8c). We have to show that if ψ is in $K * \phi$, then $K * \phi$ is a subset of $K * \phi \wedge \psi$.

Suppose that ψ and χ are in $K * \phi = Consol(\mathcal{G} \circ \langle \phi \rangle)$. That is, ψ and χ are in $Cn(G')$ for all G' in $(\mathcal{G} \circ \langle \phi \rangle) \parallel \perp$.

We want to show that χ is in $K * \phi \wedge \psi$. Suppose for reductio that $\chi \notin K * \phi \wedge \psi$. Thus there is some G'' in $(\mathcal{G} \circ \langle \phi \wedge \psi \rangle) \parallel \perp$ such that $\chi \notin Cn(G'')$.

If ϕ is inconsistent, then $(\mathcal{G} \circ \langle \phi \rangle) \parallel \perp = \mathcal{G} \parallel \perp = (\mathcal{G} \circ \langle \phi \wedge \psi \rangle) \parallel \perp$, and there there cannot be a G'' in $(\mathcal{G} \circ \langle \phi \wedge \psi \rangle) \parallel \perp$ such that $\chi \notin Cn(G'')$ since $\chi \in Cn(G'')$ by supposition.

So let ϕ be consistent. Then ϕ is in every G' in $(\mathcal{G} \circ \langle \phi \rangle) \parallel \perp$, and since by supposition ψ, too, is in every such G' and every such G' is consistent, it follows that $\phi \wedge \psi$ is consistent. So $\phi \wedge \psi$ is in every G'' in $(\mathcal{G} \circ \langle \phi \wedge \psi \rangle) \parallel \perp$.

Now let

$$F = \begin{cases} G'' \cup \{\phi\} & \text{if } \phi \wedge \psi \text{ is in } G \\ (G'' - \{\phi \wedge \psi\}) \cup \{\phi\} & \text{otherwise} \end{cases}$$

We show that F is in $(\mathcal{G} \circ \langle \phi \rangle) \perp\!\!\!\perp \bot$. Suppose for reductio that there is a subset F' of G which is "better" than F (in the sense of the relation $\prec$ introduced above). Then we can choose F' from $(\mathcal{G} \circ \langle \phi \rangle) \perp\!\!\!\perp \bot$ (because the number of subsets of G is finite (!?) and $\prec$ is transitive). Since F' is better than F and F is constructed from $G'' \in (\mathcal{G} \circ \langle \phi \wedge \psi \rangle) \perp\!\!\!\perp \bot$, it holds that $\bot \in Cn(F' \cup \{\phi \wedge \psi\})$. Since ϕ is consistent, ϕ is in F', and we conclude that $\neg \psi \in Cn(F')$, contradicting the above supposition that ψ is in $Cn(G')$ for all G' in $(\mathcal{G} \circ \langle \phi \rangle) \perp\!\!\!\perp \bot$.

We conclude that F is indeed in $(\mathcal{G} \circ \langle \phi \rangle) \perp\!\!\!\perp \bot$. But then, by our supposition, χ is in $Cn(F)$, and also in $Cn(G'')$, and we have a contradiction.

Re (*8r). We have to show that $K * \phi \vee \psi$ is a subset of $Cn((K*\phi) \cup (K*\psi))$.

Suppose that $\chi \in K * \phi \vee \psi = Consol(\mathcal{G} \circ \langle \phi \vee \psi \rangle)$. That is, $\chi \in Cn(G')$ for all G' in $(\mathcal{G} \circ \langle \phi \vee \psi \rangle) \perp\!\!\!\perp \bot$, or equivalently (by Disjunctive Reasoning),

$$\bigvee \{\bigwedge G' : G' \in (\mathcal{G} \circ \langle \phi \vee \psi \rangle) \perp\!\!\!\perp \bot\} \vdash \chi$$

If $\phi \vee \psi$ is inconsistent, then so are ϕ and ψ, and we have $(\mathcal{G} \circ \langle \phi \vee \psi \rangle) \perp\!\!\!\perp \bot = (\mathcal{G} \circ \langle \phi \rangle) \perp\!\!\!\perp \bot = (\mathcal{G} \circ \langle \psi \rangle) \perp\!\!\!\perp \bot = \mathcal{G} \perp\!\!\!\perp \bot$, hence $K * \phi \vee \psi = K * \phi = K * \psi = K$ and we are done.

So let $\phi \vee \psi$ be consistent. Then $\phi \vee \psi$ is in all G' in $(\mathcal{G} \circ \langle \phi \vee \psi \rangle) \perp\!\!\!\perp \bot$, and we can equivalently write

$$(\phi \vee \psi) \wedge \bigvee \{\bigwedge G' : G' \in \mathcal{G} \perp\!\!\!\perp (\neg \phi \wedge \neg \psi)\} \vdash \chi$$

Now we want to show that $K * \phi = Consol(\mathcal{G} \circ \langle \phi \rangle)$ and $K * \psi = Consol(\mathcal{G} \circ \langle \psi \rangle)$ taken together imply χ. That is, we have to show that

$$\bigvee \{\bigwedge F : F \in (\mathcal{G} \circ \langle \phi \rangle) \perp\!\!\!\perp \bot\}$$

and

$$\bigvee \{\bigwedge F' : F' \in (\mathcal{G} \circ \langle \psi \rangle) \perp\!\!\!\perp \bot\}$$

taken together imply χ. If either ϕ or ψ are inconsistent, then $(\mathcal{G} \circ \langle \phi \vee \psi \rangle) \perp\!\!\!\perp \bot = (\mathcal{G} \circ \langle \phi \rangle) \perp\!\!\!\perp \bot$ or $(\mathcal{G} \circ \langle \phi \vee \psi \rangle) \perp\!\!\!\perp \bot = (\mathcal{G} \circ \langle \psi \rangle) \perp\!\!\!\perp \bot$, and we are done.

So let both ϕ and ψ be consistent. Then ϕ is in every $F \in (\mathcal{G} \circ \langle \phi \rangle) \perp\!\!\!\perp \bot$ and ψ is in every $F' \in (\mathcal{G} \circ \langle \psi \rangle) \perp\!\!\!\perp \bot$. What we have to show then, is that

$$(\phi \wedge \bigvee \{\bigwedge F : F \in \mathcal{G} \perp\!\!\!\perp \neg \phi\}) \wedge (\psi \wedge \bigvee \{\bigwedge F' : F' \in \mathcal{G} \perp\!\!\!\perp \neg \psi\})$$

implies χ. By propositional reformulation, this means that

$$\bigvee \{\phi \wedge \psi \wedge (\bigwedge F) \wedge (\bigwedge F') : F \in \mathcal{G} \perp\!\!\!\perp \neg \phi \text{ and } F' \in \mathcal{G} \perp\!\!\!\perp \neg \psi\}$$

implies χ. Now we take into account that for $F \in \mathcal{G} \perp\!\!\!\perp \neg \phi$ and $F' \in \mathcal{G} \perp\!\!\!\perp \neg \psi$, the conjunction $\bigwedge F \wedge \bigwedge F'$ will entail either $\neg \phi$ or $\neg \psi$, by the maximality of F and F'—unless F and F' are identical. So for F and F' which are not identical, we get that the conjunction $\phi \wedge \psi \wedge (\bigwedge F) \wedge (\bigwedge F')$ is a contradiction, and we can drop these conjunctions in the big disjunction. What we have to show, then, is that

$$\bigvee \{\phi \wedge \psi \wedge (\bigwedge F) : F \in (\mathcal{G} \parallel \neg\phi) \cap (\mathcal{G} \parallel \neg\psi)\}$$

implies χ. But now we realize that all F in $(\mathcal{G} \parallel \neg\phi) \cap (\mathcal{G} \parallel \neg\psi)$ are in $\mathcal{G} \parallel (\neg\phi \wedge \neg\psi)$. So the above disjunction implies

$$\phi \wedge \psi \wedge \bigvee \{\bigwedge F : F \in \mathcal{G} \parallel (\neg\phi \wedge \neg\psi)\}$$

which in turn is of course stronger than

$$(\phi \vee \psi) \wedge \bigvee \{\bigwedge F : F \in \mathcal{G} \parallel (\neg\phi \wedge \neg\psi)\}$$

But by our supposition, this formula implies χ. Since implication with respect to Cn is transitive, we are done.

A counterexample to the Preservation Principle has been given in the main text of Section 5.2.2. $\square$

Observation 9

Let $\mathcal{E}$ and $\mathcal{H}$ be prioritized expectation and information bases, respectively. Let $K = Consol(\mathcal{E} \circ \mathcal{H})$, and define

$$K * \phi = Consol(\mathcal{E} \circ \mathcal{H} \circ \langle\phi\rangle)$$
$$K \dot{-} \phi = Consol(\mathcal{E} \circ \mathcal{H} \circ \langle\neg\phi^*\rangle)$$

for all sentences ϕ. Then the revision and contraction operations $*$ and $\dot{-}$ over K satisfy the Levi identity with respect to consistent sentences ϕ. However, both parts of the Harper identity are violated.

Proof of Observation 9.

We again use $\mathcal{G}$ as an abbreviation for $\mathcal{E} \circ \mathcal{H}$. For the Levi identity as applied to consistent ϕ, we need to show that

$$K * \phi = Cn((K \dot{-} \neg\phi) \cup \{\phi\})$$

Now, a sentence ψ is in the LHS of this equation if and only if ψ follows from G, for all G in $\mathcal{G} \circ \langle\phi\rangle \parallel \perp$. Since ϕ is supposed to be consistent, ϕ is an element of any such G, and we can equivalently say that ψ is in the LHS of the above equation if and only if $\phi \supset \psi$ follows from $G - \{\phi\}$, for all G in $\mathcal{G} \circ \langle\phi\rangle \parallel \perp$. In the case that ϕ is not an element of $\bigcup \mathcal{G}$, this just means that $\phi \supset \psi$ is in $K \dot{-} \neg\phi$. In the case that ϕ is an element of $\bigcup \mathcal{G}$, we know that ψ is in $K \dot{-} \neg\phi$ iff it follows from G, for all G in $\mathcal{G} \circ \langle\phi\rangle \parallel \perp$, i.e., iff it is in $K * \phi$. But in that case $K \dot{-} \neg\phi$ will actually contain ϕ, so $Cn((K \dot{-} \neg\phi) \cup \{\phi\}) = K \dot{-} \neg\phi$, and we are done.

Next let us show that the following half of the Harper identity is violated.

$$K \dot{-} \phi \subseteq K \cap K * \neg\phi$$

We repeat the example given against the Inclusion condition ($\dot{-}2$) in Section 5.2.2. Let $\mathcal{G}$ be $\langle\{\neg p\}, \{p\}\rangle$. Then $K = Consol(\mathcal{G}) = Cn(p)$ and $K \dot{-} p = Consol(\mathcal{G} \circ$

$\langle\neg p^*\rangle) = Consol(\langle\{\neg p\}, \{p\}, \{\neg p^*\}\rangle) = Cn\,(\neg p)$. Hence, $K\dot{-}p$ is not a subset of K, and a fortiori, the above inclusion fails to be satisfied.

Finally we show that the other half of the Harper identity is violated.

$$K \cap K * \neg\phi \subseteq K\dot{-}\phi$$

Let $\mathcal{G}$ be $\langle\{p \vee q, p \leftrightarrow q\}\rangle$. Then $K = Consol(\mathcal{G}) = Cn\,(p \wedge q)$ and $K * \neg p = Consol(\mathcal{G}\circ\langle\neg p\rangle) = Consol(\langle\{p\vee q, p \leftrightarrow q\}, \{\neg p\}\rangle) = Cn\,(\neg p\wedge q)\cap Cn\,(\neg p\wedge\neg q) = Cn\,(\neg p)$. However, if we look at the contraction $K\dot{-}p$, we find that it equals $Consol(\mathcal{G} \circ \langle\neg p^*\rangle) = Consol(\langle\{p \vee q, p \leftrightarrow q\}, \{\neg p^*\}\rangle) = Cn\,(p \vee q) \cap Cn\,(p \leftrightarrow q) = Cn\,(\emptyset)$. Since $Cn\,(\neg p)$ is evidently not a subset of $Cn\,(\emptyset)$, this direction of the Harper identity gets violated as well. $\qquad\square$

APPENDIX C: PROOFS FOR CHAPTER 6

Condition ($\emptyset$) implies the conjunction of ($\emptyset$1) and ($\emptyset$2); conversely ($\emptyset$) is implied by the conjunction of ($\emptyset$1) and ($\emptyset$2) if the domain of σ is closed under finite intersections.

Proof. ($\emptyset$1) follows from ($\emptyset$), since if $S \subseteq S'$ and $\sigma(S') = \emptyset$, then by ($\emptyset$) $\sigma(S) = \sigma(S) \cap S' = \emptyset$. ($\emptyset$2) follows immediately from ($\emptyset$). For the converse, suppose that ($\emptyset$1) and ($\emptyset$2) hold, that $\mathcal{X}$ is closed under finite intersections and that $\sigma(S) = \emptyset$. Then we use ($\emptyset$1) and get $\sigma(S \cap S') = \emptyset$, so by ($\emptyset$2), $\sigma(S') \cap (S \cap S') = \sigma(S') \cap S = \emptyset$. $\qquad\square$

Lemma 10
Let x be in X and consider the following conditions:
(i) $x \in S$ for some $S \in \mathcal{X}$ such that $\sigma(S) = \emptyset$
(ii) $x \notin \sigma(S)$ for every $S \in \mathcal{X}$
If σ satisfies ($\emptyset$) and (i), then it satisfies (ii). If σ is 1-covering and satisfies (ii) then it satisfies (i) .

Proof of Lemma 10.
(i) implies (ii), provided σ satisfies ($\emptyset$). Suppose that there is an S such that $x \in S$ and $\sigma(S) = \emptyset$ and suppose further, for reductio, that there is an S' such that $x \in \sigma(S')$. But then $\sigma(S') \cap S \neq \emptyset$, contradicting ($\emptyset$).
(ii) implies (i), provided σ is 1-covering. Take $S = \{x\}$. $\qquad\square$

Lemma 11
If σ satisfies ($\emptyset$1) and ($\emptyset$2) and $\bigcup \{S \in \mathcal{X} : \sigma(S) = \emptyset\}$ is in the domain $\mathcal{X}$ of σ, then it holds for every S' in $\mathcal{X}$ that $\sigma(S') = \emptyset$ if and only if $S' \subseteq \bigcup \{S \in \mathcal{X} : \sigma(S) = \emptyset\}$.

Proof of Lemma 11.
If $\sigma(S') = \emptyset$, then $S' \subseteq \bigcup\{S : \sigma(S) = \emptyset\}$ holds trivially. For the converse, assume that $\overline{R} = \bigcup\{S : \sigma(S) = \emptyset\} \in \mathcal{X}$ and $S' \subseteq \overline{R}$. For all S such that $\sigma(S) = \emptyset$, it holds that $\sigma(\overline{R}) \cap S = \emptyset$, by ($\emptyset$2). Taking the big union, we get that $\sigma(\overline{R}) \cap (\bigcup\{S : \sigma(S) = \emptyset\}) = \emptyset$. But by definition, this just means that $\sigma(\overline{R}) \cap \overline{R} = \sigma(\overline{R}) = \emptyset$, so $\sigma(S') = \emptyset$, by ($\emptyset$1). $\qquad\square$

Lemma 12
A choice function satisfies ($\emptyset$) if and only if it possesses a taboo set.

Proof of Lemma 12.
For the *if* direction, suppose that σ has a taboo set $\overline{R}$ with respect to which it satisfies (t1) and (t2). Let $\sigma(S) = \emptyset$. We have to show that $\sigma(S') \cap S = \emptyset$ for

all S' in $\mathcal{X}$. But from $\sigma(S) = \emptyset$ we conclude that $S \subseteq \overline{R}$ with the help of (t2). On the other hand, we know from (t1) that $\sigma(S') \cap \overline{R} = \emptyset$. Taking the last two conditions together, we obtain $\sigma(S') \cap S = \emptyset$, as desired.

For the *only if* direction, suppose that σ satisfies $(\emptyset)$ and put $\overline{R} = X - \bigcup \{\sigma(S) : S \in \mathcal{X}\}$. We show that $\overline{R}$ satisfies (t1) and (t2). For (t1), we have to show that for all S, $\sigma(S)$ does not intersect $X - \bigcup \{\sigma(S) : S \in \mathcal{X}\}$. But this is immediate. For (t2), let $\sigma(S) = \emptyset$ and assume for reductio that S is not a subset of $\overline{R}$. This means that S intersects $\sigma(S')$ for some S' in $\mathcal{X}$. But this is forbidden, by $(\emptyset)$. $\qquad \square$

For finitely additive and subtractive domains, condition (I) is equivalent to

(I′) $\sigma(S \cup S') \subseteq \sigma(S) \cup \sigma(S')$

Proof.

Let σ be finitely additive and subtractive. To show that (I) implies (I′), let $S, S', S \cup S' \in \mathcal{X}$ and $x \in \sigma(S \cup S')$. As $S, S' \subseteq S \cup S'$ and x is either in S or in S', we get that $x \in \sigma(S) \cup \sigma(S')$, by (I), as desired. For the converse, let $S, S' \in \mathcal{X}$ and $S \subseteq S'$. By subtractivity, $S' - S \in \mathcal{X}$ as well. Using $S' = S \cup (S' - S)$ and (I′), we get $\sigma(S') \subseteq \sigma(S) \cup \sigma(S' - S)$. Intersecting both sides with S, then, we get $\sigma(S') \cap S \subseteq (\sigma(S) \cap S) \cup (\sigma(S' - S) \cap S) = \sigma(S)$, since $\sigma(S) \subseteq S$ and $\sigma(S' - S) \cap S = \emptyset$. $\qquad \square$

Condition (III) is independent of condition (II), even for finite X and in the presence of condition (I).

Proof.

Consider the following example: Let $X = \{x, y, z\}$, $\mathcal{X}$ be the power set of X minus $\emptyset$, and consider σ_1 and σ_2 defined by $\sigma_1(\{x, y\}) = \sigma_2(\{x, y\}) = \{x, y\}$, $\sigma_1(\{x, z\}) = \{x\}$, $\sigma_1(\{y, z\}) = \{z\}$, $\sigma_1(\{x, y, z\}) = \{x\}$, $\sigma_2(\{x, z\}) = \{x, z\}$, $\sigma_2(\{y, z\}) = \{y, z\}$, and $\sigma_2(\{x, y, z\}) = \{x, y\}$. Evidently, σ_1 satisfies (I) and (II) and violates (III), while σ_2 satisfies (I) and (III) and violates (II). $\qquad \square$

Condition (IV), taken together with $(\emptyset 1)$, implies both (II) and (III).

Proof.

Conditions (IV) and $(\emptyset 1)$ imply condition (II): Let $\{S_i : i \in I\} \subseteq \mathcal{X}$, $\bigcup \{S_i : i \in I\} \in \mathcal{X}$. If $\sigma(\bigcup \{S_i : i \in I\}) = \emptyset$, then by $(\emptyset 1)$ we get that $\sigma(S_i) = \emptyset$ for each $i \in I$, and (II) is vacuously satisfied. So suppose that $\sigma(\bigcup \{S_i : i \in I\}) \neq \emptyset$. Now $\sigma(\bigcup \{S_i : i \in I\}) \subseteq \bigcup \{S_i : i \in I\}$, so $\sigma(\bigcup \{S_i : i \in I\}) \cap S_j \neq \emptyset$ for some $j \in I$. So by (IV), $\bigcap \{\sigma(S_i) : i \in I\} \subseteq \sigma(S_j) \subseteq \sigma(\bigcup \{S_i : i \in I\})$.

Conditions (IV) and $(\emptyset 1)$ imply condition (III): Let $S \subseteq S'$. If $\sigma(S') = \emptyset$, then $\sigma(S) = \emptyset$, by $(\emptyset 1)$, and the claim of (III) is trivial. So let $\sigma(S') \neq \emptyset$. But in this case $\sigma(S') \subseteq S$ implies $\sigma(S') \cap S \neq \emptyset$, so clearly (III) follows from (IV). $\qquad \square$

Condition ($\emptyset 2^+$) follows from (I) and (III).

Proof.
Let $S \subseteq S'$ and $\sigma(S) = \emptyset$. First notice that by (I) $\sigma(S') \cap S \subseteq \sigma(S) = \emptyset$. Hence $\sigma(S') \subseteq S' - S$, so $\sigma(S') \subseteq \sigma(S') \cap (S' - S) \subseteq \sigma(S' - S)$ by (I) again. For the converse, we apply (III), and conclude from $\sigma(S') \subseteq S' - S$ that $\sigma(S' - S) \subseteq \sigma(S')$. $\square$

Lemma 13

Let σ be a choice function and $< \; = \mathcal{P}(\sigma)$ and $<_2 \; = \mathcal{P}_2(\sigma)$. Then
(i)   If σ is 12-covering and satisfies ($\emptyset 1$) and ($\emptyset 2$), then $x < y$ implies $x <_2 y$.
(ii) If σ satisfies (I), then $x <_2 y$ implies $x < y$.
(iii) If σ is 12-covering and satisfies ($\emptyset 1$) and (I), then $\mathcal{P}(\sigma) = \mathcal{P}_2(\sigma)$.

Proof of Lemma 13.
(i) Let $x < y$, that is, $y \notin \sigma(S)$ for every $S \in \mathcal{X}$ such that $\{x, y\} \subseteq S$, and $x \notin \overline{R}$. We have to show that $\{x, y\}$ is in $\mathcal{X}$, $x \in \sigma(\{x, y\})$ and $y \notin \sigma(\{x, y\})\}$. That $\{x, y\}$ is in $\mathcal{X}$ follows from 12-covering, and that $y \notin \sigma(\{x, y\})$ follows directly from the assumptions. Now suppose for reductio that $x \notin \sigma(\{x, y\})$. Then $\sigma(\{x, y\}) = \emptyset$. Thus, by ($\emptyset 1$), $\sigma(\{x\}) = \emptyset$, and, by ($\emptyset 2$), $x \notin \sigma(S)$ for all $S \in \mathcal{X}$. But then x is in $\overline{R}$, what contradicts our assumptions.

(ii) Let $x <_2 y$, that is, $\{x, y\}$ is in $\mathcal{X}$, $x \in \sigma(\{x, y\})$ and $y \notin \sigma(\{x, y\})\}$, and let S be an arbitrary element of $\mathcal{X}$ which contains x and y. We have to show that $y \notin \sigma(S)$. But this follows immediately from $y \notin \sigma(\{x, y\})\}$ and (I). It is also clear from our assumptions that there is an $S' \in \mathcal{X}$ such that $x \in \sigma(S')$: take $S' = \{x, y\}$.

(iii) Follows immediately from the (i) and (ii) and the fact that (I) implies ($\emptyset 2$). $\square$

Lemma 14

(i) If σ is relational then it satisfies (I) and (II$^\infty$), and thus ($\emptyset 2$), but not necessarily ($\emptyset 1$) or ($\emptyset 2^+$). If σ is in addition smooth, then it also satisfies ($\emptyset 1$) and ($\emptyset 2^+$).

(ii) If σ is 12-covering and satisfies ($\emptyset 1$), (I) and (II$^\infty$), then it is relational with respect to $\mathcal{P}_2(\sigma)$, i.e., $\sigma = \mathcal{S}(\mathcal{P}_2(\sigma))$.

(iii) If σ is 12-covering and satisfies ($\emptyset 1$), (I) and (II$^\infty$), then it is relational with respect to $\mathcal{P}(\sigma)$, i.e., $\sigma = \mathcal{S}(\mathcal{P}(\sigma))$.

(iv) If σ is 12-covering and relational and satisfies ($\emptyset 1$), then it is relational with respect to both $\mathcal{P}(\sigma)$ and $\mathcal{P}_2(\sigma)$, i.e., $\sigma = \mathcal{S}(\mathcal{P}(\sigma)) = \mathcal{S}(\mathcal{P}_2(\sigma))$.

(v) If σ is additive and satisfies ($\emptyset 2^+$), (I) and (II$^\infty$), then it is relational with respect to $\mathcal{P}(\sigma)$, i.e., $\sigma = \mathcal{S}(\mathcal{P}(\sigma))$.

Proof of Lemma 14.
(i) Immediate from the definition of relationality: If x is minimal in S' then it is so in all subsets S of S' in which it is contained. If x is minimal in every S_i then it is so in $\bigcup S_i$.

If $S \subseteq S'$ and $\sigma(S') = \emptyset$ because $S' \subseteq \overline{R}$, then $\sigma(S) = \emptyset$ as well, because $S \subseteq \overline{R}$. On the other hand, it may be that $S' \not\subseteq \overline{R}$ and $\sigma(S') = \emptyset$ due to infinite descending $<$-chains in S' which are cut, however, in a subset S of S', so that $\sigma(S) \neq \emptyset$. Then $(\emptyset 1)$ is violated. Condition $(\emptyset 2^+)$ may fail for a similar reason.

If, however, there are no infinite descending $<$-chains in S, for every S in $\mathcal{X}$, then $\sigma(S) \neq \emptyset$ for every S such that $S \not\subseteq \overline{R}$, so that $(\emptyset 1)$, $(\emptyset 2)$ and $(\emptyset 2^+)$ are trivially satisfied.

(ii) Let σ be 12-covering and satisfy $(\emptyset 1)$, (I) and (II^∞), and let $<_2$ be $\mathcal{P}_2(\sigma)$. We have to show that for every $S \in \mathcal{X}$ such that $S \not\subseteq \overline{R}$,

$$\sigma(S) = \{x \in S : y <_2 x \text{ for no } y \in S\}$$

which means that for every $S \not\subseteq \overline{R}$

$$\sigma(S) = \{x \in S : \text{for all } y \in S \text{ such that } \{x,y\} \in \mathcal{X},$$
$$\text{either } x \in \sigma(\{x,y\}) \text{ or } y \notin \sigma(\{x,y\})\}$$

$LHS \subseteq RHS$: Let $x \in \sigma(S)$ and $y \in S$. As σ is 12-covering, we always get $\{x,y\} \in \mathcal{X}$, and (I) gives us $x \in \sigma(\{x,y\})$.

$RHS \subseteq LHS$: Let $x \in S$ and assume that for every $y \in S$ such that $\{x,y\} \in \mathcal{X}$ either $x \in \sigma(\{x,y\})$ or $y \notin \sigma(\{x,y\})$. That $\{x,y\}$ is always in $\mathcal{X}$ is clear from 12-covering. If for every $y \in S$ it holds that $x \in \sigma(\{x,y\})$, then $x \in \sigma(S)$ follows from (II^∞) (for finite S, it is sufficient to have (II)). So suppose that there is a $y \in S$ such that $x \notin \sigma(\{x,y\})$. Then $y \notin \sigma(\{x,y\})$, so $\sigma(\{x,y\}) = \emptyset$. So by $(\emptyset 1)$, $\sigma(\{x\}) = \emptyset$, so by $(\emptyset 2)$ (which follows from (I)), $x \notin \sigma(\{x,y'\})$ for all $y' \in S$. Then $y' \notin \sigma(\{x,y'\})$, so $\sigma(\{x,y'\}) = \emptyset$ for all $y' \in S$. Then by $(\emptyset 2)$, $\sigma(S) = \emptyset$. But that means that S is a subset of $\overline{R}$—which is a case to which the above equality need not apply at all.

(iii) Follows from (ii) and Lemma 13 (iii).

(iv) Follows from (i), (ii), and (iii).

(v) Let σ be additive and satisfy (I) and (II^∞) and let $<$ be $\mathcal{P}(\sigma)$. We have to show that for every $S \in \mathcal{X}$ such that $S \not\subseteq \overline{R}$,

$$\sigma(S) = \{x \in S : y < x \text{ for no } y \in S\}$$

which means that for every such $S \not\subseteq \overline{R}$

$$\sigma(S) = \{x \in S : \text{ for all } y \in S \text{ it holds that } either \text{ there is an } S' \in \mathcal{X}$$
$$\text{such that } y \in S' \text{ and } x \in \sigma(S'), \text{ } or \text{ } y \in \overline{R}$$

LHS $\subseteq$ RHS: Let $x \in \sigma(S)$. Then x is trivially in the RHS; just take $S' = S$.

RHS $\subseteq$ LHS: Let $x \in S$ and let the following hold for all $y \in S$: either there is an $S' \in \mathcal{X}$ such that $y \in S'$ or $y \in \overline{R}$. Now consider the non-empty set $S - \overline{R}$ of all $y \in S$ for which there is an S'' such that $y \in \sigma(S'')$. For all these y, we know that there is an $S' \in \mathcal{X}$ such that $y \in S'$ and $x \in \sigma(S')$. Since σ is supposed to be additive, we get that the union $S^{\cup}$ of all these sets S' is in $\mathcal{X}$. By (II^{∞}), $x \in \sigma(S^{\cup})$. (For finitary σ, additivity may be replaced by finite additivity, and (II^{∞}) by (II).)

Hence $x \notin \overline{R}$. Since $S - \overline{R} \subseteq S^{\cup}$, we get from $x \in \sigma(S^{\cup}) \cap (S - \overline{R})$ that $x \in \sigma(S - \overline{R})$, by (I). But also, by the definition of the taboo set $\overline{R}$, $\sigma(S \cap \overline{R}) = \emptyset$. Finally, the principle $(\emptyset 2^+)$ gives us $x \in \sigma((S - \overline{R}) \cup (S \cap \overline{R})) = \sigma(S)$, as desired.$\square$

Lemma 16

(i) If σ is $12n$-covering and satisfies $(\emptyset 1)$ and (I), then $\mathcal{P}(\sigma) = \mathcal{P}_2(\sigma)$ is n-acyclic. If σ is ω-covering and satisfies $(\emptyset 1)$ and (I), then $\mathcal{P}(\sigma) = \mathcal{P}_2(\sigma)$ is acyclic.

(ii) If σ is 123-covering and satisfies $(\emptyset 1)$, (I) and (III), then $\mathcal{P}(\sigma) = \mathcal{P}_2(\sigma)$ is transitive.

(iii) If σ is 123-covering and satisfies $(\emptyset 1)$, (I) and (IV), then $\mathcal{P}(\sigma) = \mathcal{P}_2(\sigma)$ is modular.

(iv) If σ is finitely additive and satisfies $(\emptyset 1)$ and (IV), then $\mathcal{P}(\sigma)$ is modular.

Proof of Lemma 16

(i) Let σ be $12n$-covering and satisfy $(\emptyset 1)$ and (I). We show the claim for $< = \mathcal{P}_2(\sigma)$, which is identical with $\mathcal{P}(\sigma)$, by Lemma 13 (iii). Suppose for reductio that $x_1 < x_2 < \cdots < x_n < x_1$ for some $x_1, \ldots x_n \in X$. That is, $\sigma(\{x_i, x_{i+1}\}) = \{x_i\}$ with $+$ denoting addition modulo n. Now consider $\sigma(\{x_1, \ldots, x_n\})$ which is non-empty, by $(\emptyset 1)$. Let $x_{k+1} \in \sigma(\{x_1, \ldots, x_n\})$. But $x_{k+1} \notin \sigma(\{x_k, x_k + 1\})$. This is in contradiction with (I). The rest of (i) is trivial.

(ii) Let σ be 123-covering and satisfy $(\emptyset 1)$, (I) and (III). We show the claim for $< = \mathcal{P}_2(\sigma)$, which is identical with $\mathcal{P}(\sigma)$, by Lemma 13 (iii). Assume that $x < y$ and $y < z$. By the definition of $<$, this means that $\sigma(\{x, y\}) = \{x\}$ and $\sigma(\{y, z\}) = \{y\}$. Now consider $\sigma(\{x, y, z\})$. By (I), $y \notin \sigma(\{x, y, z\})$ and $z \notin \sigma(\{x, y, z\})$, so $\{x\} \subseteq \sigma(\{x, y, z\})$. But $\sigma(\{x, y, z\})$ must be non-empty, by $(\emptyset 1)$. So $\sigma(\{x, y, z\}) = \{x\} \subseteq \{x, z\}$. Hence, by (III), $\sigma(\{x, z\}) \subseteq \sigma(\{x, y, z\}) = \{x\}$, so $z \notin \sigma(\{x, z\})$, i.e. $x < z$, as desired.

(iii) Let σ be 123-covering and satisfy $(\emptyset 1)$, (I) and (IV). We show the claim for $< = \mathcal{P}_2(\sigma)$, which is identical with $\mathcal{P}(\sigma)$, by Lemma 13 (iii). Let $x < y$, i.e., $\{x, y\} \in \mathcal{X}$ and $\sigma(\{x, y\}) = \{x\}$. We have to show that either $x < z$ or $z < y$. Suppose for reductio that this is not the case. Since σ is 12-covering, both $\{x, z\}$ and $\{y, z\}$ are in $\mathcal{X}$, so we get

$$z \in \sigma(\{x, z\}) \text{ or } x \notin \sigma(\{x, z\}) \tag{$\dagger$}$$

and

$$y \in \sigma(\{y, z\}) \text{ or } z \notin \sigma(\{y, z\}) \tag{$\ddagger$}$$

We first verify that the first disjunct of ($\dagger$) applies. For if $z \notin \sigma(\{x, z\})$, then by ($\dagger$), also $x \notin \sigma(\{x, z\})$, so $\sigma(\{x, z\}) = \emptyset$, so, by $(\emptyset 1)$, $\sigma(\{x\}) = \emptyset$, and so, by $(\emptyset 2)$ (which follows from (I)), $x \notin \sigma(\{x, y\})$, contradicting our initial assumption. Similarly, we verify that the first disjunct of ($\ddagger$) applies. We have just shown that

$z \in \sigma(\{x, z\})$, so, by (Ø2) again, $\sigma(\{z\}) \neq \emptyset$, and so, by (Ø1), $\sigma(\{y, z\}) \neq \emptyset$, which would be the case, by (‡), if the first disjunct of (‡) did not apply. So in sum we have got that $z \in \sigma(\{x, z\})$ and $y \in \sigma(\{y, z\})$.

Now consider $\{x, y, z\}$. This set is in $\mathcal{X}$, since σ is 3-covering, and $\sigma(\{x, y, z\}) \neq \emptyset$, by (Ø1). Hence $\sigma(\{x, y, z\}) \cap \{x, z\} \neq \emptyset$ or $\sigma(\{x, y, z\}) \cap \{y, z\} \neq \emptyset$. Therefore, by (IV), $\sigma(\{x, z\}) \subseteq \sigma(\{x, y, z\})$ or $\sigma(\{y, z\}) \subseteq \sigma(\{x, y, z\})$, and so $z \in \sigma(\{x, y, z\})$ or $y \in \sigma(\{x, y, z\})$. In any case, $\sigma(\{x, y, z\}) \cap \{y, z\} \neq \emptyset$. Applying (IV) once more, we get that $\sigma(\{y, z\}) \subseteq \sigma(\{x, y, z\})$, so $y \in \sigma(\{x, y, z\})$. So by (I), we get $y \in \sigma(\{x, y\})$. On the other hand, by our initial assumption, $y \notin \sigma(\{x, y\})$, and we have a contradiction.

(iv) Let σ be finitely additive and satisfy (Ø1) and (IV), let $< \, = \mathcal{P}(\sigma)$, and let $x < y$, i.e., for all $S \in \mathcal{X}$, if $x \in S$ then $y \notin \sigma(S)$, and there is an $S' \in \mathcal{X}$ such that $x \in \sigma(S)$. We have to show that either $x < z$ or $z < y$. Suppose for reductio that this is not the case, i.e., that both

$$x \in S_1 \text{ and } z \in \sigma(S) \text{ for some } S_1 \in \mathcal{X}, \text{ or } x \notin \sigma(S') \text{ for every } S' \in \mathcal{X} \quad (\dagger)$$

and

$$z \in S_2 \text{ and } y \in \sigma(S_2) \text{ for some } S_2 \in \mathcal{X}, \text{ or } z \notin \sigma(S') \text{ for every } S' \in \mathcal{X} \quad (\ddagger)$$

But the second disjunct of (†) is false by our assumptions. So the first disjunct of (†) applies. This makes the second disjunct of (‡) false. So the second disjunct of (‡) applies.

Now consider $S_1 \cup S_2$. This set is in $\mathcal{X}$, since σ is finitely additive, and $\sigma(S_1 \cup S_2) \neq \emptyset$, by (Ø1). Hence $\sigma(S_1 \cup S_2) \cap S_1 \neq \emptyset$ or $\sigma(S_1 \cup S_2) \cap S_2 \neq \emptyset$. Therefore, by (IV), $\sigma(S_1) \subseteq \sigma(S_1 \cup S_2)$ or $\sigma(S_2) \subseteq \sigma(S_1 \cup S_2)$, and so $z \in \sigma(S_1 \cup S_2)$ or $y \in \sigma(S_1 \cup S_2)$. In any case, $\sigma(S_1 \cup S_2) \cap S_2 \neq \emptyset$. Applying (IV) once more, we get that $\sigma(S_2) \subseteq \sigma(S_1 \cup S_2)$, so $y \in \sigma(S_1 \cup S_2)$. On the other hand, we have that $x \in S_1 \cup S_2$. So by our initial assumption, $y \notin \sigma(S_1 \cup S_2)$, and we have a contradiction. $\qquad \square$

Lemma 17

(i) If $<$ is transitive and well-founded and $\mathcal{S}(<)$ satisfies (Ø1), then $\mathcal{S}(<)$ satisfies (III).

(ii) If $<$ is transitive and smooth, $\mathcal{X}$ is subtractive and $\mathcal{S}(<)$ satisfies (Ø1), then $\mathcal{S}(<)$ satisfies (III).

(iii) If $<$ is modular, then $\mathcal{S}(<)$ satisfies (IV).

Proof of Lemma 17.

(i) Let σ be relational with respect to some transitive and well-founded preference relation $<$, and let it satisfy (Ø1). Let $S, S' \in \mathcal{X}$, $S \subseteq S'$ and $\sigma(S') \subseteq S$. We want to show that $\sigma(S) \subseteq \sigma(S')$.

Suppose for reductio that this is not the case, i.e., that there is some x which is in $\sigma(S)$ but not in $\sigma(S')$. The latter means either that $S' \subseteq \overline{R}$ or there is some y_1 in S' such that $y_1 < x$. If $S' \subseteq \overline{R}$, then $S \subseteq \overline{R}$, by (Ø1), and $\sigma(S) = \emptyset$, contradicting our assumption. If $S' \not\subseteq \overline{R}$, then $y_1 \in S' - S$, since $x \in \sigma(S)$. But because $\sigma(S') \subseteq S$, $y_1 \notin \sigma(S')$. So there is some y_2 in S' such that $y_2 < y_1$. By transitivity, $y_2 < x$. So by the same reasoning as before, $y_2 \in S' - S$ and $y_2 \notin \sigma(S')$. So there is some y_3 in S' such that $y_3 < y_2$. By transitivity again, $y_3 < x$, and the same reasoning can be continued again and again. What we get is an infinite chain y_1, y_2, y_3, $\ldots$ in $S' - S$ such that $\cdots < y_3 < y_2 < y_1$. But this contradicts the well-foundedness of $<$.

(ii) Let σ be subtractive and relational with respect to some transitive and $\mathcal{X}$-smooth preference relation $<$, and let it satisfy ($\emptyset$1). We then copy the proof of (i) until we end up with an infinite chain y_1, y_2, y_3, $\ldots$ in $S' - S$ such that $\cdots < y_3 < y_2 < y_1$. But this contradicts the smoothness of $<$, by the subtractivity of σ which guarantees that $S' - S \in \mathcal{X}$.

(iii) Let σ be relational with respect to some modular preference relation $<$. Let $S, S' \in \mathcal{X}$, $S \subseteq S'$ and $x \in \sigma(S') \cap S$. Since $\sigma(S') \neq \emptyset$, if follows that $S' \not\subseteq \overline{R}$ and $x \in \sigma(S')$ says that $y < x$ for no $y \in S'$. Now let $z \in \sigma(S)$. We have to show that $z \in \sigma(S')$, i.e., by relationality, that $S' \not\subseteq \overline{R}$ and $y < z$ for no $y \in S'$. But it was already pointed out that $S' \not\subseteq \overline{R}$. And since $x \in S$ and $z \in \sigma(S)$, it is not the case that $x < z$, so since $y \not< x$ for all $y \in S'$, it follows from the modularity of $<$ that $y \not< z$ for all y in S. □

Observation 18

Let σ be a choice function which satisfie ($\emptyset$1) and can take all and only the finite subsets of a given domain X as arguments.
(i) σ is rationalizable iff it is rationalizable by the preference relation $\mathcal{P}_2(\sigma)$ defined by

$$x < y \quad \text{iff} \quad x \in \sigma(\{x, y\}) \text{ and } y \notin \sigma(\{x, y\}) \tag{$*$}$$

(ii) σ is rationalizable iff it satisfies (I) and (II).
(iii) σ is transitively rationalizable iff it satisfies (I), (II), and (III).
(iv) σ is modularly rationalizable iff it satisfies (I) and (IV).

Proof of Observation 18.
(i) Apply Lemma 14 (iv).
(ii) Apply Lemma 14 (i) and Corollary 15 (i).
(iii) Apply the results mentioned under (i) and (ii), plus Lemma 16 (ii), Lemma 17 (ii), and the fact that irreflexivity and transitivity imply that there are no infinite descending $<$-chains in any finite set, so $<$ is smooth, and the domain $\mathcal{X}$ of σ is subtractive.
(iv) Apply the results mentioned under (i) and (ii), plus Lemma 16 (iii) (alternatively, Lemma 16 (iv)) and Lemma 17 (iii). □

If σ is iterative and finitely additive, then (PI) and (PI') are equivalent.

Proof.

From (PI) to (PI'). By the commutativity of set union and two-fold application of (PI), we get that $\sigma(S \cup S') = \sigma(\sigma(S) \cup S') = \sigma(S' \cup \sigma(S)) = \sigma(\sigma(S) \cup \sigma(S'))$, that is, (PI') holds.

From (PI') to (PI). First note that by putting $S' = S$, we see that (PI') implies that σ is idempotent, i.e., $\sigma(\sigma(S)) = \sigma(S)$ for every S. Idempotence and two-fold application of (PI') gives us that $\sigma(S \cup S') = \sigma(\sigma(S) \cup \sigma(S')) = \sigma(\sigma(\sigma(S)) \cup \sigma(S')) = \sigma(\sigma(S) \cup S')$, that is, (PI) holds. $\square$

Observation 20

A finitely additive and iterative choice function satisfies Path Independence if and only if it satisfies conditions (I) and (III).

Proof of Observation 20.

(I) and (III) jointly entail (PI). From (I), we get $\sigma(S \cup S') \cap S \subseteq \sigma(S)$ and $\sigma(S \cup S') \cap S' \subseteq \sigma(S')$. Taking the unions, we find that $(\sigma(S \cup S') \cap S) \cup (\sigma(S \cup S') \cap S') \subseteq \sigma(S) \cup \sigma(S')$. Set-theoretic transformation gives us $\sigma(S \cup S') \subseteq \sigma(S) \cup \sigma(S')$. But from this and (I) and (III) we conclude that $\sigma(S \cup S') = \sigma(\sigma(S) \cup \sigma(S'))$. This is (PI'), which we know is equivalent with (PI) for finitely additive and iterative σ.

(PI) entails (I) and (III). Let $S \subseteq S'$. For (I), consider the following chain which makes use of (PI): $\sigma(S') = \sigma(S \cup (S' - S)) = \sigma(\sigma(S) \cup (S' - S)) \subseteq \sigma(S) \cup (S' - S)$. Intersecting with S gives us $\sigma(S') \cap S \subseteq (\sigma(S) \cap S) \cup ((S' - S) \cap S) = \sigma(S)$, and this is (I). In order to prove (III), let $S \subseteq S'$ and $\sigma(S') \subseteq S$. First notice that a two-fold application of (I) which we have just proved gives us $\sigma(S') \subseteq \sigma(S) \subseteq \sigma(\sigma(S))$. We use this and (PI) in establishing the following chain of equalities: $\sigma(S) = \sigma(\sigma(S)) = \sigma(\sigma(S) \cup \sigma(S')) = \sigma(S \cup S') = \sigma(S')$. But $\sigma(S) \subseteq \sigma(S')$ is precisely what we need for (III), so we are done. $\square$

For choice functions with domains closed under either finite unions or finite intersections, (I) and (IV⁺) taken together imply (Indis).

Proof.

We have to find some R such that for all S in $\mathcal{X}$, $\sigma(S) = S \cap R$. Now define R to be the set of all x in X such that there is some $S' \in \mathcal{X}$ with $x \in \sigma(S')$. Assume that $x \in \sigma(S)$; then $x \in R$ by definition, and also $x \in S$.

For the converse, assume that $x \in S \cap R$. Since $x \in R$, we can find some S' in $\mathcal{X}$ such that $x \in \sigma(S')$. If $\mathcal{X}$ is closed under finite unions, then x is also in $\sigma(S \cup S')$, by (IV⁺), and since $x \in S$, we get $x \in \sigma(S)$, by (I), as desired. If, on the other hand, $\mathcal{X}$ is closed under finite intersections, then, since $x \in \sigma(S') \cap S$, x is also in $\sigma(S \cap S')$, by (I), and $x \in \sigma(S)$, by (IV⁺), as desired. $\square$

APPENDIX D: PROOFS FOR CHAPTER 7

Lemma 21

For all sentences ϕ and all sets F and G of sentences, we have

(i) $]\phi[=]Cn(\phi)[$;
(ii) $]F[\cup]G[=]F \cup G[$;
(iii) $F \vdash \phi$ if and only if $[\![F]\!] \cap]\phi[= \emptyset$;
(iv) $]F[\subseteq]G[$ if and only if $Cn(F) \subseteq Cn(G)$.

Proof of Lemma 21.

(i), (ii) and (iii) are immediate;

(iv) From left to right. Let $Cn(F) \not\subseteq Cn(G)$, that is, there exists some ϕ such that $F \vdash \phi$ and $G \nvdash \phi$. From the latter we conclude that there is a model m of G that does not satisfy ϕ. Hence, by the former, $m \in]F[$ and $m \notin]G[$.

From right to left. $]F[\not\subseteq]G[$, i.e. there is a model m that falsifies F but satisfies G. Take some ϕ from F which is not satisfied by m. Clearly, $F \vdash \phi$ but $G \nvdash \phi$, since m satisfies G. Hence $Cn(F) \not\subseteq Cn(G)$ □

Lemma 23

(LP1) and (Faith) for δ imply that $L - R$ is a theory.
(LP1) and (Indis) for δ imply that $L - R$ is a theory.

Proof of Lemma 23.

(LP1) and (Faith) for δ imply that $L - R$ is a theory: Suppose for reductio that $L - R$ is not a theory, that is, that $Cn(L - R) \cap R \neq \emptyset$. Then, by (Faith), $\delta(Cn(L - R)) = Cn(L - R) \cap R \neq \emptyset$. Take some ϕ from $\delta(Cn(L - R))$. Since $\phi \in Cn(L - R)$ and $L - R \subseteq Cn(L - R)$, we conclude by (LP1) that $(L - R) \cap \delta(Cn(L - R)) \neq \emptyset$. But this contradicts $\delta(Cn(L - R)) = Cn(L - R) \cap R \subseteq R$.

Since (Indis) implies (Faith), (LP1) and (Indis), too, imply that $L - R$ is a theory.   □

Lemma 24

Let δ be a choice function over sentences satisfying (LP1) and (LP2), and let F be a finite set of sentences. Furthermore, let $\dot{-} = \mathcal{C}(\delta)$ be the contraction function and $\dot{-}_{()} = \mathcal{C}_{()}(\delta)$ be the pick contraction function over K that are generated by δ. Then

(i) For ϕ in $F \cap K$, we have $\phi \in K \dot{-} \langle F \rangle$ if and only if $\phi \notin \delta(F)$
(ii) $K \dot{-} \langle F \rangle = K \dot{-} \bigwedge F$.

Proof of Lemma 24.

(i) $\phi \in K \dot{-} \langle F \rangle$ iff   (by Definition 9)
$\phi \in K$ and $\phi \vee \chi \notin \delta(Cn(F))$ for every $\chi \in F$ iff   (by (LP1) and $\phi \in F$)

$\phi \in K$ and $\phi \notin \delta(Cn(F))$ iff   (by (LP2) and $\phi \in F$)
$\phi \in K$ and $\phi \notin \delta(F)$ iff   (by $\phi \in K$)
$\phi \notin \delta(F)$
(ii) $\phi \in K \doteq \langle F \rangle$ iff   (by Definition 9)
$\phi \in K$ and $\phi \vee \chi \notin \delta(Cn(F))$ for every $\chi \in F$ iff   (by (LP1))
$\phi \in K$ and $(\phi \vee \bigwedge F) \notin \delta(Cn(F))$ iff   (by (LP2))
$\phi \in K$ and $(\phi \vee \bigwedge F) \notin \delta(\{\bigwedge F, \phi \vee \bigwedge F\})$ iff   (by Definition 10)
$\phi \in K \doteq (\bigwedge F)$ $\square$

Observation 25. For every semantic choice function γ which satisfies

$$\left.\begin{cases} \overline{\text{(Faith1) wrt } [\![K]\!]} \\ \text{(Success)} \\ \text{(I)} \\ \text{(I}^-\text{)} \\ \text{(II) and is complete} \\ \text{(II}^+\text{)} \\ \text{(III)} \\ \text{(IV)} \\ \text{(IV}^+\text{)} \\ (\emptyset 1) \\ (\emptyset 2) \end{cases}\right\}$$, the contraction function $\doteq = \mathcal{C}(\gamma)$ over K based on

γ satisfies $(\doteq 1) - (\doteq 2)$, $(\doteq 5) - (\doteq 6)$ and $\left.\begin{cases} \overline{} \\ (\doteq 3) \\ (\doteq 4) \\ (\doteq 7) \\ (\doteq 7c) \\ (\doteq 8\text{vwd}) \\ (\doteq 8d) \\ (\doteq 8c) \\ (\doteq 8) \\ (\doteq 8m) \\ (\doteq 8rc) \\ (\doteq 7r) \end{cases}\right\}$, respectively.

Proof of Observation 25.
   Let γ be a semantic choice function and $\doteq$ be the contraction function $\doteq = \mathcal{C}(\gamma)$ which is based on γ.
   It is clear that $\doteq$ satisfies $(\doteq 1)$ and $(\doteq 2)$ because $K \doteq \phi$ is defined as an intersection of K with (the intersection of) a certain number of theories $\widehat{m}$. That $\doteq$ satisfies $(\doteq 6)$ also follows immediately from the construction of $\doteq$, since $Cn(\phi) = Cn(\psi)$ implies that $]\phi[=]\psi[$ (Lemma 21).

For $(\dot{-}5)$, let ψ be in K. We have to show that ψ is in $Cn((K\dot{-}\phi)\cup\{\phi\})$, that is, by $(\dot{-}1)$ which we have already proved, that $\neg\phi\vee\psi$ is in $K\dot{-}\phi$. This means, by the definition of $\dot{-}$, we have to show that $\neg\phi\vee\psi\in K$ and $\gamma(\,]\phi[\,)\subseteq[\![\neg\phi\vee\psi]\!]$. But the former is clear from $\psi\in K=Cn(K)$, and the latter is clear from $\gamma(\,]\phi[\,)\subseteq\,]\phi[\!=[\![\neg\phi]\!]\subseteq[\![\neg\phi\vee\psi]\!]$.

(Faith1) w.r.t. $[\![K]\!]$ implies $(\dot{-}3)$. Now let γ satisfy (Faith1) with respect to $[\![K]\!]$, i.e., for all ϕ such that $]\phi[\,\cap[\![K]\!]\neq\emptyset$ it holds that $\gamma(\,]\phi[\,)\subseteq[\![K]\!]$. Assume that $\phi\notin K$. We have to show that $K\subseteq K\dot{-}\phi$. By $(\dot{-}1)$, we conclude from $\phi\notin K=Cn(K)$ that there is a model m which satisfies K but falsifies ϕ. So m is in $]\phi[\,\cap[\![K]\!]$. Now by (Faith1), we get that $\gamma(\,]\phi[\,)\subseteq[\![K]\!]$. So $K\subseteq(\bigcap\widehat{\gamma}(\,]\phi[\,))$, hence $K\subseteq K\dot{-}\phi$.

(Success) implies $(\dot{-}4)$. Now let γ satisfy (Success). Let ϕ be in $K\dot{-}\phi$. We have to show that then ϕ is in $Cn(\emptyset)$. By the definition of $\dot{-}$, $\phi\in K\dot{-}\phi$ means that ϕ is in K and $\gamma(\,]\phi[\,)\subseteq[\![\phi]\!]$. But since $\gamma(\,]\phi[\,)\subseteq\,]\phi[$ and $]\phi[\,\cap[\![\phi]\!]=\emptyset$, it follows that $\gamma(\,]\phi[\,)=\emptyset$. By (Success), we get that $]\phi[\!=\emptyset$, from which we conclude that $\phi\in Cn(\emptyset)$.

(I) implies $(\dot{-}7)$. Now let γ satisfy (I). We use version (I′). So $\gamma(\,]\phi\wedge\psi[\,)=\gamma(\,]\phi[\,\cup\,]\psi[\,)\subseteq\gamma(\,]\phi[\,)\cup\gamma(\,]\psi[\,)$. Let $\chi\in K\dot{-}\phi\cap K\dot{-}\psi$, that is $\gamma(\,]\phi[\,)\subseteq[\![\chi]\!]$ and $\gamma(\,]\psi[\,)\subseteq[\![\chi]\!]$. Hence $\gamma(\,]\phi\wedge\psi[\,)\subseteq[\![\chi]\!]$, i.e., $\chi\in K\dot{-}(\phi\wedge\psi)$, as desired. So $\dot{-}$ satisfies $(\dot{-}7)$.

(I$^-$) implies $(\dot{-}7c)$. Now let γ satisfy (I$^-$). Let $\psi\in K\dot{-}(\phi\wedge\psi)$. For $(\dot{-}7c)$, we have to show that every χ in $K\dot{-}\phi$ is in $K\dot{-}(\phi\wedge\psi)$. Let $\chi\in K\dot{-}\phi$. From $\psi\in K\dot{-}(\phi\wedge\psi)$ we get, by the definition of $\dot{-}$, that ψ is in K and $\gamma(\,]\phi\wedge\psi[\,)\subseteq[\![\psi]\!]$. Since $\gamma(\,]\phi\wedge\psi[\,)\subseteq\,]\phi\wedge\psi[$, it follows that $\gamma(\,]\phi\wedge\psi[\,)\subseteq\,]\phi[$. Since $]\phi[\,\subseteq\,]\phi\wedge\psi[$, we get, by (I$^-$), that $\gamma(\,]\phi\wedge\psi[\,)\subseteq\gamma(\,]\phi[\,)$. Now $\chi\in K\dot{-}\phi$ means that $\chi\in K$ and $\gamma(\,]\phi[\,)\subseteq[\![\chi]\!]$. Hence $\gamma(\,]\phi\wedge\psi[\,)\subseteq[\![\chi]\!]$. Since $\chi\in K$, we get $\chi\in K\dot{-}(\phi\wedge\psi)$, by the definition of $\dot{-}$.

(II) implies $(\dot{-}8vwd)$. Now let γ be complete and satisfy (II). Let $\chi\in K\dot{-}(\phi\wedge\psi)=K\cap\bigcap\widehat{\gamma}(\,]\phi\wedge\psi[\,)$. Since by (II), $\gamma(\,]\phi[\,)\cap\gamma(\,]\psi[\,)\subseteq\gamma(\,]\phi\wedge\psi[\,)$, it follows that $\chi\in\bigcap(\widehat{\gamma}(\,]\phi[\,)\cap\widehat{\gamma}(\,]\psi[\,))$. Now suppose for reductio that $\chi\notin Cn(K\dot{-}\phi\cup K\dot{-}\psi)$. Then there is a model $m\in\,]\chi[$ such that m satisfies $\bigcap\widehat{\gamma}(\,]\phi[\,)$ and $\bigcap\widehat{\gamma}(\,]\psi[\,)$. Since γ is complete, we get $m\in\gamma(\,]\phi[\,)$ and $m\in\gamma(\,]\psi[\,)$, so $m\in\gamma(\,]\phi[\,)\cap\gamma(\,]\psi[\,)$. But $\chi\notin\widehat{m}$, so $\chi\notin\bigcap(\widehat{\gamma}(\,]\phi[\,)\cap\widehat{\gamma}(\,]\psi[\,))$ which gives us a contradiction. So $\dot{-}$ satisfies $(\dot{-}8vwd)$.

(II$^+$) implies $(\dot{-}8d)$. Now let γ satisfy (II$^+$). Let $\chi\in K\dot{-}(\phi\wedge\psi)$, that is, $\gamma(\,]\phi\wedge\psi[\,)=\gamma(\,]\phi[\,\cup\,]\psi[\,)\subseteq[\![\chi]\!]$. Now suppose for reductio that $\chi\notin(K\dot{-}\phi)\cup(K\dot{-}\psi)$, that is, $\gamma(\,]\phi[\,)\not\subseteq[\![\chi]\!]$ and $\gamma(\,]\psi[\,)\not\subseteq[\![\chi]\!]$. Take some model m from $\gamma(\,]\phi[\,)$ which falsifies χ and some model m' from $\gamma(\,]\psi[\,)$ which falsifies χ. By (II$^+$), either $m\in\gamma(\,]\phi[\,\cup\,]\psi[\,)$ or $m'\in\gamma(\,]\phi[\,\cup\,]\psi[\,)$. In both cases there is a model in $\gamma(\,]\phi[\,\cup\,]\psi[\,)$ which falsifies χ, contradicting $\gamma(\,]\phi[\,\cup\,]\psi[\,)\subseteq[\![\chi]\!]$. So $\dot{-}$ satisfies $(\dot{-}8d)$.

(III) implies $(\dot{-}8c)$. Now let γ satisfy (III). Let $\psi\in K\dot{-}(\phi\wedge\psi)$. For $(\dot{-}8c)$, we have to show that every χ in $K\dot{-}(\phi\wedge\psi)$ is in $K\dot{-}\phi$. Let $\chi\in K\dot{-}(\phi\wedge\psi)$. From $\psi\in K\dot{-}(\phi\wedge\psi)$ we get, by the definition of $\dot{-}$, that ψ is in K and $\gamma(\,]\phi\wedge\psi[\,)\subseteq[\![\psi]\!]$.

Since $\gamma(\,]\phi \wedge \psi[\,) \subseteq \,]\phi \wedge \psi[\,$, it follows that $\gamma(\,]\phi \wedge \psi[\,) \subseteq \,]\phi[\,$. Since $\,]\phi[\, \subseteq \,]\phi \wedge \psi[\,$, we get, by (III), that $\gamma(\,]\phi[\,) \subseteq \gamma(\,]\phi \wedge \psi[\,)$. Now $\chi \in K \dot{-}(\phi \wedge \psi)$ means that $\chi \in K$ and $\gamma(\,](\phi \wedge \psi)[\,) \subseteq [\![\chi]\!]$. Hence $\gamma(\,]\phi[\,) \subseteq [\![\chi]\!]$. Since $\chi \in K$, we get $\chi \in K \dot{-}\phi$, by the definition of $\dot{-}$.

(IV) implies ($\dot{-}8$). Now let γ satisfy (IV). Let $\phi \notin K \dot{-}(\phi \wedge \psi)$. For ($\dot{-}8$), we have to show that every χ in $K \dot{-}(\phi \wedge \psi)$ is in $K \dot{-}\phi$. Let $\chi \in K \dot{-}(\phi \wedge \psi)$. From $\phi \notin K \dot{-}(\phi \wedge \psi)$ we get, by the definition of $\dot{-}$, that ϕ is not in K or $\gamma(\,]\phi \wedge \psi[\,)$ is not a subset of $[\![\phi]\!]$. The case when ϕ is not in K has already been dealt with in general. So let ϕ be in K. Then it follows that $\gamma(\,]\phi \wedge \psi[\,) \cap \,]\phi[\, \neq \emptyset$. Since $\,]\phi[\, \subseteq \,]\phi \wedge \psi[\,$, we get, by (IV), that $\gamma(\,]\phi[\,) \subseteq \gamma(\,]\phi \wedge \psi[\,)$. Now $\chi \in K \dot{-}(\phi \wedge \psi)$ means that $\chi \in K$ and $\gamma(\,](\phi \wedge \psi)[\,) \subseteq [\![\chi]\!]$. Hence $\gamma(\,]\phi[\,) \subseteq [\![\chi]\!]$. Since $\chi \in K$, we get $\chi \in K \dot{-}\phi$, by the definition of $\dot{-}$.

(IV$^+$) implies ($\dot{-}8m$). Now let γ satisfy (IV$^+$). For ($\dot{-}8m$), we have to show that every χ in $K \dot{-}(\phi \wedge \psi)$ is in $K \dot{-}\phi$. Let $\chi \in K \dot{-}(\phi \wedge \psi)$. Then $\chi \in K$ and $\gamma(\,](\phi \wedge \psi)[\,) \subseteq [\![\chi]\!]$. Since $\,]\phi[\, \subseteq \,]\phi \wedge \psi[\,$, we get, by (IV$^+$), that $\gamma(\,]\phi[\,) \subseteq \gamma(\,]\phi \wedge \psi[\,)$. Hence $\gamma(\,]\phi[\,) \subseteq [\![\chi]\!]$. Since $\chi \in K$, we get $\chi \in K \dot{-}\phi$, by the definition of $\dot{-}$.

($\emptyset1$) implies ($\dot{-}8rc$). Now let γ satisfy ($\emptyset1$). For ($\dot{-}8rc$), we have to show that if $\phi \wedge \psi$ is in $K \dot{-}(\phi \wedge \psi)$, then ϕ is in $K \dot{-}\phi$. Let $\phi \wedge \psi \in K \dot{-}(\phi \wedge \psi)$. From this we get, by the definition of $\dot{-}$, that $\phi \wedge \psi$ is in K and $\gamma(\,]\phi \wedge \psi[\,) \subseteq [\![\phi \wedge \psi]\!]$. Since also $\gamma(\,]\phi \wedge \psi[\,) \subseteq \,]\phi \wedge \psi[\,$, we know that $\gamma(\,]\phi \wedge \psi[\,) = \emptyset$. Thus by $\,]\phi[\, \subseteq \,]\phi \wedge \psi[\,$ and ($\emptyset1$), we get that $\gamma(\,]\phi[\,) = \emptyset$, so clearly $\gamma(\,](\phi)[\,) \subseteq [\![\phi]\!]$. But K is a theory, so $\phi \wedge \psi \in K$ entails $\phi \in K$, and we get $\phi \in K \dot{-}\phi$, by the definition of $\dot{-}$.

($\emptyset2$) implies ($\dot{-}7r$). Now let γ satisfy ($\emptyset2$). For ($\dot{-}7r$), we have to show that if ϕ is in $K \dot{-}\phi$, then it is in $K \dot{-}(\phi \wedge \psi)$, too. Let $\phi \in K \dot{-}\phi$. From this we get, by the definition of γ, that ϕ is in K and $\gamma(\,]\phi[\,) \subseteq [\![\phi]\!]$. Since also $\gamma(\,]\phi[\,) \subseteq \,]\phi[\,$, we know that $\gamma(\,]\phi[\,) = \emptyset$. Thus by $\,]\phi[\, \subseteq \,]\phi \wedge \psi[\,$ and ($\emptyset2$), we get that $\gamma(\,]\phi \wedge \psi[\,) \cap \,]\phi[\, = \emptyset$, that is, $\gamma(\,]\phi \wedge \psi[\,) \subseteq [\![\phi]\!]$, and hence $\phi \in K \dot{-}(\phi \wedge \psi)$, by the definition of γ. $\quad\square$

Observation 26
Every contraction function $\dot{-}$ over a theory K which satisfies $(\dot{-}1) - (\dot{-}2)$,

$$(\dot{-}5) - (\dot{-}6) \text{ and } \left\{ \begin{array}{l} \overline{} \\ (\dot{-}3) \\ (\dot{-}4) \\ (\dot{-}7) \\ (\dot{-}7c) \\ (\dot{-}8vwd) \\ (\dot{-}8d) \\ (\dot{-}8c) \\ (\dot{-}8) \\ (\dot{-}8m) \\ (\dot{-}8rc) \\ (\dot{-}7r) \end{array} \right\} \text{ can be represented as the contraction function}$$

$\mathcal{C}(\gamma)$ based on a semantic choice function γ which is complete and satisfies

$$\text{(Faith2) w.r.t. } [\![K]\!] \text{ and } \left\{ \begin{array}{l} \overline{} \\ \text{(Faith1) w.r.t. } [\![K]\!] \\ \text{(Success)} \\ \text{(I)} \\ \text{(I}^-) \\ \text{(II)} \\ \text{(II}^+) \\ \text{(III)} \\ \text{(IV)} \\ \text{(IV}^+) \\ (\emptyset 1) \\ (\emptyset 2) \end{array} \right\}, \text{ respectively.}$$

Proof of Observation 26.
Let $\dot{-}$ satisfy $(\dot{-}1) - (\dot{-}2)$ and $(\dot{-}5) - (\dot{-}6)$. This assumption is supposed to hold throughout this proof. We take $\gamma = \mathcal{G}(\dot{-})$ in what follows.

We first show that $\dot{-}$ is determined by γ. Let $\dot{-}'$ be $\mathcal{C}(\gamma)$; we have to show that $\dot{-}' = \dot{-}$. Let ψ be in $K \dot{-}'\phi$. This means, by the definition of $\dot{-}' = \mathcal{C}(\gamma)$, that ψ is in K and $\gamma(\,]\phi[\,) \subseteq [\![\psi]\!]$. This in turn means, by the definition of $\gamma = \mathcal{G}(\dot{-})$, that

$$\psi \in K \text{ and }]\phi[\,\cap[\![K\dot{-}\phi]\!] \subseteq [\![\psi]\!] \tag{$*$}$$

We want to show that this is equivalent with $\psi \in K\dot{-}\phi$, or equivalently, by $(\dot{-}1)$, with $[\![K\dot{-}\phi]\!] \subseteq [\![\psi]\!]$. It is clear, with the help of $(\dot{-}2)$, that the latter entails $(*)$. Now let us show that $(*)$ entails $[\![K\dot{-}\phi]\!] \subseteq [\![\psi]\!]$. By recovery $(\dot{-}5)$,

$\llbracket \phi \rrbracket \cap \llbracket K \dot{-} \phi \rrbracket \subseteq \llbracket K \rrbracket$, and since $\psi \in K$, $\llbracket K \rrbracket \subseteq \llbracket \psi \rrbracket$, so $\llbracket \phi \rrbracket \cap \llbracket K \dot{-} \phi \rrbracket \subseteq \llbracket \psi \rrbracket$. But this together with $\rrbracket \phi \llbracket \cap \llbracket K \dot{-} \phi \rrbracket \subseteq \llbracket \psi \rrbracket$ yields $\llbracket K \dot{-} \phi \rrbracket \subseteq \llbracket \psi \rrbracket$, as desired.

It remains to prove the various properties of γ.

($\dot{-}2$) implies (Faith2) with respect to $\llbracket K \rrbracket$. Let ϕ be an arbitrary sentence and let m be in $\rrbracket \phi \llbracket \cap \llbracket K \rrbracket$. We have to show that m is in $\gamma(\rrbracket \phi \llbracket)$. By ($\dot{-}2$), we have $\llbracket K \rrbracket \subseteq \llbracket K \dot{-} \phi \rrbracket$, so we get that $m \in \llbracket K \dot{-} \phi \rrbracket$. So by the definition of $\gamma = \mathcal{G}(\dot{-})$, m is in $\gamma(\rrbracket \phi \llbracket)$.

($\dot{-}3$) implies (Faith1) with respect to $\llbracket K \rrbracket$. Let $\dot{-}$ in addition satisfy ($\dot{-}3$). Let ϕ be an arbitrary sentence and let $\rrbracket \phi \llbracket \cap \llbracket K \rrbracket \neq \emptyset$. Hence $\phi \notin K$. So by ($\dot{-}3$) $K \subseteq K \dot{-} \phi$, and hence $\llbracket K \dot{-} \phi \rrbracket \subseteq \llbracket K \rrbracket$. We have to show that every m in $\gamma(\rrbracket \phi \llbracket)$ is in $\llbracket K \rrbracket$. Let m be in $\gamma(\rrbracket \phi \llbracket)$. By the definition of $\gamma = \mathcal{G}(\dot{-})$, m is in $\llbracket K \dot{-} \phi \rrbracket$. Since $\llbracket K \dot{-} \phi \rrbracket \subseteq \llbracket K \rrbracket$, m is in $\llbracket K \rrbracket$, as desired.

($\dot{-}4$) implies (Success). Let $\dot{-}$ in addition satisfy ($\dot{-}4$). Let $\gamma(\rrbracket \phi \llbracket) = \emptyset$. We have to show that $\rrbracket \phi \llbracket = \emptyset$. By the definition of γ, $\gamma(\rrbracket \phi \llbracket) = \emptyset$ means that $\rrbracket \phi \llbracket \cap \llbracket K \dot{-} \phi \rrbracket = \emptyset$, or equivalently, $\llbracket K \dot{-} \phi \rrbracket \subseteq \llbracket \phi \rrbracket$. Hence, by ($\dot{-}1$), $\phi \in K \dot{-} \phi$. By ($\dot{-}4$), we conclude that $\phi \in Cn(\emptyset)$, and thus $\rrbracket \phi \llbracket = \emptyset$.

($\dot{-}7$) implies (I). We actually show that ($\dot{-}7$) implies (I'), which is equivalent to (I) because the domain of γ is subtractive: $\rrbracket \phi - \rrbracket \psi \llbracket = \rrbracket \phi \vee \neg \psi \llbracket$. So let $\dot{-}$ satisfy ($\dot{-}7$), and let $m \in \gamma(\rrbracket \phi \wedge \psi \llbracket)$. We want to show that $m \in \gamma(\rrbracket \phi \llbracket) \cup \gamma(\rrbracket \psi \llbracket)$. By $m \in \gamma(\rrbracket \phi \wedge \psi \llbracket)$, $m \in \llbracket K \dot{-} (\phi \wedge \psi) \rrbracket$, so by ($\dot{-}7$), $m \in \llbracket K \dot{-} \phi \cap K \dot{-} \psi \rrbracket$. Now suppose for reductio that $m \notin \gamma(\rrbracket \phi \llbracket) \cup \gamma(\rrbracket \psi \llbracket)$, i.e., neither $m \in \llbracket K \dot{-} \phi \rrbracket$ nor $m \in \llbracket K \dot{-} \psi \rrbracket$. Take $\chi \in (K \dot{-} \phi) - \hat{m}$ and $\chi' \in (K \dot{-} \psi) - \hat{m}$. Since both χ and χ' are falsified by m, we get that $m \in \rrbracket \chi \vee \chi' \llbracket$. But on the other hand, since $\chi \vee \chi' \in K \dot{-} \phi \cap K \dot{-} \psi$, by ($\dot{-}1$), we also get $m \in \llbracket \chi \vee \chi' \rrbracket$, so we have a contradiction. So $\mathcal{G}(\dot{-})$ satisfies (I).

($\dot{-}7c$) implies (I$^-$). Now let $\dot{-}$ satisfy ($\dot{-}7c$), and let $\rrbracket \phi \llbracket \subseteq \rrbracket \psi \llbracket$ and $\gamma(\rrbracket \psi \llbracket) \subseteq \rrbracket \phi \llbracket$. We have to show that $\gamma(\rrbracket \psi \llbracket) \subseteq \gamma(\rrbracket \phi \llbracket)$. From $\rrbracket \phi \llbracket \subseteq \rrbracket \psi \llbracket$ we conclude that $Cn(\psi) = Cn(\phi \wedge \psi) = Cn(\phi \wedge (\neg \phi \vee \psi))$, and hence, by ($\dot{-}6$), $K \dot{-} \psi = K \dot{-} (\phi \wedge (\neg \phi \vee \psi))$. Moreover, $\gamma(\rrbracket \psi \llbracket) \subseteq \rrbracket \phi \llbracket$ means, by the definition of γ, that $\rrbracket \psi \llbracket \cap \llbracket K \dot{-} \psi \rrbracket \subseteq \rrbracket \phi \llbracket$, or equivalently, $\llbracket K \dot{-} \psi \rrbracket \subseteq \rrbracket \phi \llbracket \cup \llbracket \psi \rrbracket = \llbracket \neg \phi \vee \psi \rrbracket$. Thus $\neg \phi \vee \psi \in K \dot{-} \psi = K \dot{-} (\phi \wedge (\neg \phi \vee \psi))$. Hence, by ($\dot{-}7c$), $K \dot{-} \phi \subseteq K \dot{-} (\phi \wedge (\neg \phi \vee \psi)) = K \dot{-} \psi$, and hence $\llbracket K \dot{-} \psi \rrbracket \subseteq \llbracket K \dot{-} \phi \rrbracket$. This, together with $\rrbracket \psi \llbracket \cap \llbracket K \dot{-} \psi \rrbracket \subseteq \rrbracket \phi \llbracket$, gives us $\rrbracket \psi \llbracket \cap \llbracket K \dot{-} \psi \rrbracket \subseteq \rrbracket \phi \llbracket \cap \llbracket K \dot{-} \phi \rrbracket$, that is $\gamma(\rrbracket \psi \llbracket) \subseteq \gamma(\rrbracket \phi \llbracket)$, by the definition of $\gamma = \mathcal{G}(\dot{-})$.

($\dot{-}8vwd$) implies (II). Now let $\dot{-}$ satisfy ($\dot{-}8vwd$), and let $m \in \gamma(\rrbracket \phi \llbracket) \cap \gamma(\rrbracket \psi \llbracket)$. We want to show that $m \in \gamma(\rrbracket \phi \wedge \psi \llbracket)$. Since $m \in \rrbracket \phi \llbracket \cap \rrbracket \psi \llbracket$, it is clear that $m \in \rrbracket \phi \wedge \psi \llbracket$. Our hypothesis means that $m \in \llbracket K \dot{-} \phi \cup K \dot{-} \psi \rrbracket$, we thus also have $m \in \llbracket Cn(K \dot{-} \phi \cup K \dot{-} \psi) \rrbracket$, thus by ($\dot{-}8vwd$) $m \in \llbracket K \dot{-} (\phi \wedge \psi) \rrbracket$, thus $m \in \gamma(\rrbracket \phi \wedge \psi \llbracket)$. So $\mathcal{G}(\dot{-})$ satisfies (II).

($\dot{-}8d$) implies (II$^+$). Now let $\dot{-}$ satisfy ($\dot{-}8d$), and let $m \in \gamma(\rrbracket \phi \llbracket)$ and $m' \in \gamma(\rrbracket \psi \llbracket)$. This means that $m \in \llbracket K \dot{-} \phi \rrbracket$ and $m' \in \llbracket K \dot{-} \psi \rrbracket$. We have to show that $m \in \gamma(\rrbracket \phi \llbracket \cup \rrbracket \psi \llbracket)$ or $m' \in \gamma(\rrbracket \phi \llbracket \cup \rrbracket \psi \llbracket)$. Since $\gamma(\rrbracket \phi \llbracket \cup \rrbracket \psi \llbracket) = \gamma(\rrbracket \phi \wedge \psi \llbracket)$, this means that we have to show that $m \in \llbracket K \dot{-} (\phi \wedge \psi) \rrbracket$ or $m' \in \llbracket K \dot{-} (\phi \wedge \psi) \rrbracket$. Suppose for reductio that this is not the case. Thus there is a χ in $K \dot{-} (\phi \wedge \psi)$

which is falsified by m and a χ' in $K \dot- (\phi \wedge \psi)$ which is falsified by m'. Since m satisfies $K \dot- \phi$ and m' satisfies $K \dot- \psi$, it follows that $\chi \notin K \dot- \phi$ and $\chi' \notin K \dot- \psi$. Since both $K \dot- \phi$ and $K \dot- \psi$ are closed under conjunctions, by $(\dot- 1)$, we have $\chi \wedge \chi' \notin (K \dot- \phi) \cup (K \dot- \psi)$. On the other hand, since $K \dot- (\phi \wedge \psi)$ is also closed under conjunctions, by $(\dot- 1)$ again, we have $\chi \wedge \chi' \in K \dot- (\phi \wedge \psi)$. But this contradicts $(\dot- 8d)$. So $\mathcal{G}(\dot-)$ satisfies (II^+).

$(\dot- 8c)$ *implies* (III). Now let $\dot-$ satisfy $(\dot- 8c)$, and let $]\phi[\subseteq]\psi[$ and $\gamma(]\psi[) \subseteq]\phi[$. We have to show that $\gamma(]\phi[) \subseteq \gamma(]\psi[)$. From $]\phi[\subseteq]\psi[$ we conclude that $Cn(\psi) = Cn(\phi \wedge \psi) = Cn(\phi \wedge (\neg\phi \vee \psi))$, and hence, by $(\dot- 6)$, $K \dot- \psi = K \dot- (\phi \wedge (\neg\phi \vee \psi))$. Moreover, $\gamma(]\psi[) \subseteq]\phi[$ means, by the definition of γ, that $]\psi[\cap [K \dot- \psi] \subseteq]\phi[$, or equivalently, $[K \dot- \psi] \subseteq]\phi[\cup [\psi] = [\neg\phi \vee \psi]$. Thus $\neg\phi \vee \psi \in K \dot- \psi = K \dot- (\phi \wedge (\neg\phi \vee \psi))$. Hence, by $(\dot- 8c)$, $K \dot- \psi = K \dot- (\phi \wedge (\neg\phi \vee \psi)) \subseteq K \dot- \phi$, and hence $[K \dot- \phi] \subseteq [K \dot- \psi]$. This, together with $]\phi[\subseteq]\psi[$, gives us $]\phi[\cap [K \dot- \phi] \subseteq]\psi[\cap [K \dot- \psi]$, that is $\gamma(]\phi[) \subseteq \gamma(]\psi[)$, by the definition of $\gamma = \mathcal{G}(\dot-)$.

$(\dot- 8)$ *implies* (IV). Now let $\dot-$ satisfy $(\dot- 8)$, and let $]\phi[\subseteq]\psi[$ and $\gamma(]\psi[) \cap]\phi[\neq \emptyset$. We have to show that $\gamma(]\phi[) \subseteq \gamma(]\psi[)$. From $]\phi[\subseteq]\psi[$ we conclude that $Cn(\psi) = Cn(\phi \wedge \psi)$, and hence, by $(\dot- 6)$, $K \dot- \psi = K \dot- (\phi \wedge \psi)$. Moreover, $\gamma(]\psi[) \cap]\phi[\neq \emptyset$ means, by the definition of γ, that $]\psi[\cap [K \dot- \psi] \cap]\phi[\neq \emptyset$, hence in particular $[K \dot- \psi] \cap]\phi[\neq \emptyset$. Thus $\phi \notin K \dot- \psi = K \dot- (\phi \wedge \psi)$. Hence, by $(\dot- 8)$, $K \dot- \psi = K \dot- (\phi \wedge \psi) \subseteq K \dot- \phi$, and hence $[K \dot- \phi] \subseteq [K \dot- \psi]$. This, together with $]\phi[\subseteq]\psi[$, gives us $]\phi[\cap [K \dot- \phi] \subseteq]\psi[\cap [K \dot- \psi]$, that is $\gamma(]\phi[) \subseteq \gamma(]\psi[)$, by the definition of $\gamma = \mathcal{G}(\dot-)$.

$(\dot- 8m)$ *implies* (IV^+). Now let $\dot-$ satisfy $(\dot- 8m)$, and let $]\phi[\subseteq]\psi[$. We have to show that $\gamma(]\phi[) \subseteq \gamma(]\psi[)$. From $]\phi[\subseteq]\psi[$ we conclude that $Cn(\psi) = Cn(\phi \wedge \psi)$, and by $(\dot- 6)$, $K \dot- \psi = K \dot- (\phi \wedge \psi)$. Hence, by $(\dot- 8m)$, $K \dot- \psi = K \dot- (\phi \wedge \psi) \subseteq K \dot- \phi$, and hence $[K \dot- \phi] \subseteq [K \dot- \psi]$. This, together with $]\phi[\subseteq]\psi[$, gives us $]\phi[\cap [K \dot- \phi] \subseteq]\psi[\cap [K \dot- \psi]$, that is $\gamma(]\phi[) \subseteq \gamma(]\psi[)$, by the definition of γ.

$(\dot- 8rc)$ *implies* $(\emptyset 1)$. Now let $\dot-$ satisfy $(\dot- 8rc)$, and let $]\phi[\subseteq]\psi[$ and $\gamma(]\psi[) = \emptyset$. We have to show that $\gamma(]\phi[) = \emptyset$. From $]\phi[\subseteq]\psi[$ we conclude that $Cn(\psi) = Cn(\phi \wedge \psi)$. By the definition of γ, $\gamma(]\psi[) = \emptyset$ means that $]\psi[\cap [K \dot- \psi] = \emptyset$, thus $\psi \in K \dot- \psi$, or equivalently, by $(\dot- 1)$ and $(\dot- 6)$, $\phi \wedge \psi \in K \dot- (\phi \wedge \psi)$. By $(\dot- 8rc)$, $\phi \in K \dot- \phi$, and hence $]\phi[\cap [K \dot- \phi] = \emptyset$, that is, $\gamma(]\phi[) = \emptyset$, by the definition of $\gamma = \mathcal{G}(\dot-)$.

$(\dot- 7r)$ *implies* $(\emptyset 2)$. Now let $\dot-$ satisfy $(\dot- 7r)$, and let $]\phi[\subseteq]\psi[$ and $\gamma(]\phi[) = \emptyset$. We have to show that $\gamma(]\psi[) \cap]\phi[= \emptyset$. From $]\phi[\subseteq]\psi[$ we conclude that $Cn(\psi) = Cn(\phi \wedge \psi)$. By the definition of γ, $\gamma(]\phi[) = \emptyset$ means that $]\phi[\cap [K \dot- \phi] = \emptyset$, thus $\phi \in K \dot- \phi$. By $(\dot- 7r)$, we get $\phi \in K \dot- (\phi \wedge \psi)$, or equivalently, by $(\dot- 6)$, $\phi \in K \dot- \psi$. Hence $[K \dot- \psi] \cap]\phi[= \emptyset$ and a fortiori $]\psi[\cap [K \dot- \psi] \cap]\phi[= \emptyset$. That is, $\gamma(]\psi[) \cap]\phi[= \emptyset$, by the definition of $\gamma = \mathcal{G}(\dot-)$. $\square$

Observation 35

For every syntactic choice function δ which satisfies (LP1), (LP2) and

$$\left\{\begin{array}{l} \overline{} \\ \text{(Faith1) wrt } \delta(L) \\ \text{(Faith2) wrt } \delta(L) \\ \text{(Virtual Success)} \\ \quad \text{(I)} \\ \quad \text{(I}^-\text{)} \\ \quad \text{(II)} \\ \quad \text{(II}^+\text{)} \\ \quad \text{(III)} \\ \quad \text{(IV)} \\ \quad \text{(IV}^+\text{)} \\ \quad (\emptyset 1) \\ \quad (\emptyset 2) \end{array}\right\}$$, the inference operation Inf based on δ satisfies (Ref),

(LLE), (RW), (And) and $\left\{\begin{array}{l} \overline{} \\ \text{(WRMon)} \\ \text{(WCond)} \\ \quad \text{(CP)} \\ \quad \text{(Or)} \\ \quad \text{(Cut)} \\ \text{(WDRat)} \\ \text{(DRat)} \\ \text{(CMon)} \\ \text{(RMon)} \\ \quad \text{(Mon)} \\ (\emptyset \text{CMon)} \\ (\emptyset \text{Cond)} \end{array}\right\}$, respectively.

Proof of Observation 35.

(Ref). We have to show that $\phi \hspace{1pt}\vdash\hspace{-3pt}\sim \phi$, i.e., that $\phi \supset \phi \notin \delta(Cn(\neg\phi))$. Since $\phi \supset \phi \in Cn(\emptyset)$, this follows from (LP1).

(LLE). Let $Cn(\phi) = Cn(\psi)$. We have to show that for every χ, $\phi \hspace{1pt}\vdash\hspace{-3pt}\sim \chi$ iff $\psi \hspace{1pt}\vdash\hspace{-3pt}\sim \chi$, i.e., $\phi \supset \chi \notin \delta(Cn(\neg\phi))$ iff $\psi \supset \chi \notin \delta(Cn(\neg\psi))$. Since $Cn(\neg\phi) = Cn(\neg\psi)$ and $Cn(\phi \supset \chi) = Cn(\psi \supset \chi)$, this follows from (LP1).

(RW). Let $\chi \in Cn(\psi)$. We have to show that $\phi \hspace{1pt}\vdash\hspace{-3pt}\sim \psi$ implies $\phi \hspace{1pt}\vdash\hspace{-3pt}\sim \chi$, i.e., that $\phi \supset \psi \notin \delta(Cn(\neg\phi))$ implies $\phi \supset \chi \notin \delta(Cn(\neg\phi))$. Since $\phi \supset \chi \in Cn(\phi \supset \psi)$ and $\phi \supset \psi \in Cn(\neg\phi)$, this follows from (LP1).

(And). Let $\phi \hspace{1pt}\vdash\hspace{-3pt}\sim \psi$ and $\phi \hspace{1pt}\vdash\hspace{-3pt}\sim \chi$, i.e., $\phi \supset \psi \notin \delta(Cn(\neg\phi))$ and $\phi \supset \chi \notin \delta(Cn(\neg\phi))$. We have to show that $\psi \wedge \chi \in Inf(\phi)$, i.e., $\phi \supset (\psi \wedge \chi) \notin \delta(Cn(\neg\phi))$. Since $\phi \supset (\psi \wedge \chi) \in Cn(\{\phi \supset \psi, \phi \supset \chi\})$, this follows from (LP1).

(Faith1) w.r.t. $\delta(L)$ implies (WRMon). Let $\top \not\vdash \neg\phi$, i.e., $\top \supset \neg\phi \in \delta(Cn(\neg\top))$, which means $\neg\phi \in \delta(L)$, and hence $Cn(\neg\phi) \cap \delta(L) \neq \emptyset$. We have to show that $Inf(\top) \subseteq Inf(\phi)$. Let $\top \vdash \psi$, i.e., $\top \supset \psi \notin \delta(Cn(\neg\top))$, which means $\psi \notin \delta(L)$. We have to show that $\phi \vdash \psi$, i.e., $\phi \supset \psi \notin \delta(Cn(\neg\phi))$. Since δ satisfies (Faith1) w.r.t. $\delta(L)$ and $Cn(\neg\phi) \cap \delta(L) \neq \emptyset$, we get $\delta(Cn(\neg\phi)) \subseteq \delta(L)$. From $\psi \notin \delta(L)$ we infer that $\phi \supset \psi \notin \delta(L)$, by (LP1). But then we conclude with $\delta(Cn(\neg\phi)) \subseteq \delta(L)$ that $\phi \supset \psi \notin \delta(Cn(\neg\phi))$, as desired.

(Faith2) w.r.t. $\delta(L)$ implies (WCond). Let $\phi \vdash \psi$, i.e., $\phi \supset \psi \notin \delta(Cn(\neg\phi))$. We have to show that $\psi \in Cn(Inf(\top) \cup \{\phi\})$, i.e. $\top \vdash \phi \supset \psi$, i.e., $\top \supset (\phi \supset \psi) \notin \delta(Cn(\neg\top))$, which means $\phi \supset \psi \notin \delta(L)$. Since δ satisfies (Faith2) w.r.t. $\delta(L)$, we get $Cn(\neg\phi) \cap \delta(L) \subseteq \delta(Cn(\neg\phi))$. Since $\phi \supset \psi \notin \delta(Cn(\neg\phi))$ but $\phi \supset \psi \in Cn(\neg\phi)$, we get that $\phi \supset \psi \notin \delta(L)$, as desired.

(Virtual Success) implies (CP). Let ϕ be such that $Cn(\phi) \neq L$. We have to show that $Inf(\phi) \neq L$, and we show that $\phi \not\vdash \neg\phi$. Suppose for reductio that $\phi \vdash \neg\phi$, i.e., $\phi \supset \neg\phi \notin \delta(Cn(\neg\phi))$. Since $\phi \supset \neg\phi$ is equivalent with $\neg\phi$, (LP1) tells us that $\neg\phi \notin \delta(Cn(\neg\phi))$, so $\delta(Cn(\neg\phi))$ empty, again by (LP1). Hence by (Virtual Success), $Cn(\neg\phi) \subseteq Cn(\emptyset)$, that is, $\neg\phi$ is a logical truth. But this contradicts $Cn(\phi) \neq L$.

(I) implies (Or). Let $\phi \vdash \chi$ and $\psi \vdash \chi$, i.e., $\phi \supset \chi \notin \delta(Cn(\neg\phi))$ and $\psi \supset \chi \notin \delta(Cn(\neg\psi))$. We have to show that $\phi \lor \psi \vdash \chi$, i.e., that $(\phi \lor \psi) \supset \chi \notin \delta(Cn(\neg\phi \land \neg\psi))$. Suppose for reductio that $(\phi \lor \psi) \supset \chi \in \delta(Cn(\neg\phi \land \neg\psi))$. By (LP1), it follows that either $\phi \supset \chi \in \delta(Cn(\neg\phi \land \neg\psi))$ or $\psi \supset \chi \in \delta(Cn(\neg\phi \land \neg\psi))$. By $Cn(\neg\phi) \subseteq Cn(\neg\phi \land \neg\psi)$, $Cn(\neg\psi) \subseteq Cn(\neg\phi \land \neg\psi)$ and (I), this implies that either $\phi \supset \chi \in \delta(Cn(\neg\phi))$ or $\psi \supset \chi \in \delta(Cn(\neg\psi))$, and we have a contradiction with our assumptions.

(II) implies (WDRat). Let $\phi \lor \psi \vdash \chi$, i.e, $(\phi \lor \psi) \supset \chi \notin \delta(Cn(\neg\phi \land \neg\psi))$. By (LP2) we get that $(\phi \land \psi) \supset \chi \notin \delta(Cn(\neg\phi) \cup Cn(\neg\psi))$. By (II), we get that $(\phi \land \psi) \supset \chi \notin \delta(Cn(\neg\phi)) \cap \delta(Cn(\neg\psi))$. By (LP1), this means that either $\phi \supset (\psi \supset \chi) \notin \delta(Cn(\neg\phi))$ or $\psi \supset (\phi \supset \chi) \notin \delta(Cn(\neg\psi))$. Hence either $\phi \vdash \psi \supset \chi$ or $\psi \vdash \phi \supset \chi$, as desired for (WDRat).

(III) implies (CMon). Let $\phi \vdash \psi$ and $\phi \vdash \chi$, i.e., $\phi \supset \psi \notin \delta(Cn(\neg\phi))$ and $\phi \supset \chi \notin \delta(Cn(\neg\phi))$. We have to show that $\phi \land \psi \vdash \chi$, i.e., that $(\phi \land \psi) \supset \chi \notin \delta(Cn(\neg\phi \lor \neg\psi))$. Suppose for reductio that $(\phi \land \psi) \supset \chi \in \delta(Cn(\neg\phi \lor \neg\psi))$, or equivalently, by (LP2), $(\phi \land \psi) \supset \chi \in \delta(\{(\phi \land \psi) \supset \chi, \neg\phi \lor \neg\psi \lor \neg\chi\})$. From this and $\phi \supset \psi \notin \delta(\{\phi \supset \psi, (\phi \land \psi) \supset \chi, \neg\phi \lor \neg\psi \lor \neg\chi\})$, which follows from $\phi \supset \psi \notin \delta(Cn(\neg\phi))$ by (LP2), we conclude with the help of (III) that $(\phi \land \psi) \supset \chi \in \delta(\{\phi \supset \psi, (\phi \land \psi) \supset \chi, \neg\phi \lor \neg\psi \lor \neg\chi\})$. By (LP2), we conclude that $(\phi \land \psi) \supset \chi \in \delta(\{\phi \supset \psi, \phi \supset \chi, (\phi \land \psi) \supset \chi, \neg\phi \lor \neg\psi \lor \neg\chi\})$. But, by (LP1), this contradicts $\phi \supset \chi \notin \delta(\{\phi \supset \psi, \phi \supset \chi, (\phi \land \psi) \supset \chi, \neg\phi \lor \neg\psi \lor \neg\chi\})$ which follows from our assumption $\phi \supset \chi \notin \delta(Cn(\neg\phi))$ by (LP2).

(IV) implies (RMon). Let $\phi \not\vdash \neg\psi$ and $\phi \vdash \chi$, i.e., $\phi \supset \neg\psi \in \delta(Cn(\neg\phi))$, and $\phi \supset \chi \notin \delta(Cn(\neg\phi))$. We have to show that $\phi \land \psi \vdash \chi$, i.e., that $(\phi \land \psi) \supset \chi \notin \delta(Cn(\neg\phi \lor \neg\psi))$. Suppose for reductio that $(\phi \land \psi) \supset \chi \in \delta(Cn(\neg\phi \lor \neg\psi))$. Since $\phi \supset \neg\psi \in \delta(Cn(\neg\phi))$, we know that $\delta(Cn(\neg\phi)) \cap Cn(\neg\phi \lor \neg\psi) \neq \emptyset$, so by (IV),

we get that $(\phi \wedge \psi) \supset \chi \in \delta(Cn(\neg\phi))$. Hence, by (LP1), $\phi \supset \chi \in \delta(Cn(\neg\phi))$, and we have a contradiction with one of our assumptions.

(I^-) *implies (Cut)*. Let $\phi \vdash \psi$ and $\phi \wedge \psi \vdash \chi$, i.e., $\phi \supset \psi \notin \delta(Cn(\neg\phi))$ and $\phi \wedge \psi \supset \chi \notin \delta(Cn(\neg\phi \vee \neg\psi))$. We have to show that $\phi \vdash \chi$, i.e., that $\phi \supset \chi \notin \delta(Cn(\neg\phi))$. Suppose for reductio that $\phi \supset \chi \in \delta(Cn(\neg\phi))$. By (LP1), then, either $(\phi \wedge \psi) \supset \chi \in \delta(Cn(\neg\phi))$ or $(\phi \wedge \neg\psi) \supset \chi \in \delta(Cn(\neg\phi))$. But the latter is impossible, since it would imply $\phi \supset \psi \in \delta(Cn(\neg\phi))$, by (LP1), contradicting our first assumption. So $(\phi \wedge \psi) \supset \chi \in \delta(Cn(\neg\phi))$, or equivalently, by (LP2), $(\phi \wedge \psi) \supset \chi \in \delta(\{(\phi \wedge \psi) \supset \chi, \phi \supset \psi, \neg\phi \vee \neg\psi\})$. On the other hand, we get from $\phi \supset \psi \notin \delta(Cn(\neg\phi))$ by (LP2) that $\phi \supset \psi \notin \delta(\{(\phi \wedge \psi) \supset \chi, \phi \supset \psi, \neg\phi \vee \neg\psi\})$, or equivalently, that $\delta(\{(\phi \wedge \psi) \supset \chi, \phi \supset \psi, \neg\phi \vee \neg\psi\}) \subseteq \{(\phi \wedge \psi) \supset \chi, \neg\phi \vee \neg\psi\}$. Hence, by (I^-), $\delta(\{(\phi \wedge \psi) \supset \chi, \phi \supset \psi, \neg\phi \vee \neg\psi\}) \subseteq \delta(\{(\phi \wedge \psi) \supset \chi, \neg\phi \vee \neg\psi\})$. Since $(\phi \wedge \psi) \supset \chi \in \delta(\{(\phi \wedge \psi) \supset \chi, \phi \supset \psi, \neg\phi \vee \neg\psi\})$, we get that $(\phi \wedge \psi) \supset \chi \in \delta(\{(\phi \wedge \psi) \supset \chi, \neg\phi \vee \neg\psi\})$. Hence, by (LP2), $(\phi \wedge \psi) \supset \chi \in \delta(Cn(\neg\phi \vee \neg\psi))$, and we have a contradiction with our second assumption.

(II^+) *implies (DRat)*. Let $\phi \vee \psi \vdash \chi$, that is, $(\phi \vee \psi) \supset \chi \notin \delta(Cn(\neg(\phi \vee \psi)))$. We have to show that either $\phi \vdash \chi$ or $\psi \vdash \chi$, that is, that either $\phi \supset \chi \notin \delta(Cn(\neg\phi))$ or $\psi \supset \chi \notin \delta(Cn(\neg\psi))$. Suppose for reductio that both $\phi \supset \chi \in \delta(Cn(\neg\phi))$ and $\psi \supset \chi \in \delta(Cn(\neg\psi))$. Then by (II^+), either $\phi \supset \chi \in \delta(Cn(\neg\phi) \cup Cn(\neg\psi))$ or $\psi \supset \chi \in \delta(Cn(\neg\phi) \cup Cn(\neg\psi))$. Hence, by (LP2), either $\phi \supset \chi \in \delta(Cn(\neg\phi \wedge \neg\psi))$ or $\psi \supset \chi \in \delta(Cn(\neg\phi \wedge \neg\psi))$. Hence, by (LP1), $(\phi \vee \psi) \supset \chi \in \delta(Cn(\neg\phi \wedge \neg\psi))$, and we have a contradiction with our assumption.

(IV^+) *implies (Mon)*. Let $\phi \vdash \chi$, i.e., $\phi \supset \chi \notin \delta(Cn(\neg\phi))$. We have to show that $\phi \wedge \psi \vdash \chi$, i.e., that $(\phi \wedge \psi) \supset \chi \notin \delta(Cn(\neg\phi \vee \neg\psi))$. Suppose for reductio that $(\phi \wedge \psi) \supset \chi \in \delta(Cn(\neg\phi \vee \neg\psi))$. By $Cn(\neg\phi \vee \neg\psi) \subseteq Cn(\neg\phi)$ and (IV), we get that $(\phi \wedge \psi) \supset \chi \in \delta(Cn(\neg\phi))$. Hence, by (LP1), $\phi \supset \chi \in \delta(Cn(\neg\phi))$, and we have a contradiction with our assumption.

$(\emptyset 1)$ *implies* $(\emptyset CMon)$. Let $\phi \vdash \bot$, i.e., $\phi \supset \bot \notin \delta(Cn(\neg\phi))$. We have to show that $\phi \wedge \psi \vdash \bot$, i.e., that $(\phi \wedge \psi) \supset \bot \notin \delta(Cn(\neg(\phi \wedge \psi)))$. From $\phi \supset \bot \notin \delta(Cn(\neg\phi))$ we conclude with (LP1) that $\delta(Cn(\neg\phi)) = \emptyset$. Since $Cn(\neg(\phi \wedge \psi)) \subseteq Cn(\neg\phi)$, $(\emptyset 1)$ gives us $\delta(Cn(\neg(\phi \wedge \psi))) = \emptyset$, and we are done.

$(\emptyset 2)$ *implies* $(\emptyset Cond)$. Let $\phi \wedge \psi \vdash \bot$, i.e., $(\phi \wedge \psi) \supset \bot \notin \delta(Cn(\neg(\phi \wedge \psi)))$. We have to show that $\phi \vdash \neg\psi$, i.e., that $\phi \supset \neg\psi \notin \delta(Cn(\neg\phi))$. From $(\phi \wedge \psi) \supset \bot \notin \delta(Cn(\neg(\phi \wedge \psi)))$ we conclude with (LP1) that $\delta(Cn(\neg(\phi \wedge \psi))) = \emptyset$. Since $Cn(\neg(\phi \wedge \psi)) \subseteq Cn(\neg\phi)$, $(\emptyset 2)$ gives us $Cn(\neg(\phi \wedge \psi)) \cap \delta(Cn(\neg\phi)) = \emptyset$, so $\phi \supset \neg\psi \notin \delta(Cn(\neg\phi))$, and we are done.   □

Observation 36

Every finitary inference operation *Inf* which satisfies (Ref), (LLE), (RW),

$$
\text{(And) and } \left\{
\begin{array}{c}
— \\
\text{(WCond)} \\
\text{(WRMon)} \\
\text{(CP)} \\
\text{(Or)} \\
\text{(Cut)} \\
\text{(WDRat)} \\
\text{(DRat)} \\
\text{(CMon)} \\
\text{(RMon)} \\
\text{(Mon)} \\
\text{(}\emptyset\text{CMon)} \\
\text{(}\emptyset\text{Cond)}
\end{array}
\right\} \text{ can be represented as the inference operation fini-}
$$

tarily based on a syntactic choice function δ which satisfies (LP1), (LP2) and

$$
\left\{
\begin{array}{c}
— \\
\text{(Faith2) w.r.t. } L - Inf(\top) \\
\text{(Faith1) w.r.t. } L - Inf(\top) \\
\text{(Virtual Success)} \\
\text{(I)} \\
\text{(I}^-\text{)} \\
\text{(II)} \\
\text{(II}^+\text{)} \\
\text{(III)} \\
\text{(IV)} \\
\text{(IV}^+\text{)} \\
\text{(}\emptyset 1\text{)} \\
\text{(}\emptyset 2\text{)}
\end{array}
\right\} \text{, respectively.}
$$

Proof of Observation 36.

We take $\delta = \mathcal{D}(Inf)$ for the proof. First of all we have to show that this δ in fact finitarily represents $\mathrel{\vmid\sim}$, that is, we have to show that

(Def′ $\mathrel{\vmid\sim}$) $\phi \mathrel{\vmid\sim} \psi$ iff $\phi \supset \psi \notin \delta(\{\phi \supset \psi, \phi \supset \neg\psi\})$

According to the definition of δ, the right-hand side is true if and only if $\neg(\phi \supset \psi) \lor \neg(\phi \supset \neg\psi) \mathrel{\vmid\sim} \phi \supset \psi$. By (LLE), this is equivalent to $\phi \mathrel{\vmid\sim} \phi \supset \psi$, which by (Ref), (And) and (RW) is in turn equivalent to $\phi \mathrel{\vmid\sim} \psi$.

(LP1). Let $G \subseteq F$, $\phi \in Cn(G)$ and $\phi \in \delta(F)$, i.e., $\phi \in A$ and $\neg\bigwedge F \mathrel{\not\vmid\sim} \phi$. We need to show that $G \cap \delta(F) \neq \emptyset$. Suppose for reductio that $G \cap \delta(F) = \emptyset$, i.e., $\neg\bigwedge F \mathrel{\vmid\sim} \psi$ for every ψ in G. From $\phi \in Cn(G)$ and the compactness of Cn it

follows that there is a finite subset G_0 of G such that $\phi \in Cn(G_0)$. By (And), we get $\neg\bigwedge F \hspace{1pt}\vdash\hspace{-6pt}\sim \bigwedge G_0$, whence, by $\phi \in Cn(G_0)$ and (RW), $\neg\bigwedge F \hspace{1pt}\vdash\hspace{-6pt}\sim \phi$, and we have a contradiction.

(LP2). Let $Cn(F) = Cn(G)$. We need to show that $\delta(F) \cap G = \delta(G) \cap F$. Let ϕ be in $\delta(F) \cap G$. Hence $\phi \in F$ and $\neg\bigwedge F \hspace{1pt}\not\vdash\hspace{-6pt}\sim \phi$. Since $Cn(F) = Cn(G)$, (LLE) gives us $\neg\bigwedge G \hspace{1pt}\not\vdash\hspace{-6pt}\sim \phi$, thus, since ϕ is in G, $\phi \in \delta(G)$. Thus we have $\delta(F) \cap G \subseteq \delta(G) \cap F$. The converse direction is similar.

(WCond) implies (Faith2) w.r.t. $L - Inf(\top)$. (Faith2) with respect to $L - Inf(\top)$ states that $F - Inf(\top) \subseteq \delta(F)$. Let $\phi \in F$ and $\top \not\vdash\hspace{-6pt}\sim \phi$. We have to show that $\phi \in \delta(F)$. Suppose for reductio that $\phi \notin \delta(F)$, i.e., $\neg\bigwedge F \hspace{1pt}\vdash\hspace{-6pt}\sim \phi$. By (WCond), we get $\top \hspace{1pt}\vdash\hspace{-6pt}\sim (\neg\bigwedge F) \supset \phi$. Since $\phi \in F$, we get by (RW) that $\top \hspace{1pt}\vdash\hspace{-6pt}\sim \phi$, and we have found a contradiction.

(WRMon) implies (Faith1) w.r.t. $L - Inf(\top)$. (Faith1) with respect to $L - Inf(\top)$ states that if $F \not\subseteq Inf(\top)$, then $\delta(F) \cap Inf(\top) = \emptyset$. Let $F \not\subseteq Inf(\top)$. Then by (RW), $\top \not\vdash\hspace{-6pt}\sim \bigwedge F$. Hence, by (WRMon), $Inf(\top) \subseteq Inf(\neg\bigwedge F)$. Now suppose for reductio that $\delta(F) \cap Inf(\top) \neq \emptyset$. That is, there is a $\phi \in \delta(F)$ such that $\top \hspace{1pt}\vdash\hspace{-6pt}\sim \phi$. From the latter and $Inf(\top) \subseteq Inf(\neg\bigwedge F)$, we get that $\neg\bigwedge F \hspace{1pt}\vdash\hspace{-6pt}\sim \phi$. But $\phi \in \delta(F)$ means that $\neg\bigwedge F \hspace{1pt}\not\vdash\hspace{-6pt}\sim \phi$, so we have found a contradiction.

(CP) implies (Virtual Success). Let F be finite and $\delta(F) = \emptyset$. We have to show that $F \subseteq Cn(\emptyset)$. $\delta(F) = \emptyset$ means that for each $\phi \in F$, $\neg\bigwedge F \hspace{1pt}\vdash\hspace{-6pt}\sim \phi$. It follows, by (And), that $\neg\bigwedge F \hspace{1pt}\vdash\hspace{-6pt}\sim \bigwedge F$, and hence, by (Ref), (And) and (RW), that $\neg\bigwedge F \hspace{1pt}\vdash\hspace{-6pt}\sim \chi$ for arbitrary χ. Hence, by (CP), $\chi \in Cn(\neg\bigwedge F)$ for arbitrary χ, hence $\bigwedge F \in Cn(\emptyset)$, hence $F \subseteq Cn(\emptyset)$.

(Or) implies (I). Let $F \subseteq G$, $\phi \in F$ and $\phi \in \delta(G)$, i.e., $\neg\bigwedge G \hspace{1pt}\not\vdash\hspace{-6pt}\sim \phi$. We have to show that $\phi \in \delta(F)$, i.e., that $\neg\bigwedge F \hspace{1pt}\not\vdash\hspace{-6pt}\sim \phi$. Suppose for reductio that $\neg\bigwedge F \hspace{1pt}\vdash\hspace{-6pt}\sim \phi$. By (Ref) and (RW), we also have $\bigwedge F \wedge \neg\bigwedge(G - F) \hspace{1pt}\vdash\hspace{-6pt}\sim \phi$. By (Or), this gives us $\neg\bigwedge F \vee (\bigwedge F \wedge \neg\bigwedge(G - F)) \hspace{1pt}\vdash\hspace{-6pt}\sim \phi$. Hence, by (LLE), $\neg\bigwedge F \vee \neg\bigwedge(G - F) \hspace{1pt}\vdash\hspace{-6pt}\sim \phi$, i.e., by (LLE), $\neg\bigwedge G \hspace{1pt}\vdash\hspace{-6pt}\sim \phi$, and we have a contradiction.

(WDRat) implies (II). Let $\phi \in \delta(F) \cap \delta(G)$. We need to show that $\phi \in \delta(F \cup G)$. The supposition says that $\neg\bigwedge F \hspace{1pt}\not\vdash\hspace{-6pt}\sim \phi$ and $\neg\bigwedge G \hspace{1pt}\not\vdash\hspace{-6pt}\sim \phi$. Assume for reductio that $\phi \notin \delta(F \cup G)$, i.e. that $\neg\bigwedge(F \cup G) \hspace{1pt}\vdash\hspace{-6pt}\sim \phi$, i.e., by (LLE), $(\neg\bigwedge F) \vee (\neg\bigwedge G) \hspace{1pt}\vdash\hspace{-6pt}\sim \phi$. By (WDRat), we get $\neg\bigwedge F \hspace{1pt}\vdash\hspace{-6pt}\sim (\neg\bigwedge G) \supset \phi$ or $\neg\bigwedge G \hspace{1pt}\vdash\hspace{-6pt}\sim (\neg\bigwedge F) \supset \phi$. Since ϕ is contained in both F and G, this gives us, by (RW), $\neg\bigwedge F \hspace{1pt}\vdash\hspace{-6pt}\sim \phi$ or $\neg\bigwedge G \hspace{1pt}\vdash\hspace{-6pt}\sim \phi$, and we have a contradiction.

(CMon) implies (III). Let $F \subseteq G$ and $\delta(G) \subseteq F$. We need to show that $\delta(F) \subseteq \delta(G)$. The supposition $\delta(G) \subseteq F$ says that for each $\psi \in G - F$, $\psi \notin \delta(G)$, i.e., $\neg\bigwedge G \hspace{1pt}\vdash\hspace{-6pt}\sim \psi$. Now let ϕ be in $\delta(F)$, i.e., $\neg\bigwedge F \hspace{1pt}\not\vdash\hspace{-6pt}\sim \phi$. We have to show that ϕ is in $\delta(G)$, i.e., that $\neg\bigwedge G \hspace{1pt}\not\vdash\hspace{-6pt}\sim \phi$. Suppose for reductio that $\neg\bigwedge G \hspace{1pt}\vdash\hspace{-6pt}\sim \phi$. Since $\neg\bigwedge G \hspace{1pt}\vdash\hspace{-6pt}\sim \psi$ for every $\psi \in G - F$, we get with (Ref), (And) and (RW), that $\neg\bigwedge G \hspace{1pt}\vdash\hspace{-6pt}\sim \neg\bigwedge F$. Combining this with $\neg\bigwedge G \hspace{1pt}\vdash\hspace{-6pt}\sim \phi$, we get from (CMon) that $(\neg\bigwedge G) \wedge (\neg\bigwedge F) \hspace{1pt}\vdash\hspace{-6pt}\sim \phi$, i.e., by (LLE), $\neg\bigwedge F \hspace{1pt}\vdash\hspace{-6pt}\sim \phi$, and we have a contradiction.

(RMon) implies (IV). Let $F \subseteq G$ and $\delta(G) \cap F \neq \emptyset$. We need to show that $\delta(F) \subseteq \delta(G)$. The supposition $\delta(G) \cap F \neq \emptyset$ says that for some $\phi \in F$, $\phi \in \delta(G)$, i.e., $\neg\bigwedge G \hspace{1pt}\not\vdash\hspace{-6pt}\sim \phi$. Now let ψ be in $\delta(F)$, i.e., $\neg\bigwedge F \hspace{1pt}\not\vdash\hspace{-6pt}\sim \psi$. We have to show that ψ is

in $\delta(G)$ as well, i.e., that $\neg\bigwedge G \not\hspace{-0.3em}\sim \psi$. Suppose for reductio that $\neg\bigwedge G \hspace{0.1em}\sim\hspace{-0.9em}\vert\hspace{0.4em} \psi$. Since $\neg\bigwedge G \not\hspace{-0.3em}\sim \phi$, (RW) gives us $\neg\bigwedge G \not\hspace{-0.3em}\sim \bigwedge F$. Hence, by (RMon), $(\neg\bigwedge G)\wedge(\neg\bigwedge F) \hspace{0.1em}\sim\hspace{-0.9em}\vert\hspace{0.4em} \psi$. By (LLE), we get $\neg\bigwedge F \hspace{0.1em}\sim\hspace{-0.9em}\vert\hspace{0.4em} \psi$, and we have a contradiction.

(Cut) implies (I^-). Let $F \subseteq G$ and $\delta(G) \subseteq F$. We need to show that $\delta(G) \subseteq \delta(F)$. The supposition $\delta(G) \subseteq F$ says that for each $\psi \in G - F$, $\psi \notin \delta(G)$, i.e., $\neg\bigwedge G \hspace{0.1em}\sim\hspace{-0.9em}\vert\hspace{0.4em} \psi$. Now let ϕ be in $\delta(G)$, i.e., $\neg\bigwedge G \not\hspace{-0.3em}\sim \phi$. We have to show that ϕ is also in $\delta(F)$, i.e., that $\neg\bigwedge F \not\hspace{-0.3em}\sim \phi$. ϕ is in F, since $\delta(G) \subseteq F$. Suppose for reductio that $\neg\bigwedge F \hspace{0.1em}\sim\hspace{-0.9em}\vert\hspace{0.4em} \phi$, or equivalently, by (LLE), $(\neg\bigwedge F) \wedge (\neg\bigwedge G) \hspace{0.1em}\sim\hspace{-0.9em}\vert\hspace{0.4em} \phi$. Since $\neg\bigwedge G \hspace{0.1em}\sim\hspace{-0.9em}\vert\hspace{0.4em} \psi$ for every $\psi \in G-F$, we get with (Ref), (And) and (RW), that $\neg\bigwedge G \hspace{0.1em}\sim\hspace{-0.9em}\vert\hspace{0.4em} \neg\bigwedge F$. Hence, by $(\neg\bigwedge F) \wedge (\neg\bigwedge G) \hspace{0.1em}\sim\hspace{-0.9em}\vert\hspace{0.4em} \phi$ and (Cut), $\neg\bigwedge G \hspace{0.1em}\sim\hspace{-0.9em}\vert\hspace{0.4em} \phi$, and we have a contradiction.

(DRat) implies (II^+). Let $\phi \in \delta(F)$ and $\psi \in \delta(G)$, that is, $\neg\bigwedge F \not\hspace{-0.3em}\sim \phi$ and $\neg\bigwedge G \not\hspace{-0.3em}\sim \psi$. We have to show that either ϕ or ψ is in $\delta(F \cup G)$, i.e., that either $\neg\bigwedge(F \cup G) \not\hspace{-0.3em}\sim \phi$ or $\neg\bigwedge(F \cup G) \not\hspace{-0.3em}\sim \psi$. Suppose for reductio that both $\neg\bigwedge(F \cup G) \hspace{0.1em}\sim\hspace{-0.9em}\vert\hspace{0.4em} \phi$ and $\neg\bigwedge(S \cup G) \hspace{0.1em}\sim\hspace{-0.9em}\vert\hspace{0.4em} \psi$. Applying (And) and (LLE), we conclude that $\neg(\bigwedge F) \vee \neg(\bigwedge G) \hspace{0.1em}\sim\hspace{-0.9em}\vert\hspace{0.4em} \phi \wedge \psi$. Hence, by ($II^+$), either $\neg\bigwedge F \hspace{0.1em}\sim\hspace{-0.9em}\vert\hspace{0.4em} \phi \wedge \psi$ or $\neg\bigwedge G \hspace{0.1em}\sim\hspace{-0.9em}\vert\hspace{0.4em} \phi \wedge \psi$, and so by (RW), either $\neg\bigwedge F \hspace{0.1em}\sim\hspace{-0.9em}\vert\hspace{0.4em} \phi$ or $\neg\bigwedge G \hspace{0.1em}\sim\hspace{-0.9em}\vert\hspace{0.4em} \psi$, and we have a contradiction. Hence $\hspace{0.1em}\sim\hspace{-0.9em}\vert\hspace{0.4em}$ satisfies (II^+).

(Mon) implies (IV^+). Let $F \subseteq G$. We need to show that $\delta(F) \subseteq \delta(G)$. Let ϕ be in $\delta(F)$, i.e., $\neg\bigwedge F \not\hspace{-0.3em}\sim \phi$. We have to show that ϕ is in $\delta(G)$ as well, i.e., that $\neg\bigwedge G \not\hspace{-0.3em}\sim \phi$. Suppose for reductio that $\neg\bigwedge G \hspace{0.1em}\sim\hspace{-0.9em}\vert\hspace{0.4em} \phi$. (Mon) gives us $(\neg\bigwedge G) \wedge (\neg\bigwedge F) \hspace{0.1em}\sim\hspace{-0.9em}\vert\hspace{0.4em} \phi$. By (LLE), we get $\neg\bigwedge F \hspace{0.1em}\sim\hspace{-0.9em}\vert\hspace{0.4em} \phi$, and we have a contradiction.

($\emptyset$CMon) implies ($\emptyset$1). Let $F \subseteq G$ and $\delta(G) = \emptyset$. We need to show that $\delta(F) = \emptyset$. The supposition $\delta(G) = \emptyset$ says that for each $\phi \in G$, $\neg\bigwedge G \hspace{0.1em}\sim\hspace{-0.9em}\vert\hspace{0.4em} \phi$. By (Ref), (And) and (RW), this implies that $\neg\bigwedge G \hspace{0.1em}\sim\hspace{-0.9em}\vert\hspace{0.4em} \bot$. By ($\emptyset$CMon), this is equivalent to $\neg\bigwedge G \wedge \neg\bigwedge F \hspace{0.1em}\sim\hspace{-0.9em}\vert\hspace{0.4em} \bot$. Since $\neg\bigwedge F$ entails $\neg\bigwedge G$, this is equivalent to $\neg\bigwedge F \hspace{0.1em}\sim\hspace{-0.9em}\vert\hspace{0.4em} \bot$, by (LLE). Hence, by (RW), we have $\neg\bigwedge F \hspace{0.1em}\sim\hspace{-0.9em}\vert\hspace{0.4em} \phi$ for all ϕ in F. But this just means that $\delta(F) = \emptyset$, by the definition of δ.

($\emptyset$Cond) implies ($\emptyset$2). Let $F \subseteq G$ and $\delta(F) = \emptyset$. We need to show that $F \cap \delta(G) = \emptyset$. The supposition $\delta(F) = \emptyset$ says that for each $\phi \in F$, $\neg\bigwedge F \hspace{0.1em}\sim\hspace{-0.9em}\vert\hspace{0.4em} \phi$. By (Ref), (And) and (RW), this implies that $\neg\bigwedge F \hspace{0.1em}\sim\hspace{-0.9em}\vert\hspace{0.4em} \bot$. Since $\neg\bigwedge F$ entails $\neg\bigwedge G$, this is equivalent to $\neg\bigwedge G \wedge \neg\bigwedge F \hspace{0.1em}\sim\hspace{-0.9em}\vert\hspace{0.4em} \bot$, by (LLE). By ($\emptyset$Cond), this is equivalent to $\neg\bigwedge G \hspace{0.1em}\sim\hspace{-0.9em}\vert\hspace{0.4em} \bigwedge F$. Hence, by (RW), we have $\neg\bigwedge G \hspace{0.1em}\sim\hspace{-0.9em}\vert\hspace{0.4em} \phi$ for all ϕ in F. But this just means that $F \cap \delta(G) = \emptyset$, by the definition of δ. $\qquad\square$

Observation 37

Let δ be a finitary and ω-covering syntactic choice function which satisfies (LP1) and (LP2). If δ satisfies (II) and (III), then it satisfies (II^+).

Proof of Observation 37.

Let $\phi \in \delta(F)$ and $\psi \in \delta(G)$. We want to show that either $\phi \in \delta(F \cup G)$ or $\psi \in \delta(F \cup G)$. Now consider the sets $F' = F \cup \{\psi\}$ and $G' = G \cup \{\psi\}$. Note that F' and G' are in $\mathcal{X}$ and $\{\phi, \psi\} \subseteq F \cup G = F' \cup G'$. From $\phi \in \delta(F)$ we get with (III) that

$$\text{either } \phi \in \delta(F') \text{ or } \psi \in \delta(F') \tag{\dag}$$

Similarly, we get from $\psi \in \delta(G)$ that

$$\text{either } \psi \in \delta(G') \text{ or } \phi \in \delta(G') \tag{$\ddagger$}$$

Now if in both (†) and (‡) the disjuncts concerning ϕ are true, we get $\phi \in \delta(F' \cup G') = \delta(F \cup G)$, by (II), and we are done. Similarly, if in both (†) and (‡) the disjuncts concerning ψ are true, we get $\psi \in \delta(F \cup G)$, by (II), and we are done. The difficult case is when

$$\phi \in \delta(F') \text{ and } \psi \in \delta(G')$$

(The remaining fourth case $\psi \in \delta(F')$ and $\phi \in \delta(G')$ is analogous.) From this it follows with (LP2) that $\phi \in \delta(F' \cup \{\phi \wedge \psi\})$ and $\psi \in \delta(G' \cup \{\phi \wedge \psi\})$, so by (LP1), $\phi \wedge \psi \in \delta(F' \cup \{\phi \wedge \psi\})$ and $\phi \wedge \psi \in \delta(G' \cup \{\phi \wedge \psi\})$. Hence, by (II), $\phi \wedge \psi \in \delta(F' \cup G' \cup \{\phi \wedge \psi\})$, so by (LP1) again, either $\phi \in \delta(F' \cup G' \cup \{\phi \wedge \psi\})$ or $\psi \in \delta(F' \cup G' \cup \{\phi \wedge \psi\})$. Hence, by (LP2), either $\phi \in \delta(F \cup G)$ or $\psi \in \delta(F \cup G)$, as desired. □

Observation 38

(i) Let δ be a syntactic choice function, let $\gamma = \mathcal{G}(\delta)$ and let $\gamma' = \mathcal{G}(\mathcal{C}(\delta))$. Then for all sentences ϕ

$$\gamma(\,]\phi[\,) \subseteq \gamma'(\,]\phi[\,)$$

and if δ satisfies (Faith1) and (Faith2) with respect to $L - K$, then

$$\gamma'(\,]\phi[\,) \subseteq \gamma(\,]\phi[\,)$$

(ii) Let γ be a semantic choice function, let $\delta = \mathcal{D}(\gamma)$ and let $\delta' = \mathcal{D}(\mathcal{C}(\gamma))$. Then for all finite sets $\{\phi_1, \ldots, \phi_n\}$ of sentences

$$\delta(\{\phi_1, \ldots, \phi_n\}) \subseteq \delta'(\{\phi_1, \ldots, \phi_n\})$$

and if γ satisfies (Faith2) with respect to $[\![K]\!]$, then

$$\delta'(\{\phi_1, \ldots, \phi_n\}) \subseteq \delta(\{\phi_1, \ldots, \phi_n\})$$

Proof of Observation 38.

(i) Let δ, γ and γ' be as stated, and let $\dot{-} = \mathcal{C}(\delta)$. Then, by the definition of $\gamma = \mathcal{G}(\delta)$, m is in $\gamma(\,]\phi[\,)$ iff

$$m \in\,]\phi[\text{ and } Cn(\phi) - \delta(Cn(\phi)) \subseteq \widehat{m} \tag{$\dagger$}$$

On the other hand, by the definition of $\gamma' = \mathcal{G}(\dot{-})$, m is in $\gamma'(\,]\phi[\,)$ iff m is in $]\phi[$ and $K \dot{-} \phi \subseteq \widehat{m}$, which means, by the definition of $\dot{-} = \mathcal{C}(\delta)$ that

$$m \in\,]\phi[\text{ and } \{\psi \in K : \phi \vee \psi \notin \delta(Cn(\phi))\} \subseteq \widehat{m} \tag{$\ddagger$}$$

First we show that (†) implies (‡). Assume that ψ is in K and $\phi \vee \psi \notin \delta(Cn(\phi))$. We have to show that ψ is in $\widehat{m}$. Now we know that $\phi \vee \psi \in Cn(\phi) - \delta(Cn(\phi))$.

So by (†), $\phi \vee \psi$ is in $\widehat{m}$. But we also know that m is in $]\phi[$, i.e., $\neg\phi$ is in $\widehat{m}$. So ψ is in $\widehat{m}$, as required.

Now we show that (‡) implies (†), provided that δ satisfies (Faith1) and (Faith2) with respect to $L - K$. Assume that ψ is in $Cn(\phi) - \delta(Cn(\phi))$. We have to show that ψ is in $\widehat{m}$. We distinguish two cases.

If ϕ is in K, then since $\psi \in Cn(\phi)$, ψ is in $K = Cn(K)$ as well. Since ψ is in $Cn(\phi)$ but not in $\delta(Cn(\phi))$, we get by (LP1) that $\phi \vee \psi$ is not in $\delta(Cn(\phi))$. Hence, by (‡), ψ is in $\widehat{m}$.

If ϕ is not in K, then $Cn(\phi) \cap (L - K) \neq \emptyset$. We now see that $\{\psi \in K : \phi \vee \psi \notin \delta(Cn(\phi))\}$ is identical with K. This is because for every ψ in K, $\phi \vee \psi$ is in $K = Cn(K)$ as well, or equivalently, $\phi \vee \psi \notin L - K$, so by (Faith1), $\phi\vee\psi \notin \delta(Cn(\phi))$. So (‡) entails that $K \subseteq \widehat{m}$. In order to show that (†) is satisfied, then, it suffices to show that $Cn(\phi) - \delta(Cn(\phi)) \subseteq K$. But (Faith2) tells us that $Cn(\phi) \cap (L - K) \subseteq \delta(Cn(\phi))$, from which we can get $Cn(\phi) - \delta(Cn(\phi)) \subseteq K$ by elementary set-theoretical reasoning.

(ii) Let γ, δ and δ' be as stated, and let $\dot{-} = \mathcal{C}(\gamma)$. Then, by the definition of $\delta = \mathcal{D}(\gamma)$, ϕ_i is in $\delta(\{\phi_1, \ldots, \phi_n\})$ iff

$$\phi_i \in \{\phi_1, \ldots, \phi_n\} \text{ and }]\phi_i[\cap \gamma(]\{\phi_1, \ldots, \phi_n\}[) \neq \emptyset \tag{†}$$

On the other hand, by the definition of $\delta' = \mathcal{D}(\dot{-})$, ϕ_i is in $\delta(\{\phi_1, \ldots, \phi_n\})$ iff ϕ_i is in $\{\phi_1, \ldots, \phi_n\}$ and $\phi_i \notin K \dot{-} (\phi_1 \wedge \cdots \wedge \phi_n)$, which means, by the definition of $\dot{-} = \mathcal{C}(\gamma)$ that

$$\phi_i \in \{\phi_1, \ldots, \phi_n\}, \text{ and } \phi_i \notin K \text{ or } \gamma(]\phi_1 \wedge \cdots \wedge \phi_n[) \not\subseteq [\![\phi_i]\!] \tag{‡}$$

First we show that (†) implies (‡). This is clear from $]\{\phi_1, \ldots, \phi_n\}[=]\phi_1 \wedge \cdots \wedge \phi_n[$ and $[\![\phi_i]\!] \cap]\phi_i[= \emptyset$.

Now we show that (‡) implies (†), provided that γ satisfies (Faith2) with respect to $[\![K]\!]$. We distinguish two cases.

If ϕ_i is in K, then (‡) tells us that $\gamma(]\phi_1 \wedge \cdots \wedge \phi_n[) \not\subseteq [\![\phi_i]\!]$, and (†) follows immediately from $[\![\phi_i]\!] \cup]\phi_i[= \mathcal{M}_L$.

If ϕ_i is not in $K = Cn(K)$, then $]\phi_i[\cap [\![K]\!] \neq \emptyset$. Hence, since $]\phi_i[\subseteq]\phi_1, \ldots, \phi_n[$, we also have $]\phi_i[\cap (]\phi_1, \ldots, \phi_n[\cap [\![K]\!]) \neq \emptyset$. Hence, since $]\phi_1, \ldots, \phi_n[\cap [\![K]\!] \subseteq \gamma(]\phi_1, \ldots, \phi_n[)$ by (Faith2), $]\phi_i[\cap \gamma(]\phi_1, \ldots, \phi_n[) \neq \emptyset$, so (†) is satisfied.   □

Observation 39

(i) For every semantic choice function γ, the corresponding syntactic choice function $\delta = \mathcal{D}(\gamma)$ satisfies (LP1) and (LP2).

(ii) For every syntactic choice function δ that satisfies (LP1) and (LP2), the corresponding semantic choice function $\gamma = \mathcal{D}(\delta)$ is complete.

(iii) For every syntactic choice function δ that satisfies (LP1) and (LP2), the syntactic choice function corresponding to the semantic choice function corresponding to δ is identical with δ, that is

$$\mathcal{D}(\mathcal{G}(\delta)) = \delta$$

(iv) For every semantic choice function γ, the semantic choice function corresponding to the syntactic choice function corresponding to γ is the completion γ^+ of γ, that is

$$\mathcal{G}(\mathcal{D}(\gamma)) = \gamma^+$$

(v) Corresponding choice functions lead to identical contraction functions, that is, it holds both that

$$\mathcal{C}(\mathcal{D}(\gamma)) = \mathcal{C}(\gamma)$$

and

$$\mathcal{C}(\mathcal{G}(\delta)) = \mathcal{C}(\delta)$$

Proof of Observation 39.

(i) We want to show that $\delta = \mathcal{D}(\gamma)$ satisfies (LP1) and (LP2).

For (LP1), let $\phi \in \delta(F) \cap Cn(G)$ and $G \subseteq F$. From $\phi \in \delta(F)$ it follows that $]\phi[\cap \gamma(]F[) \neq \emptyset$, and from $G \vdash \phi$ it follows that $]\phi[\subseteq]G[$. Hence $]G[\cap \gamma(]F[) \neq \emptyset$, hence, since $]G[= \bigcup \{]\psi[: \psi \in G\}$, $]\psi[\cap \gamma(]F[) \neq \emptyset$ for some $\psi \in G \subseteq F$. Hence, by the definition of δ, $\psi \in \delta(F)$ for this ψ, hence $G \cap \delta(F) \neq \emptyset$, as desired.

For (LP2), let $Cn(F) = Cn(G)$ and $\phi \in \delta(F) \cap G$. Hence, $]\phi[\cap \gamma(]F[) \neq \emptyset$. Since $Cn(F) = Cn(G)$, we have $]F[=]G[$, so $]\phi[\cap \gamma(]G[) \neq \emptyset$. Since $\phi \in G$, we get, by the definition of δ, $\phi \in \delta(G)$, as desired. The converse direction is analogous.

(ii) We want to show that $\gamma = \mathcal{D}(\delta)$ is complete.

Let $\bigcap \{\widehat{m'} : m' \in \gamma(]F[)\} \subseteq \widehat{m}$ and suppose for reductio that $m \in]F[- \gamma(]F[)$. By the definition of γ, this means that $\bigcap \{\widehat{m'} : m' \in]F[$ and $Cn(F) - m' \subseteq \delta(Cn(F))\} \subseteq \widehat{m}$, and $Cn(F) - \widehat{m} \not\subseteq \delta(Cn(F))$. Take some ϕ such that $F \vdash \phi$, m does not satisfy ϕ, and $\phi \notin \delta(Cn(F))$. From $\phi \notin \widehat{m}$ we conclude that $\phi \notin \bigcap \{\widehat{m'} : m' \in]F[$ and $Cn(F) - m' \subseteq \delta(Cn(F))\}$. Hence some $m' \in]F[$ such that $Cn(F) - \widehat{m'} \subseteq \delta(Cn(F))$ does not satisfy ϕ. Since $\phi \in Cn(F) - \widehat{m'}$, we get $\phi \in \delta(Cn(F))$, and we have a contradiction.

(iii) We want to show that $\mathcal{D}(\mathcal{G}(\delta)) = \delta$. Let $\gamma = \mathcal{G}(\delta)$ and $\delta' = \mathcal{D}(\gamma)$. We have to show that for all ϕ and F,

$$\phi \in \delta(F) \text{ iff } \phi \in \delta'(F)$$

Unfolding the definition, we get that the right-hand side means

$$\phi \in F \text{ and }]\phi[\cap \gamma(]F[) \neq \emptyset$$

and finally

$$\phi \in F \text{ and for some } m \in]\phi[\text{ it holds that } Cn(F) - \widehat{m} \subseteq \delta(Cn(F)) \qquad (\dagger)$$

(Since $m \in]\phi[$ and $\phi \in F$, we can drop the additional condition $m \in]F[$.)

$\phi \in \delta(F)$ implies (†). Let ϕ be in $\delta(F)$. Since $\delta(F) \subseteq F$, it is clear that $\phi \in F$. (LP2) gives us $\phi \in \delta(Cn(F))$. From (LP1) we then know that $Cn(F) - \delta(Cn(F))$ does not entail ϕ. Hence there is an m which satisfies $Cn(F) - \delta(Cn(F))$ but falsifies ϕ. The latter says that $m \in]\!]\phi[\![$, while the former says that $Cn(F) - \delta(Cn(F)) \subseteq \hat{m}$, and hence $Cn(F) - \hat{m} \subseteq \delta(Cn(F))$, and we are done.

(†) implies $\phi \in \delta(F)$. Let ϕ be in F and let there be an m in $]\!]\phi[\![$ such that $Cn(F) - \hat{m} \subseteq \delta(Cn(F))$. Since $F \vdash \phi$ and $\phi \notin \hat{m}$, we immediately get that $\phi \in \delta(Cn(F))$, so by (LP2) $\phi \in \delta(F)$, we are done.

(iv) We want to show that $\mathcal{G}(\mathcal{D}(\gamma)) = \gamma^+$. Let $\delta = \mathcal{D}(\gamma)$ and $\gamma' = \mathcal{G}(\delta)$. We have to show that for all m and F,

$$m \in \gamma^+(]\!]F[\![) \text{ iff } m \in \gamma'(]\!]F[\![)$$

Unfolding the definitions, we get that the left-hand side means

$$m \in]\!]F[\![\text{ and } \bigcap\{\widehat{m'} : m' \in \gamma(]\!]F[\![)\} \subseteq \hat{m}$$

or equivalently,

$$m \in]\!]F[\![\text{ and for every } \phi, \text{ if } \gamma(]\!]F[\![) \subseteq [\![\phi]\!] \text{ then } m \in [\![\phi]\!]$$

or, again equivalently, by contraposition

$$m \in]\!]F[\![\text{ and for every } \phi, \text{ if } m \in]\!]\phi[\![\text{ then }]\!]\phi[\![\cap \gamma(]\!]F[\![) \neq \emptyset \qquad (\dagger)$$

Unfolding the definitions for the right-hand side, we get that it means

$$m \in]\!]F[\![\text{ and } Cn(F) - \hat{m} \subseteq \delta(Cn(F))$$

and finally

$$m \in]\!]F[\![\text{ and for every } \phi, \text{ if } F \vdash \phi \text{ and } m \in]\!]\phi[\![, \text{ then }]\!]\phi[\![\cap \gamma(]\!]F[\![) \neq \emptyset \ (\ddagger)$$

It is now immediately clear that (†) implies (‡). In order to show that conversely (‡) implies (†), we take an arbitrary ϕ such that $m \in]\!]\phi[\![$. Then take some $\psi \in F$ such that $m \in]\!]\psi[\![$. Such a ψ exists, since $m \in]\!]F[\![$. We infer that $m \in]\!]\phi \vee \psi[\![$. As $F \vdash \phi \vee \psi$, the assumption (‡) gives us that $]\!]\phi \vee \psi[\![\cap \gamma(]\!]F[\![) \neq \emptyset$. Since $]\!]\phi \vee \psi[\![\subseteq]\!]\phi[\![$, we get that $]\!]\phi[\![\cap \gamma(]\!]F[\![) \neq \emptyset$, as desired.

(v) We want to show that corresponding choice functions lead to identical contraction functions. So we first have to show that

$$\mathcal{C}(\mathcal{D}(\gamma)) = \mathcal{C}(\gamma)$$

Let ϕ and ψ be in K. We want to know in which cases we will have $\psi \in K \dot{-} \phi$. According to $\mathcal{C}(\gamma)$, this is true just in case $\gamma(]\!]\phi[\![) \subseteq [\![\psi]\!]$.

According to $\mathcal{C}(\delta)$, with $\delta = \mathcal{D}(\gamma)$, this is true just in case $\phi \vee \psi \notin \delta(Cn(\phi))$, or

$$]\phi \vee \psi[\cap \gamma(]Cn(\phi)[) = \emptyset$$

(It is clear that $\phi \vee \psi \in Cn(\phi)$.) Since $]Cn(\phi)[=]\phi[$, this is equivalent with

$$\gamma(]\phi[) \subseteq [\![\phi \vee \psi]\!]$$

Since $\gamma(]\phi[) \subseteq]\phi[$ and $[\![\phi \vee \psi]\!] \cap]\phi[\subseteq [\![\psi]\!] \subseteq [\![\phi \vee \psi]\!]$, this is again equivalent with

$$\gamma(]\phi[) \subseteq [\![\psi]\!]$$

and we have the desired identity.

The dual equation

$$\mathcal{C}(\mathcal{G}(\delta)) = \mathcal{C}(\delta)$$

is a corollary to the results achieved so far. Applying $\mathcal{C}(\mathcal{D}(\gamma)) = \mathcal{C}(\gamma)$, we get $\mathcal{C}(\mathcal{G}(\delta)) = \mathcal{C}(\mathcal{D}(\mathcal{G}(\delta))) = \mathcal{C}(\delta)$, the latter equality being due to the equation $\mathcal{D}(\mathcal{G}(\delta)) = \delta$ which we proved in part (iii) of this observation. $\square$

For the proof of the next observation, it is convenient to have available a preparatory result:

Auxiliary Lemma

Let $]F[\subseteq]G[$ and suppose that for all $m \in]G[$ such that $Cn(G) - \hat{m} \subseteq \delta(Cn(G))$ it holds that $m \in]F[$ (i.e., for $\gamma = \mathcal{G}(\delta)$, that $\gamma(]G[) \subseteq]F[$). Then there exists a subset Γ of $Cn(G)$ such that

(i) $Cn(F \cup \Gamma) = Cn(G)$

(ii) $\Gamma \cap \delta(Cn(G)) = \emptyset$

(iii) $\Gamma \cap Cn(F) = \emptyset$.

Notice that the Lemma allows for the case that $F = \emptyset$. For this special case the lemma says: If there is no $m \in]G[$ such that $Cn(G) - \hat{m} \subseteq \delta(Cn(G))$, then there is a subset Γ of $Cn(G)$ such that (i) $Cn(\Gamma) = Cn(G)$, (ii) $\Gamma \cap \delta(Cn(G)) = \emptyset$, and (iii) $\Gamma \cap Cn(\emptyset) = \emptyset$.

Proof of the Auxiliary Lemma.

Let $]F[\subseteq]G[$ and suppose that for all $m \in]G[$ such that $Cn(G) - \hat{m} \subseteq \delta(Cn(G))$ it holds that $m \in]F[$. The latter supposition means that

for every m, if $m \in [\![F]\!] \cap]G[$ then $Cn(G) - \hat{m} \not\subseteq \delta(Cn(G))$

Now take for every $m \in [\![F]\!] \cap]G[$ some sentence ϕ_m such that $G \vdash \phi_m$, $m \in]\phi_m[$ and $\phi_m \notin \delta(Cn(G))$, and put

$$\Gamma = \{\phi_m : m \in [\![F]\!] \cap]G[\}$$

$\Gamma = \emptyset$ if and only if $]F[=]G[$, or equivalently, $Cn(G) = Cn(F)$.

(i) Since $] F [\subseteq] G [$ implies $Cn(F) \subseteq Cn(G)$ and Γ is also a subset of $Cn(G)$, it is clear that $Cn(F \cup \Gamma) \subseteq Cn(G)$. To show that $Cn(G) \subseteq Cn(F \cup \Gamma)$, let ψ be in $Cn(G)$. Suppose for reductio that ψ is not in $Cn(F \cup \Gamma)$. Then there is a model m' that satisfies F and Γ but falsifies ψ. Since m' satisfies ϕ_m for all $m \in [\![F]\!] \cap]G[$, but by construction each such m does not satisfy ϕ_m, it follows that m' is in $] F [\![\cup [\![G]\!]$. But m' cannot be in $] F [\![$, since it satisfies F. Moreover, m' cannot be in $[\![G]\!]$, since $G \vdash \psi$ and m' falsifies ψ. So we have a contradiction in either case.

(ii) $\Gamma \cap \delta(Cn(G)) = \emptyset$ is immediate from the construction of the ϕ_ms.

(iii) $\Gamma \cap Cn(F) = \emptyset$ is also clear from the construction of the ϕ_ms: Since each relevant m satisfies F, but falsifies ϕ_m, the latter cannot be a consequence of the former. If there is no relevant m, then $\Gamma \cap Cn(F) \subseteq \Gamma = \emptyset$ is trivial.   □

Observation 40

For every syntactic choice function δ which satisfies (LP1), (LP2) and

$$\left\{\begin{array}{l} \text{(Faith1) w.r.t. } \delta(L) \\ \text{(Faith2) w.r.t. } \delta(L) \\ \text{(Virtual Success)} \\ \quad (\emptyset 1) \\ \quad (\emptyset 2) \\ \quad (\text{I}) \\ \quad (\text{II}) \\ \quad (\text{III}) \\ \quad (\text{IV}) \\ \quad (\text{I}^-) \\ \quad (\text{II}^+) \\ \quad (\text{IV}^+) \end{array}\right\}, \text{ the corresponding semantic choice function } \gamma = \mathcal{G}(\delta)$$

$$\text{satisfies} \left\{\begin{array}{l} \text{(Faith1) w.r.t. } \gamma(\mathcal{M}_L) \\ \text{(Faith2) w.r.t. } \gamma(\mathcal{M}_L) \\ \quad \text{(Virtual Success)} \\ \qquad (\emptyset 1) \\ \qquad (\emptyset 2) \\ \qquad (\text{I}) \\ \qquad (\text{II}) \\ \qquad (\text{III}) \\ \qquad (\text{IV}) \\ \qquad (\text{I}^-) \\ \qquad (\text{II}^+) \\ \qquad (\text{IV}^+) \end{array}\right\}, \text{ respectively.}$$

Proof of Observation 40. Let δ satisfy (LP1) and (LP2) and let $\gamma = \mathcal{G}(\delta)$.

(Faith1). We are going to show that

$$\forall F : \text{ if }]\!]F[\![\,\cap\,\gamma(\mathcal{M}_L) \neq \emptyset \text{ then } \gamma(]\!]F[\![\,) \subseteq \gamma(\mathcal{M}_L)$$

Let $]\!]F[\![\,\cap\,\gamma(\mathcal{M}_L) \neq \emptyset$. Since $\gamma(\mathcal{M}_L) = \gamma(]\!]L[\![\,) = \{m : L - \delta(L) \subseteq \widehat{m}\}$, it follows that $Cn(F) \cap \delta(L) \neq \emptyset$. Now we apply (Faith1) for δ and get that $\delta(Cn(F)) \subseteq \delta(L)$.

We have to show that $\gamma(]\!]F[\![\,) \subseteq \gamma(\mathcal{M}_L)$, that is, that every m in $\gamma(]\!]F[\![\,)$ satisfies K. This means, by $\gamma = \mathcal{G}(\delta)$, that for all m

$$\text{if } m \in]\!]F[\![\text{ and } Cn(F) - \widehat{m} \subseteq \delta(Cn(F)) \text{ then } m \in \gamma(\mathcal{M}_L)$$

Suppose for reductio that there is an m for which the antecedent is true but the consequent is false. From the latter, we know that there is some ϕ which is not in $\delta(L)$ and is falsified by m. Since m is in $]\!]F[\![$, there is a ψ in F which is falsified by m. Thus $\phi \vee \psi$ is falsified by m, so $\phi \vee \psi$ is in $Cn(F) - \widehat{m} \subseteq \delta(Cn(F))$. Since we have shown that $\delta(Cn(F)) \subseteq \delta(L)$, we get that $\phi \vee \psi$ is in $\delta(L)$. Thus, by (LP1), ϕ must be in $\delta(L)$ as well, but this contradicts our choice of ϕ.

(Faith2). We are going to show that

$$\forall F :]\!]F[\![\,\cap\,\gamma(\mathcal{M}_L) \subseteq \gamma(]\!]F[\![\,)$$

We have to show that every m which falsifies F but verifies $L - \delta(L)$ is in $\gamma(]\!]F[\![\,)$. This means, by $\gamma = \mathcal{G}(\delta)$, that for all m

$$\text{if } m \in]\!]F[\![\,\cap\,[\![L - \delta(L)]\!] \text{ then } m \in]\!]F[\![\text{ and } Cn(F) - \widehat{m} \subseteq \delta(Cn(F))$$

Suppose for reductio that there is an m for which the antecedent is true but the consequent is false. Hence $Cn(F) - \widehat{m} \not\subseteq \delta(Cn(F))$, that is, there is a ϕ in $Cn(F) - \widehat{m}$ which is not in $\delta(Cn(F))$. Since we know from (Faith2) for δ that $Cn(F) \cap \delta(L) \subseteq \delta(Cn(F))$, we can infer that $\phi \notin \delta(L)$. Then we get from $m \in L - \delta(L)$ that m verifies ϕ. On the other hand, $\phi \in Cn(F) - \widehat{m}$ tells us that m falsifies ϕ, so we have a contradiction.

(Virtual Success) for δ implies (Success) for γ. Let $\gamma(]\!]F[\![\,) = \emptyset$, which means, by $\gamma = \mathcal{G}(\delta)$, that

$$\text{there is no } m \text{ in }]\!]F[\![\text{ such that } Cn(F) - \widehat{m} \subseteq \delta(Cn(F))$$

We can now apply the Auxiliary Lemma, with $F' = \emptyset$ and $G' = F$, remembering that there is no m contained in $]\!]\emptyset[\![$. According to the Auxiliary Lemma, there is a set $\Gamma \subseteq Cn(G')$ such that (i) $Cn(F' \cup \Gamma) = Cn(\Gamma) = Cn(G')$ and (ii) $\Gamma \cap \delta(Cn(G')) = \emptyset$ (we do not need (iii)). Since by (i) $\Gamma \subseteq Cn(G') = Cn(\Gamma)$, we know from (LP2) that $\delta(\Gamma) \subseteq \delta(Cn(G'))$, hence, by $\delta(\Gamma) \subseteq \Gamma$ and (ii), $\delta(\Gamma) = \emptyset$. But now we can show that $\delta(Cn(G')) = \emptyset$ as well. For if there were a $\phi \in \delta(Cn(G')) \subseteq Cn(G')$, then we would get from (i) that $\Gamma \vdash \phi$ and $Cn(\Gamma \cup \{\phi\}) = Cn(G')$, and so $\phi \in \delta(\Gamma \cup \{\phi\})$, by (LP2). But from this we get $\psi \in \delta(\Gamma \cup \{\phi\})$

for some ψ in Γ, by (LP1), from which in turn we get $\psi \in \delta(\Gamma)$, by (LP2). But this contradicts $\delta(\Gamma) = \emptyset$. So in fact $\delta(Cn(G')) = \emptyset$, that is

$$\delta(Cn(F)) = \emptyset$$

From this we get, by applying (Virtual Success) for δ, that $F \subseteq Cn(\emptyset)$. But there is no model falsifying an element of $Cn(\emptyset)$, so $]F[= \emptyset$, as desired.

($\emptyset$1). Let $]F[\subseteq]G[$ and $\gamma(]G[) = \emptyset$. It follows that $Cn(F) \subseteq Cn(G)$. Using the same argument as in the proof of (Success) from (Virtual Success), it follows moreover that $\delta(Cn(G)) = \emptyset$. So by ($\emptyset$1) for δ, we get $\delta(Cn(F)) = \emptyset$. But then there can be no $m \in]F[$ such that $Cn(F) - \widehat{m} \subseteq \delta(Cn(F))$. This means that $\gamma(]F[) = \emptyset$, as desired.

($\emptyset$2). Let $]F[\subseteq]G[$ and $\gamma(]F[) = \emptyset$. It follows that $Cn(F) \subseteq Cn(G)$. Using the same argument as in the proof of (Success) from (Virtual Succes), it follows moreover that $\delta(Cn(F)) = \emptyset$. So by ($\emptyset$2) for δ, we get $\delta(Cn(G)) \cap Cn(F) = \emptyset$. On the other hand, for any m that falsifies some element ϕ of F, we know that $Cn(G) - \widehat{m}$ includes ϕ, and ϕ is in $Cn(F)$. These two pieces together imply that there can be no $m \in]F[$ such that $Cn(G) - \widehat{m} \subseteq \delta(Cn(G))$. Since $]F[\subseteq]G[$, this means that $\gamma(]G[) \cap]F[= \emptyset$, as desired.

(I). Let $]F[\subseteq]G[$ and $m \in]F[\cap \gamma(]G[)$. It follows that $Cn(F) \subseteq Cn(G)$ and $Cn(G) - \widehat{m} \subseteq \delta(Cn(G))$, by $\gamma = \mathcal{G}(\delta)$. Thus we get, using (I) for δ, that $Cn(F) - \widehat{m} \subseteq (Cn(G) - \widehat{m}) \cap Cn(F) \subseteq \delta(Cn(G)) \cap Cn(F) \subseteq \delta(Cn(F))$, which means that $m \in \gamma(]F[)$, as desired.

(II). Let $m \in \gamma(]F[) \cap \gamma(]G[)$, that is, by $\gamma = \mathcal{G}(\delta)$, $m \in]F[$, $Cn(F) - \widehat{m} \subseteq \delta(Cn(F))$, $m \in]G[$ and $Cn(G) - \widehat{m} \subseteq \delta(Cn(G))$. Hence

for every $\psi \in (Cn(F) \cap Cn(G)) - \widehat{m}$: $\psi \in \delta(Cn(F)) \cap \delta(Cn(G))$

Hence, by (II) for δ,

for every $\psi \in (Cn(F) \cap Cn(G)) - \widehat{m}$: $\psi \in \delta(Cn(F) \cup Cn(G))$

so by (LP2),

for every $\psi \in (Cn(F) \cap Cn(G)) - \widehat{m}$: $\psi \in \delta(Cn(F \cup G))$ (†)

We have to show that $m \in \gamma(]F[\cup]G[)$, that is, $m \in \gamma(]F \cup G[)$, or, by $\gamma = \mathcal{G}(\delta)$,

$$Cn(F \cup G) - \widehat{m} \subseteq \delta(Cn(F \cup G))$$

Let $\phi \in Cn(F \cup G) - \widehat{m}$. Hence m does not satisfy ϕ. Now take $\chi \in F$ and $\chi' \in G$ such that m satisfies neither χ nor χ'. Such χ and χ' exist because $m \in]F[\cap]G[$. Now we have

$$\phi \vee \chi \vee \chi' \in (Cn(F) \cap Cn(G)) - \widehat{m}$$

Hence, by (†), $\phi \vee \chi \vee \chi' \in \delta(Cn(F \cup G))$. Since $\phi \vdash \phi \vee \chi \vee \chi'$, we finally get by (LP1) that $\phi \in \delta(Cn(F \cup G))$, as desired.

(III). Let $]F[\, \subseteq \,]G[$ and $\gamma(\,]G[\,) \subseteq \,]F[$. It follows that $Cn(F) \subseteq Cn(G)$.and for every $m \in]G[$ such that $Cn(G) - \widehat{m} \subseteq \delta(Cn(G))$ it holds that $m \in]F[$, by $\gamma = \mathcal{G}(\delta)$.

Now take the set Γ from the Auxiliary Lemma. We then have its clause (ii), $\Gamma \cap \delta(Cn(G)) = \emptyset$. From this we get, by clause (i) of the Auxiliary Lemma and (LP2),

$$\Gamma \cap \delta(Cn(F) \cup \Gamma) = \emptyset$$

By clause (iii) of the Auxiliary Lemma, it follows trivially that

$$\delta(Cn(F) \cup \Gamma) \subseteq Cn(F)$$

Now (III) for δ gives us

$$\delta(Cn(F)) \subseteq \delta(Cn(F) \cup \Gamma)$$

or equivalently, by $\delta(Cn(F)) \subseteq Cn(F) \subseteq Cn(G)$, clause (i) and (LP2),

$$\delta(Cn(F)) \subseteq \delta(Cn(G)) \tag{\dagger}$$

We have to show that $\gamma(\,]F[\,) \subseteq \gamma(\,]G[\,)$, that is, by $\gamma = \mathcal{G}(\delta)$, that for every $m \in]F[$, if $Cn(F) - \widehat{m} \subseteq \delta(Cn(F))$ then $Cn(G) - \widehat{m} \subseteq \delta(Cn(G))$. Now take an $m \in]F[$ such that $Cn(F) - \widehat{m} \subseteq \delta(Cn(F))$, and suppose that $\phi \in Cn(G) - \widehat{m}$. We need to show that $\phi \in \delta(Cn(G))$. Take some $\psi \in F - \widehat{m}$; such a ψ exists since $m \in]F[$. We get that $\phi \vee \psi \in Cn(F) - \widehat{m}$. Thus $\phi \vee \psi \in \delta(Cn(F))$, and also, by (†), $\phi \vee \psi \in \delta(Cn(G))$. Hence, by (LP1), $\phi \in \delta(Cn(G))$, as desired.

(IV). Let $]F[\, \subseteq \,]G[$ and $\gamma(\,]G[\,) \cap]F[\,\neq \emptyset$. It follows that $Cn(F) \subseteq Cn(G)$ and that there is some $m \in]F[$ such that $Cn(G) - \widehat{m} \subseteq \delta(Cn(G))$, by $\gamma = \mathcal{G}(\delta)$. Now take some $\phi \in F$ for which $m \in]\phi[$; such a ϕ exists because $m \in]F[$. Since $\phi \in Cn(G) - \widehat{m}$, we get $\phi \in \delta(Cn(G))$, so $\phi \in \delta(Cn(G)) \cap Cn(F) \neq \emptyset$. Hence, by (IV) for δ

$$\delta(Cn(F)) \subseteq \delta(Cn(G)) \tag{\dagger}$$

The rest of the proof is a literal copy of the end of the proof for (III).

(I⁻). Let $]F[\, \subseteq \,]G[$ and $\gamma(\,]G[\,) \subseteq \,]F[$. It follows that $Cn(F) \subseteq Cn(G)$ and for every $m \in]G[$ such that $Cn(G) - \widehat{m} \subseteq \delta(Cn(G))$ it holds that $m \in]F[$, by $\gamma = \mathcal{G}(\delta)$.

Now take the set Γ from the Auxiliary Lemma. In the same way as in the proof for (III), we show that

$$\delta(Cn(F) \cup \Gamma) \subseteq Cn(F)$$

Now (I⁻) for δ gives us

$$\delta(Cn(F) \cup \Gamma) \subseteq \delta(Cn(F)) \tag{\dagger}$$

We have to show that $\gamma(\,]G[\,) \subseteq \gamma(\,]F[\,)$, that is, by $\gamma = \mathcal{G}(\delta)$, that for every $m \in]G[$, if $Cn(G) - \widehat{m} \subseteq \delta(Cn(G))$ then $Cn(F) - \widehat{m} \subseteq \delta(Cn(F))$. Now take an

$m \in \,]G[$ such that $Cn\,(G) - \widehat{m} \subseteq \delta(Cn\,(G))$, and suppose that $\phi \in Cn\,(F) - \widehat{m}$. We need to show that $\phi \in \delta(Cn\,(F))$. Since $\phi \in Cn\,(F) - \widehat{m}$ and $Cn\,(F) \subseteq Cn\,(G)$, we get $\phi \in Cn\,(G) - \widehat{m} \subseteq \delta(Cn\,(G))$. Since $\phi \in Cn\,(F)$, we get, with the help of clause (i) of the Auxiliary Lemma and (LP2), $\phi \in \delta(Cn\,(F) \cup \Gamma)$. By (†), this finally yields $\phi \in \delta(Cn\,(F))$, as desired.

(II⁺). Let $m \in \gamma(\,]F[\,)$ and $m' \in \gamma(\,]G[\,)$. That is, by $\gamma = \mathcal{G}(\delta)$, $m \in]F[$, $Cn\,(F) - \widehat{m} \subseteq \delta(Cn\,(F))$, $m' \in]G[$ and $Cn\,(G) - \widehat{m'} \subseteq \delta(Cn\,(G))$. We have to show that either $m \in \gamma(\,]F[\cup\,]G[\,)$ or $m' \in \gamma(\,]F[\cup\,]G[\,)$, that is, since both m and m' are in $]F[\cup\,]G[\,=\,]F \cup G[$ and $\gamma = \mathcal{G}(\delta)$,

$$Cn\,(F \cup G) - \widehat{m} \subseteq \delta(Cn\,(F \cup G)) \quad \text{or} \quad Cn\,(F \cup G) - \widehat{m'} \subseteq \delta(Cn\,(F \cup G))$$

Now suppose for reductio that this is not the case. Then there are some sentences ϕ and ψ in $Cn\,(F \cup G)$ such that m does not satisfy ϕ, m' does not satisfy ψ, and neither ϕ nor ψ is in $\delta(Cn\,(F \cup G))$. Now take a sentence χ from F which is not satisfied by m and a sentence χ' from G which is not satisfied by m'. Such χ and χ' exist since $m \in]F[$ and $m' \in]G[$. Clearly, we have

$$F \vdash \phi \vee \chi \text{ and } m \text{ does not satisfy } \phi \vee \chi$$
$$G \vdash \psi \vee \chi' \text{ and } m' \text{ does not satisfy } \psi \vee \chi'$$

By $Cn\,(F) - \widehat{m} \subseteq \delta(Cn\,(F))$ and $Cn\,(G) - \widehat{m'} \subseteq \delta(Cn\,(G))$, we get that $\phi \vee \chi \in \delta(Cn\,(F))$ and $\psi \vee \chi' \in \delta(Cn\,(G))$. Applying (II⁺) for δ, we get that

$$\phi \vee \chi \in \delta(Cn\,(F) \cup Cn\,(G)) \text{ or } \psi \vee \chi' \in \delta(Cn\,(F) \cup Cn\,(G))$$

Hence, by (LP2), we have $\phi \vee \chi \in \delta(Cn\,(F \cup G))$ or $\psi \vee \chi' \in \delta(Cn\,(F \cup G))$, and (LP1) finally yields $\phi \in \delta(Cn\,(F \cup G))$ or $\psi \in \delta(Cn\,(F \cup G))$, contradicting our initial assumptions concerning ϕ and ψ.

(IV⁺). Let $]F[\subseteq\,]G[$. It follows that $Cn\,(F) \subseteq Cn\,(G)$ and hence, by (IV⁺) for δ,

$$\delta(Cn\,(F)) \subseteq \delta(Cn\,(G)) \tag{†}$$

The rest of the proof is a literal copy of the end of the proof for (III). □

Observation 41

For every semantic choice function γ which satisfies

$$\left\{\begin{array}{c} \text{(Faith1) wrt } \gamma(\mathcal{M}_L) \\ \text{(Faith2) wrt } \gamma(\mathcal{M}_L) \\ \text{(Success)} \\ (\emptyset 1) \\ (\emptyset 2) \\ (\text{I}) \\ (\text{I}^-) \\ (\text{II}^+) \\ (\text{III}) \\ (\text{IV}) \\ (\text{IV}^+) \end{array}\right\} \text{ the corresponding syntactic choice function } \delta = \mathcal{D}(\gamma)$$

satisfies (LP1) and (LP2) and $\left\{\begin{array}{c} \text{(Faith) wrt } \delta(L) \\ \text{(Faith) wrt } \delta(L) \\ \text{(Virtual Success)} \\ (\emptyset 1) \\ (\emptyset 2) \\ (\text{I}) \\ (\text{I}^-) \\ (\text{II}^+) \\ (\text{III}) \\ (\text{IV}) \\ (\text{IV}^+) \end{array}\right\}$, respectively.

However, if γ satisfies (II), it does *not* follow that the corresponding syntactic choice function $\delta = \mathcal{D}(\gamma)$ satisfies (II), even in the finite case and when γ in addition satisfies (I) and (III).

Proof of Observation 41. Let $\delta = \mathcal{D}(\gamma)$. That δ satisfies (LP1) and (LP2) is proved in Observation 39(i).

(Faith1). We are going to show that

$$\forall F : \text{ if } F \cap \delta(L) \neq \emptyset \text{ then } \delta(F) \subseteq \delta(L)$$

Let $F \cap \delta(L) = F \cap \widehat{\gamma(\mathcal{M}_L)} \neq \emptyset$. Then there exists an m in $\gamma(\mathcal{M}_L)$ such that m falsifies some element of F, that is, $]F[\,\cap\,\gamma(\mathcal{M}_L) \neq \emptyset$. Now we apply (Faith1) for γ and get that $\gamma(]F[) \subseteq \gamma(\mathcal{M}_L)$.

We have to show that $\delta(F) \subseteq \delta(L)$, that is, that every ϕ in $\delta(F)$ is falsified by some m in $\gamma(\mathcal{M}_L)$. This means, by $\delta = \mathcal{D}(\gamma)$, that for all ϕ

$$\text{if } \phi \in F \text{ and }]\phi[\,\cap\,\gamma(]F[) \neq \emptyset \text{ then } \phi \notin \gamma(\mathcal{M}_L)$$

Suppose that ϕ satisfies the antecedent of this condition. Since we have shown that $\gamma(\,]F[\,) \subseteq \gamma(\mathcal{M}_L)$, we get that $]\phi[\,\cap\gamma(\mathcal{M}_L) \neq \emptyset$. But this just means that there is an m in $\gamma(\mathcal{M}_L)$ which falsifies ϕ, i.e., that $\phi \notin \widehat{\gamma(\mathcal{M}_L)}$.

(Faith2). We are going to show that

$$\forall F : F \cap \delta(L) \subseteq \delta(F)$$

We have to show that every ϕ in F which is falsified by some m in $\gamma(\mathcal{M}_L)$ is in $\delta(F)$. This means, by $\delta = \mathcal{D}(\gamma)$, that for all ϕ

$$\text{if } \phi \in F - \widehat{\gamma(\mathcal{M}_L)} \text{ then } \phi \in F \text{ and }]\phi[\,\cap\gamma(\,]F[\,) \neq \emptyset$$

Suppose that ϕ satisfies the antecedent of this condition. So there is an m in $\gamma(\mathcal{M}_L)$ that falsifies ϕ. Since ϕ is in F, m falsifies F. Since we know from (Faith2) for γ that $]F[\,\cap\gamma(\mathcal{M}_L) \subseteq \gamma(\,]F[\,)$, we get that m is in $\gamma(\,]F[\,)$, and hence, since m is in $]\phi[$, $]\phi[\,\cap\gamma(\,]F[\,) \neq \emptyset$.

(Success) for γ implies (Virtual Success) for δ. Let $\delta(F) = \emptyset$. This means, by $\delta = \mathcal{D}(\gamma)$, that for all ϕ in F, $]\phi[\,\cap\gamma(\,]F[\,) = \emptyset$. So $\bigcup\{\,]\phi[\,\cap\gamma(\,]F[\,) : \phi \in F\} = (\bigcup\{\,]\phi[\, : \phi \in F\})\cap\gamma(\,]F[\,) =]F[\,\cap\gamma(\,]F[\,) = \gamma(\,]F[\,) = \emptyset$. Hence, by (Success) for γ, $]F[\,= \emptyset$, which means that each element of F is satisfied in all models. Hence $F \subseteq Cn(\emptyset)$.

($\emptyset 1$). Let $F \subseteq G$ and $\delta(G) = \emptyset$. From the former it follows that $]F[\,\subseteq\,]G[$. Using the same argument as in the proof of (Virtual Success) from (Success), it follows from $\delta(G) = \emptyset$ that $\gamma(\,]G[\,) = \emptyset$. So by ($\emptyset 1$) for γ, we get $\gamma(\,]F[\,) = \emptyset$. But then for all ϕ in F, $]\phi[\,\cap\gamma(\,]F[\,) = \emptyset$, which means that $\delta(F) = \emptyset$, as desired.

($\emptyset 2$). Let $F \subseteq G$ and $\delta(F) = \emptyset$. From the former it follows that $]F[\,\subseteq\,]G[$. Using the same argument as in the proof of (Virtual Success) from (Success), it follows from $\delta(F) = \emptyset$ that $\gamma(\,]F[\,) = \emptyset$. So by ($\emptyset 2$) for γ, we get $\gamma(\,]G[\,)\cap\,]F[\,= \emptyset$, or in other words, for all $\phi \in F$ ist holds that $]\phi[\,\cap\gamma(\,]G[\,) = \emptyset$. But this means that $\delta(G) \cap F = \emptyset$, as desired.

(I). Let $F \subseteq G$ and $\phi \in \delta(G) \cap F$. We have to show that $\phi \in \delta(F)$, i.e., $]\phi[\,\cap\gamma(\,]F[\,) \neq \emptyset$. From our assumptions, it follows that $]\phi[\,\subseteq\,]F[\,\subseteq\,]G[$ and $]\phi[\,\cap\gamma(\,]G[\,) \neq \emptyset$, by $\delta = \mathcal{D}(\gamma)$. Thus we get, using (I) for γ, that $\emptyset \neq]\phi[\,\cap\gamma(\,]G[\,) \subseteq\,]F[\,\cap\gamma(\,]G[\,) \subseteq \gamma(\,]F[\,)$, hence $]\phi[\,\cap\gamma(\,]F[\,) \neq \emptyset$, as desired.

(I^-). Let $F \subseteq G$ and $\delta(G) \subseteq F$. It follows that $]F[\,\subseteq\,]G[$ and for all ϕ, if $\phi \in F'$ and $]\phi[\,\cap\gamma(\,]G[\,) \neq \emptyset$, then $\phi \in F$, by $\delta = \mathcal{D}(\gamma)$. As in the case of (III) we show that $\gamma(\,]G[\,) \subseteq\,]F[$. Then ($I^-$) for γ gives us $\gamma(\,]G[\,) \subseteq \gamma(\,]F[\,)$. Hence for all $\phi \in F'$, $]\phi[\,\cap\gamma(\,]G[\,) \neq \emptyset$ implies $]\phi[\,\cap\gamma(\,]F[\,) \neq \emptyset$, which means, by $\delta = \mathcal{D}(\gamma)$, $\delta(G) \subseteq \delta(F)$, as desired.

(II^+). Let $\phi \in \delta(F)$ and $\psi \in \delta(G)$. That is, by $\delta = \mathcal{D}(\gamma)$, $\phi \in F$, $]\phi[\,\cap\gamma(\,]F[\,) \neq \emptyset$, $\psi \in G$ and $]\psi[\,\cap\gamma(\,]G[\,) \neq \emptyset$. So there is an m in $\gamma(\,]F[\,)$ which does not satisfy ϕ, and an m' in $\gamma(\,]G[\,)$ which does not satisfy ψ. Applying (II^+) for γ, we get that either m or m' is in $\gamma(\,]F[\,\cup\,]G[\,)$. Since $m \in\,]\phi[$ and $m' \in\,]\psi[$ and $]F[\,\cup\,]G[\,=\,]F \cup G[$, we get that

$$]\phi[\,\cap\gamma(\,]F \cup G[\,) \neq \emptyset \quad \text{or} \quad]\psi[\,\cap\gamma(\,]F \cup G[\,) \neq \emptyset$$

that is, since both ϕ and ψ are in $F \cup G$ and since $\delta = \mathcal{D}(\gamma)$, $\phi \in \delta(F \cup G)$ or $\psi \in \delta(F \cup G)$, as desired.

(III). Let $F \subseteq G$ and $\delta(G) \subseteq F$. It follows that $]\!] F [\![\subseteq]\!] G [\![$ and for all ϕ, if $\phi \in F'$ and $]\!] \phi [\![\cap \gamma (]\!] G [\![) \neq \emptyset$, then $\phi \in F$, by $\delta = \mathcal{D}(\gamma)$. First we show that $\gamma (]\!] G [\![) \subseteq]\!] F [\![$. If this were not true, i.e., if there were some m in $\gamma (]\!] G [\![) -]\!] F [\![$, that would mean that there is some $\phi \in F' - F$ such that $m \in]\!] \phi [\![$. Hence $]\!] \phi [\![\cap \gamma (]\!] G [\![) \neq \emptyset$. But then, by our assumption, $\phi \in F$, contradicting $\phi \in F' - F$. Second, $\gamma (]\!] G [\![) \subseteq]\!] F [\![$ and (III) for γ give us

$$\gamma (]\!] F [\![) \subseteq \gamma (]\!] G [\![)$$

We have to show that $\delta(F) \subseteq \delta(G)$, that is, by $\delta = \mathcal{D}(\gamma)$, that for all $\phi \in F$, $]\!] \phi [\![\cap \gamma (]\!] F [\![) \neq \emptyset$ implies $]\!] \phi [\![\cap \gamma (]\!] G [\![) \neq \emptyset$. But this is immediate from $\gamma (]\!] F [\![) \subseteq \gamma (]\!] G [\![)$.

(IV). Let $F \subseteq G$ and $\delta(G) \cap F \neq \emptyset$. It follows that $]\!] F [\![\subseteq]\!] G [\![$ and that there is some $\phi \in F$ such that $]\!] \phi [\![\cap \gamma (]\!] G [\![) \neq \emptyset$, by $\delta = \mathcal{D}(\gamma)$. Hence $]\!] F [\![\cap \gamma (]\!] G [\![) \neq \emptyset$. Now (IV) for γ gives us $\gamma (]\!] F [\![) \subseteq \gamma (]\!] G [\![)$. We have to show that $\delta(F) \subseteq \delta(G)$, that is, by $\delta = \mathcal{D}(\gamma)$, that for all $\phi \in F$, $]\!] \phi [\![\cap \gamma (]\!] F [\![) \neq \emptyset$ implies $]\!] \phi [\![\cap \gamma (]\!] G [\![) \neq \emptyset$. But this is immediate from $\gamma (]\!] F [\![) \subseteq \gamma (]\!] G [\![)$.

(IV$^+$). Let $F \subseteq G$. It follows that $]\!] F [\![\subseteq]\!] G [\![$, and by (IV$^+$) for γ, that $\gamma (]\!] F [\![) \subseteq \gamma (]\!] G [\![)$. We have to show that $\delta(F) \subseteq \delta(G)$, that is, by $\delta = \mathcal{D}(\gamma)$, that for all $\phi \in F$, $]\!] \phi [\![\cap \gamma (]\!] F [\![) \neq \emptyset$ implies $]\!] \phi [\![\cap \gamma (]\!] G [\![) \neq \emptyset$. But this is immediate from $\gamma (]\!] F [\![) \subseteq \gamma (]\!] G [\![)$.

Failure of the transfer of (II). Now we show that the fact that γ satisfies (II) does not imply (II) for $\delta = \mathcal{D}(\gamma)$, even in the finite case and when γ in addition satisfies (I) and (III). We do this by showing how the attempt to prove (II) for δ fails, and then we give a counterexample:

Let $\phi \in \delta(F) \cap \delta(G)$. This means that $\phi \in F \cap G$, $]\!] \phi [\![\cap \gamma (]\!] F [\![) \neq \emptyset$ and $]\!] \phi [\![\cap \gamma (]\!] G [\![) \neq \emptyset$, by $\delta = \mathcal{D}(\gamma)$. In order to show that $\phi \in \delta(F \cup G)$, we would have to show that $]\!] \phi [\![\cap \gamma (]\!] F \cup G [\![) \neq \emptyset$, or equivalently, that $]\!] \phi [\![\cap \gamma (]\!] F [\![\cup]\!] G [\![) \neq \emptyset$. This could be guaranteed, by using (II) for γ, if we had $]\!] \phi [\![\cap \gamma (]\!] F [\![) \cap \gamma (]\!] G [\![) \neq \emptyset$. But we can *not* get this from what we have, viz., from $]\!] \phi [\![\cap \gamma (]\!] F [\![) \neq \emptyset$ and $]\!] \phi [\![\cap \gamma (]\!] G [\![) \neq \emptyset$, since there need not be common elements in these two intersections.

These reflections help us to construct a counterexample. Consider the propositional language over the three variables p, q, and r. Assign numbers to the eight models of this language as follows:

1	satisfies p, q, r	5	satisfies $\neg p, q, r$
2	satisfies $p, q, \neg r$	6	satisfies $\neg p, q, \neg r$
3	satisfies $p, \neg q, r$	7	satisfies $\neg p, \neg q, r$
4	satisfies $p, \neg q, \neg r$	8	satisfies $\neg p, \neg q, \neg r$

and define a preference relation $<$ over these models in extensional form by putting $< = \{(2,6), (2,8), (3,5), (3,7)\}$.

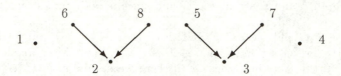

Notice that $<$ is vacuously transitive. Now let the choice function γ over sets of models be determined by the relation $<$ in the sense that always $\gamma(S) = \min_{<}(S)$. Since γ is transitively relational, it is clear that γ satisfies (I), (II) and (III). Now we see that

$$\gamma(][\{p,q\}[) \;=\; \gamma(\{3,4,5,6,7,8\}) \;=\; \{3,4,6,8\}$$
$$\gamma(][\{p,r\}[) \;=\; \gamma(\{2,4,5,6,7,8\}) \;=\; \{2,4,5,7\}$$
$$\gamma(][\{p,q,r\}[) = \gamma(\{2,3,4,5,6,7,8\}) = \;\{2,3,4\}$$

Hence clearly

$$]p[\,\cap\,\gamma(][\{p,q\}[) \neq \emptyset$$
$$]p[\,\cap\,\gamma(][\{p,r\}[) \neq \emptyset$$
$$\text{but }]p[\,\cap\,\gamma(][\{p,q,r\}[) = \emptyset$$

This means, for $\delta = \mathcal{D}(\gamma)$, that $p \in \delta(\{p,q\})$ and $p \in \delta(\{p,r\})$, but $p \notin \delta(\{p,q,r\})$. Thus δ does not satisfy (II). $\square$

Observation 42
The faithful variants of Definitions 11 and 12 satisfy

DP3 If $\phi \in K * \psi$, then $\phi \in (K * \phi) * \psi$

DP4 If $\neg\phi \notin K * \psi$, then $\neg\phi \notin (K * \phi) * \psi$

If the choice functions γ and δ satisfy (I) and (IV), then the faithful variants also satisfy

DP1 If $\phi \in Cn(\psi)$, then $(K * \phi) * \psi = K * \psi$

However, they do not satisfy

DP2 If $\neg\phi \in Cn(\psi)$, then $(K * \phi) * \psi = K * \psi$

Proof of Observation 42.
DP 3. Let $\phi \in K * \psi$. We need to show that $\phi \in (K * \phi) * \psi$.

Case 1. $\neg\psi \notin K$. Then $K * \psi = Cn(K \cup \{\psi\})$. Since $\phi \in K * \psi$, we know that $\psi \supset \phi \in K$. Hence, since K is a theory, $\neg\phi \notin K$ (otherwise $\neg\psi \in K$). So $K * \phi = Cn(K \cup \{\phi\})$. Suppose for reductio that $\neg\psi \in K * \phi = Cn(K \cup \{\phi\})$. So $\phi \supset \neg\psi \in K$. Since also $\psi \supset \phi \in K$, we get $\neg\psi \in K$, contradicting the assumption for the present case.

So $\neg\psi \notin K * \phi$, and hence $(K * \phi) * \psi = Cn((K * \phi) \cup \{\psi\}) = Cn((Cn(K \cup \{\phi\})) \cup \{\psi\}) = Cn(K \cup \{\phi, \psi\})$. Thus $K * \psi = Cn(K \cup \{\psi\}) \subseteq Cn(K \cup \{\phi, \psi\}) = (K * \phi) * \psi$, so $\phi \in (K * \phi) * \psi$.

Case 2. $\neg\psi \in K$. So $K * \psi = L * \psi$.

Case 2a. If $\neg\psi \in K * \phi$, then $(K * \phi) * \psi = L * \psi$, and we are done.

Case 2b. So let $\neg\psi \notin K * \phi$. Then $(K * \phi) * \psi = Cn((K * \phi) \cup \{\psi\})$. But since $\phi \in K * \phi$, $\phi \in Cn((K * \phi) \cup \{\psi\})$, and we are done.

DP 4. Let $\neg\phi \notin Cn(\psi)$. We need to show that $\neg\phi \notin (K * \phi) * \psi$.

Case 1. $\neg\psi \notin K$. Then $K * \psi = Cn(K \cup \{\psi\})$. Since $\neg\phi \notin K * \psi$, we know that $\psi \supset \neg\phi \notin K$. Hence, since K is a theory, $\neg\phi \notin K$, so $K * \phi = Cn(K \cup \{\phi\})$. Suppose for reductio that $\neg\psi \in K * \phi = Cn(K \cup \{\phi\})$. Then $\phi \supset \neg\psi \in K$, contradicting $\psi \supset \neg\phi \notin K$ and the fact that K is a theory. So $\neg\psi \notin K * \phi$, and hence $(K * \phi) * \psi = Cn((K * \phi) \cup \{\psi\}) = Cn((Cn(K \cup \{\phi\})) \cup \{\psi\}) = Cn(K \cup \{\phi, \psi\})$. Suppose that $\neg\phi \in (K * \phi) * \psi = Cn(K \cup \{\phi, \psi\})$. Then $(\phi \wedge \psi) \supset \neg\phi \in K$, i.e., $\neg\phi \vee \neg\psi \in K$. But this contradicts $\psi \supset \neg\phi \notin K$ and the fact that K is a theory.

Case 2. $\neg\psi \in K$. Then $K * \psi = L * \psi$.

Case 2a. If $\neg\psi \in K * \phi$, then $(K * \phi) * \psi = L * \psi$, and we are done.

Case 2b. So let $\neg\psi \notin K * \phi$. Then $(K * \phi) * \psi = Cn((K * \phi) \cup \{\psi\})$. Suppose for reductio that $\neg\phi \in (K * \phi) * \psi = Cn((K * \phi) \cup \{\psi\})$. So $\psi \supset \neg\phi \in K * \phi$. Also $\phi \in K * \phi$, so since $K * \phi$ is a theory by (*1), we get that $\neg\psi \in K * \phi$, contradicting the assumption for case 2b.

Now let the choice function γ or δ satisfy (I) and (IV).

DP1. Let $\phi \in Cn(\psi)$. We need to show that $(K * \phi) * \psi = K * \psi$.

Case 1. $\neg\psi \notin K$. Hence, since $\neg\psi \in Cn(\neg\phi)$ and K is a theory, $\neg\phi \notin K$. Hence $K * \phi = Cn(K \cup \{\phi\})$ and $K * \psi = Cn(K \cup \{\psi\})$. Since $Cn(\phi \supset \neg\psi) = Cn(\neg\phi)$, we have $\neg\psi \notin Cn(K \cup \{\phi\})$, so by definition $(Cn(K \cup \{\phi\})) * \psi = Cn((Cn(K \cup \{\phi\})) \cup \{\psi\}) = Cn(K \cup \{\phi, \psi\}) = Cn(K \cup \{\psi\})$. Thus $(K * \phi) * \psi = K * \psi$.

Case 2. $\neg\psi \in K$. Hence $K * \psi = L * \psi$.

Case 2a. If $\neg\psi \in K * \phi$, then $(K * \phi) * \psi = L * \psi$, and we are done.

Case 2b. So suppose that $\neg\psi \notin K * \phi$. Then $(K * \phi) * \psi = Cn((K * \phi) \cup \{\psi\})$. If $\neg\phi \notin K$, then $K * \phi = Cn(K \cup \{\phi\})$, so $\neg\psi \notin K * \phi$ means that $\neg\phi \vee \neg\psi \notin K$, i.e., since K is a theory, $\neg\psi \notin K$. But this contradicts the assumptions for the present case.

So $\neg\phi$ must be in K. Thus $K * \phi = L * \phi$ by definition. We need to show that $Cn((L * \phi) \cup \{\psi\}) = L * (\phi \wedge \psi)$ for all ψ such that $\neg\psi \notin L * \phi$. This holds if $*$, taken as a unary revision function for the inconsistent belief set L, satisfies $(\dot{-}7)$ and $(\dot{-}8)$. This in turn holds, by Observations 25 and 27, if the choice function γ or δ satisfies (I) and (IV).

Counterexample against DP2. We show that $\neg\phi \in Cn(\psi)$ does not imply $(K * \phi) * \psi = K * \psi$.

Define $L * \phi = Cn(\phi)$ for all ϕ. This function satisfies the AGM postulates $(\dot{-}1) - (\dot{-}8)$. Consider $K = Cn(q)$. We clearly have $(K * p) * \neg p = Cn(\neg p) \neq Cn(\neg p \wedge q) = K * \neg p$ □

APPENDIX E: PROOFS FOR CHAPTER 8

Lemma 43

The constraints (CP1) and (CP2) can be derived from the basic principles (A) – (E).

Proof of Lemma 43.

(CP1). Let $G \subseteq F$ and $\delta(F) \cap G = \emptyset$. We have to show that $\delta(F) \cap CCn(G) = \emptyset$. Principle (D) tells us that $\delta(F) \cap G = \emptyset$ means $\{\phi \in F : \phi \notin K \dot{-} \langle F \rangle\} \cap G = \emptyset$, and hence, since $G \subseteq F$, all $\phi \in G$ are in $K \dot{-} \langle F \rangle$. It follows, by principle (A), that all $\phi \in CCn(G)$ are in $K \dot{-} \langle F \rangle$. But this means that $\{\phi \in F : \phi \notin K \dot{-} \langle F \rangle\} \cap CCn(G) = \emptyset$, that is, $\delta(F) \cap CCn(G) = \emptyset$.

(CP2). Let $CCn(F) = CCn(G)$, and let ϕ be in $\delta(F) \cap G$. We have to show that ϕ is in $\delta(G)$. Principle (D) tells us that $\phi \in \delta(F) \cap G$ means $\phi \in F \cap G$ and $\phi \notin K \dot{-} \langle F \rangle$, and hence, by principle (C), $\phi \notin K \dot{-} \bigwedge F$. Since $CCn(F) = CCn(G)$, we can apply principle (B) and get that $\phi \notin K \dot{-} \bigwedge G$. Using principle (C) again, we conclude that $\phi \notin K \dot{-} \langle G \rangle$, and since ϕ is in G, principle (D) finally gives us that ϕ is in $\delta(G)$, as desired. $\qquad\qquad\square$

Substitutivity of Variants

(SV) If ϕ and ψ are $\wedge$-variants, then: $\phi < \chi$ iff $\psi < \chi$, and $\chi < \phi$ iff $\chi < \psi$

Proof of Substitutivity of Variants from the basic principles (A) – (E).

Let ϕ and ψ be $\wedge$-variants. Let $\phi < \chi$. Hence, by Definition 17, $\chi \in K \dot{-} \phi \wedge \chi$ and $\phi \notin K \dot{-} \phi \wedge \chi$. Now since ϕ and ψ are $\wedge$-variants, so are $\phi \wedge \chi$ and $\psi \wedge \chi$. By (B), then, $\chi \in K \dot{-} \psi \wedge \chi$ and $\phi \notin K \dot{-} \psi \wedge \chi$. By (A), $\phi \notin K \dot{-} (\psi \wedge \chi)$ is equivalent with $\psi \notin K \dot{-} (\psi \wedge \chi)$. Hence, by Definition 17 again, $\psi < \chi$.

Let $\chi < \phi$, i.e., by , $\phi \in K \dot{-} \phi \wedge \chi$ and $\chi \notin K \dot{-} \phi \wedge \chi$. Now since $\phi \wedge \chi$ and $\psi \wedge \chi$ are $\wedge$-variants, we get, by (B), $\phi \in K \dot{-} \psi \wedge \chi$ and $\chi \notin K \dot{-} \psi \wedge \chi$. By (A), $\phi \in K \dot{-} \psi \wedge \chi$ is equivalent with $\psi \in K \dot{-} \psi \wedge \chi$. Hence, by Definition 17 again, $\chi < \psi$. $\qquad\qquad\square$

For all admissible relations $<$, the following conditions are true.

(1) $\phi \wedge \psi < \psi$ iff $\phi < \psi$

(2) $\phi \wedge \psi < \psi \wedge \chi$ iff $\phi < \psi \wedge \chi$

(3) $\phi \wedge \psi < \chi \wedge \xi$ and $\phi \wedge \chi < \psi \wedge \xi$ iff $\phi < \psi \wedge \chi \wedge \xi$

(4) $\phi \not< \phi \wedge \psi$

Proof.

(1) $\phi \wedge \psi < \psi$ is equivalent with $\phi < \psi \wedge \psi$, by (E&C). This in turn is equivalent with $\phi < \psi$, by (SV).

(2) Let $\phi \wedge \psi < \psi \wedge \chi$. Then, by (E&C), $\phi \wedge \psi \wedge \psi < \chi$ and $\phi \wedge \psi \wedge \chi < \psi$. From the former, we get $\phi \wedge \psi < \chi$, by (SV), and from the latter, we get $\phi \wedge \chi < \psi$, by (1). Applying (E&C) again, we get $\phi < \psi \wedge \chi$, as desired. For the converse direction, let $\phi < \psi \wedge \chi$. Then $\phi < \psi \wedge \psi \wedge \chi$ by (SV), and thus $\phi \wedge \psi < \psi \wedge \chi$ by (E&C).

(3) Let $\phi \wedge \psi < \chi \wedge \xi$ and $\phi \wedge \chi < \psi \wedge \xi$. By (SV) and (2), this means that $\phi \wedge \psi \wedge \xi < \chi \wedge \xi$ and $\phi \wedge \chi \wedge \xi < \psi \wedge \xi$. By (E&C) and (SV), this is equivalent with $\phi < \psi \wedge \chi \wedge \xi$. The converse follows immediately from (E&C) and (SV).

(4) Suppose for reductio that $\phi < \phi \wedge \psi$. Then by (1) and (SV), $\phi \wedge \psi < \phi \wedge \psi$, violating Irreflexivity. $\qquad\square$

Lemma 44

Given (E&C), (ND′) is equivalent with the asymmetry of $<$.

Proof of Lemma 44.

(ND′) implies that $<$ is asymmetric: The special case $n = 2$ of (ND) tells us that either $\phi \not< \psi$ or $\psi \not< \phi$.

The asymmetry of $<$ implies (ND′): Suppose that $\phi_1 \wedge \cdots \wedge \phi_{i-1} \wedge \phi_{i+1} \wedge \cdots \wedge \phi_n < \phi_i$, for every i. From $\phi_1 \wedge \cdots \wedge \phi_{n-1} < \phi_n$ and $\phi_1 \wedge \cdots \wedge \phi_{n-2} \wedge \phi_n < \phi_{n-1}$ we get by (E&C) that $\phi_1 \wedge \cdots \wedge \phi_{n-2} < \phi_{n-1} \wedge \phi_n$. From this and $\phi_1 \wedge \cdots \wedge \phi_{n-3} \wedge \phi_{n-1} \wedge \phi_n < \phi_{n-2}$ we get, again by (E&C), that $\phi_1 \wedge \cdots \wedge \phi_{n-3} < \phi_{n-2} \wedge \phi_{n-1} \wedge \phi_n$. This process is repeated until we end up with $\phi_1 < \phi_2 \wedge \cdots \wedge \phi_n$. But this contradicts $\phi_2 \wedge \cdots \wedge \phi_n < \phi_1$ and the asymmetry of $<$. $\qquad\square$

Lemma 45

Let $<$ be an admissible relation, and let two sets of sentences F and G be given. The most general translation of $G \cap \delta(F \cup G) = \emptyset$ says that for any two sequences $F_1, \ldots, F_k$ and $G_1, \ldots, G_k$ such that $G_1 \cup \cdots \cup G_k = G$ and $F_i \cup G_i = F \cup G$ for every $i \leq k$, the following condition holds:

$$\textstyle\bigwedge(F \cup G) \not< \top \quad \text{or} \quad \bigwedge F_i < \bigwedge G_i \text{ for every } i \tag{†}$$

This translation is equivalent with the standard translation stating that

$$\textstyle\bigwedge(F \cup G) \not< \top \quad \text{or} \quad \bigwedge F < \bigwedge G \tag{‡}$$

Proof of Lemma 45.

That (†) entails (‡) is clear. For the converse, take two sequences $F_1, \ldots, F_k$ and $G_1, \ldots, G_k$ with the appropriate features.

We first prove an intermediate result for $n = 2$. Let F_1, F_2, G_1, G_2 be such that $G_1 \cup G_2 = G$ and $F_1 \cup G_1 = F_2 \cup G_2 = F \cup G$. Then we have

$$\textstyle\bigwedge F_1 < \bigwedge G_1 \text{ and } \bigwedge F_2 < \bigwedge F_2' \quad \text{iff} \quad \bigwedge(F \cup G) < \bigwedge(G_1 \cup G_2) \tag{$*$}$$

From left to right: By (E&C) and (SV), $\bigwedge F_1 < \bigwedge G_1$ and $\bigwedge F_2 < \bigwedge G_2$ are both equivalent with $\bigwedge(F \cup G) < \bigwedge G_1$ and $\bigwedge(F \cup G) < \bigwedge G_2$, and also with

$\bigwedge(F \cup G) \wedge \bigwedge G_2 < \bigwedge G_1$ and $\bigwedge(F \cup G) \wedge \bigwedge G_1 < \bigwedge G_2$, by (SV). Hence, by (E&C) and (SV), $\bigwedge(F \cup G) < \bigwedge(G_1 \cup G_2)$.

From right to left: From $\bigwedge(F \cup G) < \bigwedge(G_1 \cup G_2)$, infer with (E&C) and (SV) that $\bigwedge(F \cup G) \wedge \bigwedge G_2 < \bigwedge G_1$, which is, by $G_2 \subseteq (F \cup G)$ and (SV), $\bigwedge(F \cup G) < \bigwedge G_1$. This gives, by (SV) and (E&C), $\bigwedge F_1 < \bigwedge G_1$. The case of $\bigwedge F_2 < \bigwedge G_2$ is similar.

The technique we used for the proof of the equivalence ($*$) can be iterated, and we end up with

$$\bigwedge F_i < \bigwedge G_i \text{ for every } i \quad \text{iff} \quad \bigwedge(F \cup G) < \bigwedge(\bigcup G_i)$$

Using the fact that $\bigcup G_i = G$ and applying (E&C) and (SV) once more, we get that

$$\bigwedge F_i < \bigwedge G_i \text{ for every } i \quad \text{iff} \quad \bigwedge F < \bigwedge G$$

and we are done.   □

Observation 46. The conditions ($\emptyset 1$), ($\emptyset 2$) and (Virtual Success) for the syntactic choice function δ translate into

(E$\emptyset 1$) If $\phi < \top$, then $\phi \wedge \psi < \top$
(E$\emptyset 2$) If $\phi \wedge \psi < \top$, then $\phi < \psi$ or $\psi < \top$
(EVSuccess) If $\phi \notin Cn(\emptyset)$, then $\phi < \top$

Proof of Observation 46.
Condition ($\emptyset 1$) says that if F is a subset of G and $\delta(G)$ is empty, then $\delta(F)$ must be empty as well. Now let F and G be two sets such that F is a subset of G. Let ϕ be the conjunction of all elements in F and ψ be the conjunction of all elements in $G - F$. If we use the above translation from choices to entrenchments, then we get that $\phi \wedge \psi \not< \top$ implies $\phi \not< \top$. If we take the contraposition, we get condition (E$\emptyset 1$).

Condition ($\emptyset 2$) says that if F is a subset of G and $\delta(F)$ is empty, then all elements of F should be considered taboo when it comes to choosing in the issue G; that is, $F \cap \delta(G) = \emptyset$. Again, let F and G be two sets such that F is a subset of G. Let ϕ be short for $\bigwedge F$ and ψ be short for $\bigwedge(G - F)$. If we use our translation, then we get that $\phi \not< \top$ implies that either $\phi \wedge \psi \not< \top$ or $\psi < \phi$. If we take the contraposition, we get condition (E$\emptyset 2$).

(Virtual Success) says that $\delta(F)$ is empty only if F is a subset of $Cn(\emptyset)$. If we put $\phi = \bigwedge F$, then this means that $\phi \not< \top$ only if ϕ is in $Cn(\emptyset)$, which is (EVSuccess).   □

Any fully admissible relation validates the following conditions.

(5) If $\phi < \psi$, then $\phi < \top$
(6) $\phi \wedge \psi < \top$ if and only if $\phi < \top$ or $\psi < \top$
(7) If $\phi < \top$ and $\psi \not< \top$, then $\phi < \psi$

Proof.

For (5), let $\phi < \psi$. By (SV), this is equivalent to $\phi < \psi \wedge \top$. Hence by (E&C), $\phi \wedge \psi < \top$. By Asymmetry, we have $\psi \not< \phi$. Hence, by (E∅2), $\phi < \top$.

For (6), first assume that $\phi \wedge \psi < \top$. Then, by (E∅2), either $\phi < \psi$ or $\psi < \top$. From the former, we can infer with (5) that $\phi < \top$, so in fact either $\phi < \top$ or $\psi < \top$. For the converse, it is enough to apply (E∅1) to either $\phi < \top$ or $\psi < \top$.

For (7), let $\phi < \top$ but $\psi \not< \top$. From the former, we get $\phi \wedge \psi < \top$, by (E∅1). This, taken together with the latter, gives us $\phi < \psi$, by (E∅2). $\square$

Observation 47

The conditions (Faith1) and (Faith2) with respect to $L - K$ for the syntactic choice function δ translate into (EFaith1) and (EFaith2)

Proof of Observation 47.

(Faith1) w.r.t. $L - K$ says that if F is not a subset of K, then $\delta(F)$ does not intersect K. Since K is supposed to be a theory, this means that if F is not a subset of K, then each ψ in K is not in $\delta(F)$. Let ψ be in K and $\phi = \bigwedge(S - \{\psi\})$ be not in K. Then ψ is not in $\delta(F)$. We can translate the latter condition into $\phi \wedge \psi \not< \top$ or $\phi < \psi$. By (E∅1) and (E∅2), $\phi \wedge \psi \not< \top$ or $\phi < \psi$ is equivalent to $\phi \not< \top$, and we get (EFaith1).

(Faith2) w.r.t. $L - K$ says that $F - K$ is a subset of $\delta(F)$. So for each ϕ which is in F but not in K it holds that ϕ is in $\delta(F)$. Let $\phi \in F - K$ and $\psi = \bigwedge(F - \{\phi\})$. Then $\phi \in \delta(F)$ translates into $\phi \wedge \psi < \top$ and $\psi \not< \phi$. By (E∅1) and (E∅2), $\phi \wedge \psi < \top$ and $\psi \not< \phi$ is equivalent to $\phi < \top$ and $\psi \not< \phi$, and we get (EFaith2). $\square$

Lemma 48

Suppose the relation $<$ over L satisfies (EE1) – (EE3$^{\downarrow}$). Then it also has the following properties:

(i)   If $Cn(\phi) = Cn(\psi)$, then: $\phi < \chi$ iff $\psi < \chi$, and $\chi < \phi$ iff $\chi < \psi$

(Extensionality)

(ii)   If $\phi \wedge \chi < \psi \wedge \chi$, then $\phi < \psi$

(iii)   If $\phi < \psi$ and $\chi < \xi$, then $\phi \wedge \chi < \psi \wedge \xi$

(iv)   If $\phi < \psi$ then not $\psi < \phi$ *(Asymmetry)*

(v)   If $\phi < \psi$ and $\psi < \chi$, then $\phi < \chi$ *(Transitivity)*

(vi) $\phi < \phi \vee \psi$ iff $\phi \vee \neg\psi < \phi \vee \psi$

Proof of Lemma 48.

(i) Immediate from (EE2$^{\uparrow}$) and (EE2$^{\downarrow}$).

(ii) Let $\phi \wedge \chi < \psi \wedge \chi$. By (EE2$^{\downarrow}$), $\phi \wedge \psi \wedge \chi < \psi \wedge \chi$, so by (EE3$^{\downarrow}$) $\phi < \psi \wedge \chi$, so by (EE2$^{\uparrow}$) $\phi < \psi$.

(iii) Let $\phi < \psi$ and $\chi < \xi$. By (EE2$^{\downarrow}$), $\phi \wedge \chi < \psi$ and $\phi \wedge \chi < \xi$. Hence, by (EE3$^{\uparrow}$), $\phi \wedge \chi < \psi \wedge \xi$.

(iv) Suppose for reductio that $\phi < \psi$ and $\psi . < \phi$. Then, by (EE2$^{\downarrow}$) and (EE3$^{\uparrow}$), $\phi \wedge \psi < \phi \wedge \psi$, contradicting (EE1).

(v) Let $\phi < \psi$ and $\psi < \chi$. Then, by (iii), $\phi \wedge \psi < \chi \wedge \psi$. So $\phi < \chi$, by (ii).

(vi) From left to right: Let $\phi < \phi \vee \psi$. Then by (EE2$^{\downarrow}$) $(\phi \vee \psi) \wedge (\phi \vee \neg \psi) < \phi \vee \psi$, so by (EE3$^{\downarrow}$) $\phi \vee \neg \psi < \phi \vee \psi$. From right to left: By (EE2$^{\downarrow}$).   □

Observation 49

Suppose the relation $<$ satisfies (EE1) and (EE4). Then

(i) $<$ satisfies (EE2$^{\uparrow}$) iff it satisfies (EE2$^{\downarrow}$);

(ii) if $<$ satisfies (EE3$^{\uparrow}$), then it satisfies (EE3$^{\downarrow}$); if $<$ is asymmetric and satisfies (EE3$^{\downarrow}$), then it also satisfies (EE3$^{\uparrow}$).

Proof of Observation 49.

Let $<$ satisfy (EE1) and (EE4).

(i) *(EE2$^{\uparrow}$) implies (EE2$^{\downarrow}$).* Let $\phi < \psi$ and $\chi \vdash \phi$ and suppose for reductio that $\chi \not< \psi$. Then $\phi < \chi$, by (EE4), and thus also $\phi < \phi$, by (EE2$^{\uparrow}$). But this contradicts (EE1).

(EE2$^{\downarrow}$) implies (EE2$^{\uparrow}$). Let $\phi < \psi$ and $\psi \vdash \chi$ and suppose for reductio that $\phi \not< \chi$. Then $\chi < \psi$, by (EE4), and thus also $\psi < \psi$, by (EE2$^{\downarrow}$). But this contradicts (EE1).

(ii) *(EE3$^{\uparrow}$) implies (EE3$^{\downarrow}$).* Let $\phi \wedge \psi < \psi$ and suppose for reductio that $\phi \not< \psi$. Then $\phi \wedge \psi < \phi$, by (EE4). Thus $\phi \wedge \psi < \phi \wedge \psi$, by (EE3$^{\uparrow}$). But this contradicts (EE1).

(EE3$^{\downarrow}$) implies (EE3$^{\uparrow}$), provided $<$ is asymmetric. Let $\phi < \psi$ and $\phi < \chi$. Suppose for reductio that $\phi \not< \psi \wedge \chi$. Then $\psi \wedge \chi < \psi$ and $\psi \wedge \chi < \chi$, by (EE4). Thus $\chi < \psi$ and $\psi < \chi$, by (EE3$^{\downarrow}$). But this contradicts asymmetry.   □

Observation 51

The coherence criteria (I) – (IV), (I^{-}), (II^{+}) and (IV^{+}) for rational choice translate into the following constraints on fully admissible preference relations.

(EI)   If $\phi < \psi$, then $\phi \wedge \chi < \psi$

(EII)   If $\phi \wedge \psi < \chi$, then $\phi < \chi$ or $\psi < \chi$

(EIII)   If $\phi < \psi \wedge \chi$, then $\phi < \psi$

(EIV)   If $\phi \wedge \psi < \chi$ and $\psi \not< \phi \wedge \chi$, then $\phi < \chi$

(EI^{-})   If $\phi \wedge \psi < \chi$ and $\phi < \psi$, then $\phi \wedge \chi < \psi$

(EII^{+})   If $\phi \wedge \psi < \chi \wedge \xi$ and $\chi < \top$ and $\xi < \top$, then $\phi < \chi$ or $\psi < \xi$

(EIV^{+})   If $\phi < \top$, then $\psi \not< \phi$

Proof of Observation 51.

(I) says that if F is a subset of G and χ is in $F \cap \delta(G)$ then $\chi \in \delta(F)$. Let $\phi = \bigwedge (F - \{\chi\})$ and $\psi = \bigwedge (G - F)$. Then condition (I) translates into this one: If $\phi \wedge \psi \wedge \chi < \top$ and $\phi \wedge \psi \not< \chi$, then $\phi \wedge \chi < \top$ and $\phi \not< \chi$. By contraposition, this is equivalent to: If $\phi \wedge \psi \wedge \chi < \top$ and $(\phi \wedge \chi \not< \top$ or $\phi < \chi)$, then $\phi \wedge \psi < \chi$. But if $\phi \wedge \psi \wedge \chi < \top$ and $\phi \wedge \chi \not< \top$, then by (E02) $\psi < \phi \wedge \chi$, so by (E&C), $\phi \wedge \psi < \chi$ anyway. Hence the condition reduces to: If $\phi \wedge \psi \wedge \chi < \top$ and $\phi < \chi$, then $\phi \wedge \psi < \chi$. But if $\phi < \chi$, then $\phi \wedge \chi < \top$, by (E&C), and hence $\phi \wedge \psi \wedge \chi < \top$, by (E01). So the condition in question reduces to (EI).

(II) says that $\delta(F) \cap \delta(G)$ is a subset of $\delta(F \cup G)$. Let χ be in $F \cap G$, $\phi = \bigwedge(F - \{\chi\})$ and $\psi = \bigwedge(G - \{\chi\})$. Then condition (II) translates into this one: If $\phi \wedge \chi < \top$ and $\phi \not< \chi$ and if $\psi \wedge \chi < \top$ and $\psi \not< \chi$, then $\phi \wedge \psi \wedge \chi < \top$ and $\phi \wedge \psi \not< \chi$. By contraposition, this is equivalent to: If $\phi \wedge \chi < \top$ and $\psi \wedge \chi < \top$ and $(\phi \wedge \psi \wedge \chi \not< \top$ or $\phi \wedge \psi < \chi)$, then $\phi < \chi$ or $\psi < \chi$. By (E∅1), this is equivalent with:

If $\phi \wedge \chi < \top$ and $\psi \wedge \chi < \top$ and $\phi \wedge \psi < \chi$, then $\phi < \chi$ or $\psi < \chi$

In order to get (EII), we have to show that the following is also valid:

If $\phi \wedge \chi \not< \top$ and $\phi \wedge \psi < \chi$, then $\phi < \chi$ or $\psi < \chi$

(The case with $\psi \wedge \chi \not< \top$ in the antecedent is analogous.) So suppose that $\phi \wedge \chi \not< \top$ and $\phi \wedge \psi < \chi$. From $\phi \wedge \psi < \chi$, we get $\phi \wedge \psi \wedge \chi < \top$, by (E&C). From this and $\phi \wedge \chi \not< \top$, we get $\psi \wedge \chi < \phi \wedge \chi$, by (SV) and (E∅2). Hence, by (5), $\psi \wedge \chi < \top$. Hence, by (E∅2) again, either $\psi < \chi$ or $\chi < \top$. But the latter cannot be true, since it would entail $\phi \wedge \chi < \top$, by (E∅1), and this contradicts our supposition. Hence $\psi < \chi$, and we are done.

(III) says that if F is a subset of G and $G - F$ does not intersect $\delta(G)$, then every χ that is in $\delta(F)$ is also in $\delta(G)$. Let $\phi = \bigwedge(F - \{\chi\})$ and $\psi = \bigwedge(G - F)$. Then condition (III) translates into this one: If $(\phi \wedge \psi \wedge \chi \not< \top$ or $\phi \wedge \chi < \psi)$ and $\phi \wedge \chi < \top$ and $\phi \not< \chi$, then $\phi \wedge \psi \wedge \chi < \top$ and $\phi \wedge \psi \not< \chi$. By (E∅1) and (5), this is equivalent to: If $\phi \wedge \chi < \psi$ and $\phi \not< \chi$, then $\phi \wedge \psi \not< \chi$, or equivalently, if we take the contraposition: If $\phi \wedge \chi < \psi$ and $\phi \wedge \psi < \chi$, then $\phi < \chi$. Applying (E&C) in the antecedent gives us (EIII).

(IV) says that if F is a subset of G and F intersects $\delta(G)$, then every χ that is in $\delta(F)$ is also in $\delta(G)$. Let $\phi = \bigwedge(F - \{\chi\})$ and $\psi = \bigwedge(G - F)$. Then condition (IV) translates into this one: If $\phi \wedge \psi \wedge \chi < \top$ and $\psi \not< \phi \wedge \chi$ and $\phi \wedge \chi < \top$ and $\phi \not< \chi$, then $\phi \wedge \psi \wedge \chi < \top$ and $\phi \wedge \psi \not< \chi$. By (E∅1), this is equivalent to: If $\psi \not< \phi \wedge \chi$ and $\phi \wedge \chi < \top$ and $\phi \not< \chi$, then $\phi \wedge \psi \not< \chi$. Taking the contraposition, we get

If $\psi \not< \phi \wedge \chi$ and $\phi \wedge \chi < \top$ and $\phi \wedge \psi < \chi$, then $\phi < \chi$

From $\phi \wedge \psi < \chi$, we get that $\phi \wedge \psi \wedge \chi < \top$, by (E&C). So by (E∅2) either $\psi < \phi \wedge \chi$ or $\phi \wedge \chi < \top$. Hence $\phi \wedge \chi < \top$ follows from $\phi \wedge \psi < \chi$ and $\psi \not< \phi \wedge \chi$, and the above condition reduces to (EIV).

(I$^-$) says that if F is a subset of G and $G - F$ does not intersect $\delta(G)$, then every χ that is in $\delta(G)$ is also in $\delta(F)$. Let $\phi = \bigwedge(F - \{\chi\})$ and $\psi = \bigwedge(G - F)$. Then condition (I$^-$) translates into this one: If $(\phi \wedge \psi \wedge \chi \not< \top$ or $\phi \wedge \chi < \psi)$ and $\phi \wedge \psi \wedge \chi < \top$ and $\phi \wedge \psi \not< \chi$, then $\phi \wedge \chi < \top$ and $\phi \not< \chi$. By (E&C) and (5), this is equivalent to: If $\phi \wedge \chi < \psi$ and $\phi \wedge \psi \not< \chi$, then $\phi \not< \chi$, which gives us (EI$^-$) when we take the contraposition (and relabel the propositional variables).

(II$^+$) says that if χ is in $\delta(F)$ and ξ is in $\delta(G)$, then either χ or ξ is in $\delta(F \cup G)$. Let χ be in F, ξ be in G, and $\phi = \bigwedge(F - \{\chi\})$ and $\psi = \bigwedge(G - \{\xi\})$. Then condition (II$^+$) translates into this one: If $\phi \wedge \chi < \top$ and $\phi \not< \chi$ and if $\psi \wedge \xi < \top$

and $\psi \not< \xi$, then $\phi \wedge \psi \wedge \chi \wedge \xi < \top$ and either $\phi \wedge \psi \wedge \xi \not< \chi$ or $\phi \wedge \psi \wedge \chi \not< \xi$. Using (E∅1) and taking the contraposition gives us:

If $\phi \wedge \chi < \top$, $\psi \wedge \xi < \top$, $\phi \wedge \psi \wedge \xi < \chi$ and $\phi \wedge \psi \wedge \chi < \xi$, then either $\phi < \chi$ or $\psi < \xi$. The first clause of the antecedent of this conditional, $\phi \wedge \chi < \top$, can equivalently be rephrased as $\phi < \chi$ or $\chi < \top$ (use (E∅2), (5) and (6)). In the former case the consequent of the conditional is satisfied. So we can replace $\phi \wedge \chi < \top$ by $\chi < \top$. Similarly, we can replace the second clause of the antecedent, $\psi \wedge \xi < \top$, by $\xi < \top$. The third and the third clause of the antecedent of this conditional can be summarized as $\phi \wedge \psi < \chi \wedge \xi$, by (E&C). So we can simplify the above condition to

If $\chi < \top$, $\xi < \top$ and $\phi \wedge \psi < \chi \wedge \xi$, then either $\phi < \chi$ or $\psi < \xi$

which is precisely (EII$^+$).[373]

(IV$^+$) says that if F is a subset of G, then every χ that is in $\delta(F)$ is also in $\delta(G)$. Let $\phi = \bigwedge(F - \{\chi\})$ and $\psi = \bigwedge(G - F)$. Then condition (IV$^+$) translates into this one: If $\phi \wedge \chi < \top$ and $\phi \not< \chi$, then $\phi \wedge \psi \wedge \chi < \top$ and $\phi \wedge \psi \not< \chi$. Applying (E∅1) and (E∅2), we see that we can drop the first clause of the consequent of this conditional, and that its antecedent is equivalent with $\chi < \top$ and $\phi \not< \chi$. Taking the contraposition, we get the condition

If $\chi < \top$ and $\phi \wedge \psi < \chi$, then $\phi < \chi$ (†)

Condition (†) entails (EIV$^+$). This is seen if we take $\phi = \top$, which gives us this: If $\chi < \top$ and $\psi < \chi$, then $\top < \chi$. But this consequent can never be true (as is shown, for instance, by (5) and the Irreflexivity of $<$), so that we can conclude: If $\chi < \top$, then $\psi \not< \chi$. For the converse direction, note that condition (EIV$^+$) trivially entails (†), since (EIV$^+$) entails that the antecedent of (†) cannot be satisfied.   □

Lemma 52

(i)   Asymmetry, (E&C) and (EIV) taken together imply (EII);

(ii)   Asymmetry, (E&C) and (EIV) taken together imply (EIII);

(iii)   (EI) implies (EI$^-$);

(iv)   Asymmetry, (SV), (E&C), (E∅2) and (EII$^+$) taken together imply (EII);

(v)   (EII) and (EIII) taken together imply (EII$^+$);

(vi)   (E&C), (E∅2) and (EIV$^+$) taken together imply (EIV);

(vii)   (SV), (E&C), (EI) and (EIII) taken together imply Transitivity;

(viii) Asymmetry, (SV), (E&C), (EI) and (EIV) taken together imply Modularity;

(ix)   Asymmetry, (SV), (E&C) and Modularity taken together imply (EIV).

---

[373]It does not seem possible to remove the limiting case conditions from the antecedent of (II$^+$), if we want to use only the properties of fully admissible relations.

Proof of Lemma 52.

(i) Let $\phi \wedge \psi < \chi$ and $\psi \not< \chi$. We apply (EIII) to the latter and get that $\psi \not< \phi \wedge \chi$. (EIII) follows from Asymmetry, (E&C) and (EIV), as we show in (ii). From $\phi \wedge \psi < \chi$ and $\psi \not< \phi \wedge \chi$, we get $\phi < \chi$, as desired.

(ii) Let $\phi < \psi \wedge \chi$. Then, by (E&C), $\phi \wedge \psi < \chi$ and $\phi \wedge \chi < \psi$. From the former, we get by Asymmetry that $\chi \not< \phi \wedge \psi$. From this and $\phi \wedge \chi < \psi$, be get that $\phi < \psi$, by (EIV).

(iii) is immediate.

(iv) Let $\phi \wedge \psi < \chi$. We prove that $\phi < \chi$ or $\psi < \chi$ by cases. First, assume that $\chi < \top$. From $\phi \wedge \psi < \chi$ we get, by (SV), that $\phi \wedge \psi < \chi \wedge \chi$. Hence, by (EII$^+$), $\phi < \chi$ or $\psi < \chi$. As the second case, assume that $\chi \not< \top$. And suppose for reductio that both $\phi \not< \chi$ and $\psi \not< \chi$. Two-fold application of (E$\emptyset$2) gives us $\phi \wedge \chi \not< \top$ and $\psi \wedge \chi \not< \top$. From this it follows, by (SV) and (5) (which follows from Asymmetry, (E&C) and (E$\emptyset$2)), that $\phi \wedge \psi \wedge \chi \not< \top$. But $\phi \wedge \psi < \chi$ gives us, by (SV) and (E&C), $\phi \wedge \psi \wedge \chi < \top$, so we have a contradiction.

(v) Let $\phi \wedge \psi < \chi \wedge \xi$. We prove that either $\phi < \chi$ or $\psi < \xi$ (without exploiting the additional antecedent clauses $\chi < \top$ and $\xi < \top$). From $\phi \wedge \psi < \chi \wedge \xi$, we infer with (EII) that either $\phi < \chi \wedge \xi$ or $\psi < \chi \wedge \xi$. By (EIII), we get from the former that $\phi < \chi$ and from the latter that $\psi < \xi$, so we are done.

(vi) Let $\phi \wedge \psi < \chi$ and $\psi \not< \phi \wedge \chi$. We need to show that $\phi < \chi$. From $\phi \wedge \psi < \chi$, we get with (E&C) that $\phi \wedge \psi \wedge \chi < \top$. From this and $\psi \not< \phi \wedge \chi$, we get that $\phi \wedge \chi < \top$, by (E$\emptyset$2). On the other hand, we get from $\phi \wedge \psi < \chi$ and (EIV$^+$) that $\chi \not< \top$. Putting the two things together with the help of (E$\emptyset$2), we get that $\phi < \chi$.

(vii) Let $\phi < \psi$ and $\psi < \chi$. Then, by (EI) and (SV), $\phi \wedge \chi < \psi$ and $\phi \wedge \psi < \chi$, so $\phi < \psi \wedge \chi$, by (E&C). Therefore, by (EIII), $\phi < \chi$, as desired.

(viii) Let $\phi < \psi$. We need to show that for arbitrary χ, either $\phi < \chi$ or $\chi < \psi$. By (EI), we get $\phi \wedge \chi < \psi$. Hence, by (SV) and (EIV), either $\chi < \psi$ or $\phi < \chi \wedge \psi$. In the former case, we are done. In the latter case, we refer to part (ii) of this lemma, apply (EIII) and get $\phi < \chi$.

(ix) Let $\phi \wedge \psi < \chi$ and $\phi \not< \chi$. For (EIV), we need to show that $\psi < \phi \wedge \chi$. By Modularity, the hypothesis implies that $\phi \wedge \psi < \phi$. By (E&C), this means that $\psi < \phi$. Now by (4), we have $\psi \not< \psi \wedge \chi$, hence another application of Modularity gives us $\psi \wedge \chi < \phi$. From this and the first hypothesis, we conclude with (E&C) that $\psi < \phi \wedge \chi$, as desired. $\square$

Observation 53

The logical constraints (LP1) and (LP2) for syntactic choice functions translate into the following constraints for entrenchment relations.

(ELP1) If $\phi \in Cn(\psi)$ and $\phi \wedge \chi < \psi$, then $\psi \wedge \chi < \phi$

(ELP2) If $Cn(\phi \wedge \psi) = Cn(\phi \wedge \chi)$ and $\chi < \phi$, then $\psi < \phi$

Proof of Observation 53.

(LP1) says that if F is a subset of G which entails ϕ, and if ϕ is in $\delta(G)$, then F must intersect $\delta(G)$. This claim is trivial if ϕ is actually in F, so suppose it is not. Then let $\psi = \bigwedge F$ and $\chi = \bigwedge(G - (F \cup \{\phi\}))$. Then (LP1) translates into this condition: If $\phi \in Cn(\psi)$ and $\phi \wedge \psi \wedge \chi < \top$ and $\psi \wedge \chi \not< \phi$, then $\phi \wedge \psi \wedge \chi < \top$ and $\phi \wedge \chi \not< \psi$. Using contraposition and (E&C), we get (ELP1).

(LP2) says that if two sets F and G have the same logical consequences, then each ϕ in $\delta(F)$ which is in G must also be in $\delta(G)$. Let $\psi = \bigwedge(F - \{\phi\})$ and $\chi = \bigwedge(F - \{\phi\})$. Then (LP2) translates into this condition: If $Cn(\phi \wedge \psi) = Cn(\phi \wedge \chi)$ and $\phi \wedge \psi < \top$ and $\psi \not< \phi$, then $\phi \wedge \chi < \top$ and $\chi \not< \phi$. After contraposition, we get

If $Cn(\phi \wedge \psi) = Cn(\phi \wedge \chi)$ and $\phi \wedge \psi < \top$ and $(\phi \wedge \chi \not< \top$ or $\chi < \phi)$, then $\psi < \phi$ (†)

Next we show that

If $Cn(\phi \wedge \psi) = Cn(\phi \wedge \chi)$ and $\phi \wedge \psi < \top$ and $\phi \wedge \chi \not< \top$, then $\psi < \phi$

is valid anyway. Let $\phi \wedge \psi < \top$ and $\phi \wedge \chi \not< \top$. From the latter we get, by (E∅1), that $\phi \not< \top$. From this and the former, we get that $\psi < \phi$, by (E∅2). Hence (†) reduces to

If $Cn(\phi \wedge \psi) = Cn(\phi \wedge \chi)$ and $\phi \wedge \psi < \top$ and $\chi < \phi$, then $\psi < \phi$ (‡)

Let $Cn(\phi \wedge \psi) = Cn(\phi \wedge \chi)$ and $\chi < \phi$. In order to achieve equivalence with (ELP2), we allow ourselves to use a little help from (ELP1⁻) and conclude from the fact that $\phi \wedge \chi$ is in $Cn(\phi \wedge \psi)$ that $\phi \wedge \chi \not< \phi \wedge \psi$. On the other hand, it follows from $\chi < \phi$ that $\phi \wedge \chi < \top$, by (E&C), and thus $\phi \wedge \psi \wedge \chi < \top$, by (E∅1). So we get, by (E∅2), that $\phi \wedge \psi < \top$, which may be dropped from the antecedent of (‡), and we get in fact (ELP2).   □

Observation 54

For fully admissible relations $<$, the conjunction of (ELP1) and (ELP2) is equivalent with (Extensionality).

Proof of Observation 54.

From (ELP1) and (ELP2) to (Extensionality). Let $Cn(\phi) = Cn(\psi)$. We first show that $\phi < \chi$ iff $\psi < \chi$. But from $Cn(\phi) = Cn(\psi)$ we can conclude that $Cn(\phi \wedge \chi) = Cn(\psi \wedge \chi)$, so $\chi < \phi$ iff $\chi < \psi$ by (ELP2). We have now proved Extensionality on the LHS of $<$.

For Extensionality on the RHS of $<$, let $\chi < \phi$. We need to show that $\chi < \psi$. From $\chi < \phi$, we get $\phi \wedge \chi < \phi$, by (1). Now we remember that $Cn(\phi \wedge \chi) = Cn(\psi \wedge \chi)$ and apply Extensionality on the LHS of $<$, which gives us $\psi \wedge \chi < \phi$. From this and $\psi \in Cn(\phi)$, we can now deduce, with the help of (ELP1), that $\phi \wedge \chi < \psi$. Now we apply Extensionality on the LHS of $<$ again, which gives us $\psi \wedge \chi < \psi$. By (1) again, we get $\chi < \psi$, as desired.

From (Extensionality) to (ELP1) and (ELP2). For (ELP1), let $\phi \in Cn(\psi)$ and $\phi \wedge \chi < \psi$. We need to show that $\psi \wedge \chi < \phi$. From $\phi \in Cn(\psi)$ we conclude that $Cn(\psi) = Cn(\phi \wedge \psi)$. From $\phi \wedge \chi < \psi$ we thus get, by Extensionality on the RHS, $\phi \wedge \chi < \phi \wedge \psi$. Thus, by (2), $\chi < \phi \wedge \psi$. Hence, by (E&C), $\psi \wedge \chi < \phi$.

For (ELP2), let $Cn(\phi \wedge \psi) = Cn(\phi \wedge \chi)$ and $\chi < \phi$. We need to show that $\psi < \phi$. From $\chi < \phi$, we get $\phi \wedge \chi < \phi$, by (1), so Extensionality on the LHS gives us $\phi \wedge \psi < \phi$, which in turn gives us $\psi < \phi$, by (1) again.　　$\square$

Lemma 55

Let $<$ satisfy Irreflexivity, Extensionality and (E&C). Then it also satisfies the following properties:

(i) $\phi < \psi$ iff $\phi \wedge \psi < \psi$ (Conjunctiveness)

(ii) $\phi < \psi$ iff $\psi \supset \phi < \psi$ (Conditionalization)

(iii) If $\phi \vdash \psi$ then $\psi \not< \phi$ (GM-Dominance)

(iv) Not both $\phi \wedge \psi < \phi$ and $\phi \wedge \psi < \psi$ (GM-Conjunctiveness)

(v) If $\phi < \psi$ and $\phi < \chi$ and $\phi \vdash \psi \wedge \chi$ then $\phi < \psi \wedge \chi$

(Weak Conjunction Up)

(vi) If $\phi < \psi$ and $\psi \vdash \chi$ and $\phi \wedge \chi \vdash \psi$ then $\phi < \chi$　　(Weak Continuing Up)

(vii) If $\phi < \psi$ and $\chi \vdash \phi$ and $\phi \vdash \psi \supset \chi$ then $\chi < \psi$　(Weak Continuing Down)

Proof of Lemma 55.

(i) Conjunctiveness. By (E&C), $\phi \wedge \psi < \psi$ is equivalent with $\phi < \psi \wedge \psi$ which is equivalent with $\phi < \psi$, by Extensionality.

(ii) Conditionalization follows from Conjunctiveness and Extensionality.

(iii) GM-dominance. Let $\phi \vdash \psi$. Since $<$ is asymmetric, it follows by Extensionality that $\phi \wedge \psi \not< \phi$. So by Conjunctiveness $\psi \not< \phi$.

(iv) GM-Conjunctiveness follows from Conjunctiveness and Asymmetry.

(v) Weak Conjunction Up. Let $\phi < \psi$ and $\phi < \chi$ and $\phi \vdash \psi \wedge \chi$. By Extensionality, we get $\phi \wedge \chi < \psi$ and $\phi \wedge \psi < \chi$. Hence, by (E&C) $\phi < \psi \wedge \chi$.

(vi) Weak Continuing Up. Let $\phi < \psi$, $\psi \vdash \chi$ and $\phi \wedge \chi \vdash \psi$. By Extensionality, we get $\phi < \psi \wedge \chi$. (E&C) then gives us $\phi \wedge \psi < \chi$. By Extensionality again, we get $\phi \wedge \chi < \chi$, from which we can infer that $\phi < \chi$ by Conjunctiveness.

(vii) Weak Continuing Down. Let $\phi < \psi$ and $\chi \vdash \phi$ and $\phi \vdash \psi \supset \chi$. By Conjunctiveness, we get $\phi \wedge \psi < \psi$. By Extensionality, then, $\chi \wedge \psi < \psi$, and by Conjunctiveness once more we can infer $\chi < \psi$.　　$\square$

Observation 56

The constraints (EE1) – (EE4) on entrenchment relations translate into the following coherence criteria for rational choice.

(CC1)　$\delta(F) = \emptyset$ or $F \cap \delta(F) \neq \emptyset$

(CC2$^\uparrow$)　If $G \cap \delta(F \cup G) = \emptyset$ and $H \subseteq Cn(G)$, then $H \cap \delta(F \cup H) = \emptyset$

(CC2$^\downarrow$)　If $G \cap \delta(F \cup G) = \emptyset$ and $F \subseteq Cn(H)$, then $G \cap \delta(G \cup H) = \emptyset$

(CC3$^\uparrow$)　If $G \cap \delta(F \cup G) = \emptyset$ and $H \cap \delta(F \cup H) = \emptyset$, then
$(G \cup H) \cap \delta(F \cup G \cup H) = \emptyset$

(CC3$^\downarrow$)　If $F \cap \delta(F \cup G \cup G) = \emptyset$, then $F \cap \delta(F \cup G) = \emptyset$

(CC4)　If $G \cap \delta(F \cup G) = \emptyset$, then $H \cap \delta(F \cup H) = \emptyset$ or $G \cap \delta(G \cup H) = \emptyset$.

Proof of Observation 56.

We understand that $\phi = \bigwedge F$, $\psi = \bigwedge G$ and $\chi = \bigwedge H$.

The translations (CC1) and (CC3$^\downarrow$) of (EE1) and (EE3$^\downarrow$) are tautologies and need no comment. As regards the other conditions, the straightforward translations would always introduce accompanying limiting case conditions. We show that these are superfluous if we allow ourselves to use the basic conditions ($\emptyset$1), ($\emptyset$2), (LP1) and (LP2) for syntactic choice functions.

The translation of (EE2$^\uparrow$) literally reads: If $\delta(F \cup G) \neq \emptyset$ and $G \cap \delta(F \cup G) = \emptyset$ and $H \subseteq Cn(G)$, then $\delta(F \cup H) \neq \emptyset$ and $H \cap \delta(F \cup H) = \emptyset$. First we show that the additional consequent $\delta(F \cup H) \neq \emptyset$ follows from the antecedent anyway. If $\delta(F \cup G) \neq \emptyset$ and $G \cap \delta(F \cup G) = \emptyset$, then we get $F \cap \delta(F \cup G) \neq \emptyset$. But then, by ($\emptyset$2), $\delta(F) \neq \emptyset$, and so $\delta(F \cup H) \neq \emptyset$, by ($\emptyset$1). Now we show that the additional antecedent $\delta(F \cup G) \neq \emptyset$ is not necessary for deriving the remaining consequent. Assume that $\delta(F \cup G) = \emptyset$. If $H \subseteq Cn(G)$, then, by (LP2) and (LP1), $\delta(F \cup G \cup H) = \emptyset$, and hence, by ($\emptyset$1), $\delta(F \cup H) = \emptyset$ which of course proves the consequent. The whole translation thus reduces to this: If $G \cap \delta(F \cup G) = \emptyset$ and $H \subseteq Cn(F)$, then $H \cap \delta(F \cup H) = \emptyset$, that is (CC2$^\uparrow$).

The translation of (EE2$^\downarrow$) literally reads: If $\delta(F \cup G) \neq \emptyset$ and $G \cap \delta(F \cup G) = \emptyset$ and $F \subseteq Cn(H)$, then $\delta(G \cup H) \neq \emptyset$ and $G \cap \delta(G \cup H) = \emptyset$. First we show that that the additional consequent $\delta(G \cup H) \neq \emptyset$ follows from the antecedent anyway. If $\delta(F \cup G) \neq \emptyset$, then, by ($\emptyset$1), $\delta(F \cup G \cup H) \neq \emptyset$. If in addition $F \subseteq Cn(H)$, then we get by (LP2) and (LP1) that $\delta(G \cup H) \neq \emptyset$. Now we show that the additional antecedent $\delta(F \cup G) \neq \emptyset$ is not necessary for deriving the remaining consequent. Assume that $\delta(F \cup G) = \emptyset$. Then by ($\emptyset$1), $\delta(G) = \emptyset$, and hence, by ($\emptyset$2), $G \cap \delta(G \cup H) = \emptyset$. The whole translation thus reduces to this: If $G \cap \delta(F \cup G) = \emptyset$ and $F \subseteq Cn(H)$, then $G \cap \delta(G \cup H) = \emptyset$, and that is (CC2$^\downarrow$).

The translation of (EE3$^\uparrow$) literally reads: If $\delta(F \cup G) \neq \emptyset$, $G \cap \delta(F \cup G) = \emptyset$, $\delta(F \cup H) \neq \emptyset$, and $H \cap \delta(F \cup H) = \emptyset$, then $\delta(F \cup G \cup H) \neq \emptyset$ and $(G \cup H) \cap \delta(F \cup G \cup H) = \emptyset$. First we verify that the additional consequent $\delta(F \cup G \cup H) \neq \emptyset$ follows from the antecedent anyway. But this is clear from the antecedent $\delta(F \cup G) \neq \emptyset$ and ($\emptyset$1). Now we show that the additional antecedents $\delta(F \cup G) \neq \emptyset$ and $\delta(F \cup H) \neq \emptyset$ are not necessary for deriving the remaining consequent. Suppose, as the first case, that both $\delta(F \cup G) = \emptyset$ and $\delta(F \cup H) = \emptyset$. Then we get, by ($\emptyset$1), that both $\delta(G) = \emptyset$ and $\delta(H) = \emptyset$, and hence $(G \cup H) \cap \delta(F \cup G \cup H) = \emptyset$, by ($\emptyset$2). Suppose, as the second case that $\delta(F \cup G) = \emptyset$ and $\delta(F \cup H) \neq \emptyset$. From the latter we get, if the antecedent $H \cap \delta(F \cup H) = \emptyset$ holds, that $F \cap \delta(F \cup H) \neq \emptyset$. From the former we get, by ($\emptyset$1), that $\delta(F) = \emptyset$. But these two items contradict each other, by ($\emptyset$2). The third case, $\delta(F \cup G) \neq \emptyset$ and $\delta(F \cup H) = \emptyset$, is similar to the second one. The whole translation thus reduces to this: If $G \cap \delta(F \cup G) = \emptyset$ and $H \cap \delta(F \cup H) = \emptyset$, then $(G \cup H) \cap \delta(F \cup G \cup H) = \emptyset$, and that is (CC3$^\uparrow$).

The translation of (EE4) literally reads: If $\delta(F \cup G) \neq \emptyset$ and $G \cap \delta(F \cup G) = \emptyset$, then either $\delta(F \cup H) \neq \emptyset$ and $H \cap \delta(F \cup H) = \emptyset$, or $\delta(G \cup H) \neq \emptyset$ and $G \cap \delta(G \cup H) = \emptyset$. First we show that that the additional consequent $\delta(F \cup H) \neq \emptyset$ follows from the antecedent anyway. If $\delta(F \cup G) \neq \emptyset$ and $G \cap \delta(F \cup G) = \emptyset$, then

$F \cap \delta(F \cup G) \neq \emptyset$. From this we get that $\delta(F) \neq \emptyset$, by ($\emptyset$2), and thus $\delta(F \cup H) \neq \emptyset$, by ($\emptyset$1). So the consequent of the translation reduces to this:

$$\text{either } H \cap \delta(F \cup H) = \emptyset, \text{ or } \delta(G \cup H) \neq \emptyset \text{ and } G \cap \delta(G \cup H) = \emptyset \qquad (\dagger)$$

From this it follows that

$$\text{either } H \cap \delta(F \cup H) = \emptyset \text{ or } G \cap \delta(G \cup H) = \emptyset \qquad (\ddagger)$$

But ($\ddagger$) also entails ($\dagger$). Suppose it would not. That would mean that it is possible that $H \cap \delta(F \cup H) \neq \emptyset$ and $\delta(G \cup H) = \emptyset$ and $G \cap \delta(G \cup H) = \emptyset$. This, however, is not possible, since from $\delta(G \cup H) = \emptyset$ we can deduce that $\delta(H) = \emptyset$, by ($\emptyset$1), and from this, that $H \cap \delta(F \cup H) = \emptyset$, by ($\emptyset$2), and we have a contradiction. So the consequent of the translation further reduces to this: either $H \cap \delta(F \cup H) = \emptyset$ or $G \cap \delta(G \cup H) = \emptyset$. Finally, we can remove the antecedent clause $\delta(F \cup G) \neq \emptyset$. For if it is the case that $\delta(F \cup G) = \emptyset$, then by ($\emptyset$1), $\delta(G) = \emptyset$, so $G \cap \delta(G \cup H) = \emptyset$, by ($\emptyset$2), and the consequent is satisfied anyway. The whole translation thus reduces to this

$$\text{If } G \cap \delta(F \cup G) = \emptyset, \text{ then } H \cap \delta(F \cup H) = \emptyset \text{ or } G \cap \delta(G \cup H) = \emptyset$$

and this is (CC4). □

Observation 57

(CC1) and (CC3$^{\downarrow}$) are vacuously satisfied. Furthermore, given (LP1) and (LP2), the following implications hold true for every choice function δ:

(i) (CC2$^{\uparrow}$) is equivalent with (III);
(ii) (CC2$^{\downarrow}$) is equivalent with (I);
(iii) (CC3$^{\uparrow}$) follows from (I);
(iv) (CC4) implies (IV), and follows from the conjunction of (I) and (IV).

Proof of Observation 57.
Throughout the proof, we assume that δ satisfies (LP1) and (LP2).

(i) *(CC2$^{\uparrow}$) follows from (III).* Let $G \cap \delta(F \cup G) = \emptyset$ and $H \subseteq Cn(G)$. Then, by (LP2), $G \cap \delta(F \cup G \cup H) = \emptyset$, and thus $\delta(F \cup G \cup H) \subseteq F \cup H$. Hence, by (III), $\delta(F \cup H) \subseteq \delta(F \cup G \cup H)$. Now assume for reductio that there is a ϕ which is in $H \cap \delta(F \cup H)$. Hence ϕ is also in $H \cap \delta(F \cup G \cup H)$. Since $H \subseteq Cn(G)$, we may conclude that $G \cap \delta(F \cup G \cup H) \neq \emptyset$, by (LP1), which gives us a contradiction.

(III) follows from (CC2$^{\uparrow}$). The following is a reformulation of (III). If $F \subseteq G$ and $(G - F) \cap \delta(G) = \emptyset$ and $\{\phi\} \cap \delta(G) = \emptyset$, then $\{\phi\} \cap \delta(F) = \emptyset$. This is trivially satisfied if ϕ is not in F. So let us suppose for the following that ϕ is in F. Then we can further reformulate (III) like that:

If $F \subseteq G$ and $((G - F) \cup \{\phi\}) \cap \delta(((G - F) \cup \{\phi\}) \cup F) = \emptyset$, then $\{\phi\} \cap \delta(\{\phi\} \cup F) = \emptyset$.

Since obviously $\{\phi\} \subseteq Cn((G - F) \cup \{\phi\})$, this follows directly from (CC2$^{\uparrow}$).

(ii) *(CC2$^{\downarrow}$) follows from (I).* Let $G \cap \delta(F \cup G) = \emptyset$ and $F \subseteq Cn(H)$. From the former, we get by (I) that $G \cap \delta(F \cup G \cup H) = \emptyset$. But this entails, by $F \subseteq Cn(H)$ and (LP2) that $G \cap \delta(G \cup H) = \emptyset$.

(I) follows from (CC2$^{\downarrow}$). The following is a reformulation of (I). If $F \subseteq G$ and $\{\phi\} \cap \delta(F) = \emptyset$, then $\{\phi\} \cap F \cap \delta(G) = \emptyset$. This is trivially satisfied if ϕ is not in F. So let us suppose that ϕ is in F. Then we can further reformulate (I) like that: If $F \subseteq G$ and $\{\phi\} \cap \delta(\{\phi\} \cup F) = \emptyset$, then $\{\phi\} \cap \delta(\{\phi\} \cup G) = \emptyset$. This, however, follows directly from (CC2$^{\downarrow}$).

(iii) *(CC3$^{\uparrow}$) follows from (I).* If $G \cap \delta(F \cup G) = \emptyset$, then $G \cap \delta(F \cup G \cup H) = \emptyset$, by (I). Similarly, if $H \cap \delta(F \cup H) = \emptyset$, then $H \cap \delta(F \cup G \cup H) = \emptyset$. This proves (CC3$^{\uparrow}$).

(iv) *(CC4) follows from (I) and (IV).* Let $G \cap \delta(F \cup G) = \emptyset$, and suppose for reductio that both $H \cap \delta(F \cup H) \neq \emptyset$ and $G \cap \delta(G \cup H) \neq \emptyset$. First notice that we easily get $\delta(F \cup G \cup H) \neq \emptyset$, by ($\emptyset$1). From $H \cap \delta(F \cup H) \neq \emptyset$, we get that either $H \cap \delta(F \cup G \cup H) \neq \emptyset$ or $\delta(F \cup H) \not\subseteq \delta(F \cup G \cup H)$. The last condition yields, by (IV), that $(F \cup H) \cap \delta(F \cup G \cup H) = \emptyset$, and thus, since $\delta(F \cup G \cup H)$ is non-empty, $G \cap \delta(F \cup G \cup H) \neq \emptyset$. In any case, $(G \cup H) \cap \delta(F \cup G \cup H) \neq \emptyset$. Applying (IV) again, we get that $\delta(G \cup H) \subseteq \delta(F \cup G \cup H)$. Combining this with our supposition that $G \cap \delta(G \cup H) \neq \emptyset$ gives us $G \cap \delta(F \cup G \cup H) \neq \emptyset$. Pick some $\phi \in G \cap \delta(F \cup G \cup H)$. By (I), $(F \cup G) \cap \delta(F \cup G \cup H) \subseteq \delta(F \cup G)$, so $\phi \in G \cap \delta(F \cup G)$. Now we have got a contradiction with $G \cap \delta(F \cup G) = \emptyset$.

(IV) follows from (CC4). The following is a reformulation of (IV). If $F \subseteq G$ and $F \cap \delta(G) \neq \emptyset$ and $\{\phi\} \cap \delta(G) = \emptyset$, then $\{\phi\} \cap \delta(F) = \emptyset$. This is trivially satisfied if ϕ is not in F. So let us suppose for the following that ϕ is in F. Then we can further reformulate (IV) like that:

If $F \subseteq G$ and $\{\phi\} \cap \delta(\{\phi\} \cup G) = \emptyset$, then either $F \cap \delta(F \cup G) = \emptyset$ or $\{\phi\} \cap \delta(\{\phi\} \cup F) = \emptyset$.

But this follows directly from (CC4). $\square$

Lemma 58

Translated into choice-theoretic conditions, (EE2$^{\uparrow}$*) and (EE2$^{\downarrow}$*) become:

(CC2$^{\uparrow}$*) If $(G \cup H) \cap \delta(F \cup G \cup H) = \emptyset$, then $G \cap \delta(F \cup G) = \emptyset$

(CC2$^{\downarrow}$*) If $G \cap \delta(F \cup G) = \emptyset$, then $G \cap \delta(F \cup G \cup H) = \emptyset$

Proof of Lemma 58.

The translation of (EE2$^{\uparrow}$*) literally reads: If $\delta(F \cup G \cup H) \neq \emptyset$ and $(G \cup H) \cap \delta(F \cup G \cup H) = \emptyset$, then $\delta(F \cup G) \neq \emptyset$ and $G \cap \delta(F \cup G) = \emptyset$. First we show that the additional consequent $\delta(F \cup G) \neq \emptyset$ follows from the antecedent anyway. If $\delta(F \cup G \cup H) \neq \emptyset$ and $(G \cup H) \cap \delta(F \cup G \cup H) = \emptyset$, then we get $F \cap \delta(F \cup G \cup H) \neq \emptyset$. But then, by ($\emptyset$2), $\delta(F) \neq \emptyset$, and so $\delta(F \cup G) \neq \emptyset$, by ($\emptyset$1). Now we show that the additional antecedent $\delta(F \cup G \cup H) \neq \emptyset$ is not necessary for deriving the remaining consequent. Assume that $\delta(F \cup G \cup H) = \emptyset$. Then by ($\emptyset$1), $\delta(F \cup G) = \emptyset$ which of course proves the consequent. The whole translation thus reduces to this: If $(G \cup H) \cap \delta(F \cup G \cup H) = \emptyset$, then $G \cap \delta(F \cup G) = \emptyset$, that is (CC2$^{\uparrow}$*).

The translation of (EE2$^{\downarrow}$*) literally reads: If $\delta(F \cup G) \neq \emptyset$ and $G \cap \delta(F \cup G) = \emptyset$, then $\delta(F \cup G \cup H) \neq \emptyset$ and $G \cap \delta(F \cup G \cup H) = \emptyset$. First we need to show that that the additional consequent $\delta(F \cup G \cup H) \neq \emptyset$ follows from the antecedent anyway. But this is immediate from $\delta(F \cup G) \neq \emptyset$ and $(\emptyset 1)$. Now we show that the additional antecedent $\delta(F \cup G) \neq \emptyset$ is not necessary for deriving the remaining consequent. Assume that $\delta(F \cup G) = \emptyset$. Then by $(\emptyset 1)$, $\delta(G) = \emptyset$, and hence, by $(\emptyset 2)$, $G \cap \delta(F \cup G \cup H) = \emptyset$. The whole translation thus reduces to this: If $G \cap \delta(F \cup G) = \emptyset$, then $G \cap \delta(F \cup G \cup H) = \emptyset$, and that is $(\text{CC2}^{\downarrow *})$. □

Lemma 59
$(\text{CC2}^{\uparrow})$ is equivalent to (III) and $(\text{CC2}^{\downarrow})$ is equivalent to (I).

Proof of Lemma 59.
$(\text{CC2}^{\uparrow *})$ follows from (III). Let $(G \cup H) \cap \delta(F \cup G \cup H) = \emptyset$. Then $\delta(F \cup G \cup H) \subseteq F \subseteq F \cup G$. Hence, by (III), $\delta(F \cup G) \subseteq \delta(F \cup G \cup H)$. But then $G \cap \delta(F \cup G) \subseteq (G \cup H) \cap \delta(F \cup G \cup H) = \emptyset$.

(III) follows from $(\text{CC2}^{\uparrow *})$. The following is a reformulation of (III). If $F \subseteq G$ and $(G - F) \cap \delta(G) = \emptyset$, then $(F - \delta(G)) \cap \delta(F) = \emptyset$. We can further reformulate (III) like that:

If $F \subseteq G$ and $((G - F) \cup (F - \delta(G))) \cap \delta((G - F) \cup F \cup (F - \delta(G))) = \emptyset$, then $(F - \delta(G)) \cap \delta(F \cup (F - \delta(G))) = \emptyset$.

(Notice that clearly $(F - \delta(G)) \cap \delta((G - F) \cup F \cup (F - \delta(G))) = \emptyset$.) But this this follows directly from $(\text{CC2}^{\uparrow *})$.

$(\text{CC2}^{\downarrow *})$ follows from (I). Let $G \cap \delta(F \cup G) = \emptyset$. From this, we immediately get by (I) that $G \cap \delta(F \cup G \cup H) = \emptyset$.

(I) follows from $(\text{CC2}^{\downarrow *})$. The following is a reformulation of (I). If $F \subseteq G$ and $\{\phi\} \cap \delta(F) = \emptyset$, then $\{\phi\} \cap F \cap \delta(G) = \emptyset$. This is trivially satisfied if ϕ is not in F. So let us suppose that ϕ is in F. Then we can further reformulate (I) like that: If $F \subseteq G$ and $\{\phi\} \cap \delta(\{\phi\} \cup F) = \emptyset$, then $\{\phi\} \cap \delta(\{\phi\} \cup F \cup G) = \emptyset$. This, however, follows directly from $(\text{CC2}^{\downarrow *})$. □

Observation 60
(i) (SV) follows from $(\text{EE2}^{\uparrow})$ and $(\text{EE2}^{\downarrow})$
(ii) (E&C) follows from $(\text{EE2}^{\uparrow})$, $(\text{EE2}^{\downarrow})$, $(\text{EE3}^{\uparrow})$ and $(\text{EE3}^{\downarrow})$;
(iii) (Asymmetry) follows from (EE1), $(\text{EE2}^{\downarrow})$ and $(\text{EE3}^{\uparrow})$;
(iv) (E$\emptyset$1) follows from $(\text{EE2}^{\downarrow})$;
(v) (E$\emptyset$2) follows from $(\text{EE3}^{\downarrow})$ and (EE4);
(vi) (ELP1) follows from $(\text{EE2}^{\uparrow})$ and $(\text{EE2}^{\downarrow})$;
(vii) (ELP2) follows from $(\text{EE2}^{\downarrow})$ and $(\text{EE3}^{\downarrow})$;
(viii) (EVSuccess) follows from $(\text{EE2}^{\uparrow})$ and (EE6);
(ix) (EI) is identical with $(\text{EE2}^{\downarrow *})$ and follows from $(\text{EE2}^{\downarrow})$;
(x) (EII) follows from (EE1), $(\text{EE3}^{\uparrow})$ and (EE4);
(xi) (EIII) is identical with $(\text{EE2}^{\uparrow *})$ and follows from $(\text{EE2}^{\uparrow})$;
(xii) (EIV) follows from $(\text{EE2}^{\downarrow})$, $(\text{EE3}^{\downarrow})$ and (EE4).
(xiii) (EI^-) follows from $(\text{EE2}^{\downarrow})$;
(xiv) (EII^+) follows from (EE1), $(\text{EE2}^{\uparrow})$, $(\text{EE3}^{\uparrow})$ and (EE4);
(xv) (EIV^+) does not follow from any combination of standard postulates for epistemic entrenchment.

Proof of Observation 60. Parts (iv), (ix), (xi), (xiii) and (xv) are obvious.

(i) If ϕ and ψ are $\land$-variants, then $Cn(\phi) = Cn(\psi)$. So (SV) follows directly from (EE2$^\top$) and (EE2$^\downarrow$).

(ii) *(E&C), from left to right.* Let $\phi < \psi \land \chi$. By (EE2$^\top$), $\phi < \chi$, and then by (EE2$^\downarrow$), $\phi \land \psi < \chi$. The case $\phi \land \chi < \psi$ is analogous.

(E&C), from right to left. Let $\phi \land \psi < \chi$ and $\phi \land \chi < \psi$. By (EE2$^\downarrow$), $\phi \land \psi \land \chi < \chi$ and $\phi \land \psi \land \chi < \psi$. Therefore, by (EE3$^\top$), $\phi \land \psi \land \chi < \psi \land \chi$. So by (EE3$^\downarrow$), $\phi < \psi \land \chi$.

(iii) See Lemma 48 (iv).

(v) Let $\phi \land \psi < \top$ and $\psi \not< \top$. Then by (EE4) $\phi \land \psi < \psi$, so by (EE3$^\downarrow$) $\phi < \psi$.

(vi) Let $\phi \in Cn(\psi)$ and $\phi \land \chi < \psi$. From the former we get that $\phi \land \chi \in Cn(\psi \land \chi)$. From this and $\phi \land \chi < \psi$, we infer by (EE2$^\downarrow$) that $\psi \land \chi < \psi$, from which we get, by (EE2$^\top$), $\psi \land \chi < \phi$.

(vii) Let $Cn(\phi \land \psi) = Cn(\phi \land \chi)$ and $\psi < \phi$. Then $\phi \land \chi < \phi$, by (EE2$^\downarrow$), and hence $\chi < \phi$, by (EE3$^\downarrow$).

(viii) Let $\phi \notin Cn(\emptyset)$. Then, by (EE6), there is a ψ such that $\phi < \psi$. Hence, by (EE2$^\top$), $\phi < \top$.

(x) Let $\phi \land \psi < \chi$ and assume for reductio that $\phi \not< \chi$ and $\psi \not< \chi$. By (EE4), this implies that $\phi \land \psi < \phi$ and $\phi \land \psi < \psi$, so, by (EE3$^\top$), $\phi \land \psi < \phi \land \psi$, contradicting (EE1).

(xii) Let $\phi \land \psi < \chi$ and $\psi \not< \phi \land \chi$. We need to show that $\phi < \chi$. From $\phi \land \psi < \chi$, we get $\phi \land \psi \land \chi < \chi$, by (EE2$^\downarrow$). From $\psi \not< \phi \land \chi$, we get $\phi \land \psi \land \chi \not< \phi \land \chi$, by (EE3$^\downarrow$). Hence, by (EE4), $\phi \land \chi < \chi$, which entails, by (EE3$^\downarrow$) again, $\phi < \chi$.

(xiv) Let $\phi \land \psi < \chi \land \xi$, $\chi < \top$ and $\xi < \top$. By (EE2$^\top$), we get $\phi \land \psi < \chi$ and $\phi \land \psi < \xi$. We need to show that $\phi < \chi$ or $\psi < \xi$. Suppose this is not the case. Then $\phi \not< \chi$ and $\psi \not< \xi$. From the former and $\phi \land \psi < \chi$, we get by (EE4) that $\phi \land \psi < \phi$. From the latter and $\phi \land \psi < \xi$, we get by (EE4) that $\phi \land \psi < \psi$. So by (EE3$^\top$), $\phi \land \psi < \phi \land \psi$, contradicting (EE1). □

Observation 61

(i) Extensionality follows from (ELP1) and (ELP2);

(ii) (EE1) is Irreflexivity;

(iii) (EE2$^{\top*}$) is identical with (EIII);

(iv) (EE2$^{\downarrow*}$) is identical with (EI);

(v) (EE3$^\top$) follows from (E&C) and (EI);

(vi) (EE3$^\downarrow$) follows from (E&C) and (SV);

(vii) (EE4) follows from Irreflexivity, (SV), (E&C), (EI) and (EIV).

Proof of Observation 61.

(i) Let $Cn(\phi) = Cn(\psi)$. For Extensionality on the LHS, let $\phi < \chi$. We have $Cn(\phi \land \chi) = Cn(\psi \land \chi)$, so (ELP2) gives us $\psi < \chi$. For Extensionality on the RHS, let $\chi < \phi$. Then by (1), $\phi \land \chi < \phi$. Thus, by Extensionality on the LHS, $\psi \land \chi < \phi$, so by (ELP1), $\phi \land \chi < \psi$, thus, by Extensionality on the LHS again, $\psi \land \chi < \psi$, so by (1), $\chi < \psi$.

(ii) – (iv) are obvious.

(v) Let $\phi < \psi$ and $\phi < \chi$. By (EI), we get $\phi \wedge \chi < \psi$ and $\phi \wedge \psi < \chi$. So by (E&C), $\phi < \psi \wedge \chi$.

(vi) Let $\phi \wedge \psi < \psi$. Then $\phi < \psi \wedge \psi$ by (E&C), and $\phi < \psi$ by (SV).

(vii) This is part (viii) of Lemma 52.   □

Observation 62

For all sentences ϕ and ψ, the following conditions are equivalent, provided that ϕ is not in $Cn(\emptyset)$ and $\prec$ is stoppered:

(i) For all m in $\min_\prec \rrbracket\{\phi, \psi\}\llbracket$ it holds that $m \models \psi$, and $\min_\prec \rrbracket\{\phi, \psi\}\llbracket \neq \emptyset$;

(ii) For all m in $\rrbracket\psi\llbracket$ there is an n in $\rrbracket\phi\llbracket$ such that $n \prec m$.

Proof of Observation 62.

Let $\prec$ be stoppered and ϕ and ψ be sentences such that $\phi \notin Cn(\emptyset)$.

(i) implies (ii). Let m be in $\rrbracket\psi\llbracket$. (If $\rrbracket\psi\llbracket$ is empty, then (ii) is vacuously true.) Then m is in $\rrbracket\{\phi, \psi\}\llbracket$ and $m \not\models \psi$. Hence, by (i), m is not in $\min_\prec \rrbracket\{\phi, \psi\}\llbracket$. Thus there is an n in $\rrbracket\{\phi, \psi\}\llbracket$ such that $n \prec m$. Since $\prec$ is stoppered, we can choose n minimal in $\rrbracket\{\phi, \psi\}\llbracket$. Then, by (i) again, $n \models \psi$. But $n \in \rrbracket\{\phi, \psi\}\llbracket$, so $n \in \rrbracket\phi\llbracket$, and we have proved (ii).

(ii) implies (i). That $\min_\prec \rrbracket\{\phi, \psi\}\llbracket \neq \emptyset$ follows from the facts that ϕ is not in $Cn(\emptyset)$ and that $\prec$ is stoppered. Now let m be in $\min_\prec \rrbracket\{\phi, \psi\}\llbracket$. Suppose for reductio that $m \not\models \psi$, i.e., that m is in $\rrbracket\psi\llbracket$. But then there is an n in $\rrbracket\phi\llbracket \subseteq \rrbracket\{\phi, \psi\}\llbracket$ such that $n \prec m$, by (ii). This contradicts the assumption that m is in $\min_\prec \rrbracket\{\phi, \psi\}\llbracket$, and hence (i) is proved from (ii).   □

Observation 63

For all models m and n, the following conditions are equivalent, provided that $<$ is an admissible relation which is conversely well-founded and satisfies Extensionality, (EI), (EII), and (EIII).

(i) For all F, if m and n are in $\rrbracket F\llbracket$, then $Cn(F) - \hat{n} \not\subseteq \min_<(Cn(F))$, and there is an G such that m is in $\rrbracket G\llbracket$ and $Cn(G) - \hat{m} \subseteq \min_<(Cn(G))$;

(ii) For every ϕ falsified by m there is a ψ falsified by n such that $\phi < \psi$.

Proof of Observation 63.

Let $<$ be an an admissible, conversely well-founded relation that satisfies Extensionality, (EI), (EII), and (EIII), and let m and n be two arbitrary models.

(i) implies (ii). Let m be in $\rrbracket\phi\llbracket$. Choose some ψ that is $<$-maximal in $L - \hat{n}$. This is possible since $<$ is assumed to be conversely well-founded. The model n is in $\rrbracket\psi\llbracket$. We have that m and n are in $\rrbracket\{\phi, \psi\}\llbracket$, and hence, by (i), we get $Cn(\phi \wedge \psi) - \hat{n} \not\subseteq \min_<(Cn(\phi \wedge \psi))$. That is, there is an χ such that n is in $\rrbracket\chi\llbracket$ and χ is in $Cn(\phi \wedge \psi)$, but not minimal in this set. That χ is not minimal in $Cn(\phi \wedge \psi)$ implies, by (Extensionality) and (EI), that $\phi \wedge \psi < \chi$. From (EII), we get that $\phi < \chi$ or $\psi < \chi$. But $\psi < \chi$ is impossible, since we have chosen ψ maximal in $L - \hat{n}$. So we in fact get $\phi < \chi$, and since n is in $\rrbracket\chi\llbracket$, we have verified (ii).

(ii) implies (i). Let m and n be in $]F[$. For the first part of (i), we have to show that there is a ϕ in $Cn(F)$ such that n is in $]\phi[$ and ϕ is not in $\min_<(Cn(F))$. Now take ψ and χ from F such that $m \in]\psi[$ and $n \in]\chi[$. Then ψ is not in $\widehat{m}$, and hence, by (ii), there is a $\xi \notin \widehat{n}$ such that $\psi < \xi$. By applying (Extensionality) and (EIII), we get from $\psi < \xi$ that $\psi < \xi \wedge (\chi \vee \xi)$, and thus $\psi < \chi \vee \xi$. Summing up, we have that $\chi \vee \xi$ is in $Cn(F)$, that n is in $]\chi \vee \xi[$, and that ψ is in $Cn(F)$ with $\psi < \chi \vee \xi$. Thus $\chi \vee \xi$ is in $Cn(F) - \widehat{n}$ but not in $\min_<(Cn(F))$. It can take the role of the ϕ we set out to find for the first part of (i).

For the second part of (i), we have to show that there is a G such that m is in $]G[$ and $Cn(G) - \widehat{m} \subseteq \min_<(Cn(G))$. For this it will be enough to show (and it is in fact equivalent to show!) that there is a ϕ such that m is in $]\phi[$ and $Cn(\phi) - \widehat{m} \subseteq \min_<(Cn(\phi))$. Thus we have to find a ϕ such that $m \in]\phi[$ and for every $\psi \in Cn(\phi)$ such that $m \in]\psi[$ it holds that ψ is minimal in $Cn(\phi)$. Of course, ψ is minimal in $Cn(\phi)$ just in case there is no χ such that $\chi \in Cn(\phi)$ and $\chi < \psi$. By (Extensionality) and (EI), this just means that $\phi \not< \psi$. So we need to find a ϕ such that $m \in]\phi[$ and for every $\psi \in Cn(\phi)$ such that $m \in]\psi[$ it holds that $\phi \not< \psi$. But now we can take any ϕ which is $<$-maximal in $L - \widehat{m}$. Such ϕ's exists since we are assuming $<$ to be conversely well-founded, so we are done. $\qquad\square$

If δ satisfies (LP1) and (LP2) and $\mathcal{P}_2(\delta) = \mathcal{P}_2(\delta')$, then $\mathcal{C}(\delta) = \mathcal{C}(\delta')$.

Proof.

Let $\mathcal{P}_2(\delta) = \mathcal{P}_2(\delta')$. For $\mathcal{C}(\delta) = \mathcal{C}(\delta')$, we have to show that for all ϕ and ψ, $\phi \vee \psi \notin \delta(Cn(\phi))$ iff $\phi \vee \psi \notin \delta'(Cn(\phi))$. Or equivalently, by (LP1) and (LP2), that $\phi \vee \psi \notin \delta(\phi, \phi \vee \psi)$ iff $\phi \vee \psi \notin \delta'(\phi, \phi \vee \psi)$. With $<= \mathcal{P}_2(\delta)$ and $<'= \mathcal{P}_2(\delta')$, this means that we have show that $\phi \wedge (\phi \vee \psi) \not< \top$ or $\phi < \phi \vee \psi$ if and only if $\phi \wedge (\phi \vee \psi) \not<' \top$ or $\phi <' \phi \vee \psi$. But by hypothesis $<$ and $<'$ are identical, so this is immediate. $\qquad\square$

Observation 64

(i) Let $<$ satisfy Irreflexivity, (E&C), (Ext) and (E02), and let $<^* = \mathcal{P}_2(\mathcal{C}(<))$. Then we have $\phi <^* \psi$ if and only if $\psi \in K$ and $((\phi \notin K$ and $\phi \wedge \psi \not< \top)$ or $\phi < \psi)$.

If $<$ satisfies in addition (EFaith), then we have $\phi <^* \psi$ if and only if $\psi \in K$ and $\phi < \psi$.

If moreover $<$ satisfies (EE5'), then we have $\phi <^* \psi$ if and only if $\phi < \psi$.

(ii) Let $\dot{-}$ satisfy $(\dot{-}1) - (\dot{-}2)$, $(\dot{-}5)$ and $(\dot{-}6)$, and $\dot{-}^* = \mathcal{C}(\mathcal{P}_2(\dot{-}))$. Then we have $\psi \in K \dot{-}^* \phi$ if and only if $\psi \in K \dot{-}\phi$.

Proof of Observation 64.

(i) Let $<$ satisfy Irreflexivity, (E&C), (Ext) and (E02), and let $\dot{-} = \mathcal{C}(<)$ and $<^* = \mathcal{P}_2(\dot{-})$. Then we have

$\phi <^* \psi$

iff $\quad\quad\quad \psi \in K \dot{-}\phi \wedge \psi$ and $\phi \notin K \dot{-}\phi \wedge \psi$ $\quad\quad\quad$ (by $<^* = \mathcal{P}_2(\dot{-})$)

iff $\quad$ *either* $\quad \psi \in K \dot{-}\phi \wedge \psi$ and $\phi \notin K \dot{-}\phi \wedge \psi$ and $\phi \wedge \psi \not< \top$

$\quad\quad\quad$ *or* $\quad \psi \in K \dot{-}\phi \wedge \psi$ and $\phi \notin K \dot{-}\phi \wedge \psi$ and $\phi \wedge \psi < \top$

iff $\quad$ *either* $\quad \psi \in K$ and $\phi \notin K$ and $\phi \wedge \psi \not< \top$

$\quad\quad\quad$ *or* $\quad \psi \in K$ and $\phi \wedge \psi < (\phi \wedge \psi) \vee \psi$ and $(\phi \notin K$ or $\phi \wedge \psi \not< (\phi \wedge \psi) \vee \phi)$

$\quad\quad\quad\quad$ and $\phi \wedge \psi < \top$ $\quad\quad\quad\quad\quad\quad\quad\quad$ (by $\dot{-} = \mathcal{C}(<)$)

iff $\quad$ *either* $\quad \psi \in K$ and $\phi \notin K$ and $\phi \wedge \psi \not< \top$

$\quad\quad\quad$ *or* $\quad \psi \in K$ and $\phi \wedge \psi < \psi$ and $(\phi \notin K$ or $\phi \wedge \psi \not< \phi)$

$\quad\quad\quad\quad\quad\quad\quad\quad\quad\quad\quad$ (by Extensionality and (5))

iff $\quad$ *either* $\quad \psi \in K$ and $\phi \notin K$ and $\phi \wedge \psi \not< \top$

$\quad\quad\quad$ *or* $\quad \psi \in K$ and $\phi < \psi$ and $(\phi \notin K$ or $\psi \not< \phi)$ $\quad$ (by (E&C) and Ext.)

iff $\quad$ *either* $\quad \psi \in K$ and $\phi \notin K$ and $\phi \wedge \psi \not< \top$

$\quad\quad\quad$ *or* $\quad \psi \in K$ and $\phi < \psi$ $\quad\quad\quad\quad\quad\quad\quad\quad$ (by Asymmetry)

iff $\quad\quad\quad \psi \in K$ and $((\phi \notin K$ and $\phi \wedge \psi \not< \top)$ or $\phi < \psi)$

Let $<$ in addition satisfy (EFaith). If $\psi \in K$ and $\phi \notin K$, then by (EFaith) $\phi < \psi$, and the last condition reduces to $\psi \in K$ and $\phi < \psi$.

Let $<$ in addition satisfy (EE$'$). Then $\phi < \psi$ implies $\psi \in K$, and we finally get that $\phi <^* \psi$ if and only if $\phi < \psi$.

(ii) Let $\dot{-}$ satisfy ($\dot{-}1$) – ($\dot{-}2$), ($\dot{-}5$) and ($\dot{-}6$), and let $< = \mathcal{P}_2(\dot{-})$ and $\dot{-}^* = \mathcal{C}(<)$. Then we have

$\psi \in K \dot{-}^* \phi$

iff $\quad$ *either* $\quad \psi \in K$, and $\phi \not< \top$

$\quad\quad\quad$ *or* $\quad \psi \in K$ and $\phi < \phi \vee \psi$, and $\phi < \top$ $\quad\quad$ (by $\dot{-}^* = \mathcal{C}(<)$)

iff $\quad$ *either* $\quad \psi \in K$, and $\top \notin K \dot{-}\phi \wedge \top$ or $\phi \in K \dot{-}\phi \wedge \top$

$\quad\quad\quad$ *or* $\quad \psi \in K$ and $\top \in K \dot{-}\phi \wedge \top$ and $\phi \notin K \dot{-}\phi \wedge \top$ and $\phi \vee \psi \in$

$\quad\quad\quad\quad K \dot{-}(\phi \wedge (\phi \vee \psi))$ and $\phi \notin K \dot{-}(\phi \wedge (\phi \vee \psi))$ $\quad$ (by $< = \mathcal{P}_2(\dot{-})$)

iff $\quad$ *either* $\quad \psi \in K$, and $\phi \in K \dot{-}\phi$

$\quad\quad\quad$ *or* $\quad \psi \in K$ and $\phi \notin K \dot{-}\phi$ and $\phi \vee \psi \in K \dot{-}\phi$ $\quad$ (by ($\dot{-}1$) and ($\dot{-}6$))

iff $\quad$ *either* $\quad \psi \in K$, and $\phi \in K \dot{-}\phi$ and $K \dot{-}\phi = K$

$$or \quad \psi \in K \text{ and } \phi \notin K \dot{-} \phi \text{ and } \psi \in K \dot{-} \phi \qquad \text{(by } (\dot{-}1), (\dot{-}2) \text{ and } (\dot{-}5))$$

iff   *either* $\phi \in K \dot{-} \phi$ and $\psi \in K \dot{-} \phi$

$$or \quad \phi \notin K \dot{-} \phi \text{ and } \psi \in K \dot{-} \phi \qquad\qquad \text{(by } (\dot{-}2) \text{ and } (\dot{-}5^0))$$

iff $\psi \in K \dot{-} \phi$. $\square$

$(\dot{-}\emptyset 1)$, $(\dot{-}1)$ and $(\dot{-}7a)$ taken together imply the condition of

(Expulsiveness)   If $\phi \notin K \dot{-} \phi$ and $\psi \notin K \dot{-} \psi$, then either $\phi \notin K \dot{-} \psi$ or $\psi \notin K \dot{-} \phi$.

Proof. Let $\dot{-}$ satisfy $(\dot{-}\emptyset 1)$, $(\dot{-}1)$ and $(\dot{-}7a)$, and let $\phi \notin K \dot{-} \phi$ and $\psi \notin K \dot{-} \psi$. Then by $(\dot{-}7a)$, both $K \dot{-} \phi \subseteq K \dot{-} (\phi \wedge \psi)$ and $K \dot{-} \psi \subseteq K \dot{-} (\phi \wedge \psi)$. Suppose for reductio that both $\phi \in K \dot{-} \psi$ and $\psi \in K \dot{-} \phi$. Then, by $(\dot{-}1)$, $\phi \wedge \psi \in K \dot{-} (\phi \wedge \psi)$. But then, by $(\dot{-}\emptyset 1)$, $\phi \in K \dot{-} \phi$, and we have a contradiction. $\square$

Lemma 65

(i) If $\dot{-}$ satisfies $(\dot{-}\emptyset 2)$, $(\dot{-}1)$, $(\dot{-}2)$, $(\dot{-}5^0)$ and $(\dot{-}7a)$, then it satisfies $(\dot{-}7c)$;

(ii) If $\dot{-}$ satisfies $(\dot{-}\emptyset 1)$, $(\dot{-}1)$, $(\dot{-}2)$, $(\dot{-}5^0)$ and $(\dot{-}8)$, then it satisfies $(\dot{-}8c)$;

(iii) Severe withdrawals satisfy $(\ddot{-}9)$;

(iv) Severe withdrawals satisfy $(\ddot{-}10)$;

(v) The only severe withdrawal function $\ddot{-}$ that satisfies Recovery $(\dot{-}5)$ is the trivial one satisfying

$$K \ddot{-} \phi = \begin{cases} \overline{R} & \text{if } \phi \notin \overline{R} \\ K & \text{otherwise} \end{cases}$$

where $\overline{R} = \{\chi : \chi \in K \ddot{-} \chi\}$ is the taboo set of $\ddot{-}$. As a consequence of this, for each sentences ϕ and ψ in K which are not taboo for $\ddot{-}$, their biconditional $\phi \equiv \psi$ is taboo.

Proof of Lemma 65.

(i) Let $\dot{-}$ satisfy $(\dot{-}\emptyset 2)$, $(\dot{-}1)$, $(\dot{-}2)$, $(\dot{-}5^0)$, and $(\dot{-}7a)$, and let $\psi \in K \dot{-} (\phi \wedge \psi)$. We have to show that $K \dot{-} \phi \subseteq K \dot{-} (\phi \wedge \psi)$. If $\phi \notin K \dot{-} \phi$, this follows directly from $(\dot{-}7a)$. If $\phi \in K \dot{-} \phi$, then by $(\dot{-}\emptyset 2)$ $\phi \in K \dot{-} (\phi \wedge \psi)$, so by $(\dot{-}1)$, $\phi \wedge \psi \in K \dot{-} (\phi \wedge \psi)$. But then, by $(\dot{-}5^0)$, $K \subseteq K \dot{-} (\phi \wedge \psi)$, so $K \dot{-} \phi \subseteq K \dot{-} (\phi \wedge \psi)$ by $(\dot{-}2)$.

(ii) Let $\dot{-}$ satisfy $(\dot{-}\emptyset 1)$, $(\dot{-}2)$, $(\dot{-}5^0)$, and $(\dot{-}8)$, and let $\psi \in K \dot{-} (\phi \wedge \psi)$. We have to show that $K \dot{-} (\phi \wedge \psi) \subseteq K \dot{-} \phi$. If $\phi \notin K \dot{-} (\phi \wedge \psi)$, this follows directly from $(\dot{-}8)$. If $\phi \in K \dot{-} (\phi \wedge \psi)$, then, by $(\dot{-}1)$, $\phi \wedge \psi \in K \dot{-} (\phi \wedge \psi)$ and, by $(\dot{-}\emptyset 1)$, $\phi \in K \dot{-} \phi$, so by $(\dot{-}5^0)$, $K \subseteq K \dot{-} \phi$, so $K \dot{-} (\phi \wedge \psi) \subseteq K \dot{-} \phi$ by $(\dot{-}2)$.

(iii) Let $\ddot{-}$ be an operation of severe withdrawal, and let $\phi \notin K \ddot{-} \psi$. By (ii), $\ddot{-}$ satisfies $(\dot{-}8c)$. We have to show that $K \ddot{-} \psi \subseteq K \ddot{-} \phi$. From $\phi \notin K \ddot{-} \psi$, we infer with $(\dot{-}8c)$ that $\phi \notin K \ddot{-} (\phi \wedge \psi)$, so by $(\dot{-}8)$ $K \ddot{-} (\phi \wedge \psi) \subseteq K \ddot{-} \phi$. First assume that $\psi \notin K \ddot{-} \psi$. Then, by $(\dot{-}7a)$, $K \ddot{-} \psi \subseteq K \ddot{-} (\phi \wedge \psi)$, so $K \ddot{-} \psi \subseteq K \ddot{-} \phi$ as desired. So assume secondly that $\psi \in K \ddot{-} \psi$. Then, by $(\dot{-}5^0)$, $K \subseteq K \ddot{-} \psi$. So, since $\phi \notin K \ddot{-} \psi$, $\phi \notin K$, so $K \ddot{-} \psi \subseteq K \subseteq K \ddot{-} \phi$ by $(\dot{-}2)$ and $(\dot{-}3)$.

(iv) Let $\ddot{-}$ be an operation of severe withdrawal, and let $\phi \notin K \ddot{-} \phi$ and $\phi \in K \ddot{-} \psi$. By (ii), $\ddot{-}$ satisfies ($\dot{-}$8c). From $\phi \notin K \ddot{-}\phi$, we get $K \ddot{-} \phi \subseteq K \ddot{-}(\phi \wedge \psi)$, by ($\dot{-}$7a). First assume that $\psi \notin K \ddot{-} \psi$ (case 1). Then, by ($\dot{-}$7a), $\phi \in K \ddot{-} \psi \subseteq K \ddot{-}(\phi \wedge \psi)$, so $K \ddot{-}(\phi \wedge \psi) \subseteq K \ddot{-}\psi$, by ($\dot{-}$8c). Thus $K \ddot{-}\phi \subseteq K \ddot{-}\psi$, as desired. So assume secondly that $\psi \in K \ddot{-}\psi$ (case 2). Then, by ($\dot{-}5^0$), $K \subseteq K \ddot{-}\psi$. So $K \ddot{-}\phi \subseteq K \ddot{-}\psi$ by ($\dot{-}$2).

(v) Let $\ddot{-}$ be an operation of severe withdrawal that in addition satisfies ($\dot{-}$5). Put $\overline{R} = \{\chi : \chi \in K \ddot{-}\chi\}$. If ϕ is in $\overline{R}$, then $K \ddot{-}\phi = K$, by ($\dot{-}$2) and ($\dot{-}5^0$). So let ϕ be such that $\phi \notin K \ddot{-}\phi$. If ψ is in $\overline{R}$, then ψ is in $K \dot{-}(\phi \wedge \psi)$, by ($\dot{-}$02), and hence in $K \dot{-}\phi$, by ($\dot{-}$8c). So let ψ be such that $\psi \notin K \dot{-}\psi$. Suppose for reductio that ψ is in $K \dot{-}\phi$. Then, by ($\dot{-}$10), $K \dot{-}\psi \subseteq K \dot{-}\phi$. But by ($\dot{-}$1) and ($\dot{-}$5), $\psi \supset \phi$ is in $K \dot{-}\psi$ and thus in $K \dot{-}\phi$. Since ϕ is not in $K \dot{-}\phi$, we get, by ($\dot{-}$1), that ψ is not in $K \dot{-}\phi$, and we have a contradiction. We conclude that in fact $K \ddot{-}\phi = K$ if ϕ is in $\overline{R}$, and $K \ddot{-}\phi = \overline{R}$ otherwise.

Suppose there are sentences ϕ and ψ in K which are not taboo for $\ddot{-}$. Then $K \ddot{-}\phi = K \ddot{-}\psi = \overline{R}$. Moreover, by ($\dot{-}$1) and ($\dot{-}$5), $\phi \supset \psi$ and $\psi \supset \phi$, and thus $\phi \equiv \psi$, are in $\overline{R} = K \ddot{-}\phi = K \ddot{-}\psi$. $\qquad \square$

Lemma 66

Consider the following two conditions:

(†) $\phi \notin K \dot{-}(\phi \wedge \psi)$ and $\psi \in K \dot{-}(\phi \wedge \psi)$

(‡) $\phi \notin K \dot{-}\phi$ and $\psi \in K \dot{-}\phi$

If $\dot{-}$ satisfies ($\dot{-}$7c) and ($\dot{-}$8c), then (†) implies (‡). If $\dot{-}$ satisfies ($\dot{-}$7a) and ($\dot{-}$8c), then (‡) implies (†).

Proof of Lemma 66.

Let $\dot{-}$ satisfy ($\dot{-}$7c) and ($\dot{-}$8c), and let $\phi \notin K \dot{-}(\phi \wedge \psi)$ and $\psi \in K \dot{-}(\phi \wedge \psi)$. By, ($\dot{-}$7c) and ($\dot{-}$8c), it follows from the last condition that $K \dot{-}\phi = K \dot{-}(\phi \wedge \psi)$. Thus $\phi \notin K \dot{-}\phi$ and $\psi \in K \dot{-}\phi$.

For the converse, let $\dot{-}$ satisfy ($\dot{-}$7a) and ($\dot{-}$8c), and let $\psi \in K \dot{-}\phi$ and $\phi \notin K \dot{-}\phi$. It follows from the last condition and ($\dot{-}$7a) that $K \dot{-}\phi \subseteq K \dot{-}(\phi \wedge \psi)$. Hence $\psi \in K \dot{-}(\phi \wedge \psi)$. Thus, by ($\dot{-}$8c), also $K \dot{-}(\phi \wedge \psi) \subseteq K \dot{-}\phi$. Since $\phi \notin K \dot{-}\phi$, we get that also $\phi \notin K \dot{-}(\phi \wedge \psi)$, so we are done. $\qquad \square$

Observation 67

(i) Let $<$ satisfy Irreflexivity, (E&C) and (Ext) and let $<^* = \mathcal{P}_2(\ddot{\mathcal{C}}(<))$. Then we have $\phi <^* \psi$ if and only if $\psi \in K$ and (($\phi \notin K$ and $\phi \wedge \psi \not< \top$) or $\phi < \psi$).

If $<$ satisfies in addition (EFaith), then we have $\phi <^* \psi$ if and only if $\psi \in K$ and $\phi < \psi$.

If moreover $<$ satisfies (EE5'), then we have $\phi <^* \psi$ if and only if $\phi < \psi$.

(ii) Let $\ddot{-}$ satisfy ($\dot{-}$1) – ($\dot{-}$2), ($\dot{-}5^0$), ($\dot{-}$6), ($\dot{-}$7a) and ($\dot{-}$8c), and let $\ddot{-}^* = \ddot{\mathcal{C}}(\mathcal{P}_2(\ddot{-}))$. Then we have $\psi \in K \ddot{-}^*\phi$ if and only if $\psi \in K \ddot{-}\phi$.

Proof of Observation 67.

(i) Let $<$ satisfy Irreflexivity, (SV), (E&C) and (Ext) and let $\doteq = \ddot{\mathcal{C}}(<)$ and $<^* = \mathcal{P}_2(\dot{-})$. Then we have

$\phi <^* \psi$

iff $\qquad \psi \in K\dot{-}\phi \wedge \psi$ and $\phi \notin K\dot{-}\phi \wedge \psi$ $\qquad\qquad$ (by $<^* = \mathcal{P}_2(\dot{-})$)

iff *either* $\psi \in K\dot{-}\phi \wedge \psi$ and $\phi \notin K\dot{-}\phi \wedge \psi$ and $\phi \wedge \psi \not< \top$

$\qquad$ *or* $\psi \in K\dot{-}\phi \wedge \psi$ and $\phi \notin K\dot{-}\phi \wedge \psi$ and $\phi \wedge \psi < \top$

iff *either* $\psi \in K$ and $\phi \notin K$ and $\phi \wedge \psi \not< \top$

$\qquad$ *or* $\psi \in K$ and $\phi \wedge \psi < \psi$ and $(\phi \notin K$ or $\phi \wedge \psi \not< \phi)$

$\qquad\qquad\qquad\qquad\qquad\qquad\qquad\qquad$ (by $\doteq = \ddot{\mathcal{C}}(<)$ and (5))

iff *either* $\psi \in K$ and $\phi \notin K$ and $\phi \wedge \psi \not< \top$

$\qquad$ *or* $\psi \in K$ and $\phi < \psi$ and $(\phi \notin K$ or $\psi \not< \phi)$ $\qquad$ (by (E&C), (SV))

iff *either* $\psi \in K$ and $\phi \notin K$ and $\phi \wedge \psi \not< \top$

$\qquad$ *or* $\psi \in K$ and $\phi < \psi$ $\qquad\qquad\qquad\qquad\qquad$ (by Asymmetry)

iff $\qquad \psi \in K$ and $((\phi \notin K$ and $\phi \wedge \psi \not< \top)$ or $\phi < \psi)$

Let $<$ in addition satisfy (EFaith). If $\psi \in K$ and $\phi \notin K$, then by (EFaith) $\phi < \psi$, and the last condition reduces to $\psi \in K$ and $\phi < \psi$.

Let $<$ in addition satisfy (EE′). Then $\phi < \psi$ implies $\psi \in K$, and we finally get that $\phi <^* \psi$ if and only if $\phi < \psi$.

(ii) Let $\dot{-}$ satisfy $(\dot{-}1) - (\dot{-}2)$, $(\dot{-}5^0)$, $(\dot{-}6)$, $(\dot{-}7a)$ and $(\dot{-}8c)$, and let $< = \mathcal{P}_2(\dot{-})$ and $\dot{-}^* = \ddot{\mathcal{C}}(<)$. Then we have

$\psi \in K\dot{-}^*\phi$

iff *either* $\psi \in K$, and $\phi \not< \top$

$\qquad$ *or* $\psi \in K$ and $\phi < \psi$, and $\phi < \top$ $\qquad\qquad\qquad$ (by $\dot{-}^* = \ddot{\mathcal{C}}(<)$)

iff *either* $\psi \in K$, and $\top \notin K\dot{-}\phi \wedge \top$ or $\phi \in K\dot{-}\phi \wedge \top$

$\qquad$ *or* $\psi \in K$ and $\top \in K\dot{-}\phi \wedge \top$ and $\phi \notin K\dot{-}\phi \wedge \top$ and $\psi \in K\dot{-}(\phi \wedge \psi)$

$\qquad\qquad$ and $\phi \notin K\dot{-}(\phi \wedge \psi)$ $\qquad\qquad\qquad\qquad\qquad$ (by $< = \mathcal{P}_2(\dot{-})$)

iff *either* $\psi \in K$, and $\phi \in K\dot{-}\phi$

$\qquad$ *or* $\psi \in K$ and $\phi \notin K\dot{-}\phi$ and $\psi \in K\dot{-}(\phi \wedge \psi)$ and $\phi \notin K\dot{-}(\phi \wedge \psi)$

$\qquad\qquad\qquad\qquad\qquad\qquad\qquad\qquad$ (by $(\dot{-}1)$ and $(\dot{-}6)$)

iff   *either* $\psi \in K$, and $\phi \in K \dot{-} \phi$ and $K \dot{-} \phi = K$

      or $\psi \in K$ and $\phi \notin K \dot{-} \phi$ and $\psi \in K \dot{-} \phi$

$$\text{(by } (\dot{-}1), (\dot{-}2), (\dot{-}5^0), (\dot{-}7a) \text{ and } (\dot{-}8c))$$

iff   *either* $\phi \in K \dot{-} \phi$, and $\psi \in K \dot{-} \phi$

      or $\phi \notin K \dot{-} \phi$ and $\psi \in K \dot{-} \phi$ $\text{(by } (\dot{-}1), (\dot{-}2) \text{ and } (\dot{-}5^0))$

iff $\psi \in K \dot{-} \phi$.

Calculation of entrenchments and reconstructed contractions in the example given by Fig. 8.4.

Let $\dot{-}$ be defined as in Figure 8.4, and let $< = \mathcal{P}_2(\dot{-})$.

First of all, we have $\phi < \top$ for all ϕ in $K - Cn(\emptyset)$, since for every such ϕ, ϕ is not, but $\top$ is in $K \dot{-} (\phi \wedge \top) = K \dot{-} \phi$.

If $\phi \in Cn(\psi)$, then it can never be the case that $\phi < \psi$ because that would mean that ϕ is not, while ψ is in $K \dot{-} (\phi \wedge \psi)$ which is impossible by $(\dot{-}1)$. We now check all the other pairs in $K = Cn(\{p, q\})$, and take the following facts:

(1) $K \dot{-} (p \wedge q) = Cn(p \vee q)$
(2) $K \dot{-} p = Cn(\neg p \vee q)$
(3) $K \dot{-} q = Cn(p \vee \neg q)$
(4) $K \dot{-} (p \leftrightarrow q) = Cn(p \vee q)$

and use the Definition 17 of $\mathcal{P}_2(\dot{-})$, according to which $\phi < \psi$ iff ϕ is not, but ψ is in $K \dot{-} (\phi \wedge \psi)$. These facts settle the pairwise comparisons of the elements of K with respect to $<$. In the following table, we indicate which fact settles which comparison between beliefs in the negative. The only two positive cases are marked by a bullet and are settled by fact (1).

$<$	p	q	$p \leftrightarrow q$	$p \vee q$	$p \vee \neg q$	$\neg p \vee q$	$p \wedge q$
p	–	1	1	2	2	1	1
q	1	–	1	3	1	3	1
$p \leftrightarrow q$	1	1	–	•	4	4	1
$p \vee q$	2	3	1	–	2	3	1
$p \vee \neg q$	2	1	4	2	–	4	1
$\neg p \vee q$	1	3	4	3	4	–	1
$p \wedge q$	1	1	1	•	1	1	–

So what we essentially get in this example is $p \leftrightarrow q < p \vee q$ and $p \wedge q < p \vee q$, and nothing else.

Clearly, this relation $<$ satisfies Continuing Up, Conjunction Down and Conjunction Up, and it also satisfies Continuing Down. From the last, we conclude that $\mathcal{C}(\mathcal{D}(<)) = \mathcal{C}(<) = \dot{-}$.

Now let $\dot{-}' = \mathcal{C}(\mathcal{D}(<))$. We verify that
$K\dot{-}'(p \wedge q) = \{\phi \in K : \text{there is a } \psi \in Cn(p \wedge q) \text{ such that } \psi < (p \wedge q) \vee \phi\} = \{\phi \in K : \text{there is a } \psi \in Cn(p \wedge q) \text{ such that } \psi < \phi\} = Cn(p \vee q)$,
while
$K\dot{-}'p = \{\phi \in K : \text{there is a } \psi \in Cn(p) \text{ such that } \psi < p \vee \phi\} = Cn(\neg p \vee q)$
and
$K\dot{-}'q = \{\phi \in K : \text{there is a } \psi \in Cn(q) \text{ such that } \psi < q \vee \phi\} = Cn(p \vee \neg q)$
The contraction function $\dot{-}'$ thus violates ($\dot{-}8$vwd), as did the original $\dot{-}$. $\square$

Observation 68

(i) For every logically admissible entrenchment relation $<$, i.e., for every preference relation that satisfies Irreflexivity, Extensionality and (E&C), the contraction function $\dot{-} = \mathcal{C}(<)$ satisfies ($\dot{-}1$), ($\dot{-}2$), ($\dot{-}5$) and ($\dot{-}6$). If $<$ in addition

$$
\text{satisfies} \left\{ \begin{array}{c} Minimality\ (EE5') \\ EFaith\ wrt\ K \\ Maximality \\ Continuing\ Down\ (EE2^{\downarrow}) \\ Continuing\ Up\ (EE2^{\uparrow}) \\ (EII) \\ (EII^{+}) \\ (EIV) \end{array} \right\} \text{, then } \dot{-} \text{ satisfies} \left\{ \begin{array}{c} (\dot{-}2) \\ (\dot{-}3) \\ (\dot{-}4) \\ (\dot{-}7) \\ (\dot{-}8c) \\ (\dot{-}8wd) \\ (\dot{-}8d) \\ (\dot{-}8) \end{array} \right\}.
$$

(ii) For every contraction function $\dot{-}$ that satisfies ($\dot{-}1$) and ($\dot{-}6$), the preference relation $< = \mathcal{P}_2(\dot{-})$ satisfies Irreflexivity, Extensionality and (E&C).

$$
\text{If } \dot{-} \text{ in addition satisfies} \left\{ \begin{array}{c} (\dot{-}2) \\ (\dot{-}3) \\ (\dot{-}4) \\ (\dot{-}7) \\ (\dot{-}8c) \\ (\dot{-}7)\ and\ (\dot{-}8c) \\ (\dot{-}8wd) \\ (\dot{-}8d) \\ (\dot{-}8) \end{array} \right\} \text{, then } < \text{ satisfies}
$$

$$
\left\{ \begin{array}{c} Minimality\ (EE5') \\ EFaith\ wrt\ K \\ Maximality \\ Continuing\ Down\ (EE2^{\downarrow}) \\ Continuing\ Up\ (EE2^{\uparrow}) \\ Transitivity \\ (EII) \\ (EII^{+}) \\ (EIV) \end{array} \right\}.
$$

Proof of Observation 68.

(i) Let $<$ satisfy Irreflexivity, Extensionality and (E&C), and let $\dot{-} = \mathcal{C}(<)$. Postulates $(\dot{-}2)$ and $(\dot{-}6)$ are trivial by construction and the Extensionality of $<$.

For $(\dot{-}5)$ we have to show that for $\psi \in K$, we have $\phi \supset \psi$ is in $K \dot{-} \phi$ which means by definition that either $\phi < \phi \vee (\phi \supset \psi)$ or $\phi \not< \top$. But by Extensionality, the former condition just means that $\phi < \top$, so one of the two cases must apply.

The most interesting postulate here is $(\dot{-}1)$. In order to show that K is a theory, we use the compactness of Cn and show that if $\psi_1, \ldots, \psi_m$ are in K and $\{\psi_1, \ldots, \psi_m\} \vdash \psi$, then ψ is in K as well. So let $\psi_1, \ldots, \psi_m$ and ψ be as just stated. The limiting case $\phi \not< \top$ is trivial and can be neglected. In the principal case $\phi < \top$ we have for all i that ψ_i is in K and $\phi < \phi \vee \psi_i$. Clearly, ψ is in K, since ψ follows from $\psi_1, \ldots, \psi_m$ and all ψ_i's are in the theory K. By repeated application of Weak Conjunction Up (see Lemma 55), we get that $\phi < (\phi \vee \psi_1) \wedge \cdots \wedge (\phi \vee \psi_m)$, or equivalently, $\phi < \phi \vee (\psi_1 \wedge \cdots \wedge \psi_m)$. But now we can apply Weak Continuing Up (see again Lemma 55), thus we get that $\phi < \phi \vee \psi$, and we are done.

Now let $<$ in satisfy (Continuing Down). For $(\dot{-}7)$, suppose that χ is in $K \dot{-} \phi \cap K \dot{-} \psi$. We have to show that χ is in $K \dot{-}(\phi \wedge \psi)$. Our assumption means that χ is in K, and that both

$$\phi < \phi \vee \chi \text{ or } \phi \not< \top$$

and

$$\psi < \psi \vee \chi \text{ or } \psi \not< \top$$

We need to show that χ is in K and

$$\phi \wedge \psi < (\phi \wedge \psi) \vee \chi \text{ or } \phi \wedge \psi \not< \top$$

Case 1. Suppose that $\phi < \phi \vee \chi$ and $\psi < \psi \vee \chi$. Then, by Continuing Down, $\phi \wedge \psi < \phi \vee \chi$ and $\phi \wedge \psi < \psi \vee \chi$. Therefore, by Weak Conjunction Up (see Lemma 55), we also have $\phi \wedge \psi < (\phi \vee \chi) \wedge (\psi \vee \chi)$, and Extensionality helps us to realize that we are done.

Case 2. Suppose that $\phi < \phi \vee \chi$ and $\psi \not< \top$. If $\phi \not< \top$ as well, we can use (E$\emptyset$2) to derive $\phi \wedge \psi \not< \top$, and we are done. If on the other hand $\phi < \top$, then we can use the Top Equivalence condition (7) and get $\phi < \psi$ or equivalently, by Conjunctiveness, $\phi \wedge \psi < \psi$. We can also use Continuing Down to infer from $\phi < \phi \vee \chi$ that $\phi \wedge \psi < \phi \vee \chi$. Now we can use Weak Conjunction Up and get $\phi \wedge \psi < (\phi \vee \chi) \wedge \psi$. Finally, we can apply Weak Continuing Up (see Lemma 55), and get $\phi \wedge \psi < (\phi \wedge \psi) \vee \chi$, as desired.

Case 3, $\psi < \psi \vee \chi$ and $\phi \not< \top$, is similar.

Case 4. Suppose that $\phi \not< \top$ and $\psi \not< \top$. Then $\phi \wedge \psi \not< \top$, by (E02), and we are done.

Next let $<$ in satisfy (Continuing Up).

For ($\dot{-}$8c), suppose that ψ is in $K\dot{-}(\phi \wedge \psi)$ and that χ is in $K\dot{-}(\phi \wedge \psi)$. We have to show that χ is in $K\dot{-}\phi$. Our assumptions mean that ψ is in K, and that

$$\phi \wedge \psi < (\phi \wedge \psi) \vee \psi \text{ or } \phi \wedge \psi \not< \top$$

that χ is in K, and that

$$\phi \wedge \psi < (\phi \wedge \psi) \vee \chi \text{ or } (\phi \wedge \psi) \not< \top$$

We need to show that χ is in K and

$$\phi < \phi \vee \chi \text{ or } \phi \not< \top$$

Case 1. Suppose that $\phi \wedge \psi \not< \top$. Then, by (E$\emptyset$1), $\phi \not< \top$, and we are done.

So look at case 2 where $\phi \wedge \psi < \top$. What we have is that $\psi, \chi \in K$. By Extensionality, we further get that $\phi \wedge \psi < \psi$. We also have that $\phi \wedge \psi < (\phi \wedge \psi) \vee \chi$. So by Weak Conjunction Up, we can conclude that $\phi \wedge \psi < ((\phi \wedge \psi) \vee \chi) \wedge \psi$, or equivalently. $\phi \wedge \psi < (\phi \wedge \psi) \vee (\psi \wedge \chi)$. Now we can use Conditionalization (see Lemma 55) and get

$$((\phi \wedge \psi) \vee (\psi \wedge \chi)) \supset (\phi \wedge \psi) < (\phi \wedge \psi) \vee (\psi \wedge \chi)$$

which reduces to

$$(\psi \wedge \chi) \supset \phi < (\phi \wedge \psi) \vee (\psi \wedge \chi)$$

Now we can check that all conditions for applying Weak Continuing Down (see again Lemma 55) are satisfied, and so we get that $\phi < (\phi \wedge \psi) \vee (\psi \wedge \chi)$. Finally, we can use Continuing Up und end up with $\phi < \phi \vee \chi$, as desired, and we are done.

Now let $<$ in satisfy (EII).

For ($\dot{-}$8wd), suppose that χ is in $K\dot{-}(\phi \wedge \psi)$. We have to show that either χ is in $(K\dot{-}\phi) \cup \{\neg\psi\}$ or χ is in $(K\dot{-}\psi) \cup \{\neg\phi\}$. Since $\dot{-}$ satisfies ($\dot{-}$1), this means that we have to show that either $\psi \vee \chi$ is in $K\dot{-}\phi$ or $\phi \vee \chi$ is in $K\dot{-}\psi$.

Our assumption means that χ is in K and

$$\phi \wedge \psi < (\phi \wedge \psi) \vee \chi \text{ or } \phi \wedge \psi \not< \top$$

We know by the closure of K that both $\phi \vee \chi$ and $\psi \vee \chi$ are in K. If $\phi \wedge \psi \not< \top$, then by (E$\emptyset$1) both $\phi \not< \top$ and $\psi \not< \top$, and we are done. So suppose that

$$\phi \wedge \psi < (\phi \wedge \psi) \vee \chi$$

By Weak Continuing Up, we get that

$$\phi \wedge \psi < \phi \vee \psi \vee \chi$$

So by (EII) it follows that either $\phi < \phi \vee \psi \vee \chi$ or $\psi < \phi \vee \psi \vee \chi$ But this is just means that either $\psi \vee \chi$ is in $K\dot{-}\phi$ or $\phi \vee \chi$ is in $K\dot{-}\psi$, as desired.

Now let $<$ in satisfy (EII$^+$).

For ($\dot{-}$8d), suppose that χ is in $K\dot{-}(\phi \wedge \psi)$. We have to show that χ is either in $K\dot{-}\phi$ or in $K\dot{-}\psi$. Our assumption means that χ is in K and

$$\phi \wedge \psi < (\phi \wedge \psi) \vee \chi \text{ or } \phi \wedge \psi \not< \top$$

If $\phi \wedge \psi \not< \top$, then by (E∅1) both $\phi \not< \top$ and $\psi \not< \top$, and we are done. Now suppose that $\phi < \top$, $\psi < \top$ and

$$\phi \wedge \psi < (\phi \wedge \psi) \vee \chi$$

The latter is equivalent, by Extensionality, to

$$\phi \wedge \psi < (\phi \vee \chi) \wedge (\psi \vee \chi)$$

For ($\dot{-}$8d), it remains to show that either $\phi < \phi \vee \chi$ or $\psi < \psi \vee \chi$.

Since we have $\phi < \top$, we can apply Extensionality to get $(\phi \vee \chi) \wedge (\phi \vee \neg \chi) < \top$. Then (E∅2) gives us

$$\phi \vee \chi < \top \text{ or } \phi \vee \neg \chi < \phi \vee \chi$$

The last case gives us, with the help of (E&C), $\phi < \phi \vee \chi$, and we are ready. So suppose that $\phi \vee \chi < \top$.

Similarly, since we have $\psi < \top$, we can apply Extensionality to get $(\psi \vee \chi) \wedge (\psi \vee \neg \chi) < \top$. Then (E∅2) gives us

$$\psi \vee \chi < \top \text{ or } \psi \vee \neg \chi < \psi \vee \chi$$

The last case gives us, with the help of (E&C), $\psi < \psi \vee \chi$, and we are ready. So suppose that $\psi \vee \chi < \top$.

So we have $\phi \wedge \psi < (\phi \vee \chi) \wedge (\psi \vee \chi)$ and both $\phi \vee \chi < \top$ and $\psi \vee \chi < \top$. By (EII$^+$), it follows that either $\phi < \phi \vee \chi$ or $\psi < \psi \vee \chi$, as desired.

Now let $<$ in satisfy (EIV).

For ($\dot{-}$8), suppose that ϕ is not in $K\dot{-}(\phi \wedge \psi)$ and that χ is in $K\dot{-}(\phi \wedge \psi)$. We have to show that χ is in $K\dot{-}\phi$. Our assumptions mean that either ϕ is not in K or

$$\phi \wedge \psi \not< (\phi \wedge \psi) \vee \phi \text{ and } \phi \wedge \psi < \top$$

and that χ is in K and

$$\phi \wedge \psi < (\phi \wedge \psi) \vee \chi \text{ or } (\phi \wedge \psi) \not< \top$$

We need to show that χ is in K and

$$\phi < \phi \vee \chi \text{ or } \phi \not< \top$$

Case 1. Suppose that ϕ is not in K. As χ is in K, $\phi \vee \chi$ is in K as well, so we can apply (EFaith) and get $\phi < \phi \vee \chi$, as desired.

Case 2. Suppose that ϕ is in K. What we have is $\phi \wedge \psi \not< \phi$, $\phi \wedge \psi < \top$ and $\phi \wedge \psi < (\phi \wedge \psi) \vee \chi$. Since the latter means $\phi \wedge \psi < (\phi \vee \chi) \wedge (\psi \vee \chi)$, we can apply Weak Continuing Up and get $\phi \wedge \psi < \phi \vee \chi$. On the other hand, by (E&C) and (SV), $\phi \wedge \psi \not< \phi$ is equivalent with $\psi \not< \phi$, which by Extensionality is the same as $\psi \not< \phi \wedge (\phi \vee \chi)$. Now we can apply (EIV) and get $\phi < \phi \vee \chi$, as desired, and thus we are done.

(ii) Let $\dot{-}$ satisfy postulates $(\dot{-}1)$ and $(\dot{-}6)$, and let $< = \mathcal{P}_2(\dot{-})$.

Irreflexivity is trivial, since both $\phi \in K\dot{-}(\phi \wedge \phi)$ and $\phi \notin K\dot{-}(\phi \wedge \phi)$ at the same time is impossible.

Extensionality is trivial from $(\dot{-}1)$ and $(\dot{-}6)$.

For (E&C), let $\phi < \psi \wedge \chi$. This means that $\psi \wedge \chi$ is, but ϕ is not in $K\dot{-}\phi \wedge (\psi \wedge \chi)$. By $(\dot{-}1)$ and $(\dot{-}6)$, this implies on the one hand that χ is, but $\phi \wedge \psi$ is not in $K\dot{-}(\phi \wedge \psi) \wedge \chi$, and on the other hand that ψ is, but $\phi \wedge \chi$ is not in $K\dot{-}(\phi \wedge \chi) \wedge \psi$. That is, by definition, $\phi \wedge \psi < \chi$ and $\phi \wedge \chi < \psi$. The converse direction is equally straightforward.

Now let $\dot{-}$ satisfy $(\dot{-}2)$. Let $\phi < \psi$, which means that ϕ is not, but ψ is in $K\dot{-}(\phi \wedge \psi)$. By $(\dot{-}2)$, then, ψ is in K, as desired for (EE5′).

Now let $\dot{-}$ satisfy $(\dot{-}3)$, and let ϕ, but not ψ be in K. For (EFaith), we need to show that $\psi < \phi$, that is, that ϕ is, but ψ is not in $K\dot{-}(\phi \wedge \psi)$. But by $(\dot{-}1)$, $\phi \wedge \psi$ is not in K, so by $(\dot{-}3)$, $K\dot{-}(\phi \wedge \psi) = K$, and we are done.

Now let $\dot{-}$ satisfy $(\dot{-}4)$, and let $\nvdash \phi$. For (Maximality), we have to show that $\phi < \top$, that is that ϕ is not, but $\top$ is in $K\dot{-}(\phi \wedge \top)$. The latter belief set is identical with $K\dot{-}\phi$, by $(\dot{-}6)$. Moreover, $\top$ is in every belief set, by $(\dot{-}1)$. It remains to show that $\phi \notin K\dot{-}\phi$. But this we get from $(\dot{-}4)$, and we are done.

Now let $\dot{-}$ satisfy $(\dot{-}5)$ and $(\dot{-}7)$. For (Continuing Down), assume that $\phi < \psi$ and $\chi \vdash \phi$. We have to show that $\chi < \psi$. Our assumption means that ϕ is not, but ψ is in $K\dot{-}(\phi \wedge \psi)$. By $(\dot{-}5)$ and $(\dot{-}1)$, we know that ψ is also in $K\dot{-}(\psi \supset \chi)$. Therefore, by $(\dot{-}7)$ and $(\dot{-}6)$, ψ is in $K\dot{-}(\phi \wedge \psi \wedge \chi)$, which means, by $(\dot{-}6)$, that ψ is in $K\dot{-}(\psi \wedge \chi)$. It remains to show that χ is not in $K\dot{-}(\psi \wedge \chi)$. Suppose it were. Then by $(\dot{-}1)$, ϕ is in $K\dot{-}(\psi \wedge \chi)$ as well. So $\phi \wedge \psi \wedge \chi$ is in $K\dot{-}\psi \wedge \chi = K\dot{-}(\phi \wedge \psi \wedge \chi)$. But then, by $(\dot{-}8rc)$ which is a limiting condition that we always adopt, we also have that $\phi \wedge \psi$ is in $K\dot{-}(\phi \wedge \psi)$, contradicting our assumption. So $\chi \notin K\dot{-}(\psi \wedge \chi)$, and we are done.

Now let $\dot{-}$ satisfy $(\dot{-}8c)$. For (Continuing Up), assume that $\phi < \psi$ and $\psi \vdash \chi$. We have to show that $\phi < \chi$. Our assumption means that ϕ is not in, but ψ is in $K\dot{-}(\phi \wedge \psi)$. By $(\dot{-}6)$, we have that ψ is in $K\dot{-}(\phi \wedge \psi) = K\dot{-}(\phi \wedge \psi \wedge \chi)$. Hence, by $(\dot{-}8c)$, $K\dot{-}(\phi \wedge \psi \wedge \chi) \subseteq K\dot{-}(\phi \wedge \chi)$, and thus ψ is in $K\dot{-}(\phi \wedge \chi)$, and also χ is in $K\dot{-}(\phi \wedge \chi)$, by $(\dot{-}1)$. It remains to show that ϕ is not in $K\dot{-}(\phi \wedge \chi)$. Suppose for reductio that it is. Then $\phi \wedge \chi$ is in $K\dot{-}(\phi \wedge \chi)$, and by $(\dot{-}7r)$ we get that $\phi \wedge \chi$ is in $K\dot{-}(\phi \wedge \psi \wedge \chi)$, contradicting $\phi \notin K\dot{-}(\phi \wedge \psi \wedge \chi)$, what we had before. So we have a contradiction, and we are done.

Now let $\dot{-}$ satisfy $(\dot{-}5)$, $(\dot{-}7)$ and $(\dot{-}8c)$. Since we just showed that under these conditions, $<$ satisfies (E&C), (Continuing Up) and (Continuing Down), we can rely on Lemma 52(vii), which says that these properties entail transitivity.

Now let $\dot{-}$ satisfy $(\dot{-}8wd)$. For (EII), assume that $\phi \wedge \psi < \chi$. We have to show that either $\phi < \chi$ or $\psi < \chi$.

Our assumption means that $\phi \wedge \psi$ is not in, but χ is in $K \dot{-} (\phi \wedge \psi \wedge \chi)$.

By $(\dot{-}8wd)$, we have that χ is either in $Cn((K \dot{-} \phi \wedge \chi) \cup \{\neg(\psi \wedge \chi)\})$ or in $Cn((K \dot{-} \psi \wedge \chi) \cup \{\neg(\phi \wedge \chi)\})$. Since all contracted belief sets are closed, it follows that either $(\psi \wedge \chi) \vee \chi \in K \dot{-} (\phi \wedge \chi)$ or $(\phi \wedge \chi) \vee \chi \in K \dot{-} (\psi \wedge \chi)$. This is equivalent with the fact that either $\chi \in K \dot{-} (\phi \wedge \chi)$ or $\chi \in K \dot{-} (\psi \wedge \chi)$.

Assume without loss of generality that $\chi \in K \dot{-} (\phi \wedge \chi)$; the other case is similar.

Case 1. If $\phi \notin K \dot{-} (\phi \wedge \chi)$, then we have shown that $\phi < \chi$, and we are done.

Case 2. If $\phi \in K \dot{-} (\phi \wedge \chi)$, then we have $\phi \wedge \chi \in K \dot{-} (\phi \wedge \chi)$, so by $(\dot{-}8rc)$ and $(\dot{-}7r)$, $\chi \in K \dot{-} \chi$ and $\chi \in K \dot{-} (\psi \wedge \chi)$. For $\psi < \chi$, it remains to show that $\psi \notin K \dot{-} (\psi \wedge \chi)$. But if ψ were in $K \dot{-} (\psi \wedge \chi)$ as well, then by $(\dot{-}7r)$ we would get from $\phi \wedge \chi \in K \dot{-} (\phi \wedge \chi)$ and $\psi \wedge \chi \in K \dot{-} (\psi \wedge \chi)$, that $\phi \wedge \psi \wedge \chi \in K \dot{-} (\phi \wedge \psi \wedge \chi)$, contradicting the assumption that $\phi \wedge \psi$ is not in $K \dot{-} (\phi \wedge \psi \wedge \chi)$.

Now let $\dot{-}$ satisfy $(\dot{-}8d)$. For (EII$^+$), assume that $\phi \wedge \psi < \chi \wedge \xi$ and $\chi < \top$ and $\xi < \top$. We have to show that either $\phi < \chi$ or $\psi < \chi$.

Our main assumption means that $\phi \wedge \psi$ is not in, but $\chi \wedge \xi$ is in $K \dot{-} (\phi \wedge \psi \wedge \chi \wedge \xi)$. By $(\dot{-}8d)$, this implies that $\chi \wedge \xi$ is either in $K \dot{-} (\phi \wedge \chi)$ or in $K \dot{-} (\psi \wedge \xi)$. Assume, without loss of generality, that $\chi \wedge \xi$ is in $K \dot{-} (\phi \wedge \chi)$; the other case is similar. So by $(\dot{-}1)$, χ is in $K \dot{-} (\phi \wedge \chi)$. For $\phi < \chi$, it remains to show that ϕ is not in $K \dot{-} (\phi \wedge \chi)$. Suppose for reductio that ϕ is in this set. Then, by $(\dot{-}1)$, $\phi \wedge \chi$ is in $K \dot{-} (\phi \wedge \chi)$. and by $(\emptyset 1)$, χ is in $K \dot{-} \chi$, and we have a contradiction with $\chi < \top$.

Now let $\dot{-}$ satisfy $(\dot{-}8)$.

For (EIV), assume that $\phi \wedge \psi < \chi$ and $\psi \not< \phi \wedge \chi$. We have to show that $\phi < \chi$.

The first assumption means that $\phi \wedge \psi$ is not in, but χ is in $K \dot{-} (\phi \wedge \psi \wedge \chi)$. The second assumption means that either ψ is, or $\phi \wedge \chi$ is not in $K \dot{-} (\phi \wedge \psi \wedge \chi)$. By $(\dot{-}1)$, the two assumptions taken together are equivalent to saying that ϕ is not, but χ is in $K \dot{-} (\phi \wedge \psi \wedge \chi)$. By $(\dot{-}1)$ again, it follows that $\phi \wedge \chi$ is not in $K \dot{-} (\phi \wedge \psi \wedge \chi)$. Hence, by $(\dot{-}8)$, we get that $K \dot{-} (\phi \wedge \psi \wedge \chi) \subseteq K \dot{-} (\phi \wedge \chi)$ and that χ is in $K \dot{-} (\phi \wedge \chi)$. It remains to show that ϕ is not in $K \dot{-} (\phi \wedge \chi)$. But if it were, then $\phi \wedge \chi$ would be in $K \dot{-} (\phi \wedge \chi)$, and $(\dot{-}7r)$ would give us that $\phi \wedge \chi$ is in $K \dot{-} (\phi \wedge \chi)$, and we have a contradiction. $\square$

Observation 69

Let $<$ be an entrenchment relation satisfying Extensionality, (E&C), (EI) and (EIII). Then $\mathcal{R}(\mathcal{C}(<)) = \mathcal{R}(\ddot{\mathcal{C}}(<))$.

Proof of Observation 69.

Let $<$ satisfy Extensionality, (E&C), (EI) and (EIII), and let $\dot{-} = \mathcal{C}(<)$ and $\ddot{-} = \ddot{\mathcal{C}}(<)$. First we note that the entrenchment conditions guarantee that $K\dot{-}\phi$ and $K\ddot{-}\phi$ are theories. We show this for $K\ddot{-}\phi$. (The case for $K\dot{-}\phi$ is a bit more complicated, but essentially similar.) Let $\psi \in Cn(K\ddot{-}\phi)$. We want to show that $\psi \in K\ddot{-}\phi$, i.e., in the principal case, that $\phi < \psi$. From $\psi \in Cn(K\ddot{-}\phi)$ we know that there are $\psi_1, \ldots, \psi_m$ in $K\ddot{-}\phi$ such that $\psi \in Cn(\{\psi_1 \wedge \cdots \wedge \psi_m\})$. That ψ_j is in $K\ddot{-}\phi$ means that $\phi < \psi_j$, for all j. By (EI), $\phi \wedge \psi_1 < \psi_2$ and $\phi \wedge \psi_2 < \psi_1$, by (E&C) $\phi < \psi_1 \wedge \psi_2$. By repeated application of this method, we get that $\phi < \psi_1 \wedge \cdots \wedge \psi_m$. Hence, by (EIII) and Extensionality, $\phi < \psi$, as desired.

Now let $* = \mathcal{R}(\mathcal{C}(<))$ and $\ddot{*} = \mathcal{R}(\ddot{\mathcal{C}}(<))$. Then we have

$\psi \in K * \phi$

> iff $\psi \in Cn((K\dot{-}\neg\phi) \cup \{\phi\})$
>
> iff $\phi \supset \psi \in K\dot{-}\neg\phi$ ($K\dot{-}\neg\phi$ is closed under Cn)
>
> iff $\phi \supset \psi \in K$ and ($\neg\phi \not< \top$ or $\neg\phi < \neg\phi \vee (\phi \supset \psi)$ ($\dot{-} = \mathcal{C}(<)$)
>
> iff $\phi \supset \psi \in K$ and ($\neg\phi \not< \top$ or $\neg\phi < \phi \supset \psi$)   (Extensionality)
>
> iff $\phi \supset \psi \in K\ddot{-}\neg\phi$ ($\ddot{-} = \mathcal{C}(<)$)
>
> iff $\psi \in Cn((K\ddot{-}\neg\phi) \cup \{\phi\})$ ($K\ddot{-}\neg\phi$ is closed under Cn)
>
> iff $\psi \in K\ddot{*}\phi$ $\square$

Observation 70

Let R be a binary relation over L. Then R can be extended to an entrenchment relation iff it satisfies the following condition of *entrenchment consistency*:

(EC)   There is no finite non-empty set of pairs $\langle\phi_1, \psi_1\rangle, \ldots, \langle\phi_n, \psi_n\rangle$ in R such that $\{\phi_1, \ldots, \phi_n\} \subseteq Cn(\{\psi_1, \ldots, \psi_n\})$.

Proof of Observation 70.

Let R be a binary relation over L. Clearly, if R is extensible to a GEE-relation $<$, then it must satisfy (EC). For suppose there is a finite set $\{\langle\phi_1, \psi_1\rangle, \ldots, \langle\phi_n, \psi_n\rangle\} \subseteq R \subseteq <$ such that $\psi_1 \wedge \cdots \wedge \psi_n \vdash \phi_1 \wedge \cdots \wedge \phi_n$. Then Lemma 48 (iii) and (EE2$^\top$) give us $\phi_1 \wedge \cdots \wedge \phi_n < \phi_1 \wedge \cdots \wedge \phi_n$, contradicting (EE1).

For the converse direction, suppose that R satisfies (EC). Now define two operators Up and Down on binary relations P over L as follows (compare the conditions (EE$^\top$) and (EE$^\downarrow$) on p. 237):

> $\text{Up}(P) = \{\langle\phi, \psi\rangle : \text{there are pairs } \langle\phi, \chi_1\rangle, \ldots, \langle\phi, \chi_m\rangle \in P \text{ such that}$
> $\chi_1, \ldots, \chi_m \vdash \psi\}$
>
> $\text{Down}(P) = \{\langle\phi, \psi\rangle : \text{there is a pair } \langle\chi, \psi\rangle \in P \text{ such that } \phi, \psi \vdash \chi\}$

Then define $R_0 = R$, $R_{2i+1} = \text{Up}(R_{2i})$, $R_{2i+2} = \text{Down}(R_{2i+1})$, and $<= \bigcup_{i=1,2,3,\ldots} R_i$.

Now clearly, $R \subseteq <$, since $R_i \subseteq R_{i+1}$ for every i. Moreover, $<$ satisfies (EE$^\uparrow$) and (EE$^\downarrow$), by construction and the finitary character of (EE$^\uparrow$). Finally, for (EE1), suppose for reductio that $\phi < \phi$. Then $\langle \phi, \phi \rangle \in R_i$ for some i. This R_i then violates (EC). But as R satisfies (EC), this contradicts the fact that both Up and Down preserve (EC), as we are now going to show.

First, suppose that a relation P satisfies (EC), but Up(P) does not. The latter means that there is a finite set $\{\langle \phi_1, \psi_1 \rangle, \ldots, \langle \phi_n, \psi_n \rangle\} \subseteq \text{Up}(P)$ such that $\psi_1 \wedge \cdots \wedge \psi_n \vdash \phi_1 \wedge \cdots \wedge \phi_n$. By the definition of Up(P), then, there are finite sets $H_1, \ldots, H_n$ such that $H_i \vdash \psi_i$ and $\langle \phi_i, \chi \rangle \in P$ for every χ which is in some H_i. But since $\bigwedge H_1 \wedge \cdots \wedge \bigwedge H_n \vdash \psi_1 \wedge \cdots \wedge \psi_n \vdash \phi_1 \wedge \cdots \wedge \phi_n$, we have a violation of (EC) in P, contradicting the assumption that P satisfies (EC).

Second, suppose that a relation P satisfies (EC), but Down(P) does not. The latter means that there is a finite set $\{\langle \phi_1, \psi_1 \rangle, \ldots, \langle \phi_n, \psi_n \rangle\} \subseteq \text{Down}(P)$ such that $\psi_1 \wedge \cdots \wedge \psi_n \vdash \phi_1 \wedge \cdots \wedge \phi_n$. By the definition of Down(P), then, there are sentences $\chi_1, \ldots, \chi_n$ such that $\phi_i \vdash \psi_i \supset \chi_i$ and $\langle \chi_i, \psi_i \rangle \in P$ for every i. But then $\psi_1 \wedge \cdots \wedge \psi_n \vdash (\psi_1 \supset \chi_1) \wedge \cdots \wedge (\psi_n \supset \chi_n)$, and hence $\psi_1 \wedge \cdots \wedge \psi_n \vdash \chi_1 \wedge \cdots \wedge \chi_n$, contradicting the assumption that P satisfies (EC).   □

Note incidentally that $<$ as constructed above is the smallest GEE-relation extending R.

Lemma 71

(i) A pair $\langle \phi, \psi \rangle$ is inconsistent with an entrenchment relation $<$ if and only if $\phi \in Cn(\psi)$ or $\psi < \psi \supset \phi$.

(ii) If $<$ is a relation that satisfies (EC) and $\phi \wedge \psi$ is not in $Cn(\emptyset)$, then either $\langle \phi, \psi \rangle$ or $\langle \psi, \psi \supset \phi \rangle$ is consistent with $<$.

(iii) If $<$ is a relation that satisfies (EC) and

(**Completeness**)   For every pair of sentences ϕ and ψ such that $\phi \wedge \psi$ is not in $Cn(\emptyset)$, either $\phi < \psi$ or $\psi < \psi \supset \phi$.

then $<$ is an entrenchment relation with full comparability.

(iv) For a logically admissible relation $<$, (Completeness) is equivalent with

(**Maxichoice**)   For every pair of sentences ϕ and ψ such that ϕ is not in $Cn(\emptyset)$, either $\phi < \phi \vee \psi$ or $\phi < \phi \vee \neg\psi$.

Proof of Lemma 71.

(i) Clearly, a pair $\langle \phi, \psi \rangle$ is inconsistent with a GEE-relation $<$ if $\psi < \psi \supset \phi$. This is because every GEE-relation with $\psi < \psi \supset \phi$ and $\phi < \psi$ would would satisfy $\psi \wedge \phi < ((\psi \supset \phi) \wedge \psi)$, by Lemma 48 (iii), so $\psi \wedge \phi < \psi \wedge \phi$, by (EE2$^\uparrow$), contradicting (EE1).

Now suppose for the converse that $\langle \phi, \psi \rangle$ is inconsistent with $<$. Since $<$ is a GEE-relation, it satisfies (EC). Since $< \cup \langle \phi, \psi \rangle$ is inconsistent, however, this relation does not satisfy (EC). Hence there are $\chi_1, \ldots, \chi_n, \xi_1, \ldots, \xi_n$ such that

$\chi_i < \xi_i$ for every i and $\xi_1 \wedge \cdots \wedge \xi_n \wedge \psi \vdash \chi_1 \wedge \cdots \wedge \chi_n \wedge \phi$. If n is 0, then $\psi \vdash \phi$, i.e., $\phi \in Cn(\psi)$. If n is greater than 0, application of Lemma 48 (iii) gives us $\chi_1 \wedge \cdots \wedge \chi_n < \xi_1 \wedge \cdots \wedge \xi_n$. Hence, by $(EE^{\downarrow})$, $((\xi_1 \wedge \cdots \wedge \xi_n) \supset (\chi_1 \wedge \cdots \wedge \chi_n)) < \xi_1 \wedge \cdots \wedge \xi_n$. But on the one hand, $\psi \vdash (\xi_1 \wedge \cdots \wedge \xi_n) \supset (\chi_1 \wedge \cdots \wedge \chi_n)$. So by $(EE2^{\downarrow})$, $\psi < \xi_1 \wedge \cdots \wedge \xi_n$. And on the other hand, $\xi_1 \wedge \cdots \wedge \xi_n \vdash \psi \supset \phi$. So by $(EE2^{\uparrow})$, $\psi < \psi \supset \phi$, as desired.

(ii) Let the relation $<$ satisfy (EC). Suppose for reductio that neither $\langle \phi, \psi \rangle$ nor $\langle \psi, \psi \supset \phi \rangle$ is consistent with $<$. Then there are $\phi_1, \ldots, \phi_n$ and $\psi_1, \ldots, \psi_n$, $\phi'_1, \ldots, \phi'_m$ and $\psi'_1, \ldots, \psi'_m$ such that for all i and j, $\phi_i < \psi_i$ and $\phi'_j < \psi'_j$ and

$$\bigwedge \psi_i \wedge \psi \vdash \bigwedge \phi_i \wedge \phi \tag{$\dagger$}$$

and

$$\bigwedge \psi'_j \wedge (\psi \supset \phi) \vdash \bigwedge \phi'_j \wedge \psi \tag{$\ddagger$}$$

Condition ($\dagger$) can be rewritten as

$$\bigwedge \psi_i \vdash (\psi \supset \bigwedge \phi_i) \wedge (\psi \supset \phi)$$

and ($\ddagger$) can be rewritten as

$$\bigwedge \psi'_j \vdash ((\psi \supset \phi) \supset \bigwedge \phi'_j) \wedge ((\psi \supset \phi) \supset \psi)$$

Taking these two together, it follows that

$$\bigwedge \psi_i \wedge \bigwedge \psi'_j \vdash \bigwedge \phi_i \wedge \bigwedge \phi'_j$$

Since we have assumed that $<$ satisfies (EC), both $\{\psi_i\}$ and $\{\psi'_j\}$ must be empty. But in that case ($\dagger$) and ($\ddagger$) simplify to $\psi \vdash \phi$ and $\psi \supset \phi \vdash \psi$ what can only be the case if $\vdash \phi$ and $\vdash \psi$.

(iii) Let $<$ be a relation that satisfies (EC) and Completeness. We show that it is an entrenchment relation with full comparability, that is, it satisfies the postulates mentioned in Definition 18.

Since $<$ satisfies (EC), it is impossible that $\phi < \psi$ when ϕ is from $Cn(\emptyset)$. This makes the application of Completeness easier, since we do not have to take into account the limiting case $\phi, \psi \in Cn(\emptyset)$.

(Extensionality). Let $Cn(\phi) = Cn(\psi)$. First assume that $\phi < \chi$. We need to show that $\psi < \chi$. Suppose that not $\psi < \chi$. Then, by Completeness, $\chi < \chi \supset \psi$. Since $\phi \wedge \chi \in Cn(\chi \wedge (\chi \supset \psi))$, however, this contradicts (EC). Second, assume that $\chi < \phi$. We need to show that $\chi < \psi$. Suppose that not $\chi < \psi$. Then, by Completeness, $\psi < \psi \supset \chi$. Since $\chi \wedge \psi \in Cn(\phi \wedge (\psi \supset \chi))$, however, this also contradicts (EC).

$(EE2^{\uparrow*})$. Let $\phi < \psi \wedge \chi$. We need to show that $\phi < \psi$. Suppose that not $\phi < \psi$. Then, by Completeness, $\psi < \psi \supset \phi$. Since $\phi \wedge \psi \in Cn((\psi \wedge \chi) \wedge (\psi \supset \phi))$, however, this contradicts (EC).

$(EE2^{\downarrow*})$. Let $\phi < \psi$. We need to show that $\phi \wedge \chi < \psi$. Suppose that not $\phi \wedge \chi < \psi$. Then, by Completeness, $\psi < \psi \supset \phi \wedge \chi$. Since $\phi \wedge \psi \in Cn(\psi \wedge (\psi \supset \phi \wedge \chi))$, however, this contradicts (EC).

(EE3$^\uparrow$). Let $\phi < \psi$ and $\phi < \chi$. We need to show that $\phi < \psi \wedge \chi$. Suppose that not $\phi < \psi \wedge \chi$. Then, by Completeness, $\psi \wedge \chi < (\psi \wedge \chi) \supset \phi$. Since $\phi \wedge (\psi \wedge \chi) \in Cn(\psi \wedge \chi \wedge (\psi \wedge \chi \supset \phi))$, however, this contradicts (EC).

(EE3$^\downarrow$). Let $\phi \wedge \psi < \psi$. We need to show that $\phi < \psi$. Suppose that not $\phi < \psi$. Then, by Completeness, $\psi < \psi \supset \phi$. Since $\phi \wedge \psi \in Cn(\psi \wedge (\psi \supset \phi))$, however, this contradicts (EC).

(EE4). Let $\phi < \psi$. We need to show that either $\phi < \chi$ or $\chi < \psi$. Suppose that neither $\phi < \chi$ nor $\chi < \psi$. Then, by Completeness, $\chi < \chi \supset \phi$ and $\psi < \psi \supset \chi$. Since $\phi \wedge \chi \wedge \psi \in Cn(\psi \wedge (\chi \supset \phi) \wedge (\psi \supset \chi))$, however, this contradicts (EC).

We add that such a relation $<$ also satisfies the Maximality condition (EE6): Let $\phi \notin Cn(\emptyset)$. We need to show that $\phi < \psi$ for some ψ. Since $<$ satisfies (EC), we know that not $\phi < \phi$. Hence, by Completeness, $\phi < \phi \supset \phi$, and we are done.

(iv) Completeness implies Maxichoice. Let $\phi \not< \phi \vee \psi$. By Completeness, then we have $\phi \vee \psi < (\phi \vee \psi) \supset \phi$, or equivalently, $\phi \vee \psi < \psi \supset \phi$. Using Extensionality and (E&C), we get $(\phi \vee \psi) \wedge (\psi \supset \phi) < \psi \supset \phi$, which is again equivalent with $\phi < \phi \vee \neg\psi$, as desired.

Maxichoice implies Completeness. Let $\phi \not< \psi$, That is, by (E&C) and Extensionality, $\phi \wedge \psi \not< (\phi \wedge \psi) \vee (\neg\phi \wedge \psi)$. By Maxichoice, this implies $\phi \wedge \psi < (\phi \wedge \psi) \vee \neg(\neg\phi \wedge \psi)$, that is, by Extensionality again $\psi \wedge (\phi \vee \neg\psi) < \phi \vee \neg\psi$. Thus, by (E&C) again, $\psi < \phi \vee \neg\psi$, as desired. $\square$

Observation 72

Every entrenchment relation can be extended to an entrenchment relation with full comparability.

Proof of Observation 72.[374]

Let $<$ be a GEE-relation. $<$ satisfies (EC). Since L is supposed to be countable, we can fix an enumeration $\langle \phi_1, \psi_1 \rangle, \langle \phi_2, \psi_2 \rangle, \langle \phi_3, \psi_3 \rangle, \ldots$ of all pairs in $L \times L$, leaving out those of $Cn(\emptyset) \times Cn(\emptyset)$. We work through all of these pairs in order to make Completeness true. We begin with $<_0 = <$ and extend this initial relation step by step.

Suppose that we have reached $<_i$ and we are considering the pair $\langle \phi_{i+1}, \psi_{i+1} \rangle$ for some $i \geq 0$. If $\phi_{i+1} <_i \psi_{i+1}$, then we put $<_{i+1} = <_i$. If $\langle \phi_{i+1}, \psi_{i+1} \rangle$ is not contained in $<_i$, but $\langle \phi_{i+1}, \psi_{i+1} \rangle$ is consistent with $<_i$, then we put $<_{i+1} = <_i \cup \{\langle \phi_{i+1}, \psi_{i+1} \rangle\}$; if not, we put $<_{i+1} = <_i \cup \{\langle \psi_{i+1}, \psi_{i+1} \supset \phi_{i+1} \rangle\}$. In any case, $<_{i+1}$ satisfies (EC), by Lemma 71 (ii).

Now we put $<_\infty = \bigcup <_i$. By construction, $<_\infty$ satisfies Completeness. Since each of the $<_i$'s satisfies (EC) and (EC) is of finite character, $<_\infty$ satisfies (EC) as well. But then, by Lemma 71(iii), $<_\infty$ is an entrenchment relation with full comparability, and we are done. $\square$

---

[374] This proof replaces an incorrect proof of Rott (1992b, p. 77).

Observation 74

Every entrenchment relation $<$ is the meet of all entrenchment relations with full comparability that extend $<$.

Proof of Observation 74.

Clearly, every GEE-relation $<$ is a subset of the meet of all SEE-relations extending it. To show that it is not a proper subset, suppose that $\phi \not< \psi$. We have to show that there is a SEE-relation $<^+$ extending $<$ such that $\phi \not<^+ \psi$. To see that this is true, we first add, under preservation of (EC), $\langle \psi, \psi \supset \phi \rangle$ to $<$. This is possible, by Lemma 71 (ii). Then we extend this set to a SEE-relation $<^+$, which is possible according to Lemma 73. But by Lemma 71 (i), no GEE-relation, and a fortiori no SEE-relation, containing $\langle \psi, \psi \supset \phi \rangle$ can contain $\langle \phi, \psi \rangle$. Thus we have constructed a SEE-relation $<^+$ extending $<$ such that $\phi \not<^+ \psi$, as desired. $\square$

REFERENCES

Aizerman, Mark A.: 1985, 'New Problems in the General Choice Theory. Review of a Research Trend', *Social Choice and Welfare* **2**, 235–282.

Aizerman, Mark A., and Andrew V. Malishevski: 1981, 'General Theory of Best Variants Choice: Some Aspects', *IEEE Transactions on Automatic Control*, Vol. **AC-26**, 1030–1040.

Alchourrón, Carlos, Peter Gärdenfors and David Makinson: 1985, 'On the Logic of Theory Change: Partial Meet Contraction Functions and Their Associated Revision Functions', *Journal of Symbolic Logic* **50**, 510–530.

Alchourrón, Carlos, and David Makinson: 1982, 'On the Logic of Theory Change: Contraction Functions and Their Associated Revision Functions', *Theoria* **48**, 14–37.

Alchourrón, Carlos, and David Makinson: 1985, 'On the Logic of Theory Change: Safe Contraction', *Studia Logica* **44**, 405–422.

Alchourrón, Carlos, and David Makinson: 1986, 'Maps Between Some Different Kinds of Contraction Function: The Finite Case', *Studia Logica* **45**, 187–198.

Anderson, Alan Ross, Nuel D. Belnap and J. Michael Dunn: 1992, *Entailment. The Logic of Relevance and Necessity*, Vol. II, Princeton University Press, Princeton.

Areces, Carlos, and Veronica Becher: 2001, 'Iterable AGM Functions', in Mary-Anne Williams and Hans Rott (eds.), *Frontiers of Belief Revision*, Kluwer, Dordrecht, pp. 261–277.

Arrow, Kenneth J.: 1951, *Social Choice and Individual Values*, Wiley, New York, second edition 1963.

Arrow, Kenneth J.: 1959, 'Rational Choice Functions and Orderings', *Economica* N.S. **26**, 121–127.

Bach, Kent: 1984, 'Default Reasoning: Jumping to Conclusions and Knowing When to Think Twice', *Pacific Philosophical Quarterly* **65**, 37–58.

Bagehot, Walter: 1898, 'The Emotion of Conviction', in W.B., *Literary Studies*, Vol. 3, Longmans, Green and Co., London, pp. 190–203. (Orig. 1871)

Baier, Annette: 1979, 'Mind and Change of Mind', *Midwest Studies in Philosophy*, Vol. 4: *Metaphysics*, 157–176. Reprinted in A.B., *Postures of the Mind. Essays on Mind and Morals*, Methuen, London 1985, pp. 51–73.

Bar-Hillel, Yehoshua: 1968, 'On Alleged Rules of Detachment', in Imre Lakatos (ed.), *The Problem of Inductive Logic*, North-Holland, Amsterdam, pp. 120–128.

Bell, John L., and A.B. Slomson: 1969, *Models and Ultraproducts: An Introduction*, North-Holland, Amsterdam.

Belnap, Nuel D.: 1979, 'Rescher's Hypothetical Reasoning: An Amendment', in Ernest Sosa (ed.), *The Philosophy of Nicholas Rescher*, Reidel, Dordrecht, pp. 19–28. Reprinted, with amendments, in Alan Ross Anderson, Nuel D. Belnap and J. Michael Dunn, *Entailment. The Logic of Relevance and Necessity*, Vol. II, Princeton University Press, Princeton 1992, pp. 541–553.

Benferhat, Salem, Claudette Cayrol, Didier Dubois, Jérôme Lang and Henri Prade: 1993, 'Inconsistency Management and Prioritized Syntax-based Entailment', in R. Bajcsy (ed.), *IJCAI'93 – Proceedings of the 13th International Joint Conference on Artificial Intelligence*, Morgan Kaufmann, San Mateo, Cal., pp. 640–645.

Berger, J.O.: 1985, *Statistical Decision Theory and Bayesian Analysis*, second edition, Springer, New York.

Blair, Douglas H., Georges Bordes, Jerry S. Kelly and Kotaro Suzumura: 1976, 'Impossibility Theorems Without Collective Rationality', *Journal of Economic Theory* **13**, 361–379.

Bochman, Alexander: 2000a, 'Belief Contraction as Nonmonotonic Inference', *Journal of Symbolic Logic* **65**, 605–626.

Bochman, Alexander: 2000b, 'A Foundationalist View of the AGM Theory of Belief Change', *Artificial Intelligence* **116**, 237–263.

Bochman, Alexander: 2000c, *A Logical Theory of Nonmonotonic Inference and Belief Change*, book manuscript, Computer Science Department, Holon Academic Institute of Technology, December 2000.

BonJour, Laurence: 1985, *The Structure of Empirical Knowledge*, Harvard University Press, Cambridge, Mass.

Brewka, Gerd: 1991, 'Belief Revision in a Framework for Default Reasoning', in André Fuhrmann and Michael Morreau (eds.), *The Logic of Theory Change*, Springer LNAI **465**, Berlin etc., pp. 206–222.

Cantwell, John: 2000, 'Non-Linear Belief Revision – Foundations and Applications', PhD thesis, Acta Universitatis Upsaliensis, Faculty of Arts, Uppsala University.

Cantwell, John: 2001, 'Two Notions of Epistemic Entrenchment', in Mary-Anne Williams and Hans Rott (eds.), *Frontiers of Belief Revision*, Dordrecht: Kluwer, pp. 221–245.

Cayrol, Claudette, and Marie-Christine Lagasquie-Schiex: 1995, 'Non-monotonic Syntax-based Entailment: A Classification of Consequence Relations', in Christine Froidevaux and Jürg Kohlas (eds.), *Symbolic and Quantitative Approaches to Reasoning and Uncertainty – ECSQARU '95*, Springer LNAI **946**, Berlin etc., pp. 107–114.

Chellas, Brian: 1975, 'Basic Conditional Logic', *Journal of Philosophical Logic* **4**, 133–153.

Chernoff, Herman: 1954, 'Rational Selection of Decision Functions', *Econometrica* **22**, 423–443.

Christensen, David: 1994, 'Conservatism in Epistemology', *Noûs* **28**, 69–89.

Cohen, L. Jonathan: 1992, *An Essay on Belief and Acceptance*, Clarendon Press, Oxford.

Condorcet, Jean Antoine de: 1785, *Essai sur l'application de l'analyse à la probabilité des décisions rendues à la pluralité des voix*. Reprint Chelsea Publishing Company, New York 1972.

Cross, Charles B., and Richmond H. Thomason: 1992, 'Conditionals and Knowledge-base Update', in Peter Gärdenfors (ed.), *Belief Revision*, Cambridge University Press, Cambridge, pp. 247–275.

Curley, E.M.: 1975, 'Descartes, Spinoza and the Ethics of Belief', in Eugene Freeman and Maurice Mandelbaum (eds.), *Spinoza. Essays in Interpretation*, Open Court, La Salle, Ill., pp. 139–157.

Dalal, Mukesh: 1988, 'Investigations into a Theory of Knowledge Base Revisions: Preliminary Report', *AAAI'88 – Proceedings of the Seventh National Conference of the American Association on Artificial Intelligence*, pp. 475–479.

Darwiche, Adnan, and Judea Pearl: 1994, 'On the Logic of Iterated Belief Revision', in Ronald Fagin (ed.), *TARK'94 – Proceedings of the Fifth Conference on Theoretical Aspects of Reasoning About Knowledge*, Morgan Kaufmann, Pacific Grove, Cal., pp. 5–23.

Darwiche, Adnan, and Judea Pearl: 1997, 'On the Logic of Iterated Belief Revision', *Artificial Intelligence* **89**, 1–29.

Davidson, Donald: 1973, 'Radical Interpretation', *Dialectica* **27**, 313–328. Reprinted in D.D., *Inquiries into Truth and Interpretation*, Clarendon Press, Oxford 1984, pp. 125–139.

de Sousa, Ronald B.: 1971, 'How to Give a Piece of Your Mind, or A Logic of Belief and Assent', *Review of Metaphysics* **25**, 52–79.

del Val, Alvaro: 1994, 'On the Relation Between the Coherence and Foundations Theories of Belief Revision', *AAAI'94 – Proceedings of the National Conference of the American Association on Artificial Intelligence*, pp. 909–914.

del Val, Alvaro: 1997, 'Non-Monotonic Reasoning and Belief Revision: Syntactic, Semantic, Foundational, and Coherence Approaches', *Journal of Applied Non-Classical Logics* **7**, 213–240.

Delgrande, J.: 1987, 'A First-Order Logic for Prototypical Properties', *Artificial Intelligence* **33**, 105–130.

Dennett, Daniel: 1978, 'How to Change Your Mind', in D.D., *Brainstorms*, Harvester Press, Hassocks, pp. 300–309.

Dennett, Daniel: 1987, *The Intentional Stance*, MIT Press, Cambridge, Mass.

Descartes, René: 1984, *Meditations on First Philosophy, With Objections and Replies*, in *The Philosophical Writings of Descartes*, edited and translated by John Cottingham, Robert Stoothoff and Dugald Murdoch, Vol. 2, Cambridge University Press, Cambridge, pp. 1–397. (Orig. 1641)

Descartes, René: 1985, *Principles of Philosophy*, in *The Philosophical Writings of Descartes*, edited and translated by John Cottingham, Robert Stoothoff and Dugald Murdoch, Vol. 1, Cambridge University Press, Cambridge, pp. 193–291. (Orig. 1644)

Doyle, Jon: 1991, 'Rational Belief Revision: Preliminary Report', in J. Allen, R. Fikes and E. Sandewall (eds.), *Principles of Knowledge Representation and Reasoning. Proceedings of the 2nd International Conference*, Morgan Kaufmann, San Mateo, Cal., pp. 163–174.

Doyle, Jon: 1992, 'Reason Maintenance and Belief Revision: Foundations vs. Coherence Theories', in Peter Gärdenfors (ed.), *Belief Revision*, Cambridge University Press, Cambridge, pp. 29–51.

Dubois, Didier, and Henri Prade: 1980, *Fuzzy Sets and Systems. Theory and Applications*, Academic Press, New York.

Dubois, Didier, and Henri Prade (with the collaboration of H. Farreny, R. Martin-Clouared and C. Testemale): 1988, *Possibility Theory. An Approach to the Computerized Processing of Uncertainty*, Plenum Press, New York.

Dubois, Didier, and Henri Prade: 1991, 'Epistemic Entrenchment and Possibilistic Logic', *Artificial Intelligence* **50**, 223–239.

Ellis, Brian: 1979, *Rational Belief Systems*, Blackwell, Oxford.

Fagin, R., J.D. Ullman and M.Y. Vardi: 1983, 'On the Semantics of Updates in Databases', in *Proceedings of Second ACM SIGACT-SIGMOD, Atlanta*, pp. 352–365.

Fagin, R., J.D. Ullman, G.M. Kuper, and M.Y. Vardi: 1986, 'Updating Logical Databases', *Advances in Computing Research* **3**, 1–18.

Fermé, Eduardo: 1999, 'A Little Note About Maxichoice and Epistemic Entrenchment', *WoLLIC'99 – Proceedings of the Sixth Workshop on Logic, Language, Information and Computation*, Itatiaia, Brasil, pp. 111 -114.

Fermé, Eduardo, and Ricardo Rodriguez: 1998, 'A Brief Note About the Rott Contraction', *Logic Journal of the IGPL* **6**, 835–842.

Freund, Michael: 1993, 'Injective Models and Disjunctive Relations', *Journal of Logic and Computation* **3**, 231–247.

Freund, Michael, and Daniel Lehmann: 1993, 'Nonmonotonic Inference Operations', *Bulletin of the Interest Group in Pure and Applied Logics* **1**, 23–68. (Also compare the version 'Nonmonotonic Reasoning: From Finitary Relations to Infinitary Inference Operations', *Studia Logica* **53**, 1994, 161–201.)

Freund, Michael, and Daniel Lehmann: 1994, 'Belief Revision and Rational Inference', Technical Report TR 94-16, Leibniz Center for Research in Computer Science, Institute of Computer Science, Hebrew University of Jerusalem.

Friedman, Nir, and Joseyph Y. Halpern: 1999, 'Belief Revision: A Critique', *Journal of Logic, Language and Information* **8**, 401–420.

Fuhrmann, André: 1991, 'Theory Contraction Through Base Contraction', *Journal of Philosophical Logic* **20**, 175–203.

Fuhrmann, André, and Sven Ove Hansson: 1994, 'A Survey of Multiple Contractions', *Journal of Logic, Language and Information* **3**, 39–76.

Gabbay, Dov M.: 1985, 'Theoretical Foundations for Non-Monotonic Reasoning in Expert Systems', in Krzysztof R. Apt (ed.), *Logics and Models of Concurrent Systems*, Springer, Berlin, pp. 439–458.

Gabbay, Dov M., Christopher J. Hogger and John A. Robinson (eds.): 1994, *Handbook of Logic in Artificial Intelligence and Logic Programming, Vol. 3: Nonmonotonic Reasoning and Uncertain Reasoning*, Oxford University Press, Oxford.

Galliers, Julia Rose: 1992, 'Autonomous Belief Revision and Communication', in Peter Gärdenfors (ed.), *Belief Revision*, Cambridge University Press, Cambridge, pp. 220–246.

Gärdenfors, Peter: 1979, 'Conditionals and Changes of Belief', in Illkka Niiniluoto and Raimo Tuomela (eds.), 'The Logic and Epistemology of Scientific Change', *Acta Philosophica Fennica* **30** (1978), nos. 2–4, pp. 381–404.

Gärdenfors, Peter: 1982, 'Rules for Rational Changes of Belief', in T. Pauli (ed.), ⟨*320311*⟩: *Philosophical Essays Dedicated to Lennart Åqvist on His Fiftieth Birthday*, University of Uppsala (=University of Uppsala Philosophical Studies, No. **34**), Uppsala, pp. 88–101.

Gärdenfors, Peter: 1984, 'Epistemic Importance and Minimal Changes of Belief', *Australasian Journal of Philosophy* **62**, 136–157.

Gärdenfors, Peter: 1986, 'Belief Revisions and the Ramsey Test for Conditionals', *Philosophical Review* **95**, 81–93.

Gärdenfors, Peter: 1988, *Knowledge in Flux. Modeling the Dynamics of Epistemic States*, Bradford Books, MIT Press, Cambridge, Mass.

Gärdenfors, Peter: 1990, 'The Dynamics of Belief Systems: Foundations vs. Coherence Theories', *Revue Internationale de Philosophie* **44**, 24–46.

Gärdenfors, Peter, and David Makinson: 1988, 'Revisions of Knowledge Systems Using Epistemic Entrenchment', in Moshe Vardi (ed.), *TARK'88 – Proceedings of the Second Conference on Theoretical Aspects of Reasoning About Knowledge*, Morgan Kaufmann, Los Altos, pp. 83–95.

Gärdenfors, Peter, and David Makinson: 1994, 'Nonmonotonic Inference Based on Expectations', *Artificial Intelligence* **65**, 197–245.

Gärdenfors, Peter, and Hans Rott: 1995, 'Belief Revision', in Dov M. Gabbay, Christopher J. Hogger and John A. Robinson (eds.), *Handbook of Logic in Artificial Intelligence and Logic Programming*, Volume 4: *Epistemic and Temporal Reasoning*, Oxford University Press, Oxford, pp. 35–132.

Georgatos, Konstantinos: 1997, 'Entrenchment Relations: A Uniform Approach to Nonmonotonicity', in D. Gabbay, R. Kruse, A. Nonnengart and H.J. Ohlbach (eds.) , *Proceedings of the International Joint Conference on Qualitative and Quantitative Practical Reasoning (ESCQARU/FAPR'97)*, Springer LNAI **1244**, Berlin etc., pp. 282–297.

Gettier, Edmund, Jr.: 1963, 'Is Justified True Belief Knowledge?', *Analysis* **23**, 121–123.

Giedymin, Jerzy: 1968, 'Empiricism, Refutability, Rationality', in Imre Lakatos and Alan Musgrave (eds.), *Problems in the Philosophy of Science*, North-Holland, Amsterdam, pp. 67–78.

Ginsberg, Matthew L.: 1986, 'Counterfactuals', *Artificial Intelligence* **30**, 35–79.

Ginsberg, Matthew L. (ed.): 1987, *Readings in Nonmonotonic Reasoning*, Morgan Kaufmann, Los Altos, California.

Goodman, Nelson: 1955, *Fact, Fiction, and Forecast*, Harvard University Press, Cambridge, Mass., 4th edition 1983.

Grice, H. Paul: 1975, 'Logic and Conversation', in P. Cole and J. Morgan (eds.), *Syntax and Semantics*, Vol. 3, New York, pp. 41–58. Reprinted in H.P.G., *Studies in the Way of Words*, Harvard University Press, Cambridge, Mass. 1986, pp. 22–40.

Grove, Adam: 1988, 'Two Modellings for Theory Change', *Journal of Philosophical Logic* **17**, 157–170.

Hansson, Sven O.: 1989, 'New Operators for Theory Change', *Theoria* **55**, 114–132.

Hansson, Sven O.: 1991, 'Non-Prioritized Belief Change', in Sven O. Hansson, 'Belief Base Dynamics', Doctoral Dissertation at Uppsala University, Paper no. 8.

Hansson, Sven O.: 1992, 'In Defense of Base Contraction', *Synthese* **91**, 239–245.

Hansson, Sven O.: 1993a, 'Reversing the Levi Identity', *Journal of Philosophical Logic* **22**, 637–69.

Hansson, Sven O.: 1993b, 'Theory Contraction and Base Contraction Unified', *Journal of Symbolic Logic* **58**, 602–625.

Hansson, Sven O.: 1994a, 'Kernel Contraction', *Journal of Symbolic Logic* **59**, 845–859.

Hansson, Sven O.: 1994b, 'Taking Belief Bases Seriously', in Dag Prawitz and Dag Westerståhl (eds.), *Logic and Philosophy in Uppsala*, Synthese Library, Vol. 236, Kluwer, Dordrecht, pp. 13–28.

Hansson, Sven O.: 1996, 'Knowledge-Level Analysis of Belief Base Operations', *Artificial Intelligence* **82**, 215–235.

Hansson, Sven O.: 1997, 'Closure-invariant Rationality Postulates', in Eva Ejerhed and Sten Lindström (eds.), *Logic, Action and Cognition*, Kluwer, Dordrecht, pp. 113–136.

Hansson, Sven O.: 1999a, *A Textbook of Belief Dynamics. Theory Change and Database Updating*, Kluwer, Dordrecht.

Hansson, Sven O. (ed.): 1999b, Special Issue on Non-Prioritized Belief Revision, *Theoria* **63**, 1–134.

Hansson, Sven O., and Erik J. Olsson: 1999, 'Providing Foundations for Coherentism', *Erkenntnis* **51**, 243–265.

Harel, D.: 1984, 'Dynamic Logic', in Dov M. Gabbay and Franz Guenthner
(eds.), *Handbook of Philosophical Logic*, Vol. 2, Chapter 10, Reidel, Dordrecht, pp. 497–604.

Harman, Gilbert: 1986, *Change in View*, Bradford Books, MIT Press, Cambridge, Mass.

Herzberger, Hans G.: 1973, 'Ordinal Preference and Rational Choice', *Econometrica* **41**, 187–237.

Hilpinen, Risto: 1968, *Rules of Acceptance and Inductive Logic*, North-Holland,
Amsterdam.

Houthakker, H.S.: 1950, 'Revealed Preference and the Utility Function', *Economica* N.S. **17**, 159–174.

Hume, David: 1978, *A Treatise of Human Nature*, second edition, Peter H.
Nidditch (ed.), Oxford University Press, Oxford. (Orig. 1739)

Israel, David: 1980, 'What's Wrong With Non-Monotonic Logic?', *AAAI – Proceedings of the First National Conference on Artificial Intelligence*, pp. 99–
101. Reprinted in Matthew L. Ginsberg (ed.), *Readings in Nonmonotonic
Reasoning*, Morgan Kaufmann, Los Altos, Cal., 1987, pp. 53–55.

James, William: 1979, 'The Will to Believe', in W.J., *The Will to Believe and
Other Essays in Popular Philosophy*, The Works of William James, Vol. 6,
edited by Frederick H. Burkhardt, Fredson Bowers and Ignas K. Skrupskelis,
Cambridge, Mass., and London, Harvard University Press, pp. 13–33. (Orig.
1896)

James, William: 1975, 'Pragmatism's Conception of Truth', Lecture VI in W.J.,
Pragmatism. A New Name for Some Old Ways of Thinking, The Works of
William James, Vol. 1, edited by Fredson Bowers and Ignas K. Skrupske-
lis, Cambridge, Mass., and London, Harvard University Press, pp. 95–113.
(Orig. 1907)

Kalai, Gil, Ariel Rubinstein and Ran Spiegler: 2001, 'Comments on Rationalizing Choice Functions which Violate Rationality', manuscript January 2001.

Kaluzhny, Yuri, and Daniel Lehmann: 1995, 'Deductive Nonmonotonic Inference Operations: Antitonic Representations', *Journal of Logic and Computation* **5**, 111–122.

Kant, Immanuel: 1800, *Lectures on Logic*, translated and edited by J. Michael
Young, Cambrigde University Press, Cambridge 1992.

Katsuno, Hirofumi, and Alberto O. Mendelzon: 1991, 'Propositional Knowledge
Base Revision and Minimal Change', *Artificial Intelligence* **52**, 263–294.

Katsuno, Hirofumi, and Alberto O. Mendelzon: 1992, 'On the Difference between Updating a Knowledge Base and Revising it', in Peter Gärdenfors
(ed.), *Belief Revision*, Cambridge University Press, Cambridge, pp. 183–
203.

Kim, T., and Marcel K. Richter: 1986, 'Nontransitive-nontotal Consumer Theory', *Journal of Economic Theory* **38**, 324–363.

Kratzer, Angelika: 1981, 'Partition and Revision: The Semantics of Counterfactuals', *Journal of Philosophical Logic* **10**, 201–216.

Kraus, Sarit, Daniel Lehmann and Menachem Magidor: 1990, 'Nonmonotonic Reasoning, Preferential Models and Cumulative Logics', *Artificial Intelligence* **44**, 167–207.

Kyburg, Henry E., Jr.: 1970, 'Conjunctivitis', in Marshall Swain (ed.), *Induction, Acceptance, and Rational Belief*, Reidel, Dordrecht, pp. 55-82. Reprinted in H.E.K. Jr., *Epistemology and Inference*, University of Minnesota Press, Minneapolis 1983, pp. 232–254.

Lakatos, Imre: 1968, 'Changes in the Problem of Inductive Logic', in I.L. (ed.), *The Problem of Inductive Logic*, North-Holland, Amsterdam, pp. 315–417. Reprinted in I.L., *Philosophical Papers*, Vol. 2: *Mathematics, Science and Epistemology*, Cambridge University Press, Cambridge 1978, pp. 128–200.

Lamarre, Philippe: 1991, 'S4 and the Logic of Nonmonotonicity', in J. Allen, R. Fikes and E. Sandewall (eds.), *Principles of Knowledge Representation and Reasoning. Proceedings of the 2nd International Conference*, Morgan Kaufmann, San Mateo, Cal., pp. 357–367.

Lehmann, Erich L.: 1986, *Testing Statistical Hypothesis*, second edition, Chapman and Hall, New York and London.

Lehmann, Daniel, and Menachem Magidor: 1992, 'What Does a Conditional Knowledge Base Entail?', *Artificial Intelligence* **55**, 1–60.

Lehrer, Keith: 1975, 'Reason and Consistency', in Keith Lehrer (ed.), *Analysis and Metaphysics. Essays in Honor of R. M. Chisholm*, Reidel, Dordrecht, pp. 57–74. Reprinted in K.L., *Metamind*, Clarendon, Oxford 1990, pp. 148–166.

Lehrer, Keith: 1990, *Theory of Knowledge*, Routledge, London.

Lehrer, Keith: 1997, *Self-Trust. A Study of Reason, Knowledge, and Autonomy*, Oxford University Press, Oxford.

Lehrer, Keith: 2000, *Theory of Knowledge*, second, substantially revised edition, Westview Press, Boulder 2000.

Leibniz, Gottfried Wilhelm: 1981, *New Essays on Human Understanding*, translated and edited by Peter Remnant and Jonathan Bennett, Cambridge University Press, Cambridge. (Orig. written 1704/05, published 1765)

Levi, Isaac: 1967, *Gambling With Truth*, Knopf, New York.

Levi, Isaac: 1977, 'Subjunctives, Dispositions and Chances', *Synthese* **34**, 423–455.

Levi, Isaac: 1983, 'Truth, Fallibility and the Growth of Knowledge', in Robert S. Cohen and Marx W. Wartofsky (eds.), *Language, Logic, and Method*, Reidel, Dordrecht, pp. 153–174.

Levi, Isaac: 1984, *Decisions and Revisions. Philosophical Essays on Knowledge and Value*, Cambridge University Press, Cambridge.

Levi, Isaac: 1991, *The Fixation of Belief and Its Undoing. Changing Beliefs Through Inquiry*, Cambridge University Press, Cambridge.

Levi, Isaac: 1995, 'Closure and Consistency', in Walter Sinnott-Armstrong, Diana Raffman and Nicholas Asher (eds.), *Modality, Morality, and Belief. Essays in Honor of Ruth Barcan Marcus*, Cambridge University Press, Cambridge, pp. 215–234.

Levi, Isaac: 1998, 'Contraction and Informational Value', Manuscript Columbia University, 7th version, April 1998. Available at http://www.columbia.edu/~levi/contraction.pdf.

Lewis, David: 1973, *Counterfactuals*, Blackwell, Oxford.

Lewis, David: 1974, 'Radical Interpretation', *Synthese* **23**, 331–344. Reprinted in D.L., *Philosophical Papers*, Vol. I, Oxford University Press, New York and Oxford 1983, pp. 108–118, with a new postscript, pp. 119–121.

Lewis, David: 1981, 'Ordering Semantics and Premise Semantics for Counterfactuals', *Journal of Philosophical Logic* **10**, 217–234. Reprinted in D.L., *Papers in Philosophical Logic*, Cambridge University Press, Cambride 1998, pp. 77–96.

Lindström, Sten: 1991, 'A Semantic Approach to Nonmonotonic Reasoning: Inference Operations and Choice", Uppsala Prints and Preprints in Philosophy, Department of Philosophy, University of Uppsala, 1991:6.

Lindström, Sten, and Wlodzimierz Rabinowicz: 1991, 'Epistemic Entrenchment with Incomparabilities and Relational Belief Revision', in André Fuhrmann and Michael Morreau (eds.), *The Logic of Theory Change*, Springer LNAI **465**, Berlin etc., pp. 93–126.

Lindström, Sten, and Wlodzimierz Rabinowicz: 1992, 'Belief Revision, Epistemic Conditionals and the Ramsey Test', *Synthese* **91**, 195–237.

Luce, R. Duncan, and Howard Raiffa: 1957, *Games and Decisions*, Wiley, New York.

Makinson, David: 1985, 'How to Give it Up: A survey of Some Formal Aspects of the Logic of Theory Change', *Synthese* **62**, 347–363.

Makinson, David: 1987, 'On the Status of the Postulate of Recovery in the Logic of Theory Change', *Journal of Philosophical Logic* **16**, 383–394.

Makinson, David: 1989, 'General Theory of Cumulative Inference', in Michael Reinfrank, Johan de Kleer, Matthew L. Ginsberg and Erik Sandewall (eds.), *Non-Monotonic Reasoning. Proceedings of the 2nd International Workshop 1988*, Springer LNAI **346**, Berlin etc., pp. 1–18.

Makinson, David: 1993, 'Five Faces of Minimality', *Studia Logica* **52**, 339–379.

Makinson, David: 1994, 'General Patterns in Nonmonotonic Reasoning', in Dov M. Gabbay, Christopher J. Hogger and John A. Robinson (eds.), *Handbook of Logic in Artificial Intelligence and Logic Programming*, Vol. 3: *Nonmonotonic Reasoning and Uncertain Reasoning*, Oxford University Press, Oxford, pp. 35–110.

Makinson, David: 1997, 'On the Force of Some Apparent Counterexamples to Recovery', in E. G. Valdés et al. (eds.), *Normative Systems in Legal and Moral Theory*, Festschrift for Carlos E. Alchourrón and Eugenio Bulygin, Duncker und Humblot, Berlin, 1997, pp. 475–481.

Makinson, David, and Peter Gärdenfors: 1991, 'Relations Between the Logic of Theory Change and Nonmonotonic Logic', in André Fuhrmann and Michael Morreau (eds.), *The Logic of Theory Change*, Springer LNAI **465**, Berlin etc., pp. 185–205.

McCarthy, Gerald D. (ed.): 1986, *The Ethics of Belief Debate*, Atlanta Scholars Press, Atlanta.

McCarthy, John: 1979, 'Ascribing Mental Qualities to Machines', in Marting Ringle (ed.), *Philosophical Perspectives in Artificial Intelligence*, Harvester Press, Brighton, pp. 161–195.

Melmoth, Sebastian: 1994, 'Quantum Mutata', in S.M., *Complete Works*, Harper Collins, Glasgow, pp. 773–774. (Orig. 1881)

Meyer, Thomas Andreas, Willem Adrian Labuschagne and Johannes Heidema: 2000, 'Refined Epistemic Entrenchment', *Journal of Logic, Language and Information* **9**, 237–259.

Morreau, Michael, and Hans Rott: 1991, 'Is It Impossible to Keep Up to Date?', in Jürgen Dix, Klaus P. Jantke and Peter H. Schmitt (eds.), *Nonmonotonic and Inductive Logic*, Springer LNAI **543**, Berlin etc., pp. 233–243.

Moulin, Hervé: 1985, 'Choice Functions over a Finite Set: A Summary', *Social Choice and Welfare* **2**, 147–160.

Nayak, Abhaya C.: 1994, 'Foundational Belief Change', *Journal of Philosophical Logic* **23**, 495–533.

Nayak, Abhaya C.: 1994, 'Iterated Belief Change Based on Epistemic Entrenchment', *Erkenntnis* **41**, 353–390.

Nayak, Abhaya C., Norman Y. Foo, Maurice Pagnucco and Abdul Sattar: 1996, 'Changing Conditional Beliefs Unconditionally', in Yoav Shoham (ed.), *TARK'96 – Proceedings of the Sixth Conference on Theoretical Aspects of Rationality and Knowledge*, pp. 119–135.

Nayak, Abhaya C., Paul Nelson and Hanan Polansky: 1996, 'Belief Change as Change in Epistemic Entrenchment', *Synthese* **109**, 143–174.

Nebel, Bernhard: 1989, 'A Knowledge Level Analysis of Belief Revision', in Ronald Brachman, Hector Levesque and Ray Reiter (eds.), *Proceedings of the 1st International Conference on Principles of Knowledge Representation and Reasoning*, Morgan Kaufmann, San Mateo, Cal., pp. 301–311.

Nebel, Bernhard: 1992, 'Syntax-based Approaches to Belief Revision', in Peter Gärdenfors (ed.), *Belief Revision*, Cambridge University Press, Cambridge, pp. 52–88.

Nehring, Klaus: 1997, 'Rational Choice and Revealed Preference Without Binariness', *Social Choice and Welfare* **14**, 403–425.

Newell, Alan: 1982, 'The Knowledge Level', *Artificial Intelligence* **18**, 87–127.

Newman, John H.: 1947, *An Essay in Aid of a Grammar of Assent*, edited by C. F. Horrold, New York 1947. (Orig. 1870)

Nozick, Robert: 1981, *Philosophical Explanations*, Belknap Press, Harvard University Press, Cambridge, Mass.

Nute, Donald: 1980, *Topics in Conditional Logic*, Reidel, Dordrecht.

Nute, Donald: 1994, 'Defeasible Logic', in D.M. Gabbay, Ch.J. Hogger and J.A. Robinson (eds.), *Handbook of Logic in Artificial Intelligence and Logic Programming, Vol. 3: Nonmonotonic Reasoning and Uncertain Reasoning*, Oxford University Press, Oxford, pp. 353–395.

Olsson, Erik J.: 1998a, 'Competing for Acceptance: Lehrer's Rule and the Paradoxes of Justification', *Theoria* **64**, 34–54.

Olsson, Erik J.: 1998b, 'Making Beliefs Coherent: The Subtraction and Addition Strategies', *Journal of Logic, Language and Information* **7**, 143–163.

Olsson, Erik J.: 1999, 'Cohering With', *Erkenntnis* **50**, 273–291.

Pagnucco, Maurice: 1996, *The Role of Abductive Reasoning Within the Process of Belief Revision*, PhD Thesis, Basser Department of Computer Science, University of Sydney.

Pagnucco, Maurice, and Hans Rott: 1999, 'Severe Withdrawal – and Recovery', *Journal of Philosophical Logic* **28**, 501–547. (Corrupted printing – correct, complete reprint with original pagination in the February 2000 issue; see publisher's 'Erratum', *Journal of Philosophical Logic* **29**, 2000, 121.)

Passmore, John: 1980, 'Hume and the Ethics of Belief', Appendix in J.P., *Hume's Intentions*, third edition, Duckworth, London, pp. 160–176.

Passmore, John: 1986, 'Locke and the Ethics of Belief', in Anthony Kenny (ed.), *Rationalism, Empiricism, and Idealism*, Clarendon Press, Oxford, pp. 23–46.

Peirce, Charles S.: 1982, 'The Fixation of Belief', in Horace S. Thayer (ed.), *Pragmatism. The Classic Writings*, Hackett, Indianapolis, Cambridge, pp. 61–78. (Orig. 1877)

Peppas, Pavlos, and Mary-Anne Williams: 1995, 'Constructive Modelings for Theory Change', *Notre Dame Journal of Formal Logic* **36**, 120–133.

Piller, Christian: 1991, 'On Keith Lehrer's Belief in Acceptance', in Johannes Brandl, Wolfgang Gombocz and Christian Piller (eds.), *Metamind, Knowledge and Coherence. Essays on the Philosophy of Keith Lehrer*, Rodopi, Amsterdam (= *Grazer philosophische Studien* **40**), pp. 37–61.

Plantinga, Alvin: 1990, 'Justification in the 20th Century', *Philosophy and Phenomenological Research* **50**, Supplement Vol., 45–71.

Plato: 1994, *Meno*, in Jane M. Day (transl. and ed.), *Plato's Meno in Focus*, Routledge, London. (Orig. about 380 B.C.)

Plott, Charles R.: 1973, 'Path Independence, Rationality and Social Choice', *Econometrica* **41**, 1075–1091.

Pollock, John L.: 1976, *Subjunctive Reasoning*, Reidel, Dordrecht.

Poole, David: 1988, 'A Logical Framework for Default Reasoning', *Artificial Intelligence* **36**, 27–47.

Popper, Karl R.: 1976, *Unended Quest*, Routledge, London.

Price, Henry H.: 1954, 'Belief and Will', *Proceedings of the Aristotelian Society*, Supplement Vol. **28**, 1–26.

Price, Henry H.: 1969, *Belief*, Allen and Unwin/Humanities Press, London/New York.

Priest, Graham, Richard Routley and Jean Norman (eds.): 1989, *Paraconsistent Logic. Essays on the Inconsistent*, Philosophia Verlag, München etc.

Putnam, Hilary: 1960, 'Minds and Machines', in *Dimensions of Mind*, in Sidney Hook (ed.), State University of New York Press, Albany, N.Y., pp. 138–164; reprinted in Hilary Putnam, *Mind, Language and Reality*, Cambridge 1975, pp. 362–385;

Quine, Willard V.O., and Joseph S. Ullian: 1978, *The Web of Belief*, second edition, Random House, New York.

Ranade, R.R.: 1985, 'Rationalisable Choice Functions: An Alternative Characterisation', *Journal of Quantitative Economics* **1**, 265–272.

Ranade, R.R.: 1987, 'On the Relative Strengths of Consistency Conditions on Choice Functions', *Social Choice and Welfare* **4**, 207–212.

Reid, Thomas: 1985a, *Essays on the Intellectual Powers of Man*, in *The Works of Thomas Reid*, edited by Sir William Hamilton, eighth edition, James Thin, Edinburgh. (Orig. 1785)

Reid, Thomas: 1985b, *Essays on the Active Powers of Man*, in *The Works of Thomas Reid*, edited by Sir William Hamilton, eighth edition, Edinburgh: James Thin. (Orig. 1788)

Reiter, Raymond: 1980, 'A Logic of Default Reasoning', *Artificial Intelligence* **13**, 81–132.

Rescher, Nicholas: 1964, *Hypothetical Reasoning*, North-Holland, Amsterdam.

Rescher, Nicholas: 1973, *The Coherence Theory of Truth*, Oxford University Press, Oxford.

Rescher, Nicholas: 1976, *Plausible Reasoning*, Van Gorcum, Assen, Amsterdam.

Rescher, Nicholas: 1979, 'Reply to Belnap', in Ernest Sosa (ed.), *The Philosophy of Nicholas Rescher*, Reidel, Dordrecht, pp. 29–31.

Rescher, Nicholas: 1989, *Cognitive Economy. An Inquiry into the Economic Dimension of the Theory of Knowledge*, University of Pittsburgh Press, Pittsburgh.

Restall, Greg, and John Slaney: 1995, 'Realistic Belief Revision', in M. De Glas and Z. Pawlak (eds.), *WOCFAI'95 – Proceedings of the Second World Conference on the Fundamentals of Artificial Intelligence*, Angkor, Paris, pp. 367–378.

Richter, Marcel K.: 1966, 'Revealed Preference Theory', *Econometrica* **34**, 635–645.

Richter, Marcel K.: 1971, 'Rational Choice', in John S. Chipman, Leonid Hurwicz, Marcel K. Richter and Hugo F. Sonnenschein (eds.), *Preference, Utility and Demand*, Harcourt Brace Jovanovich, New York, pp. 29–58.

Rott, Hans: 1991a, 'Two Methods of Constructing Contractions and Revisions of Knowledge Systems', *Journal of Philosophical Logic* **20**, 149–173.

Rott, Hans: 1991b, 'A Non-monotonic Conditional Logic for Belief Revision I', in André Fuhrmann and Michael Morreau (eds.), *The Logic of Theory Change*, Lecture Notes in Computer Science **465**, Springer, Berlin etc., pp. 135–181.

Rott, Hans: 1992a, 'On the Logic of Theory Change: More Maps Between Different Kinds of Contraction Function', in Peter Gärdenfors (ed.), *Belief Revision*, Cambridge University Press, Cambridge, pp. 122–141.

Rott, Hans: 1992b, 'Preferential Belief Change Using Generalized Epistemic Entrenchment', *Journal of Logic, Language and Information* **1**, 45–78.

Rott, Hans: 1992c, 'Modellings for Belief Change: Prioritization and Entrenchment', *Theoria* **58**, 21–57.

Rott, Hans: 1993, 'Belief Contraction in the Context of the General Theory of Rational Choice', *Journal of Symbolic Logic* **58**, 1426–1450.

Rott, Hans: 1994, 'Coherent Choice and Epistemic Entrenchment (Preliminary report)', in Bernhard Nebel und Leonie Dreschler-Fischer (eds.), *KI-94: Advances in Artificial Intelligence*, Springer LNAI **861**, Berlin etc., pp. 284–295.

Rott, Hans: 1998, 'Logic and Choice', in Itzhak Gilboa (ed.), *Theoretical Aspects of Rationality and Knowledge. Proceedings of the Seventh Conference (TARK 1998, Evanston/Illinois)*, Morgan Kaufmann, San Francisco, pp. 235–248.

Rott, Hans: 1999, 'Coherence and Conservatism in the Dynamics of Belief. Part I: Finding the Right Framework', *Erkenntnis* **50**, 387–412.

Rott, Hans: 2000a, ' "Just Because": Taking Belief Bases Seriously', in Samuel R. Buss, Petr Hájek und Pavel Pudlák (eds.), *Logic Colloquium '98 – Proceedings of the Annual European Summer Meeting of the Association for Symbolic Logic held in Prague*, Lecture Notes in Logic, Vol. 13, Association for Symbolic Logic, Urbana, Ill., pp. 387–408.

Rott, Hans: 2000b, 'Two Dogmas of Belief Revision', *Journal of Philosophy* **97**, 503–522.

Rott, Hans: 2000c, 'Coherence and Conservatism in the Dynamics of Belief. Part II: Iterated Belief Change Without Dispositional Coherence', manuscript Regensburg, May 2000.

Russell, Bertrand: 1919, *Introduction to Mathematical Philosophy*, Allen and Unwin, London.

Samuelson, Paul A.: 1938, 'A Note on the Pure Theory of Consumers' Behaviour', *Economica* N.S. **5**, 61–71. Addendum *Economica* N.S. **5**, pp. 353–354.

Samuelson, Paul A.: 1947, *Foundations of Economic Analysis*, Harvard University Press, Cambridge, Mass.

Samuelson, Paul A.: 1950, 'The Problem of Integrability in Utility Theory', *Economica* **17**, 355–381.

Satoh, Ken: 1990, 'A Probabilistic Interpretation of Lazy Nonmonotonic Reasoning', *AAAI'90 – Proceedings of the Eighth National Conference of the American Association on Artificial Intelligence*, 659–664.

Schlechta, Karl: 1991a, 'Some Results on Theory Revision', in André Fuhrmann and Michael Morreau (eds.), *The Logic of Theory Change*, Springer LNAI **465**, Berlin etc., pp. 72–92.

Schlechta, Karl: 1991b, 'Theory Revision and Probability', *Notre Dame Journal of Formal Logic* **32**, 307–319.

Schlechta, K.: 1992, 'Some Results on Classical Preferential Models', *Journal of Logic and Computation* **2**, 675–686.

Schlechta, K.: 1996, 'Some Completeness Results for Stoppered and Ranked Classical Preferential Models', *Journal of Logic and Computation* **6**, 599–622.

Segerberg, Krister: 1995, 'Belief Revision from the Point of View of Doxastic Logic', *Bulletin of the Interest Group in Pure and Applied Logics* **3**, 535–553.

Segerberg, Krister: 1998, 'Irrevocable Belief Revision in Dynamic Doxastic Logic', *Notre Dame Journal of Formal Logic* **39**, 287–306.

Sen, Amartya K.: 1970, *Collective Choice Social Welfare*, Holden-Day, San Francisco.

Sen, Amartya K.: 1971, 'Choice Functions and Revealed Preference', *Review of Economic Studies* **38**, 307–317. Reprinted in A.K.S., *Choice, Welfare and Measurement*, Blackwell, Oxford 1982, pp. 41–53.

Sen, Amartya K.: 1982, *Choice, Welfare and Measurement*, Blackwell, Oxford.

Sen, Amartya K.: 1987, 'Rational Behaviour', in John Eatwell, Murray Milgate and Peter Newman (eds.), *The New Palgrave. A Dictionary of Economics*, Vol. 4, Macmillan Press, London, pp. 68–76.

Sen, Amartya K.: 1993, 'Internal Consistency of Choice', *Econometrica* **61**, 495–521.

Sen, Amartya K.: 1995, 'Is the Idea of Purely Internal Consistency of Choice Bizarre?', in J.E.J. Altham and Ross Harrison (eds.), *World, Mind, and Ethics. Essays on the Ethical Philosophy of Bernard Williams*, Cambridge University Press, Cambridge, pp. 19–31.

Sen, Amartya K.: 1997, 'Maximization and the Act of Choice', *Econometrica* **65**, 745–779.

Shackle, G.L.S.: 1961, *Decision, Order and Time in Human Affairs*, Cambridge University Press, Cambridge.

Simon, Herbert A.: 1959, 'Theories of Decision Making in Economics and Behavioral Science', *American Economic Review* **49**, 253–283.

Slote, Michael: 1989, *Beyond Optimizing*, Harvard University Press, Cambridge, Mass.

Sosa, Ernest: 1980, 'The Raft und the Pyramid: Coherence Versus Foundations in the Theory of Knowledge', *Midwest Studies in Philosophy* **5**, 3–25. Reprinted in E.S., *Knowledge in Perspective*, Cambridge University Press, Cambridge 1991, pp. 165–191.

Spohn, Wolfgang: 1988, 'Ordinal Conditional Functions', in William L. Harper and Brian Skyrms (eds.), *Causation in Decision, Belief Change, and Statistics*, Vol. II, Reidel, Dordrecht, pp. 105–134.

Stalnaker, Robert C.: 1968, 'A Theory of Conditionals', in Nicholas Rescher (ed.), *Studies in Logical Theory*, APQ Monograph Series **2**, Blackwell, Oxford, pp. 98–112.

Stalnaker, Robert C.: 1984, *Inquiry*, Bradford Books, MIT Press, Cambridge, Mass.

Stich, Stephen P.: 1978, 'Beliefs and Subdoxastic States', *Philosophy of Science* **45**, 499–518.

Suppes, Patrick: 1994, 'Qualitative Theory of Subjective Probability', in George Wright and Peter Ayton (eds.), *Subjective Probability*, Wiley, New York, pp. 17–37.

Suzumura, Kotaro: 1983, *Rational Choice, Collective Decisions, and Social Welfare*, Cambridge University Press, Cambridge.

Thayer, Horace S. (ed.): 1982, *Pragmatism. The Classic Writings*, Hackett, Indianapolis and Cambridge.

Ullmann-Margalit, Edna: 1983, 'On Presumption', *Journal of Philosophy* **80**, 143–163.

Ullmann-Margalit, Edna, and Avishai Margalit: 1992, 'Holding True and Holding As True', *Synthese* **92**, 167–187.

Ullmann-Margalit, Edna, and Sidney Morgenbesser: 1977, 'Picking and Choosing', *Social Research* **44**, 757–785.

Uzawa, H.: 1956, 'A Note on Preference and Axioms of Choice', *Annals of the Institute of Statistical Mathematics* **8**, 35–40.

Van Linder, Bernd, Wiebe van der Hoek and John-Jules Ch. Meyer: 1995, 'Actions that Make You Change Your Mind', in Armin Laux and Heinrich Wansing (eds.), *Knowledge and Belief in Philosophy and Artificial Intelligence*, Akademie Verlag, Berlin, pp. 103–146.

Veltman, Frank: 1976, 'Prejudices, Presuppositions and the Theory of Counterfactuals', in Jeroen Groenendijk and Martin Stokhof (eds.), *Amsterdam Papers of Formal Grammar*, Vol. 1, pp. 248–281.

Weintraub, Ruth: 1990, 'Decision-Theoretic Epistemology', *Synthese* **83**, 159–177.

Wernham, James C.S.: 1987, *James's Will-to-Believe Doctrine. A Heretical View*, Kingston and Montreal, McGill-Queen's University Press.

Williams, Bernard: 1970, 'Deciding to Believe', in H.E. Kiefer and M.K. Munitz (eds.), *Language, Belief, and Metaphysics*, State University of New York Press, Albany, pp. 95-111. Reprinted, with amendments, in B.W., *Problems of the Self. Philosophical Papers 1956-1972*, Cambridge University Press, Cambridge 1973, pp. 136–151.

Williams, Mary-Anne: 1994, 'On the Logic of Theory Base Change', in Craig MacNish, David Pearce and Luis M. Pereira (eds.), *Logics in Artificial Intelligence*, Springer LNAI **838**, Berlin etc., pp. 86–105.

Williams, Mary-Anne: 1995, 'Iterated Theory Base Change: A computational model', in *IJCAI'95 – Proceedings of the 14th International Joint Conference on Artificial Intelligence*, Morgan Kaufmann, San Mateo, pp. 1541–1550.

Williams, Mary-Anne, and Hans Rott (eds.): 2001, *Frontiers in Belief Revision*, Kluwer, Dordrecht.

Wittgenstein, Ludwig: 1953, *Philosophical Investigations*, translated by G. E. M. Anscombe, third edition 1967, Blackwell, Oxford.

Wolterstorff, Nicholas: 1996, *John Locke and the Ethics of Belief*, Cambridge University Press, Cambridge.

Wright, George, and Peter Ayton (eds.): 1994, *Subjective Probability*, Wiley, New York.

Zadeh, Lofti A.: 1978, 'Fuzzy Sets as a Basis for a Theory of Possibility', *Fuzzy Sets and Systems* **1**, 3–28.

INDEX OF SYMBOLS

ϕ, ψ, χ, ξ meta-variables for sentences

p, q, r atoms (propositional variables)

$[\![\phi]\!]$ set of models verifying (satisfying) ϕ

$[\![F]\!]$ set of models verifying (satisfying) all sentences ϕ in F

$]\!]\phi[\![$ set of models falsifying ϕ (satisfying $\neg\phi$)

$]\!]F[\![$ set of models falsifying some sentence ϕ in F

$\widehat{m}$ set of sentences satisfied by (true in) model m

H set of sentences (non-prioritized 'belief base')

$Cn, \vdash$ monotonic consequence operation / relation

$Inf, \mathrel{\vrule height 1.1ex depth 0pt width 0pt}\!\!\!\sim$ nonmonotonic, and possibly paraconsistent, inference operation / relation

K theory or 'belief set', closed under Cn or Inf

$\mathcal{H}$ prioritized information base $\langle H_1, \ldots, H_n \rangle$

$|\mathcal{H}|$ $\bigcup \{H_1, \ldots, H_n\}$, for the prioritized information base $\mathcal{H} = \langle H_1, \ldots, H_n \rangle$

$\mathcal{E}$ prioritized expectation base $\langle E_1, \ldots, E_m \rangle$

$\mathcal{E} \circ \mathcal{H}$ concatenation $\langle E_1, \ldots, E_m, H_1, \ldots, H_n \rangle$ of prioritized bases $\mathcal{E}$ and $\mathcal{H}$

$Consol(\mathcal{G})$ consolidation, prioritized base contraction of $\mathcal{G}$ with respect to $\bot$

$\mathcal{H} \ominus \phi$ trivial subtraction of ϕ from $\mathcal{H}$ (without constraints)

$H \dot{-} \phi,$
  $K \dot{-} \phi$ contraction of H or K with respect to ϕ (should not imply ϕ)

$K \dot{-} \langle F \rangle$ multiple contraction, pick contraction with respect to set F (retract at least one of F)

$K \doteq \phi$ severe withdrawal of ϕ from K

$\mathcal{H} \oplus \phi$ trivial addition of ϕ to $\mathcal{H}$ (without constraints)

$H * \phi,$
  $K * \phi$ revision of H or K by ϕ (should contain ϕ and be consistent)

$\mathcal{E} * \mathcal{H}$,   revision of expectations $\mathcal{E}$ or E by information $\mathcal{H}$ or h
$\quad E * h$

$H \perp \phi$ set of all maximal subsets of H that do not imply ϕ

$H *_0 \phi$ full meet base revision of H $(= \bigcap (H \perp \neg\phi) \cup \{\phi\})$

$\mathcal{H} \Downarrow \phi$ set of all "best" subsets of $\bigcup \mathcal{H}$ that do not imply ϕ

σ arbitrary choice or selection function

$x \bullet y$ compromise between, or combination of, options x and y

R absolutely satisfactory options

$\overline{R}$ taboo set (absolutely unacceptable options)

γ choice function over models

γ^+ completion of γ

δ choice function over sentences

δ^+ faithful variant of δ

ρ remainder function (complement to choice function)

$<$ arbitrary preference relation

          or preferences revealed by choice functions over sentences, $\mathcal{P}_2(\delta)$
          ('epistemic entrenchment' in a theory)

$\prec$ priorities in a belief base

          or preference relation over models

$\mathcal{C}(\dots)$ contraction function $\dot{-}$ (re-)constructed from $\dots$

$\ddot{\mathcal{C}}(<)$ withdrawal function $\ddot{-}$ constructed from entrenchment relation $<$

$\mathcal{R}(\dots)$ revision function $*$ (re-)constructed from $\dots$

$\mathcal{I}(\dots)$ inference operation Inf (re-)constructed from $\dots$

$\mathcal{S}(\dots)$ choice function γ (re-)constructed from $\dots$

$\mathcal{G}(\dots)$ semantic choice function γ (re-)constructed from $\dots$

$\mathcal{D}(\dots)$ syntactic choice function δ (re-)constructed from $\dots$

$\mathcal{P}(\dots)$ preference relation $<$ (re-)constructed from $\dots$

$\mathcal{P}_2(\sigma)$ base preference relation $<$ (re-)constructed from choice function σ

$\mathcal{P}_2(\delta)$ epistemic entrenchment relation $<$ (re-)constructed from δ

$\mathcal{P}_2(\dot{-})$ epistemic entrenchment relation $<$ (re-)constructed from $\dot{-}$

INDEX OF NAMES

SUBJECT INDEX